The Versatility of Integrability

In Memory of Igor Krichever

CONTEMPORARY MATHEMATICS

823

The Versatility of Integrability

In Memory of Igor Krichever

IDPEIS-22
Isomonodromic Deformations, Painlevé Equations, and Integrable Systems
June 27–July 1, 2022
Virtual, hosted by Columbia University

AGMPS-22
Algebraic Geometry, Mathematical Physics, and Solitons
October 7–9, 2022
Columbia University, New York, NY

Mikhail Bershtein
Anton Dzhamay
Andrei Okounkov
Editors

AMS AMERICAN MATHEMATICAL SOCIETY

2020 *Mathematics Subject Classification.* Primary 14H40, 14H70, 14K25, 14Q05, 14T20, 32G34, 33C70, 35C20, 60G55, 81R12.

Library of Congress Cataloging-in-Publication Data

Names: Bershtein, Mikhail, 1985– editor. | Dzhamay, Anton, 1970– editor. | Okounkov, Andrei, 1969– editor. | Krichever, I. M., honoree. | Isomonodromic Deformations, Painlevé Equations, and Integrable Systems (Conference) (2022 : Online) | Algebraic Geometry, Mathematical Physics, and Solitons (Conference) (2022 : Online)

Title: The versatility of integrability : in memory of Igor Krichever / Mikhail Bershtein, Anton Dzhamay, Andrei Okounkov, editors.

Description: Providence, Rhode Island : American Mathematical Society, [2025] | Series: Contemporary mathematics, 0271-4132 ; volume 823 | Isomonodromic Deformations, Painlevé Equations, and Integrable Systems, June 27–July 1, 2022, virtual, Columbia University, New York, NY, Algebraic Geometry, Mathematical Physics, and Solitons, October 7–9, 2022, Columbia University, New York, NY

Identifiers: LCCN 2025004187 | ISBN 9781470475260 (paperback) | ISBN 9781470480363 (ebook)

Subjects: LCSH: Curves, Algebraic–Congresses. | AMS: Algebraic geometry – Curves – Jacobians, Prym varieties. | Algebraic geometry – Curves – Relationships with integrable systems. | Algebraic geometry – Abelian varieties and schemes – Theta functions. | Algebraic geometry – Computational aspects in algebraic geometry – Curves. | Several complex variables and analytic spaces – Deformations of analytic structures – Moduli and deformations for ordinary differential equations (e.g. Knizhnik-Zamolodchikov equation). | Special functions – Hypergeometric functions – Other hypergeometric functions and integrals in several variables. | Partial differential equations – Representations of solutions – Asymptotic expansions. | Probability theory and stochastic processes – Stochastic processes – Point processes. | Quantum theory – Groups and algebras in quantum theory – Relations with integrable systems. | LCGFT: Conference papers and proceedings.

Classification: LCC QA565 .V47 2025 | DDC 516.3/52–dc23/eng/20250515
LC record available at https://lccn.loc.gov/2025004187

Contemporary Mathematics ISSN: 0271-4132 (print); ISSN: 1098-3627 (online)

DOI: https://doi.org/10.1090/conm/823

---

10 9 8 7 6 5 4 3 2 1   30 29 28 27 26 25

This volume is dedicated to the memory of Igor Krichever (1950–2022)
—a dear teacher, colleague, and friend.

Contents

Preface

The papers in this volume are related to two conferences that took place, under unusual and sad circumstances, in 2022:

- **IDPEIS-22**: Isomonodromic Deformations, Painlevé Equations, and Integrable Systems (June 27 – July 1, 2022);
- **AGMPS-22**: Algebraic Geometry, Mathematical Physics, and Solitons (October 7–9, 2022).

In addition to the closely related topics, these conferences were united by the figure of Igor Krichever, whose untimely passing on December 1, 2022, left a huge void that is and will be felt by many of those who knew him.

In Summer of 2021 there were good reasons to look forward to the year 2022 with a cautious optimism. The world slowly started to recover from the global COVID-19 pandemic. Students were gradually returning to classrooms, and scientific conferences began to move from virtual Zoom meetings to the traditional in-person format. And the mathematicians all over the world, but especially in Russia, were looking forward to the International Congress of Mathematicians that was scheduled to take place in Saint Petersburg in July 2022. Many people worked very hard to make ICM-2022 a success, and there was a lot of excitement and hope that this ICM would further promote international cooperation and allow young gifted students in Russia to meet people whose books and papers they were learning from.

It is customary to organize, in addition to the main Congress, a number of specialized satellite conferences. A large number of such conferences were also planned for the ICM-2022 all over Russia, to showcase the growth of mathematical research in the country, as well as the creation of several modern research centers.

At the suggestion of Igor Krichever, one such satellite conference was planned to take place at the Skolkovo Institute of Science and Technology (Skoltech) in Moscow. In one of his interviews before ICM-2022, Igor mentioned how personally important it was for him, in recognition of the guidance and support that he received from his teachers many years ago, to contribute to the education of young students in Russia and to help preserve and foster the traditions of the Soviet and Russian mathematical schools. Although Igor was a Full Professor at Columbia University since 1997, in recent years he was able to make arrangements to spend half of his time in Moscow. In 2016 Igor became the founding Director of Skoltech Center of Advanced Studies, which quickly became one of the best research centers in the world specializing in Mathematical Physics. Among its faculty were some of the best research mathematicians in the world, as well as a large group of very talented young people. The Center had an active graduate and post-graduate programs, and would have been an ideal place to have a conference on *Isomonodromic Deformations, Painlevé Equations, and Integrable Systems.*

That ICM was also very special for Igor—he was selected to be one of the plenary speakers. This great honor reflected many important contributions that Igor made to Mathematics, and in particular for his work on fruitful interaction between Integrable Systems and Algebraic Geometry.

Suddenly, everything started to fall apart. At the end of Summer 2021 Igor began having alarming health symptoms that were soon diagnosed as Amyotrophic Lateral Sclerosis (ALS), also known as the Lou Gehrig's disease—a terrible illness that results in a gradual loss of muscular functions. Then, in February 2022, Russia began a military invasion in Ukraine, which developed into the worst military conflict in Europe since the Second World War and plunged the World further into darkness. The plans to have the International Congress of Mathematicians ICM-2022 in Saint Petersburg had to be abandoned and it instead became a virtual event. Many mathematicians, especially the younger ones, had left the country, and the Skoltech Center was significantly affected by that. The enormous toll of these tragic developments, which have no clear end in sight, will be felt for many years.

Still, people tried very hard to save at least some of the planned events. Columbia University, Igor's home institution, was able to help and became a virtual host for the IDPEIS-22 conference. And Igor, after all, was able to give a wonderful plenary ICM address online. Watching this video one cannot really see the toll that the illness was taking on Igor, but things were moving fast and the prognosis was very bleak. That was the moment when the group of Igor's close colleagues and friends decided to organize a birthday conference for Igor — the original plan to have this conference to celebrate Igor's 70th birthday in 2020 in Moscow had to be abandoned because of COVID-19.

Igor Krichever at AGMPS-22: at the talks (left) and at the banquet (right)
(Photos by Andrei Okounkov, Columbia University. Reprinted with permission.)

With the help, again, of the Columbia University Department of Mathematics, we were able to organize the AGMPS-22 conference on a very short notice. Many of Igor's close friends and colleagues came to say goodbyes. Despite being very ill and confined to the wheelchair at that point, Igor attended all of the talks and was very engaged and genuinely happy. We were very glad we were able to have this conference to celebrate Igor and his life, but it was very painful to see the contrast between Igor's sharp mind and his declining body. Igor passed away shortly after the conference, on December 1, 2022.

This volume is a tribute to Igor Krichever. The range of papers in this volume, from very applied to quite abstract, gives a good perspective of the influence that Igor had on Mathematics. He is greatly missed.

Mikhail Bershtein
Anton Dzhamay
Andrei Okounkov

List of Conference Speakers

Enrico Arbarello (AGMPS-22)
Accademia Nazionale dei Lincei, Rome, Italy

Marco Bertola (IDPEIS-22)
Concordia University, Montreal, Quebec, Canada

Alexander Bobenko (IDPEIS-22)
Technische Universität Berlin, Germany

Alexei Borodin (IDPEIS-22)
Massachusetts Institute of Technology, Cambridge, MA, USA

Thomas Bothner (IDPEIS-22)
University of Bristol, Bristol, UK

Alexander Braverman (AGMPS-22)
University of Toronto, Toronto, Canada

Mattia Cafasso (IDPEIS-22)
LAREMA, Université d'Angers, Angers, France

Ivan Cherednik (AGMPS-22)
University of North Carolina, Chapel Hill, NC, USA

Philippe Di Francesco (IDPEIS-22)
University of Illinois at Urbana-Champaign, Urbana, IL, USA

Pavel Etingof (AGMPS-22)
Massachusetts Institute of Technology, Cambridge, MA, USA

Vladimir Fock (IDPEIS-22)
IRMA, Université de Strasbourg et CNRS, Strasbourg, France

Alexander Goncharov (IDPEIS-22)
Yale University, New Haven CT, USA

Alba Grassi (IDPEIS-22)
Université de Genève and CERN, Geneva, Switzerland

Tamara Grava (AGMPS-22)
University of Bristol, Bristol, UK and SISSA, Trieste, Italy

Rod Halburd (IDPEIS-22)
University College London, London, UK

Alexander Its (AGMPS-22)
Indiana University-Purdue University, Indianapolis, IL, USA

Alisa Knizel (IDPEIS-22 and AGMPS-22)
University of Chicago, Chicago, IL, USA

Peter Miller (IDPEIS-22)
University of Michigan, Ann Arbor, MI, USA

Nikita Nekrasov (IDPEIS-22 and AGMPS-22)
Simons Center for Geometry and Physics, Stony Brook University, Stony Brook, NY, USA

Masatoshi Noumi (IDPEIS-22)
Rikkyo University, Tokyo, Japan

Andrei Okounkov (AGMPS-22)
Columbia University, New York, NY, USA

Duong Phong (AGMPS-22)
Columbia University, New York, NY, USA

Axel Saenz (IDPEIS-22)
Oregon State University, Corvalis, OR, USA

Leon Takhtajan (AGMPS-22)
Stony Brook University, Stony Brook, NY, USA

Alexander Varchenko (AGMPS-22)
University of North Carolina, Chapel Hill, NC, USA

Alexander Veselov (AGMPS-22)
Loughborough University, Loughborough, UK

Paul Wiegmann (AGMPS-22)
University of Chicago, Chicago, IL, USA

Yasuhiko Yamada (IDPEIS-22)
Kobe University, Kobe, Japan

Dmitry Zakharov (AGMPS-22)
Central Michigan University, Mount Pleasan, MI, USA

Contemporary Mathematics
Volume **823**, 2025
https://doi.org/10.1090/conm/823/16489

Dimers and M-curves

Alexander I. Bobenko, Nikolai Bobenko, and Yuri B. Suris

Dedicated to Igor Krichever.

ABSTRACT. In this paper we develop a general approach to dimer models analogous to Krichever's scheme in the theory of integrable systems. We start with a Riemann surface and the simplest generic meromorphic functions on it and demonstrate how to obtain integrable dimer models. These are dimer models on doubly periodic bipartite graphs with quasi-periodic positive weights. Dimer models with periodic weights and Harnack curves are recovered as a special case. This generalization from Harnack curves to general M-curves leads to transparent algebro-geometric structures. In particular explicit formulas for the Ronkin function and surface tension as integrals of meromorphic differentials on M-curves are obtained. Furthermore we describe the variational principle for the height function in the quasi-periodic case. Based on Schottky uniformizations of Riemann surfaces we present concrete computational results including computing the weights and sampling dimer configurations with them. The computational results are in complete agreement with the theoretical predictions.

CONTENTS

The first and third authors were supported by the DFG Collaborative Research Center TRR 109 "Discretization in Geometry and Dynamics".

The second author was supported by the NCCR SwissMAP as well as by the Swiss NSF grants 200400 and 197226.

1. Introduction

Over the past two decades, dimer models became an essential part of integrable statistical mechanics, see for example [**11**, **25**, **29**]. Various aspects of integrable structure such as the invariance of the partition function under the spider move (Z-invariance), expression of the partition function, correlation functions and inverse of Dirac operators in terms of spectral curves, etc. proved to be a powerful tool in the investigation of models [**4**, **9**–**11**, **28**]. Moreover many integrable systems and systems with cluster algebra structure were identified within dimer models [**22**].

In this work we consider dimer models on planar bipartite graphs. In the foundational work [**23**] the partition function for finite graphs was shown to be the determinant of the Kasteleyn operator, which is a weighted signed adjacency matrix. Furthermore the partition function per fundamental domain was explicitly computed for periodic lattices with constant weights.

These results were generalized in [**25**] for isoradial graphs with critical weights. Further generalization for arbitrary doubly periodic weights was provided in [**29**] with the discovery of a remarkable connection to algebraic geometry. The Kasteleyn operator was used to define the spectral curve and it was shown that it is a Harnack curve, which is a classical object in algebraic geometry [**33**]. Many characteristics of dimer models were expressed purely in terms of algebro-geometric data. In particular a characterization of all ergodic Gibbs measures in terms of points in the Newton polygon was provided. Furthermore the free energy of a model with magnetic field was shown to correspond to the Ronkin function [**29**] and limit shapes for some boundary data were computed [**28**].

The solution of the inverse problem was given in [**18**] where the weights of the model were expressed in terms of algebro-geometric data. A comprehensive investigation of the corresponding models was carried out in [**10**] for spectral curves of genus 1 and in [**11**] for general genus. For any minimal graph with Fock's weights an explicit expression for inverse Kasteleyn operators was obtained yielding alternative expressions for the ergodic Gibbs measures in the doubly periodic case.

Research on dimer models and their relation to integrable systems continues to flourish. For example, an explicit description of limit shapes of the Aztec diamond as well as convergence of local fluctuations to the corresponding Gibbs measures was shown in [**4**]. The genus 1 case was studied before that in [**9**]. In [**6**] we apply the results of the present paper to obtain explicit dimer limit shapes for the Aztec diamond and the hexagon for general quasi-periodic weights. Recently a relation to circle patterns and Miquel dynamics was established in [**26**]. See also [**1**] for an elaborate investigation of variational problems of dimer limit shapes and [**13**] for an investigation of fluctuations via t-embeddings.

Somewhat similar methods based on the analysis on Riemann surfaces are known in the theory of integrable systems under the name of finite gap (or algebro-geometric) integration theory [**3**, **15**]. In 1977 Krichever [**31**] suggested an integration scheme based on the notion of Baker-Akhiezer (BA) function. This is a function on a Riemann surface which is uniquely determined by its special analytic properties, and is also expressed explicitly in terms of theta functions and Abelian integrals. From these properties one derives linear differential or difference equations satisfied by the BA function. The coefficients in these equations are also derived from explicit formulas for the BA function. Special (multiphase) solutions of many integrable systems, like Korteweg-de Vries (KdV), Kadomtsev-Petviashvili (KP), sine-Gordon etc. equations were constructed this way, see for example [**3**].

In this paper we develop a general approach to dimer models analogous to Krichever's scheme in the theory of integrable systems. We start with a Riemann surface and the simplest generic meromorphic functions on it and demonstrate how to obtain integrable dimer models (Section 3). At each vertex of a graph there is a BA-function. They satisfy certain local linear relations whose coefficients are the weights of the corresponding dimer model. Moreover, exactly as in the finite gap theory, we derive the explicit formulas for these coefficients from the formulas for the BA-functions, and they turn out to coincide with the Fock weights. It is remarkable that as a special case of our construction we obtain some fundamental algebro-geometric identities, like Fay's identity [**17, 34**] as well as the invariance of the dimer model under the spider move (Section 4). The latter is closely related to the multidimensional consistency [**7**] of the model, which is proved along the same lines (Section 5).

Whereas positivity of weights is crucial for statistical mechanics, Fock's original paper [**18**], as well as some other papers on integrable dimers associated to Riemann surfaces [**21, 22**] deal with complex Riemann surfaces that lead to complex weights. Similar problems of characterization of real and non-singular solutions were encountered in the theory of integrable systems from its early days [**3, 15**]. The class of Riemann surfaces leading to real, non-singular solutions of the KP-equation and to self-adjoint operators was identified as M-curves, [**16**]. An M-curve is a Riemann surface $\mathcal{R}$ of genus g with an anti-holomorphic involution τ and maximal number $g+1$ of real ovals $X = X_0 \cup X_1 \cup \dots X_g$. In particular, in 1985 Krichever [**32**] constructed a class of two-dimensional difference operators with positive weights, coming from M-curves.

In the context of dimer models in [**27**] it was established that doubly periodic dimer models are in one to one correspondence with Harnack curves. Their Riemann surfaces are M-curves. The above mentioned Krichever's operators can be identified with the Dirac operator with Fock's weights on the regular hexagonal lattice. For planar bipartite graphs with arbitrary faces a construction of Dirac operators with positive weights is presented in Section 6 and in [**11**][1]. In this setup one obtains quasi-periodic weights. Harnack curves appear in the special case of periodic weights (Section 7). We demonstrate that a generalization from Harnack curves to general M-curves leads to a general approach with general transparent algebro-geometric structures and self-contained presentation. It implies new results even in the classical $g = 0$ case.

An explicit formula for the free energy of a doubly periodic dimer model was derived in [**29**]. It was shown that it is given by the Ronkin function, defined on the amoeba of the corresponding Harnack curve. A remarkable generalization of all these notions was given by Krichever in [**30**] without any relation to dimer models. In particular he defined Harnack data $\mathcal{S}$ for M-curves, with the corresponding amoeba map and Ronkin function, inheriting most of the classical properties. In Section 8 we present a unified picture including diffeomorphisms of the (open) factor $\mathcal{R}_+^\circ = (\mathcal{R} \setminus X)/\tau$ to the generalized amoeba $\mathcal{A}_\mathcal{S}^\circ$ and to the generalized Newton polygon $\Delta_\mathcal{S}^\circ$. Further we present a new formula for Krichever's Ronkin function along with its Legendre dual. Our representation is given via an integral of a

[1]We discussed this construction with the authors of [**11**] in private correspondence in 2021.

holomorphic one-form. It is more convenient for investigation and computation than Krichever's original formula which is an integral of a two-form over a certain domain of $\mathcal{R}_+$. Note that for dimer models with doubly periodic weights the (Legendre) dual Ronkin function is known as surface tension [**29**].

A variational principle for the height function of a doubly periodic dimer model was obtained in [**14**]. It was shown that it is the minimizer of the surface tension functional with prescribed boundary conditions. The functional is convex which implies the uniqueness of the minimizer. In Section 9 we extend the notion of surface tension to quasi periodic weights and show that the corresponding height function is also given by its minimizer.

In Section 10 we show that all the above characteristics of dimer models including the weights, amoeba and polygon maps, surface tension as well the height function can be effectively computed. For this purpose we use the method proposed in [**5**] based on the Schottky uniformization of the corresponding M-curve. The characteristics of dimers are then represented via converging (Poincaré theta) series. This representation turns out to be useful even in the case of genus 0 which corresponds to Kenyon's isoradial weights.

Finally in Section 11 we present concrete computational results including computing general Fock weights and sampling dimer configurations with them. The computational results are in complete agreement with the theoretical predictions (See Fig. 1).

2. Dimer models on bipartite graphs

While one can consider the dimer model on an arbitrary (finite connected) graph, we will restrict our attention to graphs with certain additional structures.

Dimer model on planar graphs. We consider our graphs as *planar graphs*, i.e., graphs embedded in surfaces. Thus, the underlying graph is understood as $G = (V, E, F)$, with V being the set of vertices, E the set of edges, and F the set of faces (connected components of the complement of E in the surface). Actually, we will assume that the graph G is *simple*, so that each face of G is an n-gon with $n \geq 3$. We can introduce the *dual graph* $G^* = (V^*, E^*, F^*)$, for which $V^* = F$, $F^* = V$, and E, E^* are in a one-to-one correspondence, the corresponding edges are denoted $e \in E$, $e^* \in E^*$.

A *dimer configuration* on G is a subset $D \subset E$ such that each vertex of G is adjacent to exactly one edge from D. A purely combinatorial problem of counting all possible dimer configurations for a given graph turns into an appealing and important *dimer problem* of statistical mechanics if one supplies dimer configurations with probabilities, or Boltzmann weights. More precisely, one assumes that G has an *edge weighting* which is a function $\nu : E \to \mathbb{R}_+$, and one defines the Boltzmann weight of a dimer configuration D by

$$\nu(D) = \prod_{e \in D} \nu(e). \tag{2.1}$$

Then, the corresponding probability measure on the set of dimer configurations is given by

$$\mathbb{P}(D) = \frac{\nu(D)}{Z(G, \nu)}, \quad Z(G, \nu) = \sum_{D} \nu(D), \tag{2.2}$$

FIGURE 1. Random height function sampled according to Fock weights with a volume constraint and overlaid with theoretical prediction.[2]

where the summation in the formula for $Z(G,\nu)$ (the partition function) is extended over all possible dimer configurations on G.

We quote the famous Kasteleyn's formula for the dimers partition function for graphs embedded in the plane. For this, the following ingredients are necessary.

DEFINITION 2.1. Fix an orientation of the plane, and let the faces of the graph G embedded in the plane inherit this orientation. For each face $f \in F$, orient ∂f as the boundary of the oriented face f. This gives the *natural orientation* of the edges of ∂f. For an arbitrary orientation σ of the edges of G, denote by $N^\sigma(\partial f)$ the number of edges of the boundary ∂f for which σ induces the opposite to the

[2]This is a height function colored according to its local slope. We stress that it is practical to compute both the weights as well as the amoeba map numerically. This particular configuration was sampled on a (500×500) square in $\mathbb{Z}^2$ with a volume constraint of 0.3 volume per unit. The Schottky data used is given by $\{\beta^-, \alpha^+, \beta^+, \alpha^-\} = \{-2.4, -0.4, 0.4, 2.4\}$ and $(A, B, \mu) = (0.25 + i0.9, 0.25 - i0.9, 0.015)$.

natural orientation. An edge orientation σ is called a *Kasteleyn orientation*, if for each $f \in F$, the quantity $N^\sigma(\partial f)$ is odd.

THEOREM 2.2. *Every graph embedded in the plane admits a Kasteleyn orientation σ. If σ is such an orientation, the following formula holds true:*

$$Z(G,\nu) = |\operatorname{Pf}(A^\sigma)|, \tag{2.3}$$

where A^σ is the skew-symmetric matrix whose rows and columns are indexed by V, and the entries are given by

$$a^\sigma_{ij} = \begin{cases} \nu(e), & \text{if } \{i,j\} = e \in E, \text{ and } \sigma \text{ orients } e \text{ from } j \text{ to } i, \\ -\nu(e), & \text{if } \{i,j\} = e \in E, \text{ and } \sigma \text{ orients } e \text{ from } i \text{ to } j, \\ 0, & \text{if } \{i,j\} \notin E. \end{cases}$$

Dimer model on bipartite graphs. These results and techniques can be further specialized and refined under an additional assumption (which we adopt hereafter) that the graph G is *bipartite*, that is, the set of vertices is represented as $V = B \cup W$ (B being denoted as black vertices and W as white vertices), and each edge $e \in E$ connects a black vertex with a white one, $e = (bw)$. As an immediate corollary, each face of G is a $(2n)$-gon with $n \geq 2$.

For a bipartite graph, one can introduce an orientation of edges different from the natural one, say from the black vertex to the white one (let us call this the bw-orientation). Then it is easy to see that Definition 2.1 is equivalent to the following one.

DEFINITION 2.3. For an arbitrary orientation σ of the edges of G, denote by $M^\sigma(\partial f)$ the number of edges of the boundary ∂f which σ orients from white to black, i.e., opposite to the bw-orientation. An edge orientation σ is called a *Kasteleyn orientation*, if for each $(2n)$-gonal $f \in F$, we have $M^\sigma(\partial f) \equiv n+1 \pmod 2$.

For a bipartite graph, the matrix A^σ has (upon a suitable ordering of the index set V) a block structure:

$$A^\sigma = \begin{pmatrix} 0 & K_{WB} \\ K_{BW} & 0 \end{pmatrix},$$

where K_{WB} is a matrix whose rows are indexed by W and whose columns are indexed by B (and, of course, $K_{BW} = -K^{\mathrm{T}}_{WB}$). We call K_{WB} the *Kasteleyn matrix*. Its definition can be formulated as follows:

$$K_{wb} = \begin{cases} \nu(e), & \text{if } \{b,w\} = e \in E, \text{ and } \sigma \text{ orients } e \text{ from } b \text{ to } w, \\ -\nu(e), & \text{if } \{b,w\} = e \in E, \text{ and } \sigma \text{ orients } e \text{ from } w \text{ to } b, \\ 0, & \text{if } \{b,w\} \notin E. \end{cases}$$

Formula (2.3) takes the form

$$Z(G,\nu) = |\det(K_{WB})|, \tag{2.4}$$

An important issue is the *gauge transformations* of the dimer model. We call two weight functions ν, ν' gauge equivalent if, for each $e = \{i,j\} \in E$, we have $\nu'(e) = \nu(e)\lambda(i)\lambda(j)$ with some $\lambda : V \to \mathbb{R}_+$. One easily sees that the probability measure $\mathbb{P}(D)$ does not change under a gauge transformation of weights. For a bipartite G, one has the following characterization of gauge equivalent weights.

PROPOSITION 2.4. *Two weight functions ν, ν' are gauge equivalent iff for each $f \in F$ with $\partial f = e_1 \cup e_2 \cup \ldots \cup e_{2n}$, there holds*

$$W_f = \frac{\nu(e_1)\nu(e_3)\ldots\nu(e_{2n-1})}{\nu(e_2)\nu(e_4)\ldots\nu(e_{2n})} = \frac{\nu'(e_1)\nu'(e_3)\ldots\nu'(e_{2n-1})}{\nu'(e_2)\nu'(e_4)\ldots\nu'(e_{2n})}. \tag{2.5}$$

For the proof observe that, under the bw-orientation of edges, $\log \nu$ is a discrete 1-form on G. Equation (2.5) is equivalent to $\log \nu' - \log \nu$ being closed, which for graphs embedded in the plane is equivalent to $\log \nu' - \log \nu$ being exact, i.e., $\log \nu' - \log \nu = df$ for some $f : V \to \mathbb{R}_+$. Now set $\lambda(b) = \exp(-f(b))$ and $\lambda(w) = \exp(f(w))$ for any $b \in B$, $w \in W$.

One can extend the notion of gauge invariance by admitting complex-valued gauge functions $\lambda : V \to \mathbb{C}^*$, then the gauge equivalent weight functions are also allowed to take values in $\mathbb{C}^*$ rather than in $\mathbb{R}_+$.

It is important that the characterization of a Kasteleyn orientation for bipartite graphs is conveniently given in terms of the face gauge invariants (2.5): an edge orientation σ is a Kasteleyn orientation, if for each $f \in F$ with the boundary $(2n)$-gon $w_1b_1 \ldots w_nb_n$, the entries of the Kasteleyn matrix K_{WB} satisfy

$$\operatorname{sign}\left(\frac{K_{w_1b_1}K_{w_2b_2}\ldots K_{w_nb_n}}{K_{w_2b_1}K_{w_3b_2}\ldots K_{w_1b_n}}\right) = (-1)^{n+1}. \tag{2.6}$$

Of course, this condition will ensure the representation (2.4) for the partition function also for the case when the entries K_{wb} are from $\mathbb{C}^*$, with the understanding that the expression in the parentheses on the left-hand side must be real for each $f \in F$.

Height function. Any dimer configuration D on a planar bipartite graph G defines, via

$$\omega_D(bw) = -\omega_D(wb) = \begin{cases} 1, & \text{if } (bw) \in D, \\ 0, & \text{otherwise} \end{cases}$$

a discrete 1-form on G. This form has a special property of having divergence 1 at black vertices and -1 at white ones, where the divergence of ω at a vertex v is defined as $d^*\omega(v) = \sum_{w\sim v} \omega(vw)$. Such 1-forms are called *unit forms.*

Fix a unit form ω_0. Then $\omega = \omega_D - \omega_0$ is divergence free and therefore the dual form ω^* on the dual graph G^* is closed. Recall that the dual form ω^* is defined by $\omega^*(e^*) = \omega(e)$, where the edge $e^* \in E^*$ dual to $e \in E$ is directed so that it intersects the positively directed e from left to right. The form ω^* is closed in the sense that its integral (sum) around the boundary of any face $f^* \in F^*$ vanishes. Therefore, it is exact, $\omega^* = dh$, with a function $h : V^* = F \to \mathbb{R}$ defined, up to an additive constant, by integrating ω^* along paths in G^*. This function is called the *height function* of the dimer configuration D. Note that it depends on the choice of a unit form ω_0, but that for two dimer configurations D_1, D_2 the difference $h(D_1) - h(D_2)$ does not depend on the choice of ω_0, making this a more natural and intrinsic object than the individual height function $h(D)$.

Actually, a slight modification of this construction will be used. On the square grid, a usual choice of ω_0 is the Thurston's original one [**40**], with $\omega_0 = \frac{1}{4}$ on each positively (from black to white) directed edge. Likewise, on the hexagonal grid one can take $\omega_0 = \frac{1}{3}$ on each positively directed edge. However, with these standard choices, for a finite subgraph of a grid the boundary vertices will not have divergence ± 1. Therefore, the form ω^* will not be closed around the boundary

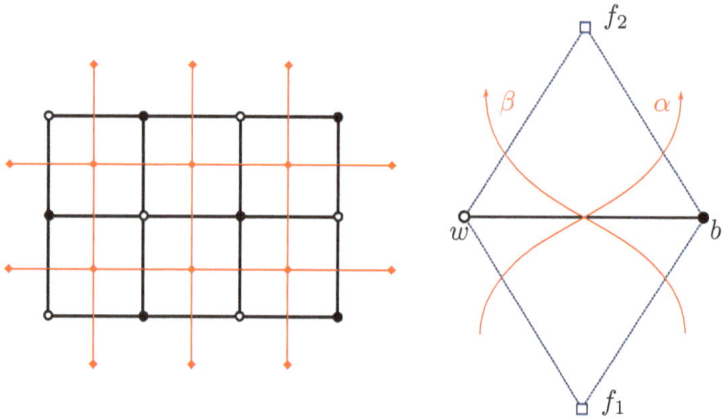

FIGURE 2. An example of a bipartite graph G and its dual graph G^*. The height function is defined on the vertices of G^* (left). A quadrilateral face of $G^\diamond$, with two directed strips through it (right).

vertices. This is usually remedied by a modification of the dual graph, in which the edges e^* dual to the boundary edges of G are considered to have distinct "remote" vertices (of valence 1) rather than to have a common vertex corresponding to the outer (unbounded) face of G, see Fig. 2. In such a modified dual graph there are no faces dual to the boundary vertices of G, i.e., no cycles around the boundary vertices of G. Note that the height function at the boundary (valence 1) vertices of G^* does not depend on the choice of the dimer configuration D.

Diamond graph. For the discussion of integrability properties of the dimer model, it is useful to introduce the concept of the *diamond graph* $G^\diamond$ associated to G and constructed as follows. The set of vertices of $G^\diamond$ is $V \sqcup V^*$. Each pair of dual edges, say $e = (bw) \in E(G)$ and $e^* = (f_1 f_2) \in E(G^*)$ define a quadrilateral $(bf_1 w f_2)$, see Fig. 2. These quadrilaterals constitute the faces of the cell decomposition (quad-graph) $G^\diamond$. Let us stress that edges of $G^\diamond$ belong neither to $E(G)$ nor to $E(G^*)$ (the latter are diagonals of the quadrilaterals).

A *strip, or a train track* in $G^\diamond$ is a sequence of quadrilateral faces $q_j \in F(G^\diamond)$ such that any pair q_{j-1}, q_j is adjacent along the edge $\mathfrak{e}_j = q_{j-1} \cap q_j$, and $\mathfrak{e}_j$, $\mathfrak{e}_{j+1}$ are opposite edges of q_j. The edges $\mathfrak{e}_j$ are called traverse edges of the strip. We visualize the strips as *strands* passing through the midpoints of the diagonals $e = (bw)$ of the quadrilaterals, and entering and leaving each quadrilateral through two opposite sides. Through the midpoint of any edge $e \in E(G)$ there pass two strips. We use the convention that these strands are directed so that white vertices lie to the left, and black vertices lie to the right of the strands. See Figure 2.

Infinite graphs. Our principal interest in this paper is towards understanding the thermodynamic limit of the dimer model on infinite regular lattices in $\mathbb{R}^2$, with the main examples being the regular square lattice and the regular hexagonal lattice. The diamond graphs for them are the regular square lattice again (rotated by 45°), resp. the regular dual kagome lattice. Both are instances of *quasicrystallic rhombic embeddings*, as introduced in [**3**, Sect. 6.8]. Those are quad-graphs embedded in the plane with all quadrilateral faces being rhombi with sides of the unit length from a finite set $\{\pm\theta_1, \dots, \pm\theta_d\} \subset \mathbb{S}^1$. As discussed in [**3**, Sect. 6.8], such quad-graph can be seen as a sort of projection of a certain two-dimensional subcomplex (stepped surface) of a multi-dimensional regular square lattice $\mathbb{Z}^d$. While the presentation in this paper will be mainly restricted to two above mentioned lattices (square and hexagonal), it admits a generalization to arbitrary quasicrystallic rhombic embeddings. A still more general class of quad-graphs for which similar theory can be developed constitute minimal isoradial immersions studied in [**10**] and [**11**].

3. Baker-Akhiezer functions and Fock weights

In this section, we recall the definition of Fock weights [**18**], which can be constructed on an arbitrary bipartite graph, with a data provided by an arbitrary compact Riemann surface. After that, we introduce *Baker-Akhiezer functions*, which are a key device in the theory of integrable systems, widely used in various contexts by I.M. Krichever (who also established this terminology).

Here are the main ingredients of the construction.

- $\mathcal{R}$ a compact Riemann surface of genus g, with a fixed homology basis $a_1, \dots, a_g, b_1, \dots, b_g$.
- $\boldsymbol{\omega} = (\omega_1, \dots, \omega_g)$ the set of normalized holomorphic differentials, with a-periods $\int_{a_j} \omega_k = \delta_{jk}$ and b-periods $\int_{b_j} \omega_k = B_{jk}$;
- $J(\mathcal{R}) = \mathbb{C}^g/(\mathbb{Z}^g + B\mathbb{Z}^g)$ the Jacobi variety of $\mathcal{R}$;
- $A : \mathcal{R} \to J(\mathcal{R})$, $P \mapsto A(P) = \int_{P_0}^{P} \boldsymbol{\omega}$ (mod $\mathbb{Z}^g + B\mathbb{Z}^g$) the Abel map.
- $\theta(z) = \theta(z|B)$ the theta function on $\mathbb{C}^g$ with the period matrix B. It is defined by

$$\theta(z) = \theta(z|B) = \sum_{m \in \mathbb{Z}^g} \exp\Big(\pi i \langle Bm, m\rangle + 2\pi i \langle z, m\rangle\Big). \tag{3.1}$$

  Its behavior under shifts of the argument by a period is given by

$$\theta(z + m + Bn) = \exp\big(-\pi i\langle n, Bn\rangle - 2\pi i \langle z, n\rangle\big)\theta(z), \quad m, n \in \mathbb{Z}^g. \tag{3.2}$$

  We will also need theta-functions with (half-integer) characteristics $\Delta = \begin{bmatrix}\Delta_1\\ \Delta_2\end{bmatrix}$, where each of Δ_1, Δ_2 is of the form $\frac{1}{2}(\delta_1, \dots, \delta_g)^{\mathrm{T}}$ with $\delta_j \in \{0, 1\}$:

$$\begin{aligned}\theta[\Delta](z) &= \sum_{m \in \mathbb{Z}^g} \exp\Big(\pi i \langle B(m + \Delta_1), m + \Delta_1\rangle + 2\pi i\langle z + \Delta_2, m + \Delta_1\rangle\Big)\\ &= \exp\Big(\pi i \langle B\Delta_1, \Delta_1\rangle + 2\pi i \langle z + \Delta_2, \Delta_1\rangle\Big)\theta(z + \Delta_2 + B\Delta_1).\end{aligned} \tag{3.3}$$

  The characteristic Δ is called even (odd), if $4\langle \Delta_1, \Delta_2\rangle \equiv 0$ (resp. $\equiv 1$ (mod 2)). The function $\theta[\Delta](z)$ is even (odd), iff the characteristic Δ is even (resp. odd).
- The prime form $E(\alpha, \beta)$, which can be expressed through a theta function with an odd characteristic Δ:

$$E(\alpha, \beta) = \frac{\theta[\Delta]\Big(\int_\alpha^\beta \boldsymbol{\omega}\Big)}{h_\Delta(\alpha) h_\Delta(\beta)} \tag{3.4}$$

  with a holomorphic spinor h_Δ (the result being independent of Δ).

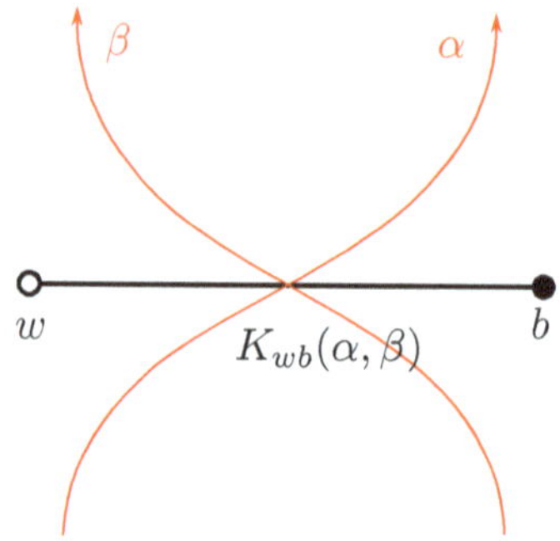

FIGURE 3. Coefficient $K_{wb}(\alpha, \beta)$ of the Dirac operator on an edge wb

Given a planar bipartite graph G (as discussed in Section 2), and a compact Riemann surface $\mathcal{R}$, we start with a *labeling* of the strips of $G^\diamond$, assigning to each (directed) strip a point $\alpha \in \mathcal{R}$. The following notions were introduced by Fock in [**18**].

DEFINITION 3.1. For a labeled graph $G^\diamond$, the *discrete Abel map* $\eta : V \cup F = V(G^\diamond) \to J(\mathcal{R})$ is defined as follows: set $\eta(v_0) = 0$ for some $v_0 \in V(G^\diamond)$, and then extend it to the whole of $V(G^\diamond)$ according to the following recursive rule (we use here notations from Fig. 2):

$$\eta(b) = \eta(f_1) - A(\beta) = \eta(f_2) - A(\alpha), \tag{3.5}$$

$$\eta(w) = \eta(f_2) + A(\beta) = \eta(f_1) + A(\alpha), \tag{3.6}$$

which also implies

$$\eta(b) = \eta(w) - A(\alpha) - A(\beta), \qquad \eta(f_2) = \eta(f_1) + A(\alpha) - A(\beta). \tag{3.7}$$

In a more informal way, this rule can be formulated as follows: if a (directed) edge of $G^\diamond$ traverses the strip labelled by α from right to left, then it contributes $A(\alpha)$, while traversing the strip from left to right contributes $-A(\alpha)$ to the discrete Abel map. This is well defined (closed along cycles in $G^\diamond$).

DEFINITION 3.2. For a labeled bipartite graph G, *Fock weights* are coefficients on (directed) edges of G depending on the labels α, β of the two strips intersecting the edge defined by the formula:

$$K_{wb}(\alpha, \beta) = \frac{E(\alpha, \beta)}{\theta(\eta(f_1) + D)\theta(\eta(f_2) + D)}, \tag{3.8}$$

where η is the discrete Abel map, and $D \in \mathbb{C}^g$ is arbitrary. Here we use notations of Fig. 3), the rule is as follows: going from w to b, and then turning left at the midpoint, one first meets the strand labeled α, then the strand labeled β.

The matrix $K_{WB} = (K_{wb}(\alpha, \beta))_{w\in W,\, b\in B}$ will play the role of the Kasteleyn matrix of a dimer model on G, as explained in Section 2, provided these weights are gauge equivalent to a system of positive weights. A construction enforcing this positivity property, based on M-curves $\mathcal{R}$, is the subject of Section 6.

In the next sections, we will be focusing on the integrability properties of the *Dirac operator*

$$K_{WB} : \mathbb{C}^B \to \mathbb{C}^W \tag{3.9}$$

defined, for $\psi \in \mathbb{C}^B$, by

$$(K_{WB}\psi)_w = \sum_{b\sim w} K_{wb}(\alpha,\beta)\psi_b. \tag{3.10}$$

The study of integrability is based on the construction of *Baker-Akhiezer functions* $(\psi_b(P))_{b\in B}$, which are functions on the Riemann surface $\mathcal{R}$. In the following their argument is denoted by $P \in \mathcal{R}$. These functions turn out to lie in the kernel of the Dirac operator K_{WB}.

We introduce more combinatorial structure for the set B. Namely, we define the *black graph* (B, E_B), by declaring $b_1, b_2 \in B$ to be neighbors, connected by an edge $(b_1 b_2) \in E_B$, if there exists $w \in W$ such that the path $b_1 w b_2$ is a part of the boundary of a face of G.

The following two analytic facts are required for introducing the Baker-Akhiezer functions.

PROPOSITION 3.3. *Let $\omega_{\alpha-\beta}$ be the unique Abelian differential of the third kind with vanishing a-periods and with exactly two poles at α and β with the residues $+1$ resp. -1 at these points. Then*

$$\exp\left(\int_Q^P \omega_{\alpha-\beta}\right) = \frac{E(P,\alpha)\,E(Q,\beta)}{E(P,\beta)\,E(Q,\alpha)}. \tag{3.11}$$

PROPOSITION 3.4. *For any $D \in J(\mathcal{R})$, $\alpha, \beta \in \mathcal{R}$, the following expression defines a genuine meromorphic function on $\mathcal{R}$:*

$$\psi(P) = \frac{\theta\big(A(P) + A(\alpha) - A(\beta) + D\big)}{\theta\big(A(P) + D\big)}\,\frac{E(P,\alpha)}{E(P,\beta)}. \tag{3.12}$$

DEFINITION 3.5. For $b \in B$, we define the *Baker-Akhiezer functions* $\psi_b : \mathcal{R} \to \mathbb{C}$ by the following rule. Fix some $b_0 \in B$ and set $\psi_{b_0}(P) = 1$. For any $b \in B$, let there be a path $b_0 b_1 \dots b_n$ in the black graph connecting b_0 to $b_n = b$ (so that all $b_k \in B$, and there exist $w_k \in W$ such that each two-edge path $b_{k-1} w_k b_k$ is a part of the boundary of a face of G). Then, in notations of Fig. 4,

$$\psi_b(P) = \frac{\theta\big(A(P) + \eta(b) + D\big)}{\theta\big(A(P) + \eta(b_0) + D\big)} \prod_{k=1}^{n} \frac{E(P,\alpha_k^-)}{E(P,\alpha_k^+)} \tag{3.13}$$

Since

$$\eta(b) - \eta(b_0) = \sum_{k=1}^{n} (A(\alpha_k^-) - A(\alpha_k^+)), \tag{3.14}$$

all $\psi_b(P)$ are, by Proposition 3.4, meromorphic functions on $\mathcal{R}$. The transition from $\psi_{b_{k-1}}(P)$ to $\psi_{b_k}(P)$ is described by the transition function

$$\frac{\psi_{b_k}(P)}{\psi_{b_{k-1}}(P)} = \frac{\theta\big(A(P) + \eta(b_k) + D\big)}{\theta\big(A(P) + \eta(b_{k-1}) + D\big)}\,\frac{E(P,\alpha_k^-)}{E(P,\alpha_k^+)}, \tag{3.15}$$

which adds a zero α_k^- and a pole α_k^+ (all $\psi_b(P)$ have additionally poles at the zeros of $\theta(A(P) + \eta(b_0) + D)$, which form a fixed effective divisor governed by the point D).

REMARK 3.6. Note that this definition is independent of the choice of path $b_0 b_1 \dots b_n$. Indeed, ψ_{b_n}/ψ_{b_0} contains terms only for every train track separating b_n and b_0.

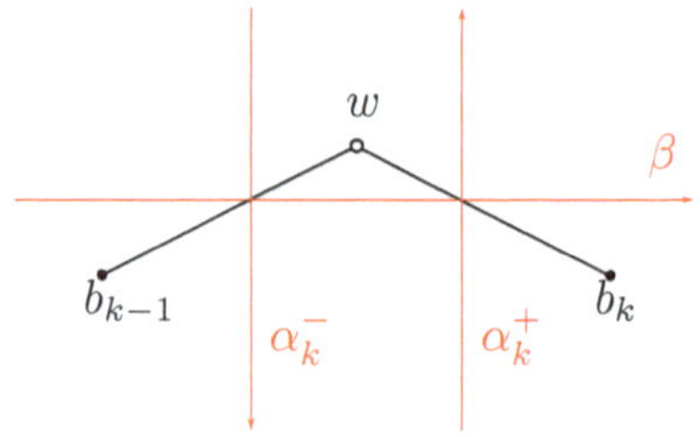

FIGURE 4. Strips along a path in G corresponding to an edge $(b_{k-1}b_k)$ in the black graph. The increment of the discrete Abel map $\eta(b_k)-\eta(b_{k-1}) = A(\alpha_k^-)-A(\alpha_k^+)$ only depends on the strands α_k^- and α_k^+ intersecting that edge

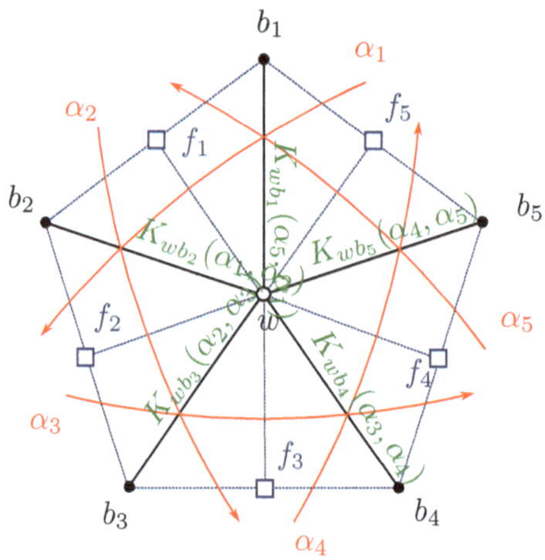

FIGURE 5. Star of a white vertex

THEOREM 3.7. *For any $P \in \mathcal{R}$, the vector $(\psi_b(P))_{b\in B} \in \mathbb{C}^B$ is in the kernel of the Dirac operator K_{WB}. In other words: consider the star of a white vertex $w \in W$, consisting of $b_1, b_2, \ldots, b_n \in B$, see Fig. 5. The associated Baker-Akhiezer functions satisfy the following relation:*

$$\sum_{k=1}^{n} K_{wb_k}(\alpha_{k-1}, \alpha_k)\psi_{b_k}(P) = 0, \tag{3.16}$$

where

$$K_{wb_k}(\alpha_{k-1}, \alpha_k) = \frac{E(\alpha_{k-1}, \alpha_k)}{\theta\big(\eta(f_{k-1}) + D\big)\theta\big(\eta(f_k) + D\big)}. \tag{3.17}$$

PROOF. For the sake of brevity, we write in this proof ψ_k for ψ_{b_k}. Since the transition from ψ_{k-1} to ψ_k adds the zero α_{k-2} and the pole α_k, and due to the corresponding cancellations, we find:

$$\frac{\psi_2(P)}{\psi_1(P)} = \frac{\theta\big(A(P)+\eta(b_2)+D\big)}{\theta\big(A(P)+\eta(b_1)+D\big)} \frac{E(P,\alpha_n)}{E(P,\alpha_2)},$$

$$\frac{\psi_k(P)}{\psi_1(P)} = \frac{\theta\big(A(P)+\eta(b_k)+D\big)}{\theta\big(A(P)+\eta(b_1)+D\big)} \frac{E(P,\alpha_n)}{E(P,\alpha_k)} \frac{E(P,\alpha_1)}{E(P,\alpha_{k-1})} \quad \text{for} \quad 3 \le k \le n-1,$$

$$\frac{\psi_n(P)}{\psi_1(P)} = \frac{\theta\big(A(P)+\eta(b_n)+D\big)}{\theta\big(A(P)+\eta(b_1)+D\big)} \frac{E(P,\alpha_1)}{E(P,\alpha_{n-1})}.$$

We determine the coefficients $c_2, \dots, c_n$ in

$$1 + \sum_{k=2}^{n} c_k \frac{\psi_k(P)}{\psi_1(P)} = 0$$

requiring that the residues at the poles $\alpha_2, \dots, \alpha_{n-1}$ vanish (this gives $n-2$ conditions), then, upon multiplication with $\theta(A(P)+\eta(b_1)+D)$, this linear combination equals a constant, and then requiring that this linear combination vanishes at α_n (the $(n-1)$-st condition), then it vanishes identically.

The point α_2 is the pole of ψ_2/ψ_1 and of ψ_3/ψ_1, with the residues

$$\frac{\theta\big(\eta(b_2)+A(\alpha_2)+D\big)}{\theta\big(\eta(b_1)+A(\alpha_2)+D\big)} E(\alpha_2,\alpha_n) \quad \text{and} \quad \frac{\theta\big(\eta(b_3)+A(\alpha_2)+D\big)}{\theta\big(\eta(b_1)+A(\alpha_2)+D\big)} \frac{E(\alpha_2,\alpha_n)}{E(\alpha_2,\alpha_3)} E(\alpha_2,\alpha_1),$$

respectively. Therefore, we kill this pole by choosing

$$\frac{c_2}{c_3} = \frac{\theta\big(\eta(b_3)+A(\alpha_2)+D\big)}{\theta\big(\eta(b_2)+A(\alpha_2)+D\big)} \frac{E(\alpha_1,\alpha_2)}{E(\alpha_2,\alpha_3)} = \frac{\theta\big(\eta(f_3)+D\big)}{\theta\big(\eta(f_1)+D\big)} \frac{E(\alpha_1,\alpha_2)}{E(\alpha_2,\alpha_3)}.$$

For $3 \le k \le n-2$, the point α_k is the pole of ψ_k/ψ_1 and of ψ_{k+1}/ψ_1, with the residues

$$\frac{\theta\big(\eta(b_k)+A(\alpha_k)+D\big)}{\theta\big(\eta(b_1)+A(\alpha_k)+D\big)} E(\alpha_k,\alpha_n) \frac{E(\alpha_k,\alpha_1)}{E(\alpha_k,\alpha_{k-1})}$$

and

$$\frac{\theta\big(\eta(b_{k+1}+A(\alpha_k)+D\big)}{\theta\big(\eta(b_1)+A(\alpha_k)+D\big)} \frac{E(\alpha_k,\alpha_n)}{E(\alpha_k,\alpha_{k+1})} E(\alpha_k,\alpha_1),$$

respectively. Therefore, we kill this pole by choosing

$$\frac{c_k}{c_{k+1}} = \frac{\theta\big(\eta(b_{k+1})+A(\alpha_k)+D\big)}{\theta\big(\eta(b_k)+A(\alpha_k)+D\big)} \frac{E(\alpha_{k-1},\alpha_k)}{E(\alpha_k,\alpha_{k+1})} = \frac{\theta\big(\eta(f_{k+1})+D\big)}{\theta\big(\eta(f_{k-1})+D\big)} \frac{E(\alpha_{k-1},\alpha_k)}{E(\alpha_k,\alpha_{k+1})}.$$

Finally, the point α_{n-1} is the pole of ψ_{n-1}/ψ_1 and of ψ_n/ψ_1, with the residues

$$\frac{\theta\big(\eta(b_{n-1})+A(\alpha_{n-1})+D\big)}{\theta\big(\eta(b_1)+A(\alpha_{n-1})+D\big)} E(\alpha_{n-1},\alpha_n) \frac{E(\alpha_{n-1},\alpha_1)}{E(\alpha_{n-1},\alpha_{n-2})}$$

and

$$\frac{\theta\big(\eta(b_n)+A(\alpha_{n-1})+D\big)}{\theta\big(\eta(b_{n-1})+A(\alpha_{n-1})+D\big)} E(\alpha_{n-1},\alpha_1).$$

Therefore, we kill this pole by choosing

$$\frac{c_{n-1}}{c_n} = \frac{\theta\big(\eta(b_n)+A(\alpha_{n-1})+D\big)}{\theta\big(\eta(b_{n-1})+A(\alpha_{n-1})+D\big)} \frac{E(\alpha_{n-2},\alpha_{n-1})}{E(\alpha_{n-1},\alpha_n)} = \frac{\theta\big(\eta(f_n)+D\big)}{\theta\big(\eta(f_{n-2})+D\big)} \frac{E(\alpha_{n-2},\alpha_{n-1})}{E(\alpha_{n-1},\alpha_n)}.$$

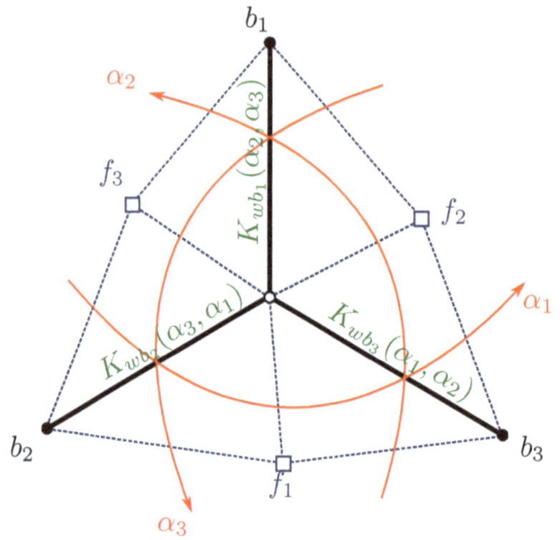

FIGURE 6. Coeffficients of the Dirac operator on a triangle

The final condition of vanishing at α_n gives:

$$1 + c_n \frac{\theta\big(\eta(b_n) + A(\alpha_n) + D\big)}{\theta\big(\eta(b_1) + A(\alpha_n) + D\big)} \frac{E(\alpha_n, \alpha_1)}{E(\alpha_n, \alpha_{n-1})} = 0,$$

or

$$c_n = \frac{\theta\big(\eta(b_1) + A(\alpha_n) + D\big)}{\theta\big(\eta(b_n) + A(\alpha_n) + D\big)} \frac{E(\alpha_{n-1}, \alpha_n)}{E(\alpha_n, \alpha_1)} = \frac{\theta\big(\eta(f_1) + D\big)}{\theta\big(\eta(f_{n-1}) + D\big)} \frac{E(\alpha_{n-1}, \alpha_n)}{E(\alpha_n, \alpha_1)}.$$

Dividing all coefficients by a common factor $\theta\big(\eta(f_n) + D\big)\theta\big(\eta(f_1) + D\big)/E(\alpha_n, \alpha_1)$, we find:

$$c_k = \frac{E(\alpha_{k-1}, \alpha_k)}{\theta\big(\eta(f_{k-1}) + D\big)\theta\big(\eta(f_k) + D\big)}, \quad k = 1, \dots, n. \qquad \square$$

REMARK 3.8. One can introduce Baker-Akhiezer "functions" on white vertices, which will lie in the kernel of the adjoint Dirac operator with the Fock weights. However, these dual objects are meromorphic 1-forms rather than functions on $\mathcal{R}$.

4. Fay identity and spider move

A particular case of Theorem 3.7 for $n = 3$ leads to a simple proof of the famous Fay formula. In this case, a different enumeration of the strips and of the faces leads to a more symmetric notation, see Fig. 6.

Moreover, one gets a symmetric expression for the discrete Abel map on the faces of G, defined by

$$\eta(f_2) - \eta(f_1) = A(\alpha_1) - A(\alpha_2),$$
$$\eta(f_3) - \eta(f_3) = A(\alpha_2) - A(\alpha_3),$$
$$\eta(f_1) - \eta(f_3) = A(\alpha_3) - A(\alpha_1),$$

if one sets

$$\eta(f_1) = A(\alpha_2) + A(\alpha_3), \quad \eta(f_2) = A(\alpha_3) + A(\alpha_1), \quad \eta(f_3) = A(\alpha_1) + A(\alpha_2).$$

This corresponds also to

$$\eta(b_1) = A(\alpha_1), \quad \eta(b_2) = A(\alpha_2), \quad \eta(b_3) = A(\alpha_3). \tag{4.1}$$

Thus, we get:

PROPOSITION 4.1. *The Baker-Akhiezer functions* $\psi_1(P)$, $\psi_2(P)$, $\psi_3(P)$ *are linearly dependent:*

$$K_{wb_1}(\alpha_2, \alpha_3)\psi_1 + K_{wb_2}(\alpha_3, \alpha_1)\psi_2 + K_{wb_3}(\alpha_1, \alpha_2)\psi_3 = 0, \tag{4.2}$$

or, more detailed:

$$\begin{aligned} &\frac{E(\alpha_2, \alpha_3)}{\theta\big(A(\alpha_1) + A(\alpha_2) + D\big)\theta\big(A(\alpha_1) + A(\alpha_3) + D\big)}\psi_1(P) \\ &+\frac{E(\alpha_3, \alpha_1)}{\theta\big(A(\alpha_2) + A(\alpha_3) + D\big)\theta\big(A(\alpha_1) + A(\alpha_2) + D\big)}\psi_2(P) \\ &+\frac{E(\alpha_1, \alpha_2)}{\theta\big(A(\alpha_1) + A(\alpha_3) + D\big)\theta\big(A(\alpha_2) + A(\alpha_3) + D\big)}\psi_3(P) = 0. \end{aligned} \tag{4.3}$$

Clearing denominators, this can be written as

$$\begin{aligned} &\theta\big(A(\alpha_2) + A(\alpha_3) + D\big)E(\alpha_2, \alpha_3)\psi_1(P) \\ &+\theta\big(A(\alpha_1) + A(\alpha_3) + D\big)E(\alpha_3, \alpha_1)\psi_2(P) \\ &+\theta\big(A(\alpha_1) + A(\alpha_2) + D\big)E(\alpha_1, \alpha_2)\psi_3(P) = 0. \end{aligned} \tag{4.4}$$

REMARK 4.2. Recall that, according to (3.15) and (4.1),

$$\begin{aligned} \frac{\psi_2(P)}{\psi_1(P)} &= \frac{\theta\big(A(P) + A(\alpha_2) + D\big)}{\theta\big(A(P) + A(\alpha_1) + D\big)} \frac{E(P, \alpha_2)}{E(P, \alpha_1)}, \\ \frac{\psi_3(P)}{\psi_1(P)} &= \frac{\theta\big(A(P) + A(\alpha_3) + D\big)}{\theta\big(A(P) + A(\alpha_1) + D\big)} \frac{E(P, \alpha_3)}{E(P, \alpha_1)}. \end{aligned} \tag{4.5}$$

Substituting (4.5) into (4.4) and clearing denominators again, we find the *Fay identity* in the following form:

$$\begin{aligned} &\theta\big(A(\alpha_2) + A(\alpha_3) + D\big)\theta\big(A(P) + A(\alpha_1) + D\big)E(\alpha_2, \alpha_3)E(P, \alpha_1) \\ &+\theta\big(A(\alpha_1) + A(\alpha_3) + D\big)\theta\big(A(P) + A(\alpha_2) + D\big)E(\alpha_3, \alpha_1)E(P, \alpha_2) \\ &+\theta\big(A(\alpha_1) + A(\alpha_2) + D\big)\theta\big(A(P) + A(\alpha_3) + D\big)E(\alpha_1, \alpha_2)E(P, \alpha_3) = 0, \end{aligned} \tag{4.6}$$

where the four points $P, \alpha_1, \alpha_2, \alpha_3$ play symmetric roles.

PROPOSITION 4.3. *Equation* (4.2) *is consistent with the spider move.*

The statement means that for the system of Dirac equations on the black graph, solutions do not change under local changes of the graph called "spider moves" and depicted on Fig. 7. This is a local statement, and to demonstrate it, one can consider suitable boundary value problems for two systems consisting of equations (4.2) on the both graphs, and show that they admit the same solution. A suitable boundary problem, defining the solution uniquely, consists in prescribing the values at any two points connected by an edge of the black graph, say at b_1 and b_2. To show is that the solutions at b_3 and b_4 for both graphs coincide.

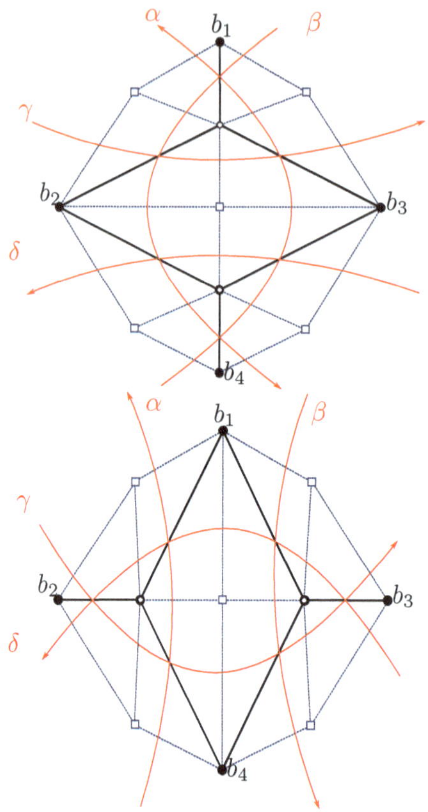

FIGURE 7. Spider move

PROOF. The argument is a combination of three observations.

- The discrete Abel maps at the black vertices of the both graphs coincide. This follows from eq. (3.14).
- The BA functions $\psi_b(P)$ provide us then with an infinite-dimensional family of (linearly independent) solutions of the system of equations (4.2) on both graphs.
- For each of the graphs, the space of solutions with prescribed initial data at b_1, b_2 is two-dimensional, therefore the solutions can be represented as linear combination of two linearly independent solutions from the infinite-dimensional family of BA functions. Therefore the solutions of the boundary value problem at b_3, b_4 coincide for both graphs. □

This argument (solutions of suitable boundary value problems for two systems of linear equations coincide, if these systems admit a common infinite-dimensional family of linearly independent solutions) works in similar situations below, for instance, for the proof of the 3D consistency of the system of Dirac equations on the square lattice in the next section.

5. Dirac operator on quad-graphs and multidimensional consistency

Consider the graph G for which the black vertices form the lattice $\mathbb{Z}^2$, and the white lattices form the lattice $(\mathbb{Z}+\frac{1}{2})^2$, compare Fig. 13.

Setting the zero of the discrete Abel map at $b_0=(0,0)$, one easily computes, with the help of eq. (3.14),

$$\eta(b_{mn})=\sum_{i=0}^{m-1}A(\alpha_i^- - \alpha_i^+)+\sum_{j=0}^{n-1}A(\beta_j^- - \beta_j^+).$$

The Baker-Akhiezer functions on $B=\mathbb{Z}^2$ are given by

$$\psi_{mn}(P)=\frac{\theta\big(A(P)+\eta(b_{mn})+D\big)}{\theta\big(A(P)+D\big)}\prod_{i=0}^{m-1}\frac{E(P,\alpha_i^-)}{E(P,\alpha_i^+)}\prod_{j=0}^{n-1}\frac{R(P,\beta_j^-)}{E(P,\beta_j^+)}. \tag{5.1}$$

On each elementary square of $B=\mathbb{Z}^2$, the BA functions $\psi_{mn}(P)$ satisfy the four-point Dirac equation (quad-equation) (3.16). This system of quad-equations in the case $g=1$ was the subject of [**8**]. In this case, if all $\alpha_i^\pm=\alpha^\pm$, $\beta_j^\pm=\beta^\pm$ are independent of i, resp. of j, and if $2A(\alpha^- - \alpha^+)=2A(\beta^- - \beta^+)\equiv 0$ in $J(\mathcal{R})$, then the coefficients of the quad-equations are 2-periodic in both directions.

Remarkably, one can extend the construction of BA functions (5.1) to a multi-dimensional lattice $\mathbb{Z}^d$. Each edge e of $\mathbb{Z}^d$ is labelled by a two-point divisor $\alpha_e^- - \alpha_e^+$, so that opposite edges of any elementary square have the same labels. Thus, now the strips of equally labeled edges are associated with hyperplanes orthogonally bisecting the edges. For any black point $b\in\mathbb{Z}^d$ connected to $b_0=0$ by a path of edges $e_1,\dots,e_p$, the discrete Abel map is given by

$$\eta(b)=\sum_{k=1}^{p}A(\alpha_{e_k}^- - \alpha_{e_k}^+).$$

Obviously, it is independent of the path. The same is true for the BA functions, which are defined by

$$\psi_b(P)=\frac{\theta\big(A(P)+\eta(b)+D\big)}{\theta\big(A(P)+D\big)}\prod_{k=1}^{p}\frac{E(P,\alpha_{e_k}^-)}{E(P,\alpha_{e_k}^+)}.$$

Quadruples of the BA functions at the vertices elementary squares of $\mathbb{Z}^d$ still satisfy four-point Dirac equations (3.16). The system of Dirac equations is multi-dimensionally consistent in the sense of [**7**]. As explained in this book, multi-dimensional consistency of quad-equations essentially follows from their 3D consistency.

Proposition 5.1. *The system of four-point Dirac equations on $\mathbb{Z}^d$ is 3D consistent.*

The statement means that a suitable boundary value problem for the system consisting of four-point Dirac equations on the six faces of a 3D cube admits a unique solution. The labelling for this system is shown on Fig. 8. A suitable boundary problem consists in prescribing four values, say at $b_0=(000)$, $b_1=(100)$, $b_2=(010)$ and $b_{13}=(101)$. To show is that the solutions at $b_{23}=(011)$ and $b_{123}=(111)$ for both graphs coincide.

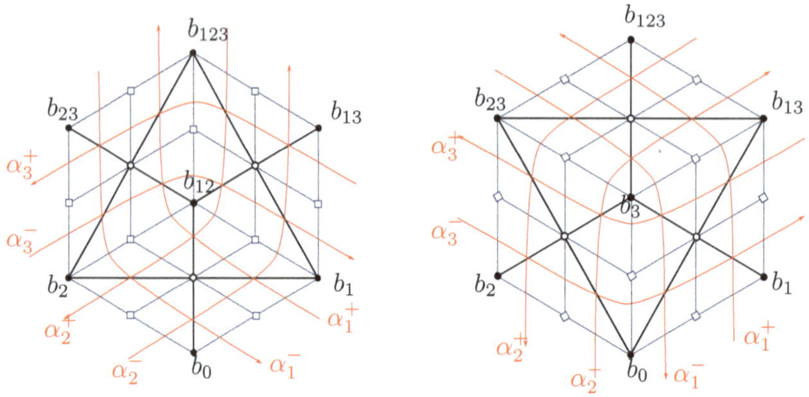

FIGURE 8. 3D consistency

PROOF. It goes along the same lines as the proof of Proposition 4.3.

- The discrete Abel maps at the black vertices of the both graphs coincide, as pointed out above.
- The BA functions $\psi_b(P)$ provide us with an infinite-dimensional family of (linearly independent) solutions of the system of Dirac equations on all six faces of the cube.
- For each of the graphs, the space of solutions with prescribed initial data at b_0, b_1, b_2 and b_{13} is four-dimensional, therefore the solutions can be represented as linear combination of four linearly independent solutions from the infinite-dimensional family of BA functions. Therefore the solutions of the boundary value problem at the common vertices $b_{12} = (110)$ and $b_{23} = (011)$ coincide for both graphs. □

Since every quasicrystallic rhombic embedding can be realized as a combinatorial surface in $\mathbb{Z}^d$ whose faces are elementary squares of $\mathbb{Z}^d$, we can restrict the above construction to any such surface. Obviously, functions $\psi_b(P)$ at the vertices of every face of the surface satisfy the four-point Dirac equation. Strips on the surface are the intersections of combinatorial coordinate hyperplanes in $\mathbb{Z}^d$ with the surface.

Such surfaces (with their respective systems of strips and Dirac equations) can be transformed one into another by local flips like on Fig. 8. These flips do not affect physical properties of the corresponding dimer model.

6. Kasteleyn property of Fock weights from M-curves

We now turn to the study of the dimer model on G with the Fock weights given by (3.8). Recall that there exists a gauge equivalent real-valued weight function whose signs define a Kasteleyn orientation, if for every face of G condition (2.6) is satisfied. In the present context, including the labeling of strands by points of $\mathcal{R}$, this condition looks as follows:

$$\text{sign}\left(\frac{K_{w_1b_1}(\alpha_1,\beta_1)}{K_{w_2b_1}(\beta_2,\alpha_1)}\cdot\frac{K_{w_2b_2}(\alpha_2,\beta_2)}{K_{w_3b_2}(\beta_3,\alpha_2)}\cdots\frac{K_{w_nb_n}(\alpha_n,\beta_n)}{K_{w_1b_n}(\beta_1,\alpha_n)}\right)=(-1)^{n+1}. \tag{6.1}$$

for any face of the graph G, see Fig. 9.

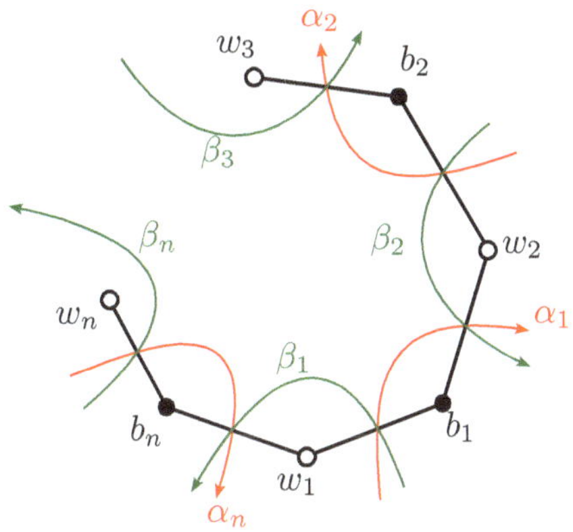

FIGURE 9. A face of the graph G

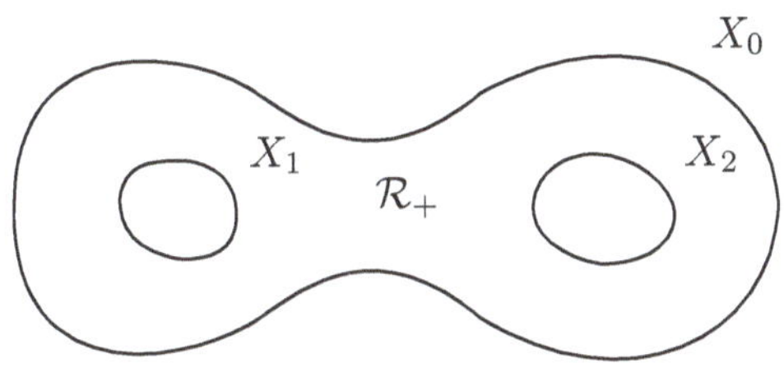

FIGURE 10. Upper half $\mathcal{R}_+$ of an M-curve $\mathcal{R}$.

We show how construct the Fock weights with the Kasteleyn property from M-curves.

DEFINITION 6.1. A Riemann surface $\mathcal{R}$ of genus g with an anti-holomorphic involution $\tau : \mathcal{R} \to \mathcal{R}$ is called an *M-curve* if it has the maximal possible number $g+1$ of real ovals $X_0, \ldots, X_g$, see Fig. 10.

Let $\mathcal{R}$ be an M-curve. Choose the canonical base of cycles: $a_1, \ldots, a_g, b_1, \ldots, b_g$ with the intersection numbers $a_j \circ a_k = b_j \circ b_k = 0$, $a_j \circ b_k = \delta_{jk}$ as follows: $a_1 = X_1, \ldots, a_g = X_g$, so that $\tau(a_j) = a_j$, and $\tau(b_j) = -b_j$. Choose a basis of holomorphic differentials $\omega_1, \ldots, \omega_g$ on $\mathcal{R}$ normalized by the condition $\int_{a_k} \omega_j = \delta_{jk}$, then $\overline{\tau^* \omega_j} = \omega_j$ on a_k, while the b-periods $B_{jk} = \int_{b_k} \omega_j$ are purely imaginary. As usual, the matrix $B = (B_{jk})_{j,k=1}^g$ is symmetric with a positive definite imaginary part (which is in this case just the matrix $-iB$).

LEMMA 6.2.

- *For any $x \in \mathbb{R}^g$, we have $\theta(x) > 0$.*
- *For any $x \in \mathbb{R}^g$ and for any half-integer characteristic Δ, we have $\theta[\Delta](x) \in \mathbb{R}$.*

PROOF. The fact that $\theta(x) \in \mathbb{R}$ for $x \in \mathbb{R}^g$ follows from $\overline{\theta(z)} = \theta(-\bar{z}) = \theta(\bar{z})$ (the latter equality due to $\theta(z)$ being an even function). To prove the stronger positivity statement, we use the multi-dimensional analog of the imaginary Jacobi transformation:

$$\theta(x|B) = \frac{1}{\sqrt{\det(-iB)}} \exp\big(-\pi i \langle x, B^{-1}x\rangle\big)\theta(B^{-1}x| - B^{-1}).$$

Here the first factor is positive, since the matrix $-iB$ is positive definite, the second factor is positive, since the matrix $-iB^{-1}$ is real, and all terms in the defining series of the third factor,

$$\theta(B^{-1}x|-B^{-1}) = \sum_{m\in\mathbb{Z}^g} \exp\Big(-\pi\langle iB^{-1}m,m\rangle + 2\pi\langle iB^{-1}x,m\rangle\Big),$$

are positive for the same reason. To prove the second statement, we derive from (3.3), (3.2) that $\overline{\theta[\Delta](z)} = \theta[\Delta](\bar{z})$. □

Choose $\alpha_i, \beta_i \in X_0$ (all points associated to strips are on the real oval which does not belong to the chosen homology basis of $\mathcal{R}$). Choose the initial integration point of the Abel map $P_0 \in X_0$, as well. Then all $A(\alpha_i), A(\beta_i) \in \mathbb{R}^g$. This implies that the discrete Abel map η is $\mathbb{R}^g$-valued. Further, choose $D \in \mathbb{R}^g$. Then all theta-functions $\theta(\eta(f)+D)$ in the denominators of the Fock weights are positive, according to Lemma 6.2. The Kasteleyn condition (6.1) turns into

$$\begin{aligned} &\operatorname{sign}\left(\frac{E(\alpha_1,\beta_1)E(\alpha_2,\beta_2)\cdots E(\alpha_n,\beta_n)}{E(\beta_2,\alpha_1)E(\beta_3,\alpha_2)\cdots E(\beta_1,\alpha_n)}\right) \\ &= \operatorname{sign}\left(\frac{\theta[\Delta]\Big(\int_{\alpha_1}^{\beta_1}\omega\Big)\theta[\Delta]\Big(\int_{\alpha_2}^{\beta_2}\omega\Big)\cdots\theta[\Delta]\Big(\int_{\alpha_n}^{\beta_n}\omega\Big)}{\theta[\Delta]\Big(\int_{\beta_2}^{\alpha_1}\omega\Big)\theta[\Delta]\Big(\int_{\beta_3}^{\alpha_2}\omega\Big)\cdots\theta[\Delta]\Big(\int_{\beta_1}^{\alpha_n}\omega\Big)}\right) = (-1)^{n+1}. \end{aligned} \tag{6.2}$$

The expression in the parentheses on the l.h.s. is real by Lemma 6.2.

THEOREM 6.3. *Consider an orientation preserving homeomorphism $X_0 \to S^1$, and denote by $\hat{\alpha}_i, \hat{\beta}_i \in S^1$ the images of α_i, β_i under this map. Then the Kasteleyn condition is equivalent to*

$$\operatorname{sign}\left(\frac{(\hat{\beta}_1-\hat{\alpha}_1)(\hat{\beta}_2-\hat{\alpha}_2)\cdots(\hat{\beta}_n-\hat{\alpha}_n)}{(\hat{\alpha}_1-\hat{\beta}_2)(\hat{\alpha}_2-\hat{\beta}_3)\cdots(\hat{\alpha}_n-\hat{\beta}_1)}\right) = (-1)^{n+1} \tag{6.3}$$

for all faces f of G.

PROOF. The expression on the l.h.s. of (6.2) does not change sign if one moves α_i, β_i on X_0 preserving their cyclic order. Indeed, by such transformations $\theta[\Delta]\Big(\int_{\alpha_i}^{\beta_i}\omega\Big)$ never vanish. One can move all the points (preserving their cyclic order) in a small neighborhood of some $P \in X_0$, covered by some coordinate chart z of $\mathcal{R}$ with $\tau^*z = \bar{z}$. Setting $\widetilde{\alpha}_i = z(\alpha_i), \widetilde{\beta}_i = z(\beta_i) \in \mathbb{R}$, we obtain for the expression in question a formula like (6.3) with hats replaces by tildes. But this multi-ratio is, first, Möbius invariant and second, does not change sign under moving the points on S^1 while preserving their cyclic order. □

6.1. Example: regular hexagonal lattice. See Fig. 11 for notations. Considering a typical hexagon, we see that the Kasteleyn condition reads:

$$\operatorname{sign}\frac{(\hat{\gamma}_1-\hat{\beta}_1)(\hat{\alpha}_2-\hat{\gamma}_2)(\hat{\beta}_2-\hat{\alpha}_1)}{(\hat{\beta}_1-\hat{\alpha}_2)(\hat{\gamma}_2-\hat{\beta}_2)(\hat{\alpha}_1-\hat{\gamma}_1)} = 1. \tag{6.4}$$

This condition is satisfied if the points $\hat{\alpha}_i, \hat{\beta}_i, \hat{\gamma}_i$ lie on S^1 in three clusters, so that, with respect to the cyclic order on S^1,

$$\hat{\alpha}_i < \hat{\beta}_j < \hat{\gamma}_k < \hat{\alpha}_i \quad \text{for all} \quad i,j,k, \tag{6.5}$$

see Fig. 12.

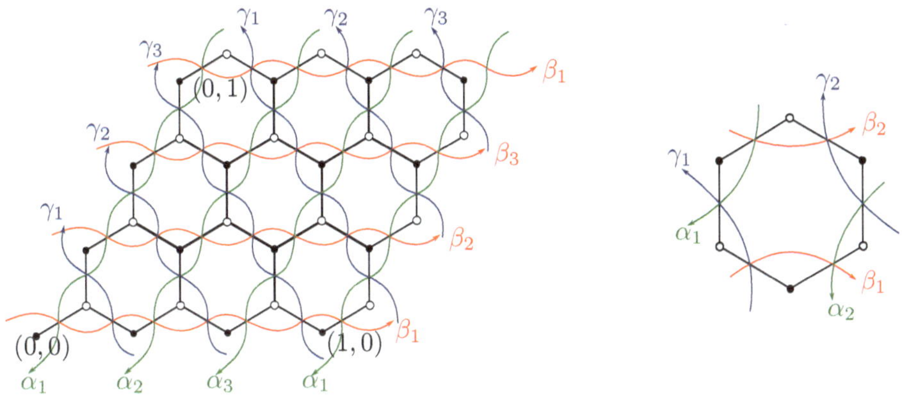

FIGURE 11. Regular hexagonal lattice on a 3 × 3 fundamental domain (left) and the strips for a typical hexagonal face (right).

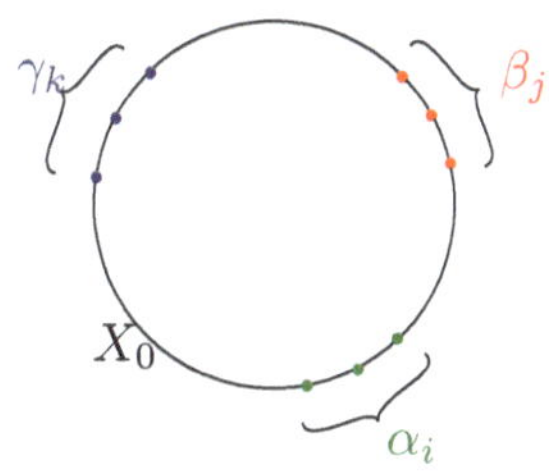

FIGURE 12. Clusters of points on $X_0 \simeq S^1$ corresponding to the strips of the regular hexagonal lattice, ensuring the Kasteleyn property

6.2. Example: regular square lattice. See Fig. 13 for notations. There are two types of elementary squares. For the first one, the Kasteleyn condition reads:

$$\operatorname{sign} \frac{(\hat{\alpha}_i^+ - \hat{\beta}_k^-)(\hat{\alpha}_j^- - \hat{\beta}_k^+)}{(\hat{\beta}_k^- - \hat{\alpha}_j^-)(\hat{\beta}_k^+ - \hat{\alpha}_i^+)} = -1. \tag{6.6}$$

For the second one, the Kasteleyn condition reads:

$$\operatorname{sign} \frac{(\hat{\beta}_j^- - \hat{\alpha}_i^+)(\hat{\beta}_k^+ - \hat{\alpha}_i^-)}{(\hat{\alpha}_i^+ - \hat{\beta}_k^+)(\hat{\alpha}_i^- - \hat{\beta}_j^-)} = -1. \tag{6.7}$$

Both conditions are satisfied if the points $\hat{\alpha}_i^\pm, \hat{\beta}_i^\pm$ lie on S^1 in four clusters, so that, with respect to the cyclic order on S^1,

$$\hat{\alpha}_i^+ < \hat{\beta}_j^+ < \hat{\alpha}_k^- < \hat{\beta}_l^- < \hat{\alpha}_i^+ \quad \text{for all} \quad i, j, k, l, \tag{6.8}$$

see Fig. 14.

7. Graphs on a torus and Harnack curves

Starting from here, we consider a $\mathbb{Z}^2$-periodic graph G on the plane $\mathbb{R}^2$. We can consider such a graph as a lift to $\mathbb{R}^2$ of the graph $G_1 = G/\mathbb{Z}^2$ on the torus $\mathbb{R}^2/\mathbb{Z}^2$.

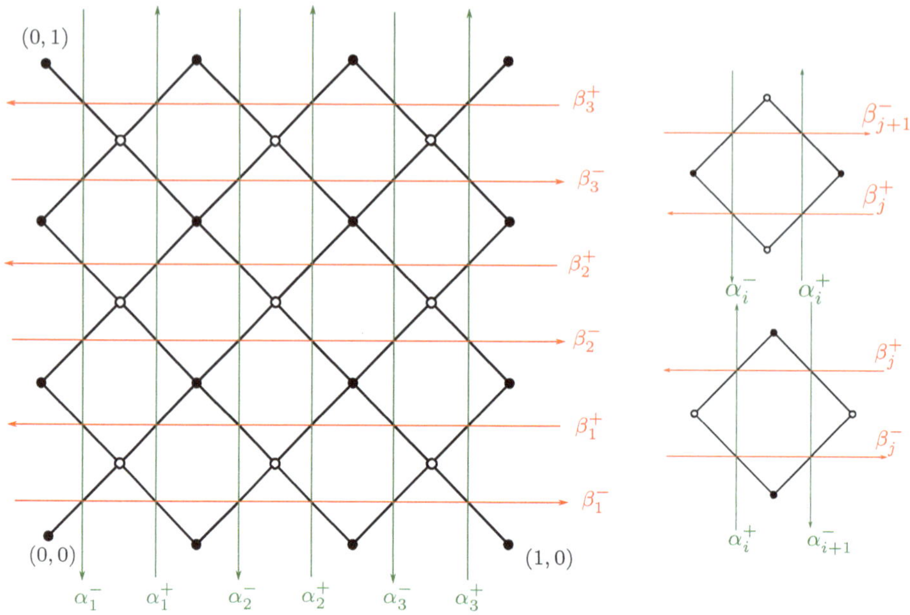

FIGURE 13. Regular square lattice on a 3×3 fundamental domain (left) and the strips for two types of square faces (right).

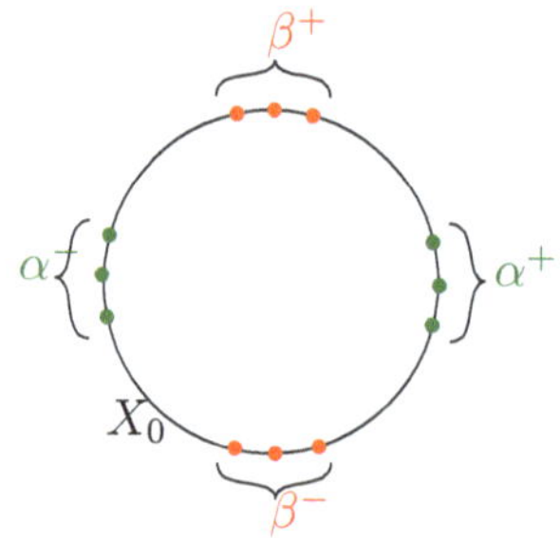

FIGURE 14. Clusters of points on $X_0 \simeq S^1$ corresponding to the strips of the regular square lattice, ensuring the Kasteleyn property

Every strip of G projects to a directed closed curve (loop) on the torus. Let (m_σ, n_σ) be the homology class of a loop σ, so that $\gamma_1 = [(0,0),(1,0)] \in H_1(\mathbb{T}^2, \mathbb{Z})$ intersects the loop σ from left to right n_σ times, while $\gamma_2 = [(0,0),(0,1)] \in H_1(\mathbb{T}^2, \mathbb{Z})$ intersects the loop σ from right to left m_σ times. One has

$$\sum_\sigma n_\sigma = 0, \quad \sum_\sigma m_\sigma = 0. \tag{7.1}$$

As seen in **[22]**, one can draw the vectors (m_σ, n_σ) in $\mathbb{Z}^2$ in a uniquely defined order so that

- the initial point of each vector is the end point of the previous one, and
- they build the boundary of a convex polygon, the *Newton polygon* $N(G_1)$.

Parallel strips correspond to vectors aligned along the same side of the Newton polygon. The criterium for the Kasteleyn property for the periodic case can be formulated as follows.

THEOREM 7.1. *A sufficient condition for the Fock weights defined by an M-curve to satisfy the Kasteleyn property is that the points α_σ are ordered on X_0 in the same cyclic order as the corresponding directed edges of the Newton polygon $N(G_1)$, with clusters of $\alpha_\sigma \in X_0$ corresponding to the sides of $N(G_1)$.*

One easily sees:

PROPOSITION 7.2. *The discrete Abel map η, and therefore the discrete Dirac operator on G are $\mathbb{Z}^2$-periodic iff*

$$\sum_\sigma n_\sigma A(\alpha_\sigma) = 0 \text{ in } J(\mathcal{R}), \quad \sum_\sigma m_\sigma A(\alpha_\sigma) = 0 \text{ in } J(\mathcal{R}), \tag{7.2}$$

i.e., if both divisors $\sum_\sigma n_\sigma \alpha_\sigma$ and $\sum_\sigma m_\sigma \alpha_\sigma$ are principal (are divisors of certain meromorphic functions on $\mathcal{R}$).

Moreover, we see from (3.13) that these meromorphic functions are nothing but the monodromies of the Baker-Akhiezer functions along the loops γ_1, γ_2:

(7.3)

$$z(P) := \frac{\psi_{(1,0)}(P)}{\psi_{(0,0)}(P)} = \prod_\sigma (E(P,\alpha_\sigma))^{-n_\sigma}, \quad w(P) := \frac{\psi_{(0,1)}(P)}{\psi_{(0,0)}(P)} = \prod_\sigma (E(P,\alpha_\sigma))^{m_\sigma}.$$

By the general theory, the points $(z(P), w(P)) \in \mathbb{C}P^1 \times \mathbb{C}P^1$ lie on an algebraic curve $\mathcal{P}(z,w) = 0$, called the *spectral curve* and parametrized by the Riemann surface $\mathcal{R}$ and having therefore geometric genus g (but usually a much higher arithmetic genus due to numerous singularities).

For an arbitrary Dirac operator $K : \mathbb{C}^B \to \mathbb{C}^W$ on a bipartite $\mathbb{Z}^2$-periodic graph with $\mathbb{Z}^2$-periodic coefficients, the equation $\mathcal{P}(z,w) = 0$ of the spectral curve can be obtained as follows. Choose a fundamental domain $\Omega \subset \mathbb{R}^2$ (so that the translations $\Omega + \mathbb{Z}^2$ are disjoint and their union is $\mathbb{R}^2$). Set $\widetilde{B} = B \cap \Omega$ and $\widetilde{W} = W \cap \Omega$. These sets represent the black and the white vertices of the graph G_1 on $\mathbb{T}^2$. We now define an operator $\widetilde{K}(z,w) : \mathbb{C}^{\widetilde{B}} \to \mathbb{C}^{\widetilde{W}}$ with coefficients depending on z, w such that

$$\mathcal{P}(z,w) = \det \widetilde{K}(z,w). \tag{7.4}$$

For each $w \in \widetilde{W}$, consider its star in $\mathbb{R}^2$ consisting of $b_1, b_2, \ldots, b_n \in B$ and the corresponding Dirac equation (3.16). For those black vertices b_k lying outside Ω (i.e., for such that the edge (wb_k) intersects the boundary $\partial\Omega$), determine $(i,j) \in \mathbb{Z}^2$ so that $\widetilde{b}_k := b_k - (i,j) \in \Omega$. Set in the Dirac equation $\psi_{b_k} = z^i w^j \psi_{\widetilde{b}_k}$. This is equivalent to multiplying the weight K_{wb_k} by $z^i w^j$. Thus, the Dirac equation becomes

$$\sum_{\widetilde{b}_k \sim w} \widetilde{K}_{w\widetilde{b}_k} \psi_{\widetilde{b}_k} = 0, \quad \text{where} \quad \widetilde{K}_{w\widetilde{b}_k} = z^i w^j K_{wb_k},$$

and the incidence $\widetilde{b}_k \sim w$ is understood relative to the graph G_1 on $\mathbb{T}^2$. A different choice of the fundamental domain leads to the multiplication of $\det(\widetilde{K})$ by a factor $z^k w^\ell$ and does not chage the spectral curve.

According to [**27**] if the coefficients of the Dirac operator satisfy the Kasteleyn condition then the spectral curve (7.4) is a *Harnack curve*. Moreover all Harnack

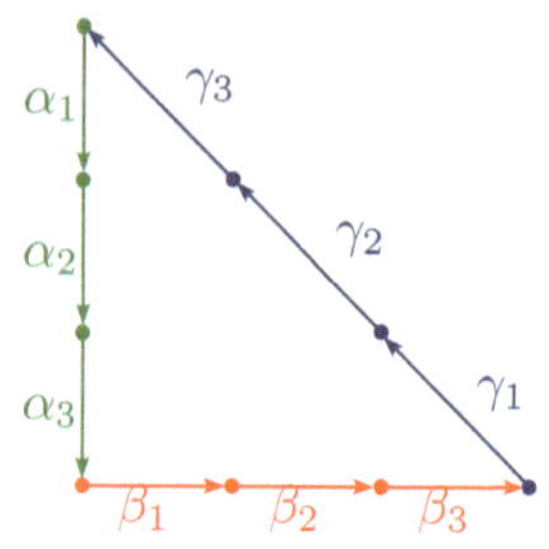

FIGURE 15. Newton polygon for a 3×3 fundamental domain of the hexagonal lattice

curves can be obtained this way. We use this characterizatiuon of the Harnack curves as M-curve $\mathcal{R}$ with two principal divisors $\sum_\sigma n_\sigma \alpha_\sigma$ and $\sum_\sigma m_\sigma \alpha_\sigma$ satisfying clustering conditions.

7.1. Example: regular hexagonal lattice. Consider the case of a $d \times d$ piece of the regular hexagonal lattice as the fundamental domain (in Fig. 11 $d = 3$ and one should assume $\alpha_4 = \alpha_1$, $\beta_4 = \beta_1$, $\gamma_4 = \gamma_1$; the coordinate directions are defined by the shifts $(0,0) \to (1,0)$ resp. $(0,0) \to (0,1)$). The homology class of all α_i-strips is $(0,-1)$, the homology class of all β_j-strips is $(1,0)$, and the homology class of all γ_k-strips is $(-1,1)$. The Newton polygon is shown in Fig. 15.

The monodromies are

$$z(P) = \frac{\psi_{(1,0)}(P)}{\psi_{(0,0)}(P)} = \prod_{k=1}^{d} \frac{E(P,\alpha_k)}{E(P,\gamma_k)}, \quad w(P) = \frac{\psi_{(0,1)}(P)}{\psi_{(0,0)}(P)} = \prod_{k=1}^{d} \frac{E(P,\beta_k)}{E(P,\gamma_k)}. \tag{7.5}$$

The periodicity condition of the Dirac operator is

$$\sum_{k=1}^{d} (A(\alpha_k) - A(\gamma_k)) = 0 \ \text{ in } J(\mathcal{R}), \quad \sum_{k=1}^{d} (A(\beta_k) - A(\gamma_k)) = 0 \ \text{ in } J(\mathcal{R}), \tag{7.6}$$

so that both divisors $\sum_{k=1}^{d} (\alpha_k - \gamma_k)$ and $\sum_{k=1}^{d} (\beta_k - \gamma_k)$ are principal. Thus, the problem consists in finding three equivalent divisors $\sum_{k=1}^{d} \alpha_k$, $\sum_{k=1}^{d} \beta_k$ and $\sum_{k=1}^{d} \gamma_k$ of points on the oval X_0 of the M-curve $\mathcal{R}$ satisfying clustering condition (6.5).

7.2. Example: regular square lattice. Consider the case of a $d \times d$ piece of the regular square lattice as the fundamental domain (in Fig. 13 $d = 3$; the coordinate directions are defined by the shifts $(0,0) \to (1,0)$ resp. $(0,0) \to (0,1)$). The homology class of all $\alpha_i^{\pm}$-strips is $(0, \pm 1)$, the homology class of all $\beta_j^{\pm}$-strips is $(\mp 1, 0)$. The Newton polygon is shown in Fig. 16.

The monodromies are

$$z(P) = \frac{\psi_{(1,0)}(P)}{\psi_{(0,0)}(P)} = \prod_{k=1}^{d} \frac{E(P,\alpha_k^-)}{E(P,\alpha_k^+)}, \quad w(P) = \frac{\psi_{(0,1)}(P)}{\psi_{(0,0)}(P)} = \prod_{k=1}^{d} \frac{E(P,\beta_k^-)}{E(P,\beta_k^+)}. \tag{7.7}$$

The periodicity condition of the Dirac operator is

$$\sum_{k=1}^{d} (A(\alpha_k^-) - A(\alpha_k^+)) = 0 \ \text{ in } J(\mathcal{R}), \quad \sum_{k=1}^{d} (A(\beta_k^-) - A(\beta_k^+)) = 0 \ \text{ in } J(\mathcal{R}), \tag{7.8}$$

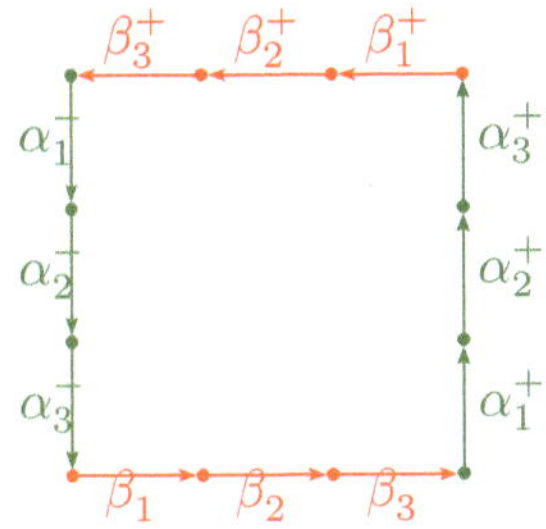

FIGURE 16. Newton polygon for a 3×3 fundamental domain of the square lattice

so that both divisors $\sum_{k=1}^{d}(\alpha_k^- - \alpha_k^+)$ and $\sum_{k=1}^{d}(\beta_k^- - \beta_k^+)$ are principal. Thus, the problem consists in finding two pairs of equivalent divisors $\sum_{k=1}^{d}\alpha_k^{\pm}$ and $\sum_{k=1}^{d}\beta_k^{\pm}$ of points on the oval X_0 of the M-curve $\mathcal{R}$ satisfying clustering condition (6.8).

7.3. Existence for Harnack curves over a given M-curve. In this Section we will show that on any M-curve there exists a collection of equivalent clustered divisors on X_0, as required by Theorem 7.1 and Proposition 7.2. These are three equivalent clustered divisors for the hexagon lattice, or two pairs of equivalent clustered divisors for the square lattice. As explained above , the corresponding functions (7.3) define a Harnack curve $\mathcal{P}(z,w) = 0$. We formulate the following result for the square lattice for the sake of simplicity only, it is actually of a quite general nature.

We first formulate the following Lemma, which we will use also in Section 9.

LEMMA 7.3. *Let $\ell^{\pm}$ be two arcs in X_0. Then there exists $L \in \mathbb{N}$ such that for any $R \in T^g = \mathbb{R}^g/\mathbb{Z}^g = \operatorname{Re}(\operatorname{Jac}(\mathcal{R}))$ there exists a divisor $\sum_j^L(s_j^- - s_j^+)$ with $s_j^{\pm} \in \ell^{\pm}$ and*

$$\sum_{j=1}^{L} A(s_j^- - s_j^+) = R.$$

PROOF. Any arc supports a non-special divisor, since they are generic. The set

$$S_g(\ell^{\pm}) = \{\sum_{i=1}^{g} A(t_i^- - t_i^+) \in \mathbb{R}^g | t_i^{\pm} \in \ell^{\pm}\}$$

is an open set in $\mathbb{R}^g$. There follows that the set $NS_g(\ell^{\pm}) \subset \mathbb{R}^g$ (the homothetic image of $S_g(\ell^{\pm})$ scaled by N) for a sufficiently large N contains the torus T^g. This completes the proof for $L = Ng$. □

THEOREM 7.4. *Let $\mathcal{R}$ be an M-curve. There exists $N_0 \in \mathbb{N}$ such that for any $m, n \geq N_0$ there exist principal divisors $\sum_{i=1}^{m}(\alpha_i^- - \alpha_i^+), \sum_{j=1}^{n}(\beta_j^- - \beta_j^+)$ satisfying clustering conditions* (6.8). *Functions* (7.7) *define a Harnack curve with the Riemann surface $\mathcal{R}$.*

PROOF. Choose four disjoint arcs satisfying clustering condition, see Fig. 17 and apply Lemma 7.3 with $R = 0$. □

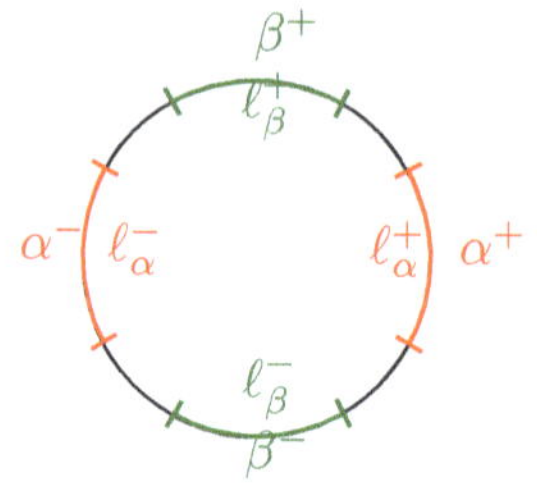

FIGURE 17. Distjoint clustering arcs for the square grid.

8. The Ronkin function

In this and the following section for simplicity we consider the case of the square grid only. General considerations are similar.

Let $\mathcal{R}$ be an M-curve, τ its anti-holomorphic involution and $X = \bigcup_{i=0}^{g} X_i$ the set of real ovals. X decomposes $\mathcal{R}$ into two components $\mathcal{R}_+$ and $\mathcal{R}_- = \tau(\mathcal{R}_+)$. We have $\mathcal{R}_+ = \mathcal{R}/\tau$ and $\mathcal{R}_+ \cap \mathcal{R}_- = X$. We also denote the corresponding open components by $\mathcal{R}_\pm^\circ := \mathcal{R}_\pm \setminus X$.

Let X_0 be a distinguished real oval and $\{\alpha, \beta\} := \{\alpha_1^\pm, \dots, \alpha_m^\pm, \beta_1^\pm, \dots, \beta_n^\pm\} \subset X_0$ be points on it clustered as in (6.8). They separate X_0 into $2m + 2n$ arcs δ_i.

We also choose a canonical basis of cycles as in Section 6 , i.e. $a_i = X_i$, $i = 1, \dots, g$, and $\tau a_i = a_i$, $\tau b_i = -b_i$.

8.1. The Ronkin function on Harnack data. Following Krichever [**30**]

$$\mathcal{S} = \{\mathcal{R}, \{\alpha, \beta\}\}$$

is called *Harnack data* for the M-curve $\mathcal{R}$.

Denote by $d\zeta^{\alpha_i}, d\zeta^{\beta_i}$ the Abelian differential of the third kind with simple poles at $\alpha_i^\pm$ or respectively $\beta_i^\pm$ and

$$\operatorname{res}_{\alpha_i^-} d\zeta^{\alpha_i} = -\operatorname{res}_{\alpha_i^+} d\zeta^{\alpha_i} = 1 = \operatorname{res}_{\beta_i^-} d\zeta^{\beta_i} = -\operatorname{res}_{\beta_i^+} d\zeta^{\beta_i}, \tag{8.1}$$

and imaginary periods. Introduce two differentials

$$d\zeta_1 = \sum_{i=1}^{m} d\zeta^{\alpha_i}, \quad d\zeta_2 = \sum_{i=1}^{n} d\zeta^{\beta_i}. \tag{8.2}$$

They are real on X, moreover $d\zeta_k(\tau P) = \overline{d\zeta_k(P)}$. The corresponding integrals

$$\zeta_k(P) = \int_Q^P d\zeta_k$$

with $Q \in X_0$ have jumps by $2\pi i$ on some paths $[\alpha_i^-, \alpha_i^+]$ and $[\beta_j^-, \beta_j^+]$. We chose these paths to lie in $\mathcal{R}_-$. Since the periods may be only imaginary, all periods of ζ_k over the real ovals vanish, and we obtain functions ζ_k well defined on $\mathcal{R}_+ \setminus \{\alpha, \beta\}$.

Denote the real and imaginary parts of ζ_k by

$$\zeta_k(P) = x_k + iy_k, \quad i = 1, 2.$$

DEFINITION 8.1. The map

$$\mathcal{A}(P) := (x_1, x_2) = (\operatorname{Re}\zeta_1(P), \operatorname{Re}\zeta_2(P))$$

is called the *amoeba map*, and its image

$$\mathcal{A}(\mathcal{R}_+ \setminus \{\alpha, \beta\}) =: \mathcal{A}_\mathcal{S} \subset \mathbb{R}^2$$

is called the *amoeba*.

Similarly the map

$$\Delta(P) = (s_1, s_2) = J(y_1, y_2) := \frac{1}{\pi}(-y_2, y_1) = \frac{1}{\pi}(-\operatorname{Im}\zeta_2(P), \operatorname{Im}\zeta_1(P))$$

we call the *polygon map*, and its image

$$\Delta(\mathcal{R}_+ \setminus \{\alpha, \beta\}) =: \Delta_{\mathcal{S}} \subset \mathbb{R}^2$$

is called the *Newton polygon*.

REMARK 8.2. The scaling and rotation J in the definition of Δ is introduced so that it matches the classical Newton polygon. It will often be more convenient to work directly with (y_1, y_2) as it produces more symmetric formulas.

Under the amoeba map X_0 is mapped to the concave boundary of $\mathcal{A}_{\mathcal{S}}$, where the points $\{\alpha, \beta\}$ are mapped to $2m + 2n$ infinite tentacles of the amoeba. The ovals $X_1, \ldots, X_g$ are mapped to g convex ovals bounding interior islands in the amoeba. Under the polygon map the (interior) ovals X_i are mapped to the interior points of $\Delta_{\mathcal{S}}$ given by the b-periods

$$\int_{b_i} d\zeta_k = iy_{k|X_i}. \tag{8.3}$$

The singular points $\{\alpha, \beta\}$ separate X_0 into $2m + 2n$ connected components $\delta_1, \ldots, \delta_{2m+2n}$. Each δ_l is mapped under the amoeba map to an external boundary component of $\mathcal{A}_{\mathcal{S}}$, and to an exterior vertex of Δ_S under the polygon map. See Fig. 18 for an illustration.

The differential $d\zeta_1$ has $2g - 2 + 2m$ zeros, and $d\zeta_2$ has $2g - 2 + 2n$ zeros. We observe that we know locations of all these zeros.

LEMMA 8.3. *The differentials $d\zeta_1, d\zeta_2$ have two zeros on each real oval $X_i, i = 1, \ldots, g$. Each of the components $(\alpha_i^+, \alpha_{i+1}^+), (\alpha_i^-, \alpha_{i+1}^-)$ of X_0 contains exactly one of the remaining $2m-2$ zeros of $d\zeta_1$. Each of the components $(\beta_j^+, \beta_{j+1}^+), (\beta_j^-, \beta_{j+1}^-)$ of X_0 contains exactly one of the remaining $2n - 2$ zeros of $d\zeta_2$. Moreover, $2g + (2m - 2) + (2n - 2)$ zeros of the differential $d\zeta_{\alpha,\beta} := \alpha d\zeta_1 + \beta d\zeta_2$ for any $\alpha, \beta \in \mathbb{R}_*$ are also located in X as described above; and the remaining two zeros lie in two of the four arcs $(\alpha^\pm, \beta^\pm)$.*

PROOF. $d\zeta_k$ must have two zeros on each $X_i, i = 1, \ldots, g$ since they are real and the periods over X_i vanish. The facts about the zeros on X_0 follow from the continuity of $d\zeta_k$ on the corresponding component δ_i and the fact that it has poles of the same sign at both ends of the component. The arguments for $d\zeta_{\alpha,\beta}$ are the same. □

See Fig. 18 for an illustration of the zeros.

$\zeta_{\alpha,\beta}$ are periodic functions on X_i with exactly two critical points. Also on each $\delta_k \in X_0$ the integral $\zeta_{\alpha,\beta}$ has at most one critical point. We conclude that the lines $\alpha x_1 + \beta x_2 = \gamma$ intersect the components of X at most in two points.

COROLLARY 8.4. *The images $\mathcal{A}(X_i) \subset \mathbb{R}^2, i = 1, \ldots, g$ are convex ovals. The images $\mathcal{A}(\delta_k) \subset \mathbb{R}^2, k = 1, \ldots, 2m + 2n$ are infinite concave arcs.*

COROLLARY 8.5. $R := \frac{d\zeta_1}{d\zeta_2}$ *is a holomorphic function without zeros on* $\mathcal{R}_+^\circ$.

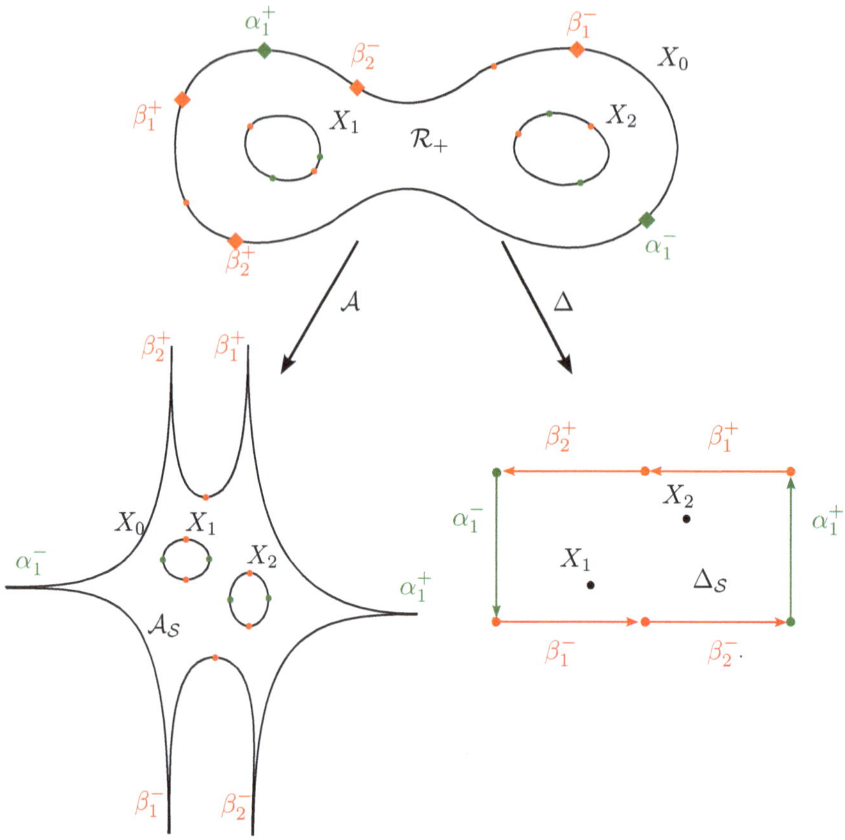

FIGURE 18. The two diffeomorphisms $\mathcal{A}$ and Δ coming from the M-curve $\mathcal{R}$. Amoeba $\mathcal{A}_\mathcal{S}$ (left) and Newton polygon $\Delta_\mathcal{S}$ (right). Zeros of $d\zeta_1, d\zeta_2$ as green and red circles respectively. Note that the X_i do not need to be mapped to integer points in $\Delta_\mathcal{S}$ as opposed to the algebraic case.

The identity $d\zeta_1 = Rd\zeta_2$ is equivalent to

$$1 + i\frac{\partial y_1}{\partial x_1} = iR\frac{\partial y_2}{\partial x_1}, \quad i\frac{\partial y_1}{\partial x_2} = R\left(1 + i\frac{\partial y_2}{\partial x_2}\right).$$

Equalizing the real and imaginary parts in these identities one easily arrives at the following formulas for the derivatives:

$$\frac{\partial y_2}{\partial x_1} = -\frac{1}{\operatorname{Im} R}, \quad \frac{\partial y_2}{\partial x_2} = -\frac{\partial y_1}{\partial x_1} = \frac{\operatorname{Re} R}{\operatorname{Im} R}, \quad \frac{\partial y_1}{\partial x_2} = \frac{|R|^2}{\operatorname{Im} R}. \tag{8.4}$$

We denote the open sets

$$\mathcal{A}_\mathcal{S}^\circ := \mathcal{A}(\mathcal{R}_+^\circ), \quad \Delta_\mathcal{S}^\circ := \Delta(\mathcal{R}_+^\circ).$$

$\Delta_\mathcal{S}^\circ$ is the interior of a polygon with g points removed. The latter are the images under Δ of the real ovals $X_1, \ldots, X_g$. The following result is due to Krichever [**30**]. We present its proof for completeness.

LEMMA 8.6. *The amoeba map* $\mathcal{A} : \mathcal{R}_+^\circ \to \mathcal{A}_\mathcal{S}^\circ$ *and the polygon map* $\Delta : \mathcal{R}_+^\circ \to \Delta_\mathcal{S}^\circ$ *are diffeomorphisms.*

PROOF. First we show that the set of critical points of the amoeba map coincides with the set of real ovals X. At regular points (x_1, x_2) are local coordinates, which yields $dx_1 \wedge dx_2 \neq 0$. The representation $d\zeta_2 = i\frac{\partial y_2}{\partial x_1}dx_1 + (1 + i\frac{\partial y_2}{\partial x_2})dx_2$ implies $d\zeta_2 \wedge d\bar{\zeta_2} = 2i\frac{\partial y_2}{\partial x_1}dx_1 \wedge dx_2$. Using (8.4) for the partial derivative we obtain

$$2dx_1 \wedge dx_2 = i \operatorname{Im} R \, d\zeta_2 \wedge d\bar{\zeta_2}.$$

We know that all zeros of $d\zeta_2$ are in X, therefore a point $P_0 \in \mathcal{R}_+^\circ$ is critical if and only if $R(P_0) \in \mathbb{R}$. Obviously $d\zeta := d\zeta_1 - R(P_0)d\zeta_2$ vanishes at P_0. On the other hand, all $2g+2$ zeros of $d\zeta$ must lie in X. Indeed, each oval $X_i, i = 1, \ldots, g$ contains 2 zeros because of the vanishing periods, and X_0 also contains 2 zeros between the poles with the residues of the same sign. Thus there are no critical points in $\mathcal{R}_+^\circ$.

The map $\mathcal{A}$ is injective on X_0 as well as on each $X_i, i = 1, \ldots, g$ because of the monotonicity of $d\zeta_i$ on the corresponding components of X, see Lemma 8.3. Due to Corollary 8.4 the set $\{(x_1, x_2) \in \mathcal{A}_\mathcal{S}^\circ | x_1 = a\}$ consists of disjoint intervals. Since $\mathcal{A}$ is a local diffeomorphism on $\mathcal{R}_+^\circ$, the derivative of x_2 on the segments of the level line $x_1^{-1}(a) \subset \mathcal{R}_+^\circ$ cannot vanish, thus the function x_2 is monotonic. This proves the injectivity of $\mathcal{A}$.

For the polygon map the proof is the same. □

We see that both $(x_1, x_2) \in \mathcal{A}_\mathcal{S}$ and $(y_1, y_2) \in J^{-1}\Delta_\mathcal{S}$ as well as $(s_1, s_2) \in \Delta_\mathcal{S}$ can be treated as global variables on $\mathcal{R}_+^\circ$, and one obtains well defined functions

$$y_1(x_1, x_2), y_2(x_1, x_2); \quad x_1(y_1, y_2), x_2(y_1, y_2).$$

Note that ζ_k and therefore x_k are defined up to real constants that correspond to choice of Q. This is just a global shift of the amoeba map.

Let us introduce the function

$$h(P) := \frac{1}{\pi} \operatorname{Im} \int_\ell \zeta_2 d\zeta_1 = \frac{1}{2\pi i} \int_\mathcal{L} \zeta_2 d\zeta_1$$

on $\mathcal{R}_+ \setminus \{\alpha, \beta\}$. Here $\ell \subset \mathcal{R}_+$ as a path from the boundary $Q \in X_0$ to P. The path $\mathcal{L} = -\tau\ell + \ell$ from $\tau P \in \mathcal{R}_-$ to $P \in \mathcal{R}_+$ has the symmetry $\tau\mathcal{L} = -\mathcal{L}$. This is a real valued function. It vanishes on the external boundary component $\delta_\mathcal{L} \ni Q$ of $\mathcal{A}_\mathcal{S}$ which is crossed by $\mathcal{L}$. Functions h that correspond to different choices of $\delta_\mathcal{L}$ differ by constants.

Considering (x_1, x_2) as variables on $\mathcal{R}_+ \setminus \{\alpha, \beta\}$ we obtain

$$d\zeta_1 = \left(1 + i\frac{\partial y_1}{\partial x_1}\right) dx_1 + i\frac{\partial y_1}{\partial x_2} dx_2.$$

This implies the following formulas for the derivatives of h:

$$\frac{\partial h}{\partial x_1} = \frac{1}{\pi} \operatorname{Im}\left((x_2 + iy_2)\left(1 + i\frac{\partial y_1}{\partial x_1}\right)\right) = \frac{1}{\pi}\left(x_2\frac{\partial y_1}{\partial x_1} + y_2\right),$$
$$\frac{\partial h}{\partial x_2} = \frac{1}{\pi} x_2 \frac{\partial y_1}{\partial x_2}.$$

Similarly in the coordinates (y_1, y_2) one obtains

$$d\zeta_1 = \left(i + \frac{\partial x_1}{\partial y_1}\right) dy_1 + \frac{\partial x_1}{\partial y_2} dy_2.$$

and thus

$$\frac{\partial h}{\partial y_1} = \frac{1}{\pi}\left(y_2 \frac{\partial x_1}{\partial y_1} + x_2\right),$$
$$\frac{\partial h}{\partial y_2} = \frac{1}{\pi} y_2 \frac{\partial x_1}{\partial y_2}.$$

Introducing

$$\rho(x_1, x_2) := -h(P) + \frac{1}{\pi} x_2 y_1 \tag{8.5}$$

and its Legendre dual

$$\sigma(y_1, y_2) := h(P) - \frac{1}{\pi} x_1 y_2 = -\rho(x_1, x_2) + \frac{1}{\pi}(x_2 y_1 - x_1 y_2), \tag{8.6}$$

we get

$$\begin{aligned} \frac{\partial \rho}{\partial x_1} &= -\frac{1}{\pi} y_2, & \frac{\partial \rho}{\partial x_2} &= \frac{1}{\pi} y_1, \\ \frac{\partial \sigma}{\partial y_1} &= \frac{1}{\pi} x_2, & \frac{\partial \sigma}{\partial y_2} &= -\frac{1}{\pi} x_1, \end{aligned} \tag{8.7}$$

More symmetric representations for ρ and σ are given by

$$\rho(x_1, x_2) = H(P) + \frac{1}{2\pi}(x_2 y_1 - x_1 y_2) \tag{8.8}$$

$$\sigma(y_1, y_2) = -H(P) + \frac{1}{2\pi}(x_2 y_1 - x_1 y_2) = -\rho(x_1, x_2) + \frac{1}{\pi}(x_2 y_1 - x_1 y_2) \tag{8.9}$$

with

$$H(P) = \frac{1}{4\pi i}\int_{\mathcal{L}} \zeta_1 d\zeta_2 - \zeta_2 d\zeta_1 = \frac{1}{2\pi}\operatorname{Im}\int_{\ell} \zeta_1 d\zeta_2 - \zeta_2 d\zeta_1.$$

Note that in (s_1, s_2) coordinates this recovers the standard Legendre duality

$$\sigma(s_1, s_2) = -\rho(x_1, x_2) + x_1 s_1 + x_2 s_2. \tag{8.10}$$

THEOREM 8.7. *Both $\rho(x_1, x_2)$ and $\sigma(y_1, y_2)$ are strictly convex functions on $\mathcal{A}_{\mathcal{S}}^{\circ}$ and $J^{-1}\Delta_{\mathcal{S}}^{\circ}$ respectively. Their Hessians are equal to*

$$\operatorname{Hess}(\rho) = \operatorname{Hess}(\sigma) = \frac{1}{\pi \operatorname{Im} R}\begin{pmatrix} 1 & -\operatorname{Re} R \\ -\operatorname{Re} R & |R|^2 \end{pmatrix}.$$

PROOF. Formulas for the coefficients of

$$\operatorname{Hess}(\rho) = \begin{pmatrix} \frac{\partial^2 \rho}{\partial x_1^2} & \frac{\partial^2 \rho}{\partial x_1 \partial x_2} \\ \frac{\partial^2 \rho}{\partial x_1 \partial x_2} & \frac{\partial^2 \rho}{\partial x_2^2} \end{pmatrix}.$$

follow directly from (8.7) and (8.4).

We omit the computation for the Legendre dual, which is the same. □

PROPOSITION 8.8. *The function $\rho : \mathcal{A}_{\mathcal{S}} \to \mathbb{R}$ can be continuously differentiably extended to the whole $\mathbb{R}^2$ plane (i.e. to the exterior of the amoeba and its islands) by affine functions. It vanishes on the exterior component of $\mathbb{R}^2 \setminus \mathcal{A}_{\mathcal{S}}$ bounded by $\mathcal{A}(\delta_{\mathcal{L}})$, which is crossed by $\mathcal{L}$.*

PROOF. Since both differentials $d\zeta_k$ are real on X the imaginary components y_1, y_2 are constant on every arc δ of X_0 built by the points α, β. Wenn we pass one of these points the corresponding value of y_1 or y_2 changes by π. On the oval $X_1, \ldots, X_g$ the imaginary components y_1, y_2 are also constant since they are the b-periods (8.3). Thus the gradient of ρ on each component of the boundary $\mathcal{A}(X)$ of amoeba is constant (8.7), which completes the proof. □

DEFINITION 8.9. The function $\rho : \mathbb{R}^2 \to \mathbb{R}$ is called the *Ronkin function* of the Harnack data $\mathcal{S}$.

The notion of a Ronkin function of Harnack data was originally introduced by Krichiver in [**30**] as an integral of a two-form over the lower left quadrant in $\mathbb{R}^2$ with the corner at (x_1, x_2). Our representation via an integral of a holomorphic one form is more convenient for investigation and computation (see Section 11).

8.2. Doubly periodic weights and identification with free energy. Consider the case when the weights are doubly periodic. The graph G_1 on a torus contains m pairs of (horizontal) $\alpha_i^{\pm}$ train tracks and n pairs of (vertical) $\beta_j^{\pm}$ train tracks. The periodicity condition (7.8) reads as

$$\sum_{i=1}^{m}(A(\alpha_i^-) - A(\alpha_i^+)) = 0, \quad \sum_{j=1}^{n}(A(\beta_j^-) - A(\beta_j^+)) = 0$$

The differentials $d\zeta_1, d\zeta_2$ are given by

$$d\zeta_1 = \frac{dz}{z}, \quad d\zeta_2 = \frac{dw}{w},$$

where z and w are meromorphic functions (7.7) on $\mathcal{R}$.

The Kasteleyn operator $K(z, w)$ with monodromies z, w determines the spectral curve $\mathcal{P}(z, w) = 0$, where

$$\mathcal{P}(z, w) = \det K(z, w).$$

Its Riemann surface is $\mathcal{R}$.

Consider the graph G_N on a torus which is the $N \times N$ covering of G_1 (see Fig. 19) with the same weights ν as of G_1. The thermodynamic limit of the free energy is defined as

$$\log Z(\nu) := \lim_{n \to \infty} \frac{1}{N^2} \log Z(G_N, \nu).$$

This limit exists and depends on the weights ν on G_1. In our case we denote it

$$E(\mathcal{R}, \{\alpha, \beta\}) = \log Z(\nu)$$

since it is a function of the Harnack data. Note that it is independent of the free vector $Z \in J(\mathcal{R})$ which enters the formulas (3.8) for the weights.

In [**29**] it was shown that for doubly-periodic weights $E = F(0, 0)$, where

$$F(x_1, x_2) := \frac{1}{(2\pi i)^2} \int_{\mathbb{T}^2} \log |\mathcal{P}(e^{x_1} z, e^{x_2} w)| \frac{dz}{z} \frac{dw}{w}.$$

Here the integration is taken over the torus

$$\mathbb{T}^2 = \{(z, w) \in \mathbb{C}^2 | \, |z| = |w| = 1\}.$$

The curve $\mathcal{P}(z, w) = 0$ is a Harnack curve. Its Riemann surface is an M-curve $\mathcal{R}$.

The classical amoeba map $\mathcal{R}_+^\circ \to \mathbb{R}^2$ is given by $(\log|z|, \log|w|)$. Comparing to the one from Definition 8.1 it is shifted by the values of z and w at the starting integration point Q of ζ_k:

$$\log|z(P)| - \log|z(Q)| = \zeta_1(P), \quad \log|w(P)| - \log|w(Q)| = \zeta_2(P).$$

Its image $\mathcal{A}$ is $\mathcal{A}_\mathcal{S}$ shifted by the same constant.

Generally $F(x_1, x_2)$ is interpreted as the free energy of the dimer model on a torus with magnetic field. The behaviour of the model qualitatively depends on the value of the magnetic field:

(1) For $(x_1, x_2) \in \mathcal{A}_\mathcal{S}$ the model is in the *liquid* phase.
(2) For (x_1, x_2) in the infinite exterior domain of the amoeba the model is in the *solid* phase.
(3) For (x_1, x_2) in the amoeba islands the model is in the *gas* phase.

The function $F(x_1, x_2)$ is the classical *Ronkin function of a Harnack curve*. It is a continuously differentiable convex function on $\mathbb{R}^2$, which is piecewise affine on $\mathbb{R}^2 \setminus \mathcal{A}$ and smooth and strictly convex on $\mathcal{A}$.

THEOREM 8.10. *$F(x_1, x_2) = \rho(x_1 - c_1, x_2 - c_2) + \log|\mathcal{P}(0,0)|$, $c_1 = \log|z(Q)|, c_2 = \log|w(Q)|$.*

PROOF. As it was shown in [**35**], the Hessian of $F(x_1, x_2)$ computed at the point P, i.e. at $(x_1, x_2) = (\log|z(P)|, \log|w(P)|)$ is given by $\mathrm{Hess}(\rho(P))$, see also [**30**]. Thus the difference $F - \rho$ is an affine function $a_1x_1 + a_2x_2 + a_3$. With our choice the integration contour $\mathcal{L}$ intersecting the component (α_m^-, β_1^-), ρ identically vanishes on the infinite domain of $\mathbb{R}^2 \setminus \mathcal{A}_\mathcal{S}$ bounded by $\mathcal{A}((\alpha_m^-, \beta_1^-))$. In the limit we have

$$\lim_{\substack{x_1 \to -\infty \\ x_2 \to -\infty}} F(x_1, x_2) = \log|\mathcal{P}(0,0)|.$$

Thus $a_1 = a_2 = 0$. □

REMARK 8.11. We note that $\mathcal{P}(0,0)$ is the partition function of the finite planar graph we get if for fixed i, j we remove the edges crossing the train tracks α_i^+ or β_j^+ from the graph G_1 on a torus. This partition function is independent of the choice of i, j but does depend on the cut more generally. In this sense ρ is a correction term from the finite planar case to the thermodynamic limit on a torus.

From (8.2) we obtain explicit formulas for ρ and σ in the form of a sum over the edges of G_1. Note the similarity of these functions when represented in terms of the Riemann surface.

(8.11)
$$\begin{aligned}\rho(x_1, x_2) &= \sum_{i=1}^m \sum_{j=1}^n \left(-\frac{1}{\pi}\operatorname{Im}\int_\ell \zeta^{\beta_j} d\zeta^{\alpha_i} + \frac{1}{\pi}\operatorname{Re}\zeta^{\beta_j}(P)\operatorname{Im}\zeta^{\alpha_i}(P)\right) \\ &= \sum_{i=1}^m \sum_{j=1}^n \left(-\frac{1}{2\pi}\operatorname{Im}\int_\ell \zeta^{\beta_j} d\zeta^{\alpha_i} - \zeta^{\alpha_i} d\zeta^{\beta_j} + \frac{1}{2\pi}\left(\operatorname{Re}\zeta^{\beta_j}(P)\operatorname{Im}\zeta^{\alpha_i} - \operatorname{Re}\zeta^{\alpha_i}(P)\operatorname{Im}\zeta^{\beta_j}(P)\right)\right)\end{aligned}$$

(8.12)
$$\begin{aligned}\sigma(s_1, s_2) &= \sum_{i=1}^m \sum_{j=1}^n \left(\frac{1}{\pi}\operatorname{Im}\int_\ell \zeta^{\beta_j} d\zeta^{\alpha_i} - \frac{1}{\pi}\operatorname{Re}\zeta^{\alpha_i}(P)\operatorname{Im}\zeta^{\beta_j}(P)\right) \\ &= \sum_{i=1}^m \sum_{j=1}^n \left(\frac{1}{2\pi}\operatorname{Im}\int_\ell \zeta^{\beta_j} d\zeta^{\alpha_i} - \zeta^{\alpha_i} d\zeta^{\beta_j} + \frac{1}{2\pi}\left(\operatorname{Re}\zeta^{\beta_j}(P)\operatorname{Im}\zeta^{\alpha_i} - \operatorname{Re}\zeta^{\alpha_i}(P)\operatorname{Im}\zeta^{\beta_j}(P)\right)\right).\end{aligned}$$

9. Height Function

Let G be an infinite $\mathbb{Z}^2$-invariant bipartite planar graph with doubly periodic dimer weights and a finite fundamental domain G_1 which is embedded in the unit square and let G_N be its $N \times N$ covering. A Gibbs measure on G is a probability measure $\mu(G)$ on infinite dimer configurations with the property that if we fix a matching on a topological annulus the matching inside it is independent of the matching outside. Furthermore the measure restricted to the inside of the annulus is the Boltzmann measure defined in (2.2).

It was shown in [**38**] that for any $(s_1, s_2) \in \Delta_S$ there exists a unique ergodic Gibbs measure $\mu_{(s_1,s_2)}$ and that the measures $\mu(G_N)$ conditioned on having height change

$$(h(f+(0,N))-h(f), h(f+(-N,0))-h(f)) = (\lfloor Ns_1 \rfloor, \lfloor Ns_2 \rfloor) \tag{9.1}$$

converge to $\mu_{(s_1,s_2)}$ as N goes to infinity.

Let now $\mathcal{M}_{(s_1,s_2)}(G_N)$ be the set of all dimer configurations with this given height change on G_N and

$$Z_{(s_1,s_2)}(G_N) = \sum_{M \in \mathcal{M}_{(s_1,s_2)}(G_N)} \prod_{e \in M} \nu(e)$$

be the partition function of $\mu_{(s_1,s_2)}(G_N)$. The surface tension is then defined as a function of (s_1, s_2) through the free energy per fundamental domain of $\mu_{(s_1,s_2)}$:

$$-\varsigma(s_1,s_2) = \log Z_{(s_1,s_2)} = \lim_{N\to\infty} \frac{1}{N^2} \log Z_{(s_1,s_2)}(G_N).$$

It was shown in [**29**] that this surface tension is the Legendre dual of the Ronkin function F. That is

$$F(x_1,x_2) = \max_{(s_1,s_2)} \left(-\varsigma(s_1,s_2) + x_1 s_1 + x_2 s_2\right).$$

REMARK 9.1. For clarity we note here that our setup differs from that of [**29**] by a rotation of all coordinates by π. In their work the monodromies z, w go in the directions $(0,-1), (1,0)$ and the height function change is considered in the directions $(1,0), (0,1)$.

9.1. Regularized convergence of the surface tension. We have seen in Theorem 8.10 that $F = \rho$ up to a shift of the argument and an additive constant in the case of doubly periodic weights. Since ς, σ are Legendre duals of F, ρ respectively we obtain that they coincide up to an affine function

$$\varsigma(s_1,s_2) = \sigma(s_1,s_2) + b_1 s_1 + b_2 s_2 + b_3. \tag{9.2}$$

This affine function dissapears in the variational description of the height function (see Remark 9.4), and one obtains an equivalence class $[\sigma]$ of surface tension functions that differ by affine functions. We consider particular representatives in the equivalence class to formulate the regularized convergence results for σ.

Let ζ_1, ζ_2 in general (quasi-periodic) case be given by

$$\zeta_1 = \sum_{i=1}^{m} \zeta^{\alpha_i}, \ \zeta_2 = \sum_{j=1}^{n} \zeta^{\beta_j}.$$

Let N be the parameter of the thermodynamic limit and L fixed (See Lemma 7.3) with $(\gamma_1^\pm, \dots, \gamma_L^\pm) \in \ell_\alpha^\pm$, $(\delta_1^\pm, \dots, \delta_L^\pm) \in \ell_\beta^\pm$ clustered points (See Fig. 17) satisfying

$$N \sum_{i=1}^{m} (\alpha_i^- - \alpha_i^+) + \sum_{l=1}^{L} (\gamma_l^- - \gamma_l^+) = 0$$

$$N \sum_{j=1}^{m} (\beta_i^- - \beta_i^+) + \sum_{l=1}^{L} (\delta_l^- - \delta_l^+) = 0$$

Now let us consider the regularized graph $\hat{G}_N$ with the Fock weights defined by the train tracks $\alpha, \beta, \gamma, \delta$ (See Fig. 19). Note that the weights on various copies of G_1 in $\hat{G}_N$ generically do not coincide (they do under the periodicity condition (7.8)).

The integrals

$$\zeta_1^N = N \sum_{i=1}^{m} \zeta^{\alpha_i} + \sum_{l=1}^{L} \zeta^{\gamma_l},$$

$$\zeta_2^N = N \sum_{j=1}^{n} \zeta^{\beta_j} + \sum_{l=1}^{L} \zeta^{\delta_l}.$$

correspond to periodic weights on $\hat{G}_N$. Denote their real and imaginary parts by

$$\zeta_k^N(P) = x_k^N + i y_k^N, \; i = 1, 2$$

and again introduce $s_1^N, s_2^N := J(y_1^N, y_2^N)$. The surface tension on the periodic graph $\hat{G}_N$ is given by

$$\sigma^N(s_1^N, s_2^N) = \frac{1}{\pi} \operatorname{Im} \int_\ell \zeta_2^N d\zeta_1^N + x_1^N s_1^N.$$

This function is defined on the Newton polygon $\hat{\Delta}_N^\circ$. Observe that the Legendre dual Ronkin function σ of G_1 is defined (8.12) on the Newton polygon of G_1, $\Delta^\circ \subset \frac{1}{N}\hat{\Delta}_N^\circ$. We define the regularized surface tension as a function on $\frac{1}{N}\hat{\Delta}_N^\circ$:

$$\hat{\sigma}^N\left(\frac{s_1^N}{N}, \frac{s_2^N}{N}\right) := \frac{1}{N^2} \sigma^N(s_1^N, s_2^N). \tag{9.3}$$

LEMMA 9.2. *The regularized surface tension function $\hat{\sigma}^N$ converges on Δ° to σ given by* (8.12) *in the supremum norm. That is*

$$\sup_{\Delta^\circ} |\hat{\sigma}^N - \sigma| \xrightarrow{N \to \infty} 0, \text{ and } \quad \hat{\Delta}_N^\circ \xrightarrow{N \to \infty} \Delta^\circ.$$

PROOF. A point $P \in \mathcal{R}_+^\circ$ corresponds to $(s_1, s_2) = J(\operatorname{Im} \zeta_1, \operatorname{Im} \zeta_2)(P) \in \Delta^\circ$ and to $\frac{1}{N}(s_1^N, s_2^N) = \frac{1}{N} J(\operatorname{Im} \zeta_1^N, \operatorname{Im} \zeta_2^N)(P) \in \hat{\Delta}_N^\circ$.

We have

$$|\hat{\sigma}^N(s_1, s_2) - \sigma(s_1, s_2)| \le |\frac{1}{N^2}\sigma^N(Ns_1, Ns_2) - \frac{1}{N^2}\sigma^N(s_1^N, s_2^N)| + |\frac{1}{N^2}\sigma^N(s_1^N, s_2^N) - \sigma(s_1, s_2).| \tag{9.4}$$

Observe that the arguments of the second term correspond to the same point $P \in \mathcal{R}_+^\circ$ and thus all the terms corresponding to the edges labelled by (α, β) cancel. Thus this term consists of contributions of N edges (α_i, δ_l), N edges (β_j, γ_l) and one edge $\gamma_{l'}, \delta_l$ for all i, j, l, l'. All these entries are of the form (8.12) and are bounded. Thus the second term in (9.4) is $\mathcal{O}\left(\frac{1}{N}\right)$.

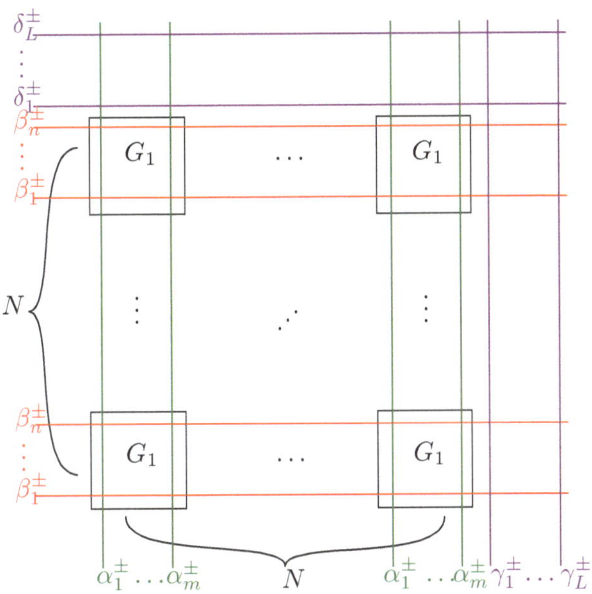

FIGURE 19. Graph G_N and regularized graph $\hat{G}_N$. G_N is composed of $N \times N$ copies of G_1 with doubly periodic train track parameters and thus quasi periodic weights. $\hat{G}_N$ is the graph G_N extended by two strips with L trains tracks each. $\hat{G}_N$ has zero monodromy for its Baker-Akhiezer functions and therefore defines doubly periodic weights. Each graph G_1 is a diagonal square lattice as in Fig. 13.

The first term in (9.4) can be easily estimated through the derivatives. We obtain two terms of the form

$$\frac{1}{N^2}\left|\frac{\partial \sigma^N}{\partial s_k^N}\right|\left|s_k^N - N s_k\right|.$$

For $k = 1$ this gives (and the same estimate for $k = 2$)

$$\frac{1}{N^2}|x_1^N|\left|\sum_{l=1}^{L} \operatorname{Im} \zeta^{\delta_l}\right| \leq \frac{1}{N}\left(\sum_{i=1}^{m}\sum_{l=1}^{L}|\operatorname{Re} \zeta^{\alpha_i}(P)||\operatorname{Im} \zeta^{\delta_l}(P)|\right) + \mathcal{O}(\frac{1}{N^2}),$$

where we have used (8.6).

To show that this term is $\mathcal{O}(\frac{1}{N})$ one observes that $|\operatorname{Re} \zeta^{\alpha_i}(P)||\operatorname{Im} \zeta^{\delta_l}(P)|$ remain bounded at the singularities $\alpha_i^\pm$. Indeed at $z \to \alpha_i$ these terms have the following behaviour:

$$|\operatorname{Re} \zeta^{\alpha_i}(P)| \approx \log|z - \alpha_i|, \quad |\operatorname{Im} \zeta^{\delta_l}(P)| \approx |\operatorname{Im}(z - \alpha_i)| \leq |z - \alpha_i|.$$

The product can be estimated by $|z - \alpha_i| \log|z - \alpha_i| \to 0$. Finally we obtain that both terms in (9.4) are uniformly bounded by $\mathcal{O}\left(\frac{1}{N}\right)$. □

9.2. Regularized convergence of the height function. Surface tension is important because it defines a functional that is minimized by the limiting height function of the corresponding dimer model.

For a domain $\Omega \subset \mathbb{R}^2$ and a set $\Delta \subset \mathbb{R}^2$ let us denote the space of functions

$$\mathrm{Lip}_\Delta(\Omega) := \{f \in \mathrm{Lip}(\Omega, \mathbb{R}) \mid \nabla f \in \Delta \text{ a.e.}\}, \quad \mathrm{Lip}_\Delta(\Omega, h_b) := \left\{f \in \mathrm{Lip}_\Delta(\Omega) \mid \ f|_{\partial\Omega} = h_b\right\}.$$

Note that the existence of the gradient is guaranteed almost everywhere due to Rademacher's theorem. This is the natural family of spaces for limiting height functions with Newton polygon Δ. We omit the index Δ when it is clear from context.

The following is a result from [**14**] and can also be found in [**38**].

THEOREM 9.3. *Let $\Omega \subset \mathbb{R}^2$ be a simply connected region with piecewise smooth boundary and h_b be a function in $\mathrm{Lip}_{[0,m]\times[0,n]}(\Omega)$ restricted to $\partial\Omega$. Let $G_k \subset \frac{1}{k}\left(\frac{1}{m}\mathbb{Z} \times \frac{1}{n}\mathbb{Z}\right)$ be a sequence of graphs that approximate (Ω, h_b) in the following sense:*

- *$G_k \subset \Omega$ for all k.*
- *The Hausdorff distance between the boundary of G_k and $\partial\Omega$ tends to 0.*
- *G_k permits dimer configurations and the height function h_k on the boundary approximates the boundary conditions. That is for $x \in \partial\Omega$ if we define $h_k(x)$ as h_k of the face closest to the point x then $\|\frac{h_k}{k} - h_b\|_\infty \to 0$.*

Fix a fundamental domain of size $(m \times n)$ with some fixed dimer weights on it and populate all of G_k doubly periodically with those weights. Let $\mathbb{P}_k$ be the corresponding dimer Boltzmann measures. Let σ be the surface tension corresponding to these weights. Then for the limiting height function defined as the minimizer

$$h := \underset{h \in \mathrm{Lip}(\Omega, h_b)}{\arg\min} \int_\Omega \sigma(\nabla h) \tag{9.5}$$

we have the large deviations principle

$$\mathbb{P}_k(\|\frac{h_k}{k} - h\|_\infty > \varepsilon) \approx e^{-k^2\varepsilon}.$$

The surface tension minimizer is unique due to the strict convexity of σ.

REMARK 9.4. Surface tension functions that differ by an affine function $b_1 y_1 + b_2 y_2 + b_3$ lead to the same variational description. Indeed, by Stokes theorem we have

$$\int_\Omega \nabla h \ dudv = \int_{\partial\Omega} h \ n \ ds,$$

where n is the outward normal to the boundary curve $\partial\Omega$. The last expression is given in terms of the boundary data.

We will now use Theorem 9.3 for doubly periodic weights to establish convergence of height functions in a regularized sense for quasi periodic weights.

To that end we consider a sequence of graphs $G_k \subset \frac{1}{k}\left(\frac{1}{m}\mathbb{Z} \times \frac{1}{n}\mathbb{Z}\right)$ that approximate a domain Ω with boundary height h_b as above.

Let $\alpha_i^\pm, i \leq m$, $\beta_j^\pm, j \leq n$ be the train track angles repeating periodically. As before we consider a fundamental domain $\hat{G}_N$ of size $(Nm+L) \times (Nn+L)$ with $4L$ added regularizing train tracks $\gamma_k^\pm, \delta_l^\pm$ such that the periodicity condition (7.8) is satisfied on this domain. We denote by σ^N the corresponding surface tension and $\hat{\sigma}^N(\frac{s_1^N}{N}, \frac{s_2^N}{N}) := \frac{1}{N^2}\sigma^N(s_1^N, s_2^N)$ its normalized version defined on $\hat{\Delta}_N^\circ$. Henceforth we denote the surface tension functional as $S_N(h) := \int \hat{\sigma}^N(\nabla h)$ and $S(h) := \int \sigma(\nabla h)$.

For each of these fundamental domains we are in the doubly periodic case so by a scaled version of Theorem 9.3 we have the limiting average height functions

$$h_N = \operatorname*{arg\,min}_{h \in \mathrm{Lip}_{\hat{\Delta}_N^\circ}(\Omega, h_b)} S_N(h).$$

It was shown in [**14**] that the functional S is lower semicontinuous with respect to the supremum norm. This is the key ingredient to establishing convergence of the minimizing functions h_N.

THEOREM 9.5. *The limiting height functions h_N converge to the minimizer of the surface tension σ. That is*

$$h_N \xrightarrow{\|\cdot\|_\infty} h := \operatorname*{arg\,min}_{f \in \mathrm{Lip}_{\Delta^\circ}(\Omega, h_b)} S(f).$$

PROOF. We have seen that $\hat{\sigma}^N \xrightarrow{\|\cdot\|_\infty} \sigma$. Since $h_N \in \mathrm{Lip}_{\hat{\Delta}_N^\circ}(\Omega, h_b) \supset \mathrm{Lip}_{\Delta^\circ}(\Omega, h_b)$ for all N we get by Arzela-Ascoli the existence of a convergent subsequence $h_{N_k} \xrightarrow{k\to\infty} g \in \mathrm{Lip}_{\Delta^\circ}(\Omega, h_b)$. We show that this limit minimizes $S(\cdot)$.

Suppose that $S(g') = S(g) - \delta$ for some $\delta > 0$. Then for $\varepsilon = |\Omega| \|\sigma - \hat{\sigma}^N\|_\infty$ and due to lower semicontinuity of S we have

$$\begin{aligned} S_N(g') &\le S(g') + \varepsilon = S(g) + \varepsilon - \delta \\ &\le S(h_N) + \varepsilon + \varepsilon' - \delta \\ &\le S_N(h_N) + 2\varepsilon + \varepsilon' - \delta \\ &< S_N(h_N) \qquad \text{for N large enough.} \end{aligned}$$

This contradicts h_N being the minimizer of S_N. Therefore g is indeed a minimizer of S. Since σ is stricly convex, the minimizer is unique. Therefore in the sequence h_N any subsequence has a subsequence converging to the same limit h. This yields convergence $h_N \xrightarrow{\|\|\|_\infty} h$. □

We thus have a limiting height function h under the regularized limit. We conjecture that this convergence also holds without a regularization step in a direct sense akin to Theorem 9.3.

Imposing a volume constraint of $V \in \mathbb{R}$ on the minimization problem (9.5) yields

$$h := \operatorname*{arg\,min}_{h \in \mathrm{Lip}(\Omega, h_b)} \int_\Omega \sigma(\nabla h), \quad \text{s.t.} \int_\Omega h = V. \tag{9.6}$$

It is well known (See e.g. [**29**] for details and further references) that the minimizer h satisfies the Euler-Lagrange equation

$$\operatorname{div}(\nabla\sigma \circ \nabla h(x)) = \lambda$$

with some Lagrange multiplier $\lambda \in \mathbb{R}$ (depending on V and h_b) where h is C^2, and σ is smooth at $h(x)$. Here $\lambda = 0$ corresponds to the unconstrained minimization problem.

Since ρ is the Legendre dual of σ we have

$$\operatorname{div}(\nabla\sigma \circ \nabla\rho) = \operatorname{div}(\mathrm{Id}) = 2.$$

The Ronkin function thus satisfies (9.6) for some volume V and its own boundary conditions. See Figure 1 for an example of this match of theoretical prediction and simulation and Section 11 for a description of the computational aspects.

10. Schottky Uniformization

Formulas for weights, surface tension and other characteristics of dimer models presented in this paper are parametrized by compact Riemann surfaces. This seems to be a rather complicated parametrization if one wants to investigate these dimer models or just plot them. Fortunately this problem can be solved using the Schottky uniformization of Riemann surfaces. This method was first suggested in [**5**] (see also [**3**]) for a similar problem of computation of algebro-geometric solutions of nonlinear integrable equations. We shortly present it here.

Let $\{C_1, C_1', \dots, C_g, C_g'\}$ be a collection of $2g$ mutually disjoint Jordan curves that form the boundary of a $2g$-connected domain $\Pi \subset \mathbb{C} \cup \{\infty\}$. The fractional-linear transformation $\sigma \in \mathrm{PSL}(2, \mathbb{C})$ given by

$$\frac{\sigma_n(z) - B_n}{\sigma_n(z) - A_n} = \mu_n \frac{z - B_n}{z - A_n}, \qquad |\mu_n| < 1.$$

maps the outside of a boundary curve C_n onto the inside of the boundary curve C_n', $\sigma_n C_n = C_n'$. Its fixed point A_n lies inside C_n and B_n lies inside C_n'.

The freely generated group $G =< \sigma_1, \dots, \sigma_g >$ is called a *Schottky group*. If all boundary curves C_n, C_n' are circles the Schottky group is called *classical*.

Let Ω be the discontinuity set of a Schottky group G. It is a classical result (see e.g. [**19**]) that $\mathcal{R} = \Omega/G$ is a compact Riemann surface of genus g. Note that G acts freely on Ω and that $\Omega/G \approx \Pi(G) := \mathbb{C} \setminus \{2g \text{ disks}\}$. Here $\Pi(G)$ is a fundamental domain of G. Observe that in fact the topology is that of a sphere with $2g$ holes glued in pairs, thus a surface with g handles.

Note that given a Riemann surface $\mathcal{R}$, the choice of Schottky group representing it is not unique. The Schottky uniformization theorem states that any compact Riemann surface $\mathcal{R}$ with a choice of homologically independent simple disjoint loops $v_1, \dots, v_g$ can be realized as $\mathcal{R} = \Omega(G)/G$, with G Schottky. It is, however, an open question whether G can be chosen to be classical.

The uniformization parameters $\{A_n, B_n, \mu_n\}$ and $\{\tau A_n, \tau B_n, \mu_n\}$ that differ by a fractional-linear mapping τ represent the same Riemann surface. The complex dimension of the uniformization space is thus 1 for $g = 1$ and $3g - 3$ otherwise. This corresponds to the complex dimension of the moduli space of compact Riemann surfaces of genus g as one would expect from the Schottky uniformization theorem.

In order to compute the period matrix and Abel map on $\mathcal{R}$, we need to fix a basis of cycles. Given a Schottky group G we chose a *canonical homology basis* $a_1, \dots, a_g, b_1, \dots, b_g$ such that the a_n are positively oriented cycles around the B_n, b_n are paths from some $z_n \in \sigma_n^{-1}(a_n)$ to $\sigma_n(z_n) \in a_n$ such that the intersection indices are $a_k \circ b_l = \delta_{kl}$ and $a_k \circ a_l = b_k \circ b_l = 0 \ \forall k, l$.

Any element in G can be represented as a word consisting of its generators and inverses

$$G \ni \sigma = \sigma_{i_1}^{j_1} \dots \sigma_{i_k}^{j_k}, \quad i_1, \dots i_k \in \{1, \dots, g\}, \ j_1, \dots, j_k \in \mathbb{Z}.$$

This representation is unique if we impose that $i_k \neq i_{k+1}$. We furthermore denote by $G/G_n := \{\sigma \mid i_k \neq n\}$ the elements whose words do not end in σ_n, and similarly $G_m \setminus G/G_n := \{\sigma \mid i_1 \neq m, i_k \neq n\}$. The following theorem gives expressions for the holomorphic differentials and the period matrix in terms of series over the group G, see [**12**].

THEOREM 10.1. *Given a Schottky group G with a canonical homology basis. If*

$$\omega_n(z) = \frac{1}{2\pi i} \sum_{\sigma \in G/G_n} \left(\frac{1}{z - \sigma B_n} - \frac{1}{z - \sigma A_n} \right) dz$$

converges absolutely, then $\omega_1, \ldots \omega_g$ forms a basis of holomorphic differentials dual to the given homology basis.

Furthermore the period matrix is then given by

$$B_{nm} = \frac{1}{2\pi i} \sum_{\sigma \in G_m \backslash G/G_n} \log \{B_m, \sigma B_n, A_m, \sigma A_n\}, \quad \text{if } n \neq m$$

$$B_{nn} = \frac{1}{2\pi i} \log \mu_n + \frac{1}{2\pi i} \sum_{\sigma \in G_n \backslash G/G_n, \sigma \neq I} \log \{B_n, \sigma B_n, A_n, \sigma A_n\}$$

where the curly brackets indicate the cross-ratio $\{z_1, z_2, z_3, z_4\} = \frac{z_1 - z_2}{z_2 - z_3} \frac{z_3 - z_4}{z_4 - z_1}$.

To see this, first observe that ω_n is indeed invariant under action in G and thus well defined on $\mathcal{R}$. Furthermore it has no poles in the fundamental domain and is thus holomorphic. The normalization $\int_{a_m} \omega_n = \delta_{nm}$ is true by Cauchy's residue theorem. A slightly more involved calculation yields the period matrix.

In general the question of absolute convergence of these Poincaré series is open. The convergence is known [**2, 12**] for sufficiently small $C_1, C_1', \ldots, C_g, C_g'$ or, equivalently, for sufficiently small $|\mu|$'s. Another classical result [**20, 37**] is that the convergence is guaranteed in the circle-decomposable case, i.e. when the fundamental domain $\Pi(G)$ of a classical Schottky group can be decomposed into non-intersecting pairs of pants by circles. See [**3**] for more details and [**36**] for error estimates.

Schottky uniformization of M-curves. The problem of convergence of the Poincaré series, as well as the problem of the description of the Schottky space

$$S = \{A_1, B_1, \mu_1, \ldots, A_g, B_g, \mu_g\},$$

can be solved for M-curves, and more general for real Riemann surfaces. In this section we describe two Schottky uniformizations of M-curves, called U1 and U2, that are useful in different regimes, see [**3**, Chapter 5].

Let $\mathcal{R}$ be an M-curve with antiholomorphic involution τ that is separated into two parts $\mathcal{R}_+, \mathcal{R}_-$ by the real ovals $X_0, \ldots, X_g$.

The surface $\mathcal{R}_+$ can be conformally mapped to the upper half plane with g round holes, so that X_0 being mapped to the real axis. Applying the antiholomorphic symmetry $\tau z = \bar{z}$ we obtain a fundamental domain of a classical Schottky group, see Fig. 20. The boundary circles C_n and C_n' are mutually complex conjugated C_n and C_n', and represent other real ovals $X_n, n = 1, \ldots, g$. We call this U2-uniformization of the M-curve $\mathcal{R}$.

The parameters of this Schottky group satisfy

$$A_n = \overline{B_n},\ \mathrm{Im}(A_n) > 0,\ 0 < \mu_n < 1,\ \forall n.$$

Furthermore it is easy to deduce inequalities for the discs bounded by C_n to be disjoint:

$$\left| \frac{A_n - \mu_n \bar{A}_n}{1 - \mu_n} - \frac{A_m - \mu_m \bar{A}_m}{1 - \mu_m} \right| > \frac{2 \,\mathrm{Im}\, A_n}{1 - \mu_n} + \frac{2 \,\mathrm{Im}\, A_m}{1 - \mu_m}, \ \forall m, n.$$

This gives a complete description of the Schottky space S. Note that in general this uniformization is not circle decomposable and convergence for the differentials is not guaranteed. However it is circle decomposable for $g \leq 2$ as well as for sufficiently small μ which corresponds to amoebas with small hole sizes.

The meromorphic differentials $d\zeta_i$ can be written in terms of Poincaré series as well. The formulas in for the U2-uniformization are especially simple.

PROPOSITION 10.2. *Let G be a Schottky group in U2 form. Then the differentials $d\zeta_k$ and integrals ζ_k are given by the following series (provided they converge):*

(10.1)

$$d\zeta^{\alpha_i}(z) = \sum_{\sigma\in G}\left(\frac{1}{z-\sigma\alpha_i^-} - \frac{1}{z-\sigma\alpha_i^+}\right)dz, \quad d\zeta^{\beta_j}(z) = \sum_{\sigma\in G}\left(\frac{1}{z-\sigma\beta_j^-} - \frac{1}{z-\sigma\beta_j^+}\right)dz,$$

(10.2)

$$\zeta^{\alpha_i}(z) = \sum_{\sigma\in G}\log\{z,\sigma\alpha_i^-,z_0,\sigma\alpha_i^+\}, \quad \zeta^{\beta_j}(z) = \sum_{\sigma\in G}\log\{z,\sigma\beta_j^-,z_0,\sigma\beta_j^+\}.$$

Here $z_0 \in \mathbb{R}$ is the starting integration point and the Harnack data are cyclicaly ordered $\beta_j^- < \alpha_i^+ < \beta_j^+ < \alpha_l^- \in \mathbb{R}$.

PROOF. It is easy to check that the differentials are invariant with respect to G, and clearly all a-periods vanish. Because of the complex conjugation symmetry all b-periods are imaginary, so the differentials given by formulas (10.1) are indeed the differentials $d\zeta_k$. □

Another natural Schottky uniformization for arbitrary M-curves comes from the Fuchsian uniformization H/G of $\mathcal{R}_+$. Here H is the upper half-plane and G is the Fuchsian group of the second kind. In this case all A_n, B_n are real valued, and are in nested pairs with non-intersecting circles, forming a fundamental domain of G. Extending its action to the lower half-plane we obtain a classical Schottky group uniformizing $\mathcal{R}$. It follows both from circle decomposability as well as from general theory of Fuchsian groups that in this uniformization, which we call U1, the series in Theorem 10.1 always converge.

11. Computation

For computations we use the U2 uniformization.

We use the jtem library introduced in [**36**] for calculation of Abelian differentials and integrals coming from a given Schottky uniformization as well as corresponding theta functions in order to get the weights. Its functionality was extended to compute the differentials $d\zeta_k$ and their integrals ζ_k thus in particular yielding the amoeba map and generalized Newton polygon.

The amoeba map is given by the real part of (10.2):

$$\begin{aligned}\mathcal{A}(z) &= (\mathrm{Re}(\zeta_1), \mathrm{Re}(\zeta_2)) \\ &= \left(\sum_i\sum_{\sigma\in G}\log\left|\{z,\sigma\alpha_i^-,z_0,\sigma\alpha_i^+\}\right|, \sum_j\sum_{\sigma\in G}\log\left|\{z,\sigma\beta_j^-,z_0,\sigma\beta_j^+\}\right|\right).\end{aligned}$$

We choose the starting integration point Q between α^- and β^- so we set $z_0 = \infty$.

The group's elements $\sigma \in G$ can be written as words consisting of the generators $(\sigma_1, \ldots, \sigma_g)$. By limiting the maximal word length L we have a finite sum approximation of $d\zeta_k$. For approximation error estimates and more sophisticated

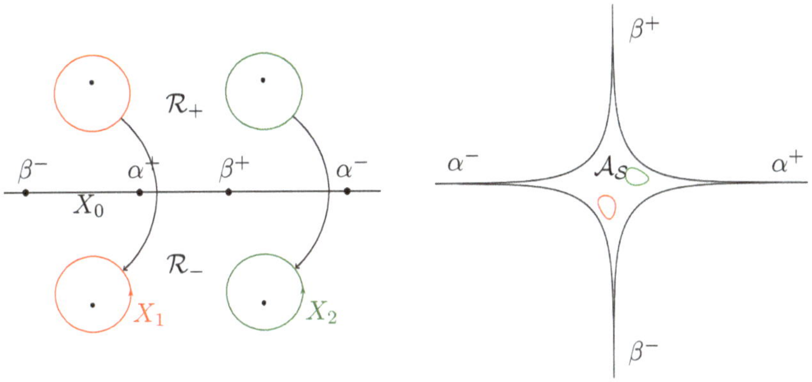

FIGURE 20. Schottky uniformization U2 and the corresponding amoeba.

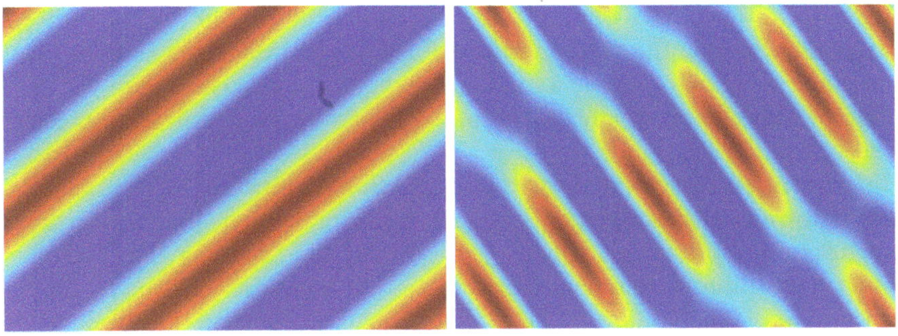

FIGURE 21. Two examples W_f as a continuous function of $\eta(f)$ (11.1) with g=1 and g=2. Computed for Schottky data with $\{\beta^-, \alpha^+, \beta^+, \alpha^-\} = \{-2.4, -0.4, 0.4, 2.4\}$ and $(A, B, \mu) = (0.1 + i, 0.1 - i, 0.02)$(left) and $(A_1, B_1, \mu_1, A_2, B_2, \mu_2) = (1.2 + i1.3, 1.2 - i1.3, 0.08, -0.4 + i0.6, -0.4 - i0.6, 0.03)$(right).

approximation schemes we refer the reader to [**36**]. Fig. 20 presents the image of the amoeba map of the real ovals.

We now take an $N \times N$ square grid with repeating train track angles $\alpha^-, \beta^-, \alpha^+, \beta^+$. The face weight W_f defined in (2.5) of a face f surrounded by faces f_n, f_w, f_s, f_e is then given by

$$W_f = \frac{\theta[\Delta](\int_{\alpha^\pm}^{\beta^\pm} \omega)}{\theta[\Delta](\int_{\beta^\pm}^{\alpha^\mp} \omega)} \frac{\theta[\Delta](\int_{\alpha^\mp}^{\beta^\mp} \omega)}{\theta[\Delta](\int_{\beta^\mp}^{\alpha^\pm} \omega)} \frac{\theta(\eta(f_e) + D)\theta(\eta(f_w) + D)}{\theta(\eta(f_n) + D)\theta(\eta(f_s) + D)}. \tag{11.1}$$

The choice of signs here depends on the order of traintracks surrounding the face f. Note that the ratio of prime forms simplifies to a ratio of theta functions with odd characteristics because the holomorphic spinors in (3.4) cancel. Hence, to compute W_f we need to compute the period matrix, the Abel maps and the corresponding theta functions. See [**36**] for a description of both pointwise and uniform approximation schemes as well as error estimates and numerically stable decompositions.

Intuitively the weights represent a non-linear interaction of g waves analogous to multiphase solutions of nonlinear integrable equations (see, for example, [**3**]). Fig. 21 illustrates W_f in (11.1) for two given Harnack data as a continuous function of $\eta(f)$. The face weights on $\mathbb{Z}^2$ are then given by evaluating this function on a periodic lattice. The periodicity condition (7.8) is equivalent to the fact that the periods of the lattice and of this function match.

We sample a random height function on the square grid with Fock weights via the Metropolis-Hastings algorithm to illustrate the convergence results in Section 9.[3] A dimer configuration on the square grid can be encoded via 4-bit incidence variables n_f at every face indicating which edges are part of the dimer configuration with (n, w, s, e) corresponding to $(1, 2, 4, 8)$. If north and south are present, i.e. $n_f = 5$ or east and west are present, i.e. $n_f = 10$ we can perform a local flip operation by updating n_f and the neighboring values. This can be done efficiently via bitwise XOR operations with appropriate filters.

In [**39**] it was shown that any dimer configuration can be reached via local flips from any other. Since our state space is connected we can run the Metropolis-Hastings algorithm on the random walk defined by the local flip operation weighted by the dimer weights to sample a random height function according to the measure induced by the chosen weights. See also [**24**] for more detail and extension to GPU-based parallelization.

In order to get behaviour of the type of the Ronkin function we need to impose a volume constraint as described in Section 9. This can be done by first starting with a configuration of prescribed volume $\int h = V$ and then running the Markov chain under the constraint of volume preservation. This is easiest achieved by flipping in pairs that cancel the volume change. Note that this construction of flipping two faces in opposite directions is not as amenable to parallelization as sampling from the unconditioned measure.

One could get similar results by sampling from a volume weighted measure. That is for some $q < 1$ the probability of a configuration M becomes $\mathbb{P}(M) = \frac{1}{Z} q^{|V_M - V|} \prod_{e \in M} w_e$ where V_M is the total volume of the height function corresponding to M. This is naively parallelizable and has similar behaviour to the volume constrained approach. See [**38**] for more detail.

As we saw in Section 9 the average height function under a volume constraint converges to the minimizer of the corresponding surface tension under a volume constraint. The volume constraint constant as well as boundary values of the height function can be tricky to choose to match the Ronkin function in the general case but we show an example hereof in Figs. 1 and 22.

[3] All code and further examples can be found at `github.com/nikolaibobenko/FockDimerSimulation`.

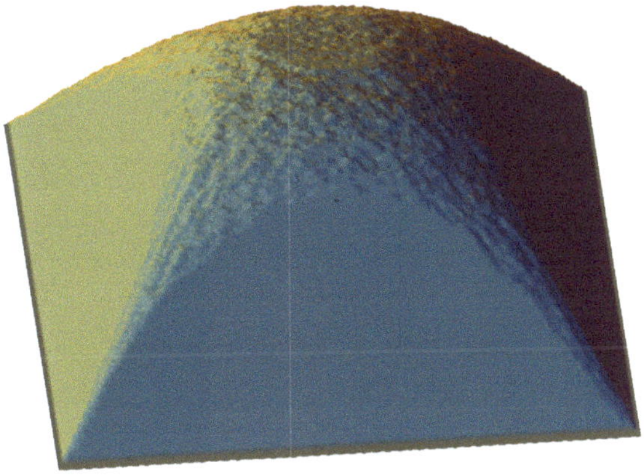

FIGURE 22. Sampled height function. This is same matching as seen in Fig. 1.

References

[1] Kari Astala, Erik Duse, István Prause, and Xiao Zhong, *Dimer Models and Conformal Structures*, September 2023, `arXiv:2004.02599` [math-ph].

[2] H. F. Baker, *Abelian functions*, Cambridge Mathematical Library, Cambridge University Press, Cambridge, 1995. Abel's theorem and the allied theory of theta functions; Reprint of the 1897 original; With a foreword by Igor Krichever. MR1386644

[3] E. D. Belokolos, Alexander I. Bobenko, V.Z. Enolskii, A.R. Its, and V.B. Matveev (eds.), *Algebro-geometric approach to nonlinear integrable equations*, Springer series in nonlinear dynamics, Springer-Verlag, Berlin ; New York, 1994.

[4] I. W. van Spiegel, *Geometry of aggregates*, Van Gorcum & Co. N. V.—G. A. Hak & Dr. H. J. Prakke, Assen, 1957. MR86322

[5] A. I. Bobenko, *Schottky uniformization and finite-gap integration* (Russian), Dokl. Akad. Nauk SSSR **295** (1987), no. 2, 268–272; English transl., Soviet Math. Dokl. **36** (1988), no. 1, 38–42. MR902679

[6] Alexander I. Bobenko and Nikolai Bobenko, *Dimers and M-curves: Limit shapes from Riemann surfaces*, 2024, `arXiv:2407.19462` [math-ph] (to appear in Duke Math. J.).

[7] A. I. Bobenko and Y. B. Suris, *Discrete differential geometry: Integrable structure*, Graduate Studies in Mathematics, vol. 98, American Mathematical Society, Providence, RI, 2008, DOI 10.1090/gsm/098. MR2467378

[8] ———, *Linear integrable systems on quad-graphs*, Int. Math. Res. Not. **2022** (2022), no. 19, 14639–14674 (English).

[9] A. Borodin and M. Duits, *Biased 2×2 periodic Aztec diamond and an elliptic curve*, Probab. Theory Related Fields **187** (2023), no. 1-2, 259–315, DOI 10.1007/s00440-023-01195-8. MR4634340

[10] C. Boutillier, D. Cimasoni, and B. de Tilière, *Elliptic dimers on minimal graphs and genus 1 Harnack curves*, Comm. Math. Phys. **400** (2023), no. 2, 1071–1136, DOI 10.1007/s00220-022-04612-6. MR4589721

[11] C. Boutillier, D. Cimasoni, and B. de Tilière, *Minimal bipartite dimers and higher genus Harnack curves*, Probab. Math. Phys. **4** (2023), no. 1, 151–208, DOI 10.2140/pmp.2023.4.151. MR4567400

[12] W. Burnside, *On a Glass of Automorphic Functions*, Proc. Lond. Math. Soc. **23** (1891/92), 49–88, DOI 10.1112/plms/s1-23.1.49. MR1576383

[13] D. Chelkak, B. Laslier, and M. Russkikh, *Dimer model and holomorphic functions on t-embeddings of planar graphs*, Proc. Lond. Math. Soc. (3) **126** (2023), no. 5, 1656–1739, DOI 10.1112/plms.12516. MR4588721

[14] H. Cohn, R. Kenyon, and J. Propp, *A variational principle for domino tilings*, J. Amer. Math. Soc. **14** (2001), no. 2, 297–346, DOI 10.1090/S0894-0347-00-00355-6. MR1815214
[15] B. A. Dubrovin, *Theta-functions and nonlinear equations* (Russian), Uspekhi Mat. Nauk **36** (1981), no. 2(218), 11–80. With an appendix by I. M. Krichever. MR616797
[16] B. A. Dubrovin and S. M. Natanzon, *Real theta-function solutions of the Kadomtsev-Petviashvili equation* (Russian), Izv. Akad. Nauk SSSR Ser. Mat. **52** (1988), no. 2, 267–286, 446, DOI 10.1070/IM1989v032n02ABEH000759; English transl., Math. USSR-Izv. **32** (1989), no. 2, 269–288. MR941677
[17] J. D. Fay, *Theta functions on Riemann surfaces*, Lecture Notes in Mathematics, Vol. 352, Springer-Verlag, Berlin-New York, 1973. MR335789
[18] V. V. Fock, *Inverse spectral problem for GK integrable system*, March 2015, `arXiv:1503.00289` [math].
[19] L. R. Ford, *Automorphic functions*, 2nd ed ed., AMS Chelsea Pub, Providence, R.I, 2004 (eng).
[20] R. Fricke and F. Klein, *Vorlesungen über die Theorie der automorphen Funktionen. Band 1: Die gruppentheoretischen Grundlagen. Band II: Die funktionentheoretischen Ausführungen und die Andwendungen* (German), Bibliotheca Mathematica Teubneriana, Bande 3, vol. 4, Johnson Reprint Corp., New York; B. G. Teubner Verlagsgesellschaft, Stuttgart, 1965. MR183872
[21] T. George, *Spectra of biperiodic planar networks*, Comm. Math. Phys. **405** (2024), no. 1, Paper No. 10, 47, DOI 10.1007/s00220-023-04878-4. MR4691860
[22] A. B. Goncharov and R. Kenyon, *Dimers and cluster integrable systems* (English, with English and French summaries), Ann. Sci. Éc. Norm. Supér. (4) **46** (2013), no. 5, 747–813, DOI 10.24033/asens.2201. MR3185352
[23] P. W. Kasteleyn, *Graph theory and crystal physics*, Graph Theory and Theoretical Physics, Academic Press, London-New York, 1967, pp. 43–110. MR253689
[24] D. Keating and A. Sridhar, *Random tilings with the GPU*, J. Math. Phys. **59** (2018), no. 9, 091420, 17, DOI 10.1063/1.5038732. MR3858689
[25] R. Kenyon, *The Laplacian and Dirac operators on critical planar graphs*, Invent. Math. **150** (2002), no. 2, 409–439, DOI 10.1007/s00222-002-0249-4. MR1933589
[26] R. Kenyon, W. Y. Lam, S. Ramassamy, and M. Russkikh, *Dimers and circle patterns* (English, with English and French summaries), Ann. Sci. Éc. Norm. Supér. (4) **55** (2022), no. 3, 865–903, DOI 10.24033/asens.2507. MR4553657
[27] R. Kenyon and A. Okounkov, *Planar dimers and Harnack curves*, Duke Math. J. **131** (2006), no. 3, 499–524, DOI 10.1215/S0012-7094-06-13134-4. MR2219249
[28] R. Kenyon and A. Okounkov, *Limit shapes and the complex Burgers equation*, Acta Math. **199** (2007), no. 2, 263–302, DOI 10.1007/s11511-007-0021-0. MR2358053
[29] R. Kenyon, A. Okounkov, and S. Sheffield, *Dimers and amoebae*, Ann. of Math. (2) **163** (2006), no. 3, 1019–1056, DOI 10.4007/annals.2006.163.1019. MR2215138
[30] I. Krichever, *Amoebas, Ronkin function, and Monge-Ampère measures of algebraic curves with marked points*, Topology, geometry, integrable systems, and mathematical physics, Amer. Math. Soc. Transl. Ser. 2, vol. 234, Amer. Math. Soc., Providence, RI, 2014, pp. 265–278, DOI 10.1090/trans2/234/13. MR3307152
[31] I. M. Krichever, *Methods of algebraic geometry in the theory of nonlinear equations* (Russian), Uspehi Mat. Nauk **32** (1977), no. 6(198), 183–208, 287. MR516323
[32] I. M. Krichever, *Two-dimensional periodic difference operators and algebraic geometry* (Russian), Dokl. Akad. Nauk SSSR **285** (1985), no. 1, 31–36. MR812140
[33] G. Mikhalkin, *Real algebraic curves, the moment map and amoebas*, Ann. of Math. (2) **151** (2000), no. 1, 309–326, DOI 10.2307/121119. MR1745011
[34] D. Mumford, *Tata lectures on theta. I*, Modern Birkhäuser Classics, Birkhäuser Boston, Inc., Boston, MA, 2007. With the collaboration of C. Musili, M. Nori, E. Previato and M. Stillman; Reprint of the 1983 edition, DOI 10.1007/978-0-8176-4578-6. MR2352717
[35] M. Passare and H. Rullgård, *Amoebas, Monge-Ampère measures, and triangulations of the Newton polytope*, Duke Math. J. **121** (2004), no. 3, 481–507, DOI 10.1215/S0012-7094-04-12134-7. MR2040284
[36] Markus Schmies, *Computational Methods for Riemann Surfaces and Helicoids with Handles*, Ph.D. thesis, Technische Universität Berlin, 2005.

[37] F. Schottky, *Ueber eine specielle Function, welche bei einer bestimmten linearen Transformation ihres Arguments unverändert bleibt* (German), J. Reine Angew. Math. **101** (1887), 227–272, DOI 10.1515/crll.1887.101.227. MR1580127
[38] S. Sheffield, *Random surfaces* (English, with English and French summaries), Astérisque **304** (2005), vi+175. MR2251117
[39] W. P. Thurston, *Conway's tiling groups*, Amer. Math. Monthly **97** (1990), no. 8, 757–773, DOI 10.2307/2324578. MR1072815
[40] W.P. Thurston, *Groups, tilings and finite state automata: Summer 1989 AMS Colloquium Lectures*, Research report GCG, Geometry Computing Group, 1989.

Institute of Mathematics, Technische Universität Berlin
Email address: bobenko@math.tu-berlin.de

Department of Mathematics, University of Geneva
Email address: nikolai.bobenko@unige.ch

Institute of Mathematics, Technische Universität Berlin
Email address: suris@math.tu-berlin.de

Contemporary Mathematics
Volume **823**, 2025
https://doi.org/10.1090/conm/823/16490

A matroidal perspective on the tropical Prym variety

Felix Röhrle and Dmitry Zakharov

to the memory of Igor Krichever

ABSTRACT. We associate a matroid $M(\widetilde{\Gamma}/\Gamma)$ to a harmonic double cover $\pi : \widetilde{\Gamma} \to \Gamma$ of metric graphs. The matroid $M(\widetilde{\Gamma}/\Gamma)$ is a geometric interpretation of Zaslavsky's signed graphic matroid. We show that the principalization $\mathrm{Prym}_p(\widetilde{\Gamma}/\Gamma)$ of the tropical Prym variety of the double cover can be reconstructed from $M(\widetilde{\Gamma}/\Gamma)$, equipped with certain additional decorations. We describe the simplification of the matroid $M(\widetilde{\Gamma}/\Gamma)$ and show that the Prym variety does not change under simplification.

CONTENTS

1. Introduction

The two most-studied objects in algebraic geometry are curves and abelian varieties. The construction of the Jacobian variety $\mathrm{Jac}(C)$ of an algebraic curve C, in the guise of abelian integrals, goes back to the foundations of algebraic geometry in the 19th century. In moduli, the association $C \mapsto \mathrm{Jac}(C)$ is called the *Torelli morphism* $\tau_J : \mathcal{M}_g \to \mathcal{A}_g$, where $\mathcal{M}_g$ and $\mathcal{A}_g$ are the moduli spaces of smooth genus g curves and dimension g principally polarized abelian varieties (ppavs), respectively. The Torelli morphism is injective, and an effective description of its image is the still-unsolved *Schottky problem*.

There is another natural way to associate a ppav to an algebraic curve. Given an étale double cover $\widetilde{C} \to C$ of smooth curves of genera $2g-1$ and g, respectively, the *Prym variety* $\mathrm{Prym}(\widetilde{C}/C)$ is the connected component of the identity of the kernel of the norm map $\mathrm{Jac}(\widetilde{C}) \to \mathrm{Jac}(C)$. Prym varieties have been extensively studied since

2020 *Mathematics Subject Classification.* Primary 14T20, 14H40.
Key words and phrases. Tropical Prym variety, signed graphic matroid.

their reintroduction by Mumford in [**Mum74**]. The Prym variety carries a principal polarization that is half the polarization induced from $\mathrm{Jac}(\widetilde{C})$, and the assignment $(\widetilde{C} \to C) \mapsto \mathrm{Prym}(\widetilde{C}/C)$ defines the *Prym–Torelli map* $\tau_P : \mathcal{R}_g \to A_{g-1}$, where $\mathcal{R}_g$ is the moduli space of connected étale double covers of smooth genus g curves. Unlike τ_J, the morphism τ_P is never injective, and it is likewise an open problem to describe its image.

Tropical geometry is a modern branch of mathematics that studies certain combinatorial, piecewise-linear objects, which mimic concepts from algebraic geometry. A *tropical curve* of genus g is a metric graph Γ with $h_1(\Gamma) = g$, a *harmonic morphism* of tropical curves is a piecewise-linear map satisfying a certain balancing condition, and a *tropical abelian variety* is a real torus carrying an auxiliary integral structure. By varying lengths and combinatorial types, we obtain moduli spaces $\mathcal{M}_g^{\mathrm{trop}}$ and $\mathcal{A}_g^{\mathrm{trop}}$ of tropical curves and tropical ppavs, respectively.

We can naturally associate a *tropical Jacobian* $\mathrm{Jac}(\Gamma)$ to a tropical curve Γ (see [**MZ08**]). The corresponding tropical Torelli map $\tau_J^{\mathrm{trop}} : \mathcal{M}_g^{\mathrm{trop}} \to \mathcal{A}_g^{\mathrm{trop}}$ (which is not injective, unlike τ_J) was investigated in [**CV10**] and [**BMV11**]. The key insight in these works is that the tropical Jacobian $\mathrm{Jac}(\Gamma)$ is in fact entirely governed by the graphic matroid $M(\Gamma)$ of Γ. It is therefore possible to study τ_J, and in particular describe its fibers, from a purely matroidal perspective, and similarly to give a matroidal interpretation of the image of τ_J^{trop}. Indeed, the relevance of the graphic matroid in this context was well-known before the advent of tropical geometry, see e.g. [**Ger82**].

The tropical analogue of an étale double cover is a harmonic morphism $\widetilde{\Gamma} \to \Gamma$ of metric graphs of degree two. The *tropical Prym variety* $\mathrm{Prym}(\widetilde{\Gamma}/\Gamma)$ of $\widetilde{\Gamma} \to \Gamma$ was defined in [**JL18**], in analogy with the algebraic setting, and is a polarized tropical abelian variety of dimension $g(\widetilde{\Gamma}) - g(\Gamma)$. The tropical Prym variety was further investigated in [**LU21**], [**LZ22**], [**RZ22**], and [**GZ23**].

The purpose of this paper is to develop a matroidal perspective on the tropical Prym variety, with the aim of studying the tropical Prym–Torelli map. We show how to associate to a harmonic double cover $\pi : \widetilde{\Gamma} \to \Gamma$ a dual pair of matroids $M(\widetilde{\Gamma}/\Gamma)$ and $M^*(\widetilde{\Gamma}/\Gamma)$ (see Definition 3.3). These matroids turn out to be *signed (co-)graphic matroids* in the sense of [**Zas82**], and the precise translation of Zaslavsky's definition into the language of double covers is given in Proposition 3.9. In Proposition 3.10 we give interpretations of the cryptomorphic descriptions of $M(\widetilde{\Gamma}/\Gamma)$ in terms of the double cover $\widetilde{\Gamma} \to \Gamma$. In Section 4 we prove the main result of our paper and show that the matroid $M(\widetilde{\Gamma}/\Gamma)$, suitably decorated, recovers all of the information of the Prym variety $\mathrm{Prym}(\widetilde{\Gamma}/\Gamma)$ except for its polarization. We canonically associate a tropical ppav to $\mathrm{Prym}(\widetilde{\Gamma}/\Gamma)$, the *principalization* $\mathrm{Prym}_p(\widetilde{\Gamma}/\Gamma)$, and signed cographic matroid recovers this object:

THEOREM A (Theorem 4.9). *Let $\pi : \widetilde{\Gamma} \to \Gamma$ be a harmonic double cover of metric graphs, and let $M = M(\widetilde{\Gamma}/\Gamma)$ be the signed graphic matroid equipped with index function* $\mathrm{ind} : \mathcal{I}(M^*) \to \mathbb{Z}_{>0}$ *(Definition 4.3), orientation $\overrightarrow{M}$ on M induced by an orientation of π (Section 4.4), and edge length function $\ell : E(\Gamma) \to \mathbb{R}$. Then we can associate a tropical ppav* $\mathrm{Prym}(M)$ *to the matroidal data such that*

$$\mathrm{Prym}_p(\widetilde{\Gamma}/\Gamma) \cong \mathrm{Prym}(M).$$

It follows that two double covers with isomorphic decorated matroidal data have isomorphic Prym varieties, see Corollary 4.11. Example 4.12 shows that distinct double covers may have identical matroidal data and hence isomorphic Prym varieties.

For any matroid M, we can construct its *simplification* M_{sim}, having no circuits of size 1 or 2. In [**CV10**] and [**BMV11**] it is shown that simplifying the cographic matroid of Γ (contracting bridges and parallel edges, with an appropriate redistribution of edge lengths) does not change the Jacobian $\mathrm{Jac}(\Gamma)$. In Section 5, we describe an analogous simplification procedure for double covers, and show that it does not change the Prym variety.

THEOREM B (Theorem 5.3). *Let* $\pi : \widetilde{\Gamma} \to \Gamma$ *be a double cover of metric graphs and let* $\pi_{\mathrm{sim}} : \widetilde{\Gamma}_{\mathrm{sim}} \to \Gamma_{\mathrm{sim}}$ *be a simplification of* π *(see Definition 5.2). Then* $\mathrm{Prym}_p(\widetilde{\Gamma}/\Gamma) \cong \mathrm{Prym}_p(\widetilde{\Gamma}_{\mathrm{sim}}/\Gamma_{\mathrm{sim}})$.

In contrast to the case of metric graphs, we show by example that double covers with non-isomorphic simple matroids may have isomorphic Prym varieties.

In hindsight, it is clear that the signed graphic matroid $M(\widetilde{\Gamma}/\Gamma)$ and its dual ought to have a close relation to the Prym variety $\mathrm{Prym}(\widetilde{\Gamma}/\Gamma)$. For example, the papers [**LZ22**] and [**GZ23**] proved a formula for the volume of $\mathrm{Prym}(\widetilde{\Gamma}/\Gamma)$ as a sum over certain subsets of $E(\Gamma)$ called *odd genus one decompositions* or *ogods*, and these turn out to be the bases of $M^*(\widetilde{\Gamma}/\Gamma)$ (see Section 2.8 for details).

A natural question is to investigate the tropical Prym–Torelli morphism τ_P^{trop} that associates to a harmonic double cover $\widetilde{\Gamma} \to \Gamma$ its Prym variety $\mathrm{Prym}(\widetilde{\Gamma}/\Gamma)$. There are two related issues with defining τ_P^{trop}. First, the polarization type of $\mathrm{Prym}(\widetilde{\Gamma}/\Gamma)$ may change when the double cover $\widetilde{\Gamma} \to \Gamma$ is deformed, specifically if the number of connected components of the dilation subgraph changes. Second, the behavior of $\mathrm{Prym}(\widetilde{\Gamma}/\Gamma)$ is not continuous under such deformations. In future work, we plan to give an alternative definition of the tropical Prym variety that solves these issues, and investigate the redefined Prym–Torelli map from a matroidal perspective. The above Theorem shows that some of the non-injectivity of τ_P^{trop} is purely matroidal in nature, and describing the remaining non-injectivity is the subject of ongoing research.

We end by outlining a series of tropical problems inspired by Krichever's solution to the algebraic Schottky problems, which we now recall. Let X be a ppav of dimension g with principal polarization Θ, and let $K(X)$ be the *Kummer variety* of X, the image of X under the map $\phi_{2\Theta} : X \to \mathbb{P}^{2^g-1}$. Fay's trisecant identity [**Fay06**] states that if $X = \mathrm{Jac}(C)$ is a Jacobian, then $K(X)$ admits a four-parameter family of trisecant lines, indexed by quadruples of points of $C \subset \mathrm{Jac}(C)$. Gunning [**Gun82**] showed that the existence of such a family characterizes Jacobians among all ppavs. Welters [**Wel83**], [**Wel84**] strengthened this result by showing that Jacobians are characterized by the existence of a formal one-parameter family of flexes of $K(X)$ (limit trisecants when the three points are equal), and conjectured that the existence of a single trisecant of the Kummer variety characterizes Jacobians among ppavs. After a number of progressively stronger results, Krichever proved Welters's conjecture in [**Kri06**] and [**Kri10**]. An analogous solution to the Prym–Schottky problem was given by Grushevsky and Krichever in [**GK10**]. Beauville and Debarre [**BD87**] showed that if $X = \mathrm{Prym}(\widetilde{C}/C)$ is a Prym variety, then $K(X)$ admits a four-parameter family of quadrisecant planes, and it was shown in [**GK10**] that the existence of a symmetric pair of quadrisecants (but not a single quadrisecant) characterizes Prym varieties among all ppavs.

It is natural to formulate an analogous series of questions in the tropical setting. A theory of tropical theta functions has already been developed in [**MZ08**] and [**FRSS18**]. Fay's trisecant identity is based on an elementary property of divisorial rank (see Proposition 11.9.1 in [**BL04**]). It turns out that the analogous statement for tropical Baker–Norine rank is false, so it is not immediately clear what happens to Fay's identity under tropicalization. Nevertheless it is natural to ask whether the theta functions of a tropical Jacobian satisfy any identities, whether these identities have a geometric manifestation, whether they characterize tropical Jacobians among all tropical ppavs, and whether this characterization is computationally effective. Finally, all these questions can be asked about the tropical Prym variety as well.

2. Setup and definitions

We begin by reviewing a number of basic definitions concerning graphs, metric graphs, harmonic morphisms, double covers, and tropical abelian varieties.

2.1. Graphs, harmonic morphisms, and double covers. A *graph* G consists of a set of *vertices* $V(G)$, a set of *half-edges* $H(G)$, a *root map* $r : H(G) \to V(G)$, and a fixed-point-free involution $\iota : H(G) \to H(G)$. The *edges* of G are the orbits of ι, with each edge consisting of two half-edges, and the set of edges is denoted $E(G)$. We allow graphs with loops and multiedges. The *tangent space* to a vertex $v \in V(G)$ is the set $T_vG = r^{-1}(v)$ of half-edges rooted at v. The *genus* $g(G)$ of a graph G is its first Betti number:

$$g(G) = |E(G)| - |V(G)| + 1.$$

Unless otherwise specified, we consider only finite connected graphs. Given a set of edges $F \subset E(G)$, we denote by $G\backslash F$ the (possibly disconnected) graph obtained by removing the edges in F, but retaining all vertices, even isolated ones. We denote $G[F]$ the minimal subgraph of G containing all edges in F, in other words $G[F]$ is obtained from $G\backslash\big(E(G)\backslash F\big)$ by removing all isolated vertices.

A *morphism* of graphs $f : \widetilde{G} \to G$ is a pair of maps $f : V(\widetilde{G}) \to V(G)$ and $f : H(\widetilde{G}) \to H(G)$ commuting with the root and involution maps, and thus inducing a map $f : E(\widetilde{G}) \to E(G)$ on the edges. A *harmonic morphism* of graphs is a pair (f, d_f) consisting of a morphism of graphs $f : \widetilde{G} \to G$ and a *local degree* function $d_f : V(\widetilde{G}) \cup H(\widetilde{G}) \to \mathbb{Z}_{>0}$, such that $d_f(\widetilde{h}) = d_f(\widetilde{h}')$ for any edge $\widetilde{e} = \{\widetilde{h}, \widetilde{h}'\} \in E(\widetilde{G})$ and such that the local balancing condition

$$d_f(\widetilde{v}) = \sum_{\widetilde{h} \in T_{\widetilde{v}}\widetilde{G} \cap f^{-1}(h)} d_f(\widetilde{h}) \tag{1}$$

holds for each vertex $\widetilde{v} \in V(\widetilde{G})$ and each half-edge $h \in T_{f(\widetilde{v})}G$. A harmonic morphism $f : \widetilde{G} \to G$ to a connected target graph G is surjective and has a well-defined *global degree* given by

$$\deg f = \sum_{\widetilde{v} \in f^{-1}(v)} d_f(\widetilde{v}) = \sum_{\widetilde{h} \in f^{-1}(h)} d_f(\widetilde{h})$$

for any $v \in V(G)$ or any $h \in H(G)$.

A harmonic morphism $p : \widetilde{G} \to G$ of global degree two is called a *double cover*, and we describe its structure in detail. A vertex $v \in V(G)$ is called *dilated* if it has a unique preimage that we label $p^{-1}(v) = \{\widetilde{v}\}$, with local degree $d_p(\widetilde{v}) = 2$. Similarly, a half-edge $h \in H(G)$ is *dilated* if $p^{-1}(h) = \{\widetilde{h}\}$ with $d_p(\widetilde{h}) = 2$, and a dilated edge $e = \{h, h'\}$ (consisting of dilated half-edges) has a unique edge $\widetilde{e} = \{\widetilde{h}, \widetilde{h}'\}$ lying over it. The root vertex of a dilated half-edge is dilated, so the dilated set forms a subgraph $G_{\mathrm{dil}} \subset G$ called the *dilation subgraph*, which is isomorphic to its preimage $p^{-1}(G_{\mathrm{dil}})$. We say that p is a *free* double cover if $G_{\mathrm{dil}} = \varnothing$ and *dilated* otherwise. We say that p is *edge-free* if G_{dil} consists of only vertices. The *dilation index* of a double cover is defined to be

$$d(\widetilde{G}/G) = \begin{cases} \text{number of connected components of } G_{\mathrm{dil}}, & \text{if } p \text{ is dilated,} \\ 1, & \text{if } p \text{ is free.} \end{cases}$$

A vertex $v \in V(G)$ having two preimages is called *undilated*, and we label its preimages $p^{-1}(v) = \{\widetilde{v}^+, \widetilde{v}^-\}$, where the local degrees are $d_p(\widetilde{v}^\pm) = 1$. Similarly, a half-edge $h \in H(G)$ is *undilated* if $p^{-1}(h) = \{\widetilde{h}^+, \widetilde{h}^-\}$ with $d_p(\widetilde{h}^\pm) = 1$. If the root vertex $v = r(h)$ is undilated, we label the preimages of h so that $r(\widetilde{h}^\pm) = \widetilde{v}^\pm$ (if v is dilated then $r(\widetilde{h}^\pm) = \widetilde{v}$ regardless of labeling). An undilated edge $e \in E(G)$ is called *free* if both root vertices of e (which may be the same) are undilated. The *free subgraph* $G_{\mathrm{fr}} \subset G$ consists of the undilated vertices and the free edges, and the restriction $p|_{p^{-1}(G_{\mathrm{fr}})} : p^{-1}(G_{\mathrm{fr}}) \to G_{\mathrm{fr}}$ of p to the free subgraph is a degree two covering space in the topological sense.

The structure of the graph $\widetilde{G}$ is determined by the involution map on the preimages $\widetilde{h}^\pm$ of the undilated half-edges, in other words by how the $\widetilde{h}^\pm$ pair up into edges. If the edge $e = \{h, \iota(h)\} \in E(G)$ is undilated but not free, then we can label the preimages of the half-edges in such a way that the edges of $\widetilde{G}$ lying above e are $\widetilde{e}^+ = \{\widetilde{h}^+, \widetilde{\iota(h)}^+\}$ and $\widetilde{e}^- = \{\widetilde{h}^-, \widetilde{\iota(h)}^-\}$, in other words the involution is $\iota(\widetilde{h}^\pm) = \widetilde{\iota(h)}^\pm$. If e is free, however, then we record the involution by

$$\sigma(e) = \begin{cases} +1, & \text{if } \iota(\widetilde{h}^\pm) = \widetilde{\iota(h)}^\pm, \\ -1, & \text{if } \iota(\widetilde{h}^\pm) = \widetilde{\iota(h)}^\mp. \end{cases}$$

Given an orientation on G, we always give $\widetilde{G}$ the induced orientation. We label the oriented preimages of an oriented edge $e = (h, \iota(h))$ as $\widetilde{e}^+ = (\widetilde{h}^+, \iota(\widetilde{h}^+))$ and $\widetilde{e}^- = (\widetilde{h}^-, \iota(\widetilde{h}^-))$.

The double cover $p : \widetilde{G} \to G$ is thus determined by G, the choice of a dilation subgraph $G_{\mathrm{dil}} \subset G$ that also determines G_{fr}, and a parity assignment $\sigma : E(G_{\mathrm{fr}}) \to \{\pm 1\}$, in other words the structure of a *signed graph* on G_{fr} (in fact, as we shall see in Proposition 3.9, the set of undilated edges $E(G) \backslash E(G_{\mathrm{dil}})$ carries the structure of a signed graph, in the more general sense of Zaslavsky [**Zas82**]). Exchanging the preimages of an undilated vertex $v \in V(G_{\mathrm{fr}})$ swaps the signs of all free edges rooted at it, so the parity assignment σ is determined up to *vertex switching equivalence*. In modern terminology, the switching equivalence class of a parity assignment σ is a well-defined element of the simplicial cohomology group $H^1(G_{\mathrm{fr}}, \mathbb{Z}/2\mathbb{Z})$, which is identified with the set of $\mathrm{Hom}\left(\pi_1(G_{\mathrm{fr}}), \mathbb{Z}/2\mathbb{Z}\right)$ of free double covers by covering space theory and the universal coefficient theorem.

For another perspective on double covers note that $p : \widetilde{G} \to G$ determines an involution $i : \widetilde{G} \to \widetilde{G}$ that changes signs on undilated elements and fixes dilated elements. Conversely, given an involution $i : \widetilde{G} \to \widetilde{G}$, the quotient map $p : \widetilde{G} \to \widetilde{G}/i$ naturally has the structure of a double cover, with the dilation subgraph of G corresponding to the fixed subgraph of $\widetilde{G}$.

2.2. Edge contractions. We extensively use the *edge contraction* operation, which is a graph-theoretic deformation that allows us to move through the moduli spaces of graphs and double covers. Let G be a graph, let $F \subset E(G)$ be a set of edges, and let $H = G[F]$ be the subgraph generated by these edges. We denote by G_F the graph obtained by contracting each connected component of H to a separate vertex. An edge contraction induces a natural surjective contraction homomorphism $H_1(G, \mathbb{Z}) \to H_1(G_F, \mathbb{Z})$.

An important property of contraction is that it can be performed for a harmonic morphism. Let $f : \widetilde{G} \to G$ be a harmonic morphism of graphs, let $F \subset E(G)$ be a set of edges, and let $\widetilde{H} = \widetilde{G}[\widetilde{F}]$ be the subgraph of $\widetilde{G}$ generated by $\widetilde{F} = f^{-1}(F)$. Consider the map $f_F : \widetilde{G}_{\widetilde{F}} \to G_F$, where the vertex $\widetilde{v}_i \in V(\widetilde{G}_{\widetilde{F}})$ corresponding to a connected component $\widetilde{H}_i$ of $\widetilde{H}$ is mapped to the vertex corresponding to the connected component $f(\widetilde{H}_i)$ of H. It is elementary to check that f_F is harmonic, provided that we set the local degree of f_F at the vertex $\widetilde{v}_i$ to be the global degree of f on the subgraph $\widetilde{H}_i$.

EXAMPLE 2.1. Let $p : \widetilde{G} \to G$ be a double cover, let $F \subset E(G)$ be a set of edges, and assume for simplicity that the subgraph $H = G[F]$ generated by F is connected. If H has no dilated vertices, then the restricted cover $p|_{p^{-1}(H)} : p^{-1}(H) \to H$ is a free double cover, and $p^{-1}(H)$ has two connected components if and only if this double cover is trivial. In this case, the vertex $v \in V(G_F)$ corresponding to H is undilated. However, if $p|_{p^{-1}(H)}$ is free but nontrivial, or if H has at least one dilated vertex, then the vertex v is dilated. In particular, the contraction of a free double cover need not be free.

EXAMPLE 2.2. We will frequently make use of the following construction, visualized in Figure 1. Let $p : \widetilde{G} \to G$ be a dilated double cover. We construct the edge-free double cover $p_{\text{ef}} : \widetilde{G}_{\text{ef}} \to G_{\text{ef}}$ by contracting all dilated edges, in other words by contracting each connected component G_i of G_{dil} to a dilated vertex v_i. We then resolve p_{ef} into a free double cover $p_{\text{free}} : \widetilde{G}_{\text{free}} \to G_{\text{free}}$ by adding a loop e_i at each dilated vertex v_i of G_{ef}, and replacing the preimage vertex $\widetilde{v}_i$ with two vertices $\widetilde{v}_i^{\pm}$ joined by two parallel edges $\widetilde{e}_i^{\pm}$ mapping to the loop e_i. The cover p_{free} is not defined canonically (unlike p_{ef}), since we must choose how to reattach the edges of $\widetilde{G}_{\text{ef}}$ to the new vertices $\widetilde{v}_i^{\pm}$. The cover p_{ef} is obtained from p_{free} by contracting the new loops e_i.

2.3. Metric graphs. Let G be a graph and let $\ell : E(G) \to \mathbb{R}_{>0}$ be an *edge length* assignment. The pair (G, ℓ) defines a *metric graph* Γ, which is the topological space obtained by viewing each edge $e \in E(G)$ as a closed interval of length $\ell(e)$ and attaching the intervals according to the structure of G. We endow Γ with the shortest-path metric. The pair (G, ℓ) is called a *model* for Γ, and the genus of Γ is the genus of any underlying model. We will often abuse notation and refer to the edges and vertices of a metric graph Γ, with respect to an implied choice of model.

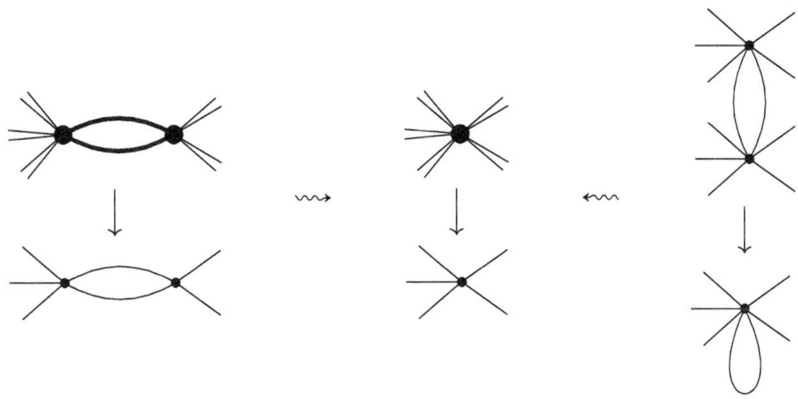

FIGURE 1. Local picture of a dilated double cover p (left) with the edge contractions producing p_{ef} (middle). On the right: p_{free} with its contraction back to p_{ef}.

A *harmonic morphism of metric graphs* $\phi : \widetilde{\Gamma} \to \Gamma$ is a continuous, piecewise-linear map whose slopes $d_f(\widetilde{e})$ along the edges $\widetilde{e} \in E(\widetilde{G})$ are positive integers satisfying the slope-balancing condition (1). In other words, there exist models $(\widetilde{G}, \widetilde{\ell})$ and (G, ℓ) of $\widetilde{\Gamma}$ and Γ, respectively, and a harmonic morphism of graphs $f : \widetilde{G} \to G$ modeling ϕ, such that $d_\phi(\widetilde{e}) = d_f(\widetilde{e})$ for all $\widetilde{e} \in E(\widetilde{\Gamma})$. The continuity of ϕ imposes the restriction

$$\ell(\phi(\widetilde{e})) = d_\phi(\widetilde{e})\widetilde{\ell}(\widetilde{e})$$

for all edges $\widetilde{e} \in E(\widetilde{G})$. If $\phi : \widetilde{\Gamma} \to \Gamma$ is a double cover of metric graphs, then ϕ is a factor two dilation along a dilated edge and an isometry along an undilated edge, explaining the terminology.

2.4. Divisor theory on metric graphs. For the reader's convenience, we briefly recall the divisor theory on a metric graph Γ. The divisor group $\mathrm{Div}(\Gamma)$ is the free abelian group on the points of Γ. A rational function on Γ is a real-valued piecewise-linear function with integer slopes, and the group of rational functions is denoted $\mathrm{Rat}(\Gamma)$. The divisor of a rational function is

$$\mathrm{div} : \mathrm{Rat}(\Gamma) \longrightarrow \mathrm{Div}(\Gamma), \qquad f \longmapsto \sum_{x\in\Gamma} (\text{sum of incoming slopes at } x) \cdot x$$

The Picard group $\mathrm{Pic}(\Gamma)$ is the cokernel of div. The degree map

$$\deg : \mathrm{Div}(\Gamma) \longrightarrow \mathbb{Z}, \qquad \sum_{x\in\Gamma} a_x \cdot x \longmapsto \sum_{x\in\Gamma} a_x$$

descends to $\mathrm{Pic}(\Gamma)$, and we denote $\mathrm{Pic}_d(\Gamma)$ the set of equivalence classes of degree d divisors.

2.5. Tropical abelian varieties. A *real torus with integral structure* Σ of dimension g, or an *integral torus* for short, is defined by a triple $(\Lambda, \Lambda', [\cdot,\cdot])$, where Λ and Λ' are free abelian groups of rank g and $[\cdot,\cdot] : \Lambda \times \Lambda' \to \mathbb{R}$ is a nondegenerate pairing. The torus itself is $\Sigma = \mathrm{Hom}(\Lambda, \mathbb{R})/\Lambda'$, where Λ' is embedded in $\mathrm{Hom}(\Lambda, \mathbb{R})$ via the assignment $\lambda' \mapsto [\cdot, \lambda']$. The integral lattice $\mathrm{Hom}(\Lambda, \mathbb{Z})$ in the universal cover $\mathrm{Hom}(\Lambda,\mathbb{R})$ of Σ may be viewed as a tropical analogue of a complex structure on a real torus.

Let $\Sigma_1=(\Lambda_1,\Lambda_1',[\cdot,\cdot]_1)$ and $\Sigma_2=(\Lambda_2,\Lambda_2',[\cdot,\cdot]_2)$ be integral tori, and let $f^{\#}:\Lambda_2\to\Lambda_1$ and $f_{\#}:\Lambda_1'\to\Lambda_2'$ be homomorphisms satisfying the relation

$$\left[f^{\#}(\lambda_2),\lambda_1'\right]_1=\left[\lambda_2,f_{\#}(\lambda_1')\right]_2$$

for all $\lambda_2\in\Lambda_2$ and $\lambda_1'\in\Lambda_1'$. The Hom-dual $\mathrm{Hom}(\Lambda_1,\mathbb{R})\to\mathrm{Hom}(\Lambda_2,\mathbb{R})$ of $f^{\#}$ restricts to $f_{\#}$ on Λ_1', hence the pair $f=(f^{\#},f_{\#})$ defines a *homomorphism of integral tori* $f:\Sigma_1\to\Sigma_2$. The maps $f^{\#}$ and $f_{\#}$ have the same rank, which is denoted $\mathrm{rk}\,f$ and which is equal to the dimension of $f(\Sigma_1)$.

Let $f:\Sigma_1\to\Sigma_2$ be a homomorphism of integral tori. The group-theoretic kernel $\mathrm{Ker}\,f$ is the direct sum of a finite abelian group and an integral torus $(\mathrm{Ker}\,f)_0$, which we call the *kernel torus* of f. It is defined by the triple

$$(\mathrm{Ker}\,f)_0=\left((\mathrm{Coker}\,f^{\#})^{\mathrm{tf}},\mathrm{Ker}\,f_{\#},[\cdot,\cdot]_K\right),$$

where for an abelian group A we denote A^{tf} the quotient by its torsion subgroup, and $[\cdot,\cdot]_K$ is the pairing $(\mathrm{Coker}\,f^{\#})^{\mathrm{tf}}\times\mathrm{Ker}\,f_{\#}\to\mathbb{R}$ induced by $[\cdot,\cdot]_1$. The pair $i=(i^{\#},i_{\#})$, where $i_{\#}:\mathrm{Ker}\,f_{\#}\hookrightarrow\Lambda_1'$ is the natural inclusion and $i^{\#}:\Lambda_1\twoheadrightarrow(\mathrm{Coker}\,f^{\#})^{\mathrm{tf}}$ is the quotient map, defines the natural injective homomorphism of real tori $i:(\mathrm{Ker}\,f)_0\to\Sigma_1$.

A *polarization* on an integral torus $\Sigma=(\Lambda,\Lambda',[\cdot,\cdot])$ is a group homomorphism $\xi:\Lambda'\to\Lambda$ with the property that the induced bilinear form $(\cdot,\cdot)=[\xi(\cdot),\cdot]:\Lambda'_{\mathbb{R}}\times\Lambda'_{\mathbb{R}}\to\mathbb{R}$ is symmetric and positive definite. A polarization is necessarily injective, and its *type* is determined by the invariant factors of its Smith normal form. A polarization is *principal* if it is bijective, in other words if its type is $(1,\ldots,1)$. An integral torus with a principal polarization is called a *principally polarized tropical abelian variety*, or *pptav* for short. Given an integral torus $\Sigma=(\Lambda,\Lambda',[\cdot,\cdot])$ with a polarization $\xi:\Lambda'\to\Lambda$, we define the *principalization* to be the integral torus $\Sigma_p=(\mathrm{Im}\,\xi,\Lambda',[\cdot,\cdot])$ with the principal polarization $\xi:\Lambda'\to\mathrm{Im}\,\xi\cong\Lambda'$. The injection $j:\mathrm{Im}\,\xi\to\Lambda$ and the identity map $\mathrm{Id}:\Lambda'\to\Lambda'$ define a homomorphism $(j,\mathrm{Id}):\Sigma\to\Sigma_p$, which is a *dilation* in the sense of Definition 4.6 of [**RZ22**], so in particular it is a bijection on the underlying tori.[1]

Let $f:\Sigma_1\to\Sigma_2$ be a homomorphism of integral tori, and let ξ_2 be a polarization on Σ_2. If f is *finite*, in other words if $\mathrm{rk}\,f=\dim\Sigma_1$, then the homomorphism $\xi_1=f^{\#}\circ\xi_2\circ f_{\#}$ determines an *induced polarization* on Σ_1. As in the algebraic setting, a polarization induced from a principal polarization is not necessarily principal; we shall see that the Prym variety of a double cover is an example of this.

Let $\Sigma=(\Lambda,\Lambda',[\cdot,\cdot])$ be an integral torus of dimension n with polarization ξ. The bilinear form $(\cdot,\cdot)=[\xi(\cdot),\cdot]$ on Λ' extends to an inner product on the vector space $\mathrm{Hom}(\Lambda,\mathbb{R})\supset\Lambda'$, which is the universal cover of $\Sigma=\mathrm{Hom}(\Lambda,\mathbb{R})/\Lambda'$, and hence defines a translation-invariant Riemannian metric on Σ. The *volume of* Σ is the volume of a fundamental parallelotope and is given by the Grammian

[1]In Lemma 4.11 of [**RZ22**] we introduced a different notion of principalization Σ^{pp} which naturally comes with a dilation $\Sigma^{\mathrm{pp}}\to\Sigma$. The combined morphism $\Sigma^{\mathrm{pp}}\to\Sigma_p$ is of degree 2^g and is given by global multiplication by 2. In this article, we work with Σ_p because it seems more natural from the matroidal perspective, but the reconstruction of the Prym variety in Section 4 can be phrased for $\mathrm{Prym}^{\mathrm{pp}}$ as well, see also Remark 4.10.

determinant

$$\mathrm{Vol}^2(\Sigma) = \det\left((\lambda'_i, \lambda'_j)\right)_{i,j}, \tag{2}$$

where $\lambda'_1, \ldots, \lambda'_n$ is any basis of Λ'.

2.6. The tropical Jacobian and the Abel–Jacobi map. Let Γ be a metric graph, and choose an oriented model (G, ℓ). The simplicial chain group $C_1(G, \mathbb{Z})$ is the free abelian group on $E(G)$ and contains the simplicial homology group $H_1(G, \mathbb{Z})$ which is of rank $g(\Gamma)$. The chain groups naturally fit into a directed system with respect to refinements of models, and the images of the homology groups agree. We denote the direct limits $C_1(\Gamma, \mathbb{Z})$ and $H_1(\Gamma, \mathbb{Z})$, and refer the reader to [**BF11**] for details.

It is conceptually convenient to introduce the group Ω^1_Γ of *tropical 1-forms* on Γ. This group is canonically isomorphic to $H_1(\Gamma, \mathbb{Z})$, and a cycle $\gamma = \sum \gamma(e) \cdot e \in H_1(\Gamma, \mathbb{Z})$ is written $\sum \gamma(e) de$ when it is viewed as a 1-form. There is a natural *integration pairing*

$$[\cdot, \cdot] : \Omega^1_\Gamma \times C_1(\Gamma, \mathbb{Z}) \longrightarrow \mathbb{R}, \qquad \left[\sum \gamma(e) de, \sum \delta(e) \cdot e\right] = \int_\gamma \omega = \sum \gamma(e)\delta(e)\ell(e), \tag{3}$$

which restricts to a perfect pairing $\Omega^1_\Gamma \times H_1(\Gamma, \mathbb{Z}) \to \mathbb{R}$. The *Jacobian variety* of Γ is the $g(\Gamma)$-dimensional pptav

$$\mathrm{Jac}(\Gamma) = \left(\Omega^1_\Gamma,\, H_1(\Gamma, \mathbb{Z}),\, [\cdot, \cdot]\right) = \mathrm{Hom}(\Omega^1_\Gamma, \mathbb{R}) \,/\, H_1(\Gamma, \mathbb{Z}),$$

where the principal polarization is the trivial isomorphism $H_1(\Gamma, \mathbb{Z}) = \Omega^1_\Gamma$.

REMARK 2.3. We note that the integration pairing (3), and hence the induced inner product on the universal cover $\mathrm{Hom}(\Omega^1_\Gamma, \mathbb{R})$ of $\mathrm{Jac}(\Gamma)$, has a physical peculiarity: it is linear, rather than quadratic, in the edge lengths $\ell(e)$. Hence the units that are used to measure edge lengths on Γ become units of area from the point of view of the Riemannian geometry of $\mathrm{Jac}(\Gamma)$.

Let $q \in \Gamma$ be a base point, and for $p \in \Gamma$ let $\gamma_p \in C_1(\Gamma, \mathbb{Z})$ denote any path from q to p. The *Abel–Jacobi map* $\phi_q : \Gamma \to \mathrm{Jac}(\Gamma)$ relative to q is given by

$$p \longmapsto \left(\omega \longmapsto \int_{\gamma_p} \omega\right).$$

The map ϕ_q extends to symmetric powers of Γ and thus to divisors. The map respects linear equivalence, and hence induces a map $\mathrm{Pic}_0(\Gamma) \to \mathrm{Jac}(\Gamma)$ that does not depend on the choice of q. The tropical Abel–Jacobi theorem (see Theorem 6.3 in [**MZ08**]) states that this map is an isomorphism. Under this identification, the Abel–Jacobi map sends p to the divisor class $p - q$.

2.7. The tropical Prym and the Abel–Prym map. Let $\phi : \widetilde{\Gamma} \to \Gamma$ be a harmonic morphism of metric graphs. We consider the natural pushforward and pullback homomorphisms

$$\phi^* = \mathrm{Nm}^\# : \Omega^1_\Gamma \longrightarrow \Omega^1_{\widetilde{\Gamma}}, \qquad \phi_* = \mathrm{Nm}_\# : H_1(\widetilde{\Gamma}, \mathbb{Z}) \longrightarrow H_1(\Gamma, \mathbb{Z}),$$

$$\sum_{e \in E(\Gamma)} \gamma(e) de \longmapsto \sum_{e \in E(\Gamma)} \gamma(e) \sum_{\tilde{e} \in \phi^{-1}(e)} d_\phi(\tilde{e})\, d\tilde{e}, \qquad \sum_{\tilde{e} \in E(\widetilde{\Gamma})} \gamma(\tilde{e}) \cdot \tilde{e} \longmapsto \sum_{\tilde{e} \in E(\widetilde{\Gamma})} \gamma(\tilde{e}) \cdot \phi(\tilde{e}).$$

The pair $\mathrm{Nm} = (\mathrm{Nm}^{\#}, \mathrm{Nm}_{\#}) = (\phi^*, \phi_*)$ defines the surjective *norm homomorphism* $\mathrm{Nm} : \mathrm{Jac}(\widetilde{\Gamma}) \to \mathrm{Jac}(\Gamma)$ of pptavs, which corresponds, under the identification of Jac with Pic_0, to the direct image homomorphism

$$\mathrm{Nm} : \mathrm{Pic}_0(\widetilde{\Gamma}) \longrightarrow \mathrm{Pic}_0(\Gamma), \qquad \sum_{\widetilde{x}\in\widetilde{\Gamma}} a_{\widetilde{x}}\, \widetilde{x} \longmapsto \sum_{\widetilde{x}\in\widetilde{\Gamma}} a_{\widetilde{x}}\, \phi(\widetilde{x}).$$

We now consider a double cover $\pi : \widetilde{\Gamma} \to \Gamma$ of metric graphs of genera $\widetilde{g} = g(\widetilde{\Gamma})$ and $g = g(\Gamma)$, respectively. Let

$$\pi_* : H_1(\widetilde{\Gamma}, \mathbb{Z}) \longrightarrow H_1(\Gamma, \mathbb{Z}), \qquad \pi^* : \Omega^1_\Gamma \longrightarrow \Omega^1_{\widetilde{\Gamma}}$$

denote the pushforward and pullback homomorphisms, denote

$$K = (\mathrm{Coker}\, \pi^*)^{\mathrm{tf}}, \qquad K' = \mathrm{Ker}\, \pi_*,$$

and denote $[\cdot,\cdot]_K : K \times K' \to \mathbb{R}$ the pairing induced by the integration pairing on $\mathrm{Jac}(\widetilde{\Gamma})$.

DEFINITION 2.4. The *tropical Prym variety* $\mathrm{Prym}(\widetilde{\Gamma}/\Gamma)$ of the double cover $\pi : \widetilde{\Gamma} \to \Gamma$ is the kernel torus of the norm homomorphism:

$$\mathrm{Prym}(\widetilde{\Gamma}/\Gamma) = (\mathrm{Ker}\, \mathrm{Nm})_0 = \big(K, K', [\cdot,\cdot]_K\big) = \frac{\mathrm{Hom}(\mathrm{Coker}\, \pi^*, \mathbb{R})}{\mathrm{Ker}\, \pi_*}.$$

The tropical Prym variety $\mathrm{Prym}(\widetilde{\Gamma}/\Gamma)$ has dimension $h = \widetilde{g} - g$ and is the connected component of the identity of the kernel Ker Nm of the norm homomorphism $\mathrm{Nm} : \mathrm{Jac}(\widetilde{\Gamma}) \to \mathrm{Jac}(\Gamma)$. The structure of the kernel is described by Proposition 6.1 in [**JL18**]. Let $i : \widetilde{\Gamma} \to \widetilde{\Gamma}$ denote the involution associated to π, then any element in Ker Nm has a representative (viewed as an element of $\mathrm{Pic}_0(\widetilde{\Gamma})$) of the form $E - i(E)$, where E is an effective divisor E on $\widetilde{\Gamma}$. If π is a free double cover, then the parity of $\deg E$ does not depend on the choice of E, and Ker Nm has two connected components distinguished by the parity. If π is dilated, however, then Ker Nm is connected.

The tropical Prym variety $\mathrm{Prym}(\widetilde{\Gamma}/\Gamma)$ has a polarization induced from the principal polarization on $\mathrm{Jac}(\widetilde{\Gamma})$. Its type was computed in Theorem 1.5.7 of [**LU21**] and Proposition 4.21 of [**RZ22**] and is $(1, \ldots, 1, 2, \ldots, 2)$ with $d-1$ many 1's, where d is the dilation index of the double cover. For the most part of this article we will work with the principalization of the Prym variety, which we denote by $\mathrm{Prym}_p(\widetilde{\Gamma}/\Gamma)$ and refer to as *Prym variety* as well. For later reference, we spell out the defining data of $\mathrm{Prym}_p(\widetilde{\Gamma}/\Gamma)$ and introduce the following notation. Any element of $\mathrm{Ker}\, \pi_*$ is contained in the $\mathbb{Z}$-span of elements $\widetilde{e}^+ - \widetilde{e}^-$ for e ranging over the undilated edges of Γ. Hence we can view elements $\gamma \in \mathrm{Ker}\, \pi_*$ as *chains* on Γ and write

$$\gamma = \sum_{e\in E(G)} \gamma(e) \cdot (\widetilde{e}^+ - \widetilde{e}^-) = \sum_{e\in E(G)} \gamma(e) \cdot e \in C_1(G, \mathbb{Z}). \tag{4}$$

Then $\mathrm{Prym}_p(\widetilde{\Gamma}/\Gamma)$ is the pptav defined by the triple $(\mathrm{Ker}\, \pi_*, \mathrm{Ker}\, \pi_*, [\cdot,\cdot])$, where the pairing is induced from $H_1(\widetilde{\Gamma}, \mathbb{Z})$ and is given by

$$[\gamma, \gamma'] = \sum_{\widetilde{e}\in E(\widetilde{\Gamma})} \gamma(\widetilde{e})\gamma'(\widetilde{e})\ell(\widetilde{e}) = \sum_{e\in E_{\mathrm{ud}}(\Gamma)} 2\gamma(e)\gamma'(e)\ell(e) \tag{5}$$

for all $\gamma, \gamma' \in \mathrm{Ker}\, \pi_*$, where the second equality uses $\ell(\widetilde{e}^+) = \ell(\widetilde{e}^-) = \ell(e)$.

We also define the tropical Abel–Prym map. Let $q \in \widetilde{\Gamma}$ be a base point, and for $p \in \widetilde{\Gamma}$ let $\gamma_p \in C_1(\widetilde{\Gamma}, \mathbb{Z})$ denote any path from q to p. Let $i_*\gamma_p$ denote the image of this path under the involution. The *Abel–Prym map* $\psi_q : \widetilde{\Gamma} \to \operatorname{Prym}(\widetilde{\Gamma}/\Gamma)$ is given by

$$p \longmapsto \left(\omega \longmapsto \int_{\gamma_p} \omega - \int_{i_*\gamma_p} \omega\right).$$

We note that $\psi_q = p - i(p) - (q - i(q))$ when viewed as an element of $\operatorname{Pic}_0(\widetilde{\Gamma})$, hence lies in the connected component of the identity of $\operatorname{Ker} \operatorname{Nm}$ (which is not true of $p - i(p)$ if the double cover is free).

2.8. Volumes of the Jacobian and the Prym. One of the motivations for this paper are the formulas for the volume of the tropical Jacobian $\operatorname{Jac}(\Gamma)$ found in [**ABKS14**] and the tropical Prym variety $\operatorname{Prym}(\widetilde{\Gamma}/\Gamma)$ found in [**LZ22**] and [**GZ23**]. In both cases, the square of the volume is a polynomial in the lengths of the edges of Γ. We shall see in the next section that the monomials are in fact indexed by the bases of an appropriate matroid on the set of edges of Γ.

We recall the results of [**ABKS14**], [**LZ22**] and [**GZ23**].

THEOREM 2.5 (Theorem 5.2 of [**ABKS14**]). *Let Γ be a metric graph of genus g with model (G, ℓ). The volume of the Jacobian of Γ is given by*

$$\operatorname{Vol}^2\left(\operatorname{Jac}(\Gamma)\right) = \sum_{F \subset E(G)} \prod_{e \in F} \ell(e), \tag{6}$$

where the sum is taken over all g-element subsets $F \subset E(G)$ such that $G \backslash F$ is a tree.

The dimensional peculiarity noted in Remark 2.3 makes its appearance: the square of the volume of $\operatorname{Jac}(\Gamma)$ (which is a Riemannian manifold of dimension g) is a polynomial in the edge lengths of degree g, not the expected $2g$. The volume of the tropical Prym variety of a free and dilated double cover was computed in [**LZ22**] and [**GZ23**], respectively.

THEOREM 2.6 (Theorem 3.4 of [**LZ22**] and Theorem 3.3 of [**GZ23**]). *Let $\pi : \widetilde{\Gamma} \to \Gamma$ be a double cover of metric graphs of genera $\widetilde{g}$ and g, respectively, and let $h = \widetilde{g} - g$. The volume of the Prym variety of π is given by*

$$\operatorname{Vol}^2\left(\operatorname{Prym}(\widetilde{\Gamma}/\Gamma)\right) = 2^{1-d(\widetilde{\Gamma}/\Gamma)} \sum_{F \subset E_{\mathrm{ud}}(G)} 4^{\operatorname{ind}(F)-1} \prod_{e \in F} \ell(e). \tag{7}$$

Here the sum is taken over all h-element subsets of undilated edges $F \subset E_{\mathrm{ud}}(G)$ called ogods (see Definition 3.7), the index $\operatorname{ind}(F)$ of an ogod F is the number of connected components of $G \backslash F$, and $d(\widetilde{\Gamma}/\Gamma)$ is the dilation index.

3. The matroid of a double cover

The papers [**CV10**] and [**BMV11**] studied the tropical Torelli map that associates to a metric graph Γ its Jacobian $\operatorname{Jac}(\Gamma)$, and showed that its fibers are most conveniently understood in terms of a combinatorial object, the *graphic matroid* $M(\Gamma)$ (we note that many of these results earlier appeared in German and without the use of the tropical language in [**Ger82**]). In this section, we describe an analogous object, the *signed graphic matroid* $M(\widetilde{\Gamma}/\Gamma)$ of a double cover

$\pi : \widetilde{\Gamma} \to \Gamma$, which plays a role in describing the fibers of the tropical Prym–Torelli map $(\widetilde{\Gamma} \to \Gamma) \mapsto \mathrm{Prym}(\widetilde{\Gamma}/\Gamma)$ in Section 5. This matroid was originally defined by Zaslavsky [**Zas82**] in the framework of signed graphs, and we translate his definition into the language of double covers. We begin with the basic definitions.

3.1. Matroids and graph theory. A *matroid* M consists of a finite set $E(M)$, called the *ground set* of M, and a family of subsets $\mathcal{I}(M)$ of $E(M)$, called the *independent sets*, satisfying the following axioms:

(1) $\varnothing \in \mathcal{I}(M)$.
(2) If $A \subset B$ and $B \in \mathcal{I}(M)$, then $A \in \mathcal{I}(M)$.
(3) If $A, B \in \mathcal{I}(M)$ and A has more elements than B, then there exists $x \in A\backslash B$ such that $B \cup \{x\} \in \mathcal{I}(M)$.

The maximal elements of $\mathcal{I}(M)$ are the *bases* of the matroid, and the set of bases is denoted $\mathcal{B}(M)$. All bases have the same cardinality, called the *rank* of M. A subset of $E(M)$ that is not independent is called *dependent*, and the set of dependent sets is denoted $\mathcal{D}(M)$. A minimal dependent set is a *circuit*, and the set of circuits is denoted $\mathcal{C}(M)$. Each of the sets $\mathcal{I}(M)$, $\mathcal{B}(M)$, $\mathcal{D}(M)$, and $\mathcal{C}(M)$ determines the others and satisfies a set of properties that can be used to give an equivalent definition of a matroid (see Chapter 1 in [**Oxl11**]).

Starting with a matroid M, we construct several related matroids. The *dual matroid* M^* has the same ground set $E(M)$, and $B^* \subset E(M) = E(M^*)$ is a basis of M^* if and only if $E(M)\backslash B^*$ is a basis of M. Given a subset $X \subset E(M)$, the *restriction* of M to X is the matroid $M|_X$ with ground set X and independent set $\mathcal{I}(M|_X) = \{I \in \mathcal{I}(M) : I \subset X\}$. The *deletion* of X from M is the restriction $M\backslash X = M|_{E(M)-X}$ to the complement. Finally, the *contraction* of M along X is the matroid $M/X = (M^*\backslash X)^*$.

An *n-circuit* of a matroid M is a circuit with n elements (a 1-circuit is sometimes called a *loop*, but we avoid this terminology). Two elements $e, f \in E(M)$ are *parallel* if $\{e, f\}$ is a 2-circuit, and a *parallel class* of M is a maximal subset of $F \subset E(M)$ containing no 1-circuits and such that any 2-element subset is a 2-circuit. A parallel class is *trivial* if it consists of a single element, and a matroid is called *simple* if it has no non-trivial parallel classes. Given a matroid M, the *simplification* of M is the simple matroid M_{sim} constructed by deleting all 1-circuits, and deleting all but one element from each parallel class. The simplification procedure is non-canonical, since it involves choosing a representative from each parallel class, but any two simplifications are isomorphic as matroids.

EXAMPLE 3.1. Let G be a connected graph of genus g. The *graphic matroid* $M(G)$ has ground set $E(G)$, and a subset $F \subset E(G)$ is independent if the subgraph $G[F]$ generated by F is a forest (equivalently, if the homology group $H_1(G[F], \mathbb{Z})$ is trivial). The bases of $M(G)$ are the spanning trees of G, the dependent sets are subgraphs having a nontrivial cycle, and the circuits are subgraphs consisting of a simple cycle. Given a set of edges $F \subset E(G)$, the deletion $M(G)\backslash F$ and contraction $M(G)/F$ are the graphic matroids of the graphs $G\backslash F$ and G_F, respectively. A 1-circuit of $M(G)$ is a loop of G, and a parallel class is the set of edges between two vertices. The matroid $M(G)$ is simple if and only if the graph G is simple, and $M(G)_{\mathrm{sim}}$ is the graphic matroid of the simplification of the graph G, obtained by deleting all loops, and deleting all but one edge in each set of multiedges between any pair of vertices.

The dual matroid $M^*(G)$ is called the *cographic* or *bond matroid.* The independent sets of $M^*(G)$ are the sets of edges $F \subset E(G)$ whose removal does not disconnect G (equivalently, such that the reduced homology group $\widetilde{H}_0(G\backslash F, \mathbb{Z})$ is trivial). The bases of $M^*(G)$ are the complements of the spanning trees of G, and the rank of $M^*(G)$ is equal to g. A 1-circuit of $M^*(G)$ is a bridge edge. A parallel class consists of a set of n edges whose removal disconnects G into n connected components that are cyclically linked in G by the removed edges. The simplification of $M^*(G)$ is obtained by contracting the bridge edges and all but one edge from each parallel class (since $M^*(G)$ is the dual, matroid-theoretic deletion is graph-theoretic contraction).

3.2. The matroid of a double cover. In the seminal paper [**Zas82**], Zaslavsky introduces the *signed graphic matroid* $M(G, \sigma)$ on the edge set $E(G)$ of a graph G equipped with an edge parity assignment $\sigma : E(G) \to \{\pm 1\}$. As described in Section 2.1, the pair (G, σ) determines a free double cover $p : \widetilde{G} \to G$, and Zaslavsky shows that the matroid $M(G, \sigma)$ is switching-equivalent, in other words depends only on the cover p (Corollary 5.4 in [**Zas82**]). He then considers the contraction of the matroid $M(G, \sigma)$ along a subset $F \subset E(G)$. Since the contracted double cover $p_F : \widetilde{G}_{\widetilde{F}} \to G_F$ may no longer be free, in order to stay within the framework of signed graphs, Zaslavsky is forced to introduce a more complicated category of graphs, having *half-arcs* and *free loops* in addition to ordinary edges (see Proposition 3.9 for details).

In this section, we give an elementary geometric description of the signed graphic matroid and its dual in terms of double covers, which works equally well in the free and dilated cases. In addition, the description via double covers allows us to define an auxiliary function on the independent sets called the *index* (see Definition 4.3), which is crucial for reconstructing the Prym variety and which cannot be seen from Zaslavsky's approach. To motivate our definition, we recall that a set of edges of a graph G is independent for the cographic matroid $M^*(G)$ if removing it does not disconnect G. We give an analogous definition for double covers.

DEFINITION 3.2. Let $p : \widetilde{G} \to G$ be a double cover of graphs, with G possibly disconnected. We say that p is *relatively connected* if the preimage of each connected component of G is connected in $\widetilde{G}$.

A double cover $p : \widetilde{G} \to G$ of a connected graph G is relatively connected if and only if it is not the trivial free double cover. If $e \in E(G)$ is a dilated edge, then $G\backslash\{e\}$ may or may not be connected, but each connected component has a dilated vertex, and hence the double cover obtained from p by deleting e from G and its preimage from $\widetilde{G}$ is still relatively connected. Removing an undilated edge, however, may relatively disconnect p. Hence we make the following definition.

DEFINITION 3.3. Let $p : \widetilde{G} \to G$ be a nontrivial double cover of a connected graph G. We define the matroid $M^*(\widetilde{G}/G)$ of p as follows:

(1) The ground set of $M^*(\widetilde{G}/G)$ is $E_{\mathrm{ud}}(G) = E(G)\backslash E(G_{\mathrm{dil}})$, the set of undilated edges of G.
(2) A set of undilated edges $F \subset E_{\mathrm{ud}}(G)$ is independent for the matroid $M^*(\widetilde{G}/G)$ if the double cover $p|_{\widetilde{G}\backslash p^{-1}(F)} : \widetilde{G}\backslash p^{-1}(F) \to G\backslash F$ obtained by removing these edges from G and their preimages from $\widetilde{G}$ is relatively connected. Equivalently, F is independent if the double cover

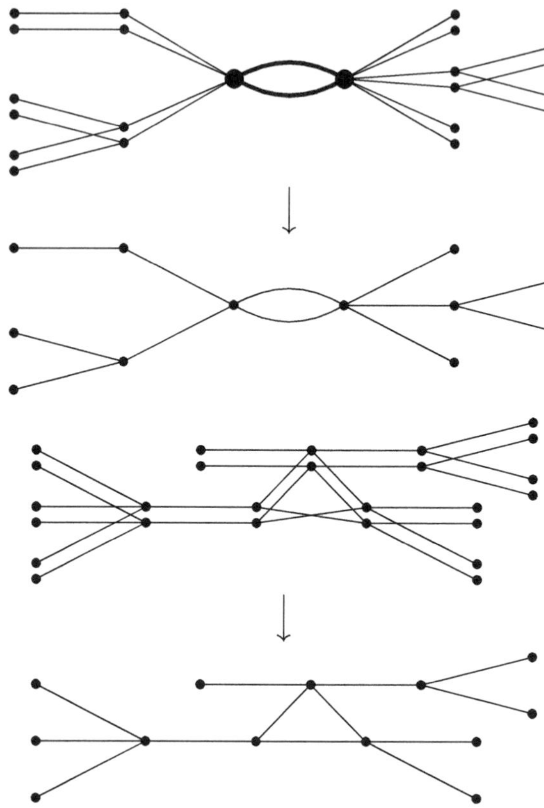

FIGURE 2. Elementary double covers of the first and second types.

$p|_{\widetilde{G}\backslash p^{-1}(F)}$ induces an injective map on the reduced homology groups $\widetilde{H}_0(\widetilde{G}\backslash p^{-1}(F),\mathbb{Z}) \to \widetilde{H}_0(G\backslash F,\mathbb{Z})$.

We show in Proposition 3.9 that $M^*(\widetilde{G}/G)$ is a matroid, by identifying it with the signed graphic matroid of Zaslavsky. We first investigate when it is possible to remove a single undilated edge without relatively disconnecting the double cover. To simplify notation, given a double cover $p:\widetilde{G}\to G$ and an undilated edge $e\in E_{\mathrm{ud}}(G)$, we denote $G^e = G\backslash\{e\}$, $\widetilde{G}^e = \widetilde{G}\backslash p^{-1}(e)$, and $p^e:\widetilde{G}^e\to G^e$ the restriction of p. We first determine the relatively connected double covers that do not admit the removal of an edge.

DEFINITION 3.4. An *elementary double cover* $p:\widetilde{G}\to G$ is one of the following two types (see Figure 2):

(1) p is dilated, G and G_{dil} are connected, and $g(G_{\mathrm{dil}}) = g(G)$.
(2) p is free, G is connected of genus one, and p is not the trivial free double cover.

LEMMA 3.5. *Let $p:\widetilde{G}\to G$ be an elementary double cover. Then p is relatively connected, but for any undilated edge $e\in E(G)$ the double cover $p^e:\widetilde{G}^e\to G^e$ is not relatively connected.*

PROOF. If p is of the first type, then G consists of the dilation subgraph G_{dil} together with a number of undilated trees attached to it. Hence for any undilated edge e, the graph G^e has two connected components, one of which is a tree with no dilation. Hence p^e is not relatively connected since a tree has no nontrivial free double cover.

Similarly, if p is free and $g(G) = 1$, then G^e is either a tree or a disjoint union of a genus one graph and a tree, and in both cases p^e is not relatively connected. □

The next lemma shows that no other double covers have this property.

LEMMA 3.6. *Let $p : \widetilde{G} \to G$ be a double cover of connected graphs of any of the following three types:*

(1) *p is dilated and G_{dil} is not connected.*
(2) *p is dilated, G_{dil} is connected, and $g(G) > g(G_{\text{dil}})$.*
(3) *p is free and $g(G) \geqslant 2$.*

There exists an undilated edge $e \in E(G)$ such that $p^e : \widetilde{G}^e \to G^e$ is relatively connected.

PROOF. If G_{dil} has two distinct connected components, choose a path between them consisting of undilated edges, and let e be any edge on this path. The graph G^e may be disconnected (if e is a bridge), but each of its connected components has dilation, hence p^e is relatively connected.

If G_{dil} is connected but $g(G) > g(G_{\text{dil}})$, then the graph obtained by contracting G_{dil} to a single vertex has genus $g(G) - g(G_{\text{dil}}) > 0$ and hence a nontrivial cycle. Let e be an (undilated) edge on this cycle. Removing e from the original graph G does not disconnect it, hence p^e is relatively connected since G^e has dilation.

Finally, if p is free, then there is a simple cycle C on G such that the restriction of p to C is not trivial. If $g(G) \geqslant 2$, we can also find a non-bridge edge $e \in G$ that does not lie on C. Removing e does not disconnect G, and the cover p^e remains relatively connected. □

We now prove that $M^*(\widetilde{G}/G)$ is a matroid by showing that it is the dual of the signed graphic matroid of Zaslavsky. For the convenience of the reader, we give the bases of the matroid $M^*(\widetilde{G}/G)$ as a separate definition.

DEFINITION 3.7. Let $p : \widetilde{G} \to G$ be a nontrivial double cover of a connected graph G, let $F \subset E_{\text{ud}}(G)$ be a set of undilated edges, let $G\backslash F = G_1 \cup \cdots \cup G_k$ be the decomposition into connected components, and let $p_i : p^{-1}(G_i) \to G_i$ be the restrictions of p. The set F is called an *odd genus one decomposition*, or *ogod* for short, if each p_i is an elementary double cover.

REMARK 3.8. This terminology was introduced in [**LZ22**] in the context of free double covers. If $p : \widetilde{G} \to G$ is free, then each $p_i : p^{-1}(G_i) \to G_i$ is a nontrivial double cover of a genus one graph by a genus one graph, corresponding to the nontrivial (i.e. *odd*) element of $H^1(G_i, \mathbb{Z}/2\mathbb{Z}) = \mathbb{Z}/2\mathbb{Z}$. If p is edge-free, then each G_i may also be viewed as a genus one graph by assigning each dilated vertex an intrinsic genus. Hence the acronym "ogod". The terminology breaks down for double covers with edge dilation, however, we generally do not consider such covers (see Lemma 4.1).

PROPOSITION 3.9. *Let $p : \widetilde{G} \to G$ be a nontrivial double cover of a connected graph G. Then $M^*(\widetilde{G}/G)$ is a matroid of rank $g(\widetilde{G}) - g(G)$ on the set $E_{\mathrm{ud}}(G)$ of undilated edges of G, whose bases are the ogods of p.*

PROOF. It follows immediately from Lemmas 3.5 and 3.6 that a subset $F \subset E_{\mathrm{ud}}(G)$ is a basis, in other words a maximal independent set of $M^*(\widetilde{G}/G)$, if and only if each p_i is an elementary double cover. To show that $M^*(\widetilde{G}/G)$ is a matroid, we show that its bases are the complements of the bases of the matroid $M(\Sigma)$ of a signed graph Σ defined by Zaslavsky in [**Zas82**]. We recall the definitions. Consider a tuple $\Sigma = (V, E, A, F, \sigma)$ of the following kind:

(1) V and E are respectively the vertex and edge sets of a graph G', which is not necessarily connected. The set of connected components of G' is denoted $\pi(\Sigma)$.
(2) A is a set of *half-arcs*, equipped with a root map $r : A \to V(G')$. An element $e \in A$ is an edge having one end rooted on the graph G' and one loose end. When speaking of connected components of G', we include the half-arcs rooted on it.
(3) F is a set of *free loops*, which are edges with no root vertices.
(4) $\sigma : E \cup F \to \{\pm 1\}$ is a partial parity assignment on the edges, taking values $\sigma(f) = +1$ on all $f \in F$.

The set of *arcs* of Σ is $E \cup A \cup F$ and it is the ground set of the matroid $M(\Sigma)$. A closed edge path $e_1 \cdots e_n$ in G' is called *balanced* if $\sigma(e_1) \cdots \sigma(e_n) = 1$. A set of arcs is called *balanced* if it contains no half-arcs and if every closed path in it is balanced. The set of connected components of G' whose arc set is balanced is denoted $\pi_b(\Sigma)$.

The bases of the matroid $M(\Sigma)$ are defined by Theorem 5.1 in [**Zas82**] (we note that part (g) of Theorem 5.1, which specifically defines bases, is inaccurately stated, and was subsequently corrected in [**Zas83**]). A set $S \subset E \cup A \cup F$ is a basis if the following conditions hold:

(1) For each balanced component $B \in \pi_b(\Sigma)$, the edges of S in B form a spanning tree for B.
(2) Each unbalanced connected component of the subgraph induced by S is either a tree together with a unique half-arc, or a genus one graph with unbalanced cycle and no half-arcs.
(3) S contains no free loops.

Now let $p : \widetilde{G} \to G$ be a double cover. We define the associated tuple $\Sigma = (V, E, A, F, \sigma)$ as follows:

(1) The graph G' is the free subgraph $G_{\mathrm{fr}} \subset G$, and $\sigma : E \to \{\pm 1\}$ is the parity assignmend defining the free double cover $p|_{p^{-1}(G_{\mathrm{fr}})} : p^{-1}(G_{\mathrm{fr}}) \to G_{\mathrm{fr}}$ with respect to some choice of vertex labeling (see Section 2.1 for details).
(2) The half-arcs A are the undilated edges having one free root vertex on G_{fr} and one dilated root vertex (which we discard).
(3) The free loops F are the undilated edges both of whose root vertices are dilated.

In other words, we delete the dilated vertices and edges from G, and leave dangling any undilated edges with missing root vertices, so that $E_{\mathrm{ud}}(G) = E \cup A \cup F$ according to whether an undilated edge loses none, one, or both root vertices.

Suppose further that G is connected. If p is free and nontrivial, then $G' = G$ is the unique (unbalanced) connected component, and there are no half-arcs or free loops. If p is dilated, then each connected component of G' has at least one half-arc (an undilated edge with one dilated and one undilated root vertex) or is a free loop. We see that $\pi_b(\Sigma)$ is empty in both cases.

Now let $F \subset E_{\mathrm{ud}}(G)$ be a basis of $M^*(\widetilde{G}/G)$, let $G\backslash F = G_1 \cup \cdots G_k$ be the decomposition into connected components, and let $p_i : p^{-1}(G_i) \to G_i$ be the corresponding elementary double covers. If p_i is of the first type, then G_i consists of its dilation subgraph $(G_i)_{\mathrm{dil}}$ with a number of attached undilated trees. Removing $(G_i)_{\mathrm{dil}}$, each tree acquires a unique half-arc, and no free loops are formed. If p_i is of the second type, then G_i is a genus one graph and its unique cycle is unbalanced because the free double cover p_i is nontrivial. Hence $E_{\mathrm{ud}}(G)\backslash F$ is a basis of the signed graphic matroid $M(\Sigma)$. Reversing this construction, we see that all bases are obtained in this way. Hence $M^*(\widetilde{G}/G)$ is the dual matroid to $M(\Sigma)$.

To complete the proof, we show that the rank of $M^*(\widetilde{G}/G)$ is equal to $g(\widetilde{G}) - g(G)$. We use the construction of Example 2.2. Let $p_{\mathrm{ef}} : \widetilde{G}_{\mathrm{ef}} \to G_{\mathrm{ef}}$ be the edge-free double cover obtained from $p : \widetilde{G} \to G$ by contracting each of the n connected components of G_{dil} to a separate dilated vertex. The matroids $M^*(\widetilde{G}/G)$ and $M^*(\widetilde{G}_{\mathrm{ef}}/G_{\mathrm{ef}})$ are canonically isomorphic. Indeed, G and G_{ef} have the same set of undilated edges, and for any $F \subset E_{\mathrm{ud}}(G) = E_{\mathrm{ud}}(G_{\mathrm{ef}})$, the connected componets of $G_{\mathrm{ef}}\backslash F$ are obtained from the connected components of $G\backslash F$ by contracting the dilated edges. Hence F is independent for $M^*(\widetilde{G}/G)$ if and only if it is independent for $M^*(\widetilde{G}_{\mathrm{ef}}/G_{\mathrm{ef}})$. Furthermore, $g(\widetilde{G}_{\mathrm{ef}}) - g(G_{\mathrm{ef}}) = g(\widetilde{G}) - g(G)$, since both $\widetilde{G}$ and G lose genus equal to the total genera of the components of G_{dil}.

Now let $p_{\mathrm{free}} : \widetilde{G}_{\mathrm{free}} \to G_{\mathrm{free}}$ be the free double cover obtained by resolving the n dilated vertices of G_{ef}. Then $g(G_{\mathrm{free}}) = g(G_{\mathrm{ef}}) + n$ and $g(\widetilde{G}_{\mathrm{free}}) = g(\widetilde{G}_{\mathrm{ef}}) + n$, so $g(\widetilde{G}_{\mathrm{free}}) - g(G_{\mathrm{free}}) = g(\widetilde{G}_{\mathrm{ef}}) - g(G_{\mathrm{ef}})$. Furthermore, under the natural inclusion $E_{\mathrm{ud}}(G_{\mathrm{ef}}) \subset E_{\mathrm{ud}}(G_{\mathrm{free}})$, any basis of $M^*(\widetilde{G}_{\mathrm{ef}}/G_{\mathrm{ef}})$ is also a basis of $M^*(\widetilde{G}_{\mathrm{free}}/G_{\mathrm{free}})$. Hence it is enough to prove the assertion for the free double cover $p_{\mathrm{free}} : \widetilde{G}_{\mathrm{free}} \to G_{\mathrm{free}}$. Let $F \subset E(G_{\mathrm{free}}) = E_{\mathrm{ud}}(G_{\mathrm{free}})$ be a basis of $M^*(\widetilde{G}_{\mathrm{free}}/G_{\mathrm{free}})$, then the connected components $G_{\mathrm{free}}\backslash F = G_1 \cup \cdots \cup G_k$ all have genus one. Since the quantity $g(\cdot) - 1 = |E(\cdot)| - |V(\cdot)|$ is additive in connected components, it follows that

$$\begin{aligned}\operatorname{rk} M^*(\widetilde{G}/G) = |F| &= g(G_{\mathrm{free}}) - 1 - \sum_{i=1}^{k} \bigl(g(G_i) - 1\bigr)\\ &= g(G_{\mathrm{free}}) - 1 = g(\widetilde{G}_{\mathrm{free}}) - g(G_{\mathrm{free}}) = g(\widetilde{G}) - g(G).\end{aligned}$$

□

As an illustration, and for future reference, we describe the 1-circuits and 2-circuits of the signed cographic matroid $M^*(\widetilde{G}/G)$. A 1-circuit is an undilated edge $e \in E_{\mathrm{ud}}(G)$ whose removal relatively disconnects the double cover. Figure 3 shows the two topological types of 1-circuits of $M^*(\widetilde{G}/G)$, depending on whether or not removing e disconnects the target graph G. In the second diagram, the edge e may or may not be a loop.

Similarly, a 2-circuit is a subset $\{e, f\} \subset E_{\mathrm{ud}}(G)$ whose removal relatively disconnects p but such that neither $\{e\}$ nor $\{f\}$ is a 1-circuit. There are four topological

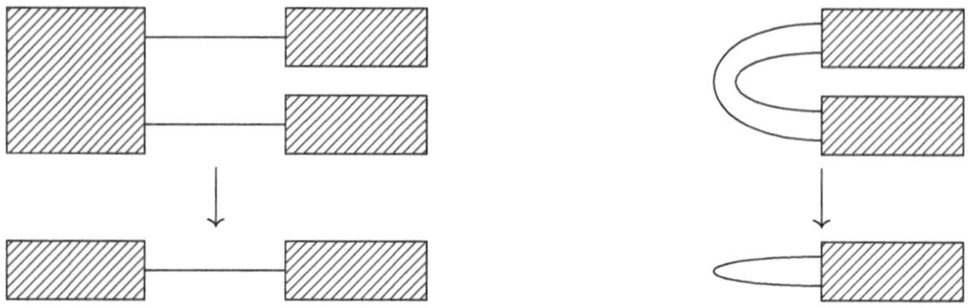

FIGURE 3. 1-circuits of $M^*(\widetilde{G}/G)$. Boxes denote arbitrary connected graphs. The curved edge on the target may or may not be a loop.

types of 2-circuits, illustrated on Figure 4, where the curved edges may or may not represent loops. We label the edges of the 2-circuit with multiplicities that are used to define the 2-simplification of the matroid $M^*(\widetilde{G}/G)$ (see Definition 5.1 for details, or simply note that an edge is labeled by a 2 if it is a bridge and 1 otherwise).

3.3. The dual matroid $M(\widetilde{G}/G)$. In this section, we describe the *signed graphic matroid* $M(\widetilde{G}/G)$, which is the dual of the signed cographic matroid $M^*(\widetilde{G}/G)$, in terms of the double cover $\widetilde{G} \to G$. We first give a description of its bases, independent sets, and circuits. As motivation, recall that a set of edges $F \subset E(G)$ of a graph G is independent for the graphic matroid if the subgraph $G[F]$ induced by F is a forest, or equivalently if the group $H_1(G[F], \mathbb{Z})$ is trivial. We now show that a similar statement holds for $M(\widetilde{G}/G)$ as well.

PROPOSITION 3.10. *Let $p : \widetilde{G} \to G$ be a double cover and let $F \subset E_{\mathrm{ud}}(G)$ be a set of undilated edges. Let $G[F] = G_1 \cup \cdots \cup G_k$ be the connected component decomposition of the subgraph $G[F]$ generated by F, and let $p[F] : p^{-1}(G[F]) \to G[F]$ and $p_i : p^{-1}(G_i) \to G_i$ denote the restrictions of p.*

(1) *F is a basis of $M(\widetilde{G}/G)$ if and only if $G[F]$ contains all undilated vertices of G and each p_i is an elementary double cover.*
(2) *F is independent if and only if the map $p[F]_* : H_1(p^{-1}(G[F]), \mathbb{Z}) \to H_1(G[F], \mathbb{Z})$ induced by $p[F]$ is injective.*
(3) *F is a circuit if and only if $G[F] = G_1$ is connected, has no undilated vertices of valency one, and either $g(G[F]) = 1$ and $p[F]$ is the trivial free double cover (type I on Figure 5), or $p[F]$ is one of types II-VI shown on Figure 5.*

PROOF. The bases of $M(\widetilde{G}/G)$ are the complements of the bases of $M^*(\widetilde{G}/G)$ in the set of undilated edges $E_{\mathrm{ud}}(G)$, so Part (1) is simply a restatement of Proposition 3.9. To prove Part (2), we note that $(p_i)_*$ is injective if p_i is an elementary double cover, and that this property is preserved if additional edges are removed. Hence $p[F]_*$ is injective for any independent set F. Conversely, it is easy to see (for example, from the explicit description of the kernel given in Propositions 4.19 and 4.20 in [**RZ22**]) that if $(p_i)_*$ is not injective, then either p_i is a trivial double cover of a graph G_i of genus $g(G_i) \geqslant 1$, or else it is one of the three types listed in Lemma 3.6. In either case, it is clear that F is not contained in a basis of $M(\widetilde{G}/G)$, hence it is dependent.

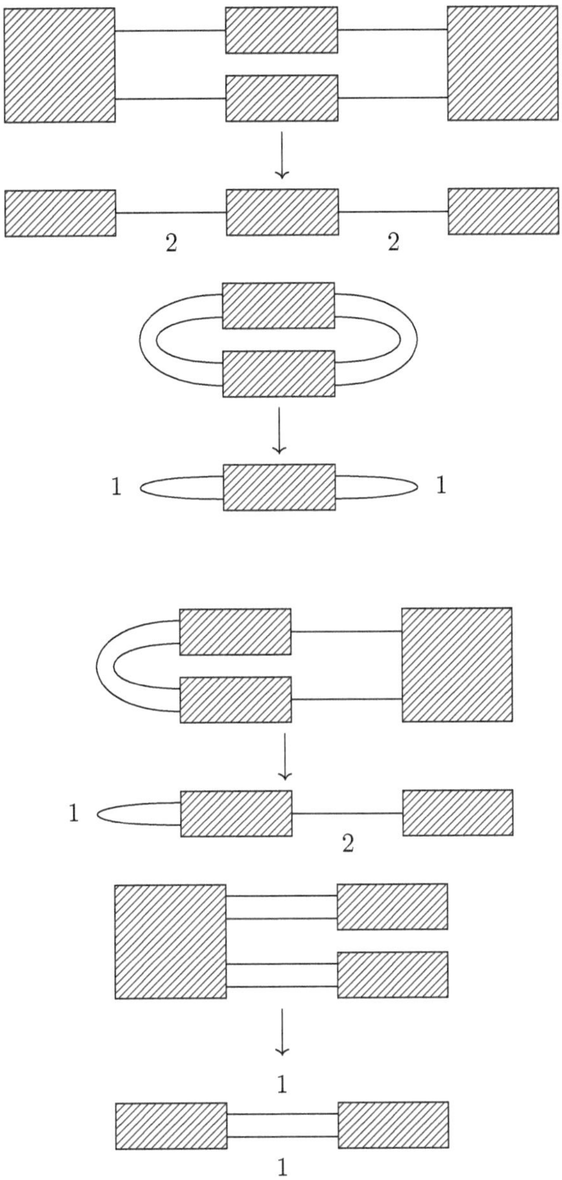

FIGURE 4. 2-circuits of $M^*(\widetilde{G}/G)$. Boxes denote arbitrary connected graphs. Numbers indicate multiplicities (see Definition 5.1).

We now describe the circuits of $M(\widetilde{G}/G)$. Looking at each of the six double covers shown in Figure 5, we see that the pushforward map on homology is not injective, but becomes so if any edge is removed. Hence each of the double covers represents a minimal dependent set of $M(\widetilde{G}/G)$, in other words a circuit.

Conversely, let F be a dependent set of $M(\widetilde{G}/G)$. We remove edges from F to find a minimal dependent set. Since $p[F]_*$ is not injective, we can assume without loss of generality that $(p_1)_*$ is not injective and replace F by $E(G_1)$. If $G[F]$ has an undilated cycle with disconnected preimage, then we can remove all other edges and obtain a circuit of type I. If it does not, then it is one of the three types listed in Lemma 3.6, and it is then elementary to verify that the circuits of types II-VI are the minimal double covers of the types given in the lemma. □

Figure 5 shows the six topological types of circuits of the signed graphic matroid $M(\tilde{G}/G)$. Undilated vertices of valency 2 are not shown, so a single edge in the picture may represent a chain of edges in G. Bold dots represent dilated vertices. In each type, the pushforward map $p[C]_* : H_1(p^{-1}(G[C]),\mathbb{Z}) \to H_1(G[C],\mathbb{Z})$ has one-dimensional kernel, and the indicated cycle $\gamma_C \in H_1(p^{-1}(G[C]),\mathbb{Z})$ is a generator.

4. Recovering the Prym variety from the matroid

In this section, we explain how to reconstruct the principalized Prym variety $\operatorname{Prym}_p(\tilde{\Gamma}/\Gamma)$ of a double cover $\pi : \tilde{\Gamma} \to \Gamma$ of metric graphs from the signed graphic matroid equipped with certain additional decorations. We always implictly choose a model for the double cover, and by abuse of notation denote its signed graphic and cographic matroid by $M(\tilde{\Gamma}/\Gamma)$ and $M^*(\tilde{\Gamma}/\Gamma)$. To motivate the reconstruction procedure, we first recall how to reconstruct the Jacobian variety of a metric graph from the graphic matroid.

4.1. Matroidal reconstruction of Jacobian and Prym: an overview. Let Γ be a metric graph. Choose an orientation for Γ. Given a circuit C of the graphic matroid $M(\Gamma)$, there exists a simple cycle $\delta_C = \sum \delta_C(e) \cdot e$ in $H_1(\Gamma,\mathbb{Z})$ supported on C with $\delta_C(e) = \pm 1$ for all edges $e \in C$. The cycle δ_C is unique up to sign. While the support of δ_C can be determined from $M(\Gamma)$, the signs of its coefficients depend on the chosen orientation and are determined by the condition that δ_C is a closed cycle. This information is not contained in $M(\Gamma)$.

To make the construction fully matroid-theoretic, we use the chosen orientation on Γ to define the structure of an *oriented matroid* $\overrightarrow{M}(\Gamma)$ on $M(\Gamma)$ (see Example 4.6 or [**BLVS**$^+$**99**, Section 1.1]). The coefficients of the cycles δ_C are then simply the signs with which the edges of C occur in an oriented circuit $\overrightarrow{C}$ of $\overrightarrow{M}(\Gamma)$ with underlying (unoriented) circuit C.

It is an elementary fact that the cycles $\{\delta_C : C \in \mathcal{C}(M(\Gamma))\}$ are a spanning set for $H_1(\Gamma,\mathbb{Z})$. Moreover, we can choose a basis as follows. Let $B = \{e_1,\ldots,e_g\}$ be a basis of the cographic matroid $M^*(\Gamma)$, so that $T = \Gamma\backslash B$ is a spanning tree. For each $i = 1,\ldots,g$ let C_i be the unique circuit of $M(\Gamma)$ supported on $T \cup \{e_i\}$, then the cycles δ_{C_i} form a basis for $H_1(\Gamma,\mathbb{Z})$. Therefore, we can associate a pptav $\operatorname{Jac}(\overrightarrow{M}(\Gamma))$ to the oriented matroid $\overrightarrow{M}(\Gamma)$ equipped with the edge length function $\ell : E(\Gamma) \to \mathbb{R}_{>0}$, and $\operatorname{Jac}(\overrightarrow{M}(\Gamma))$ is isomorphic to the Jacobian $\operatorname{Jac}(\Gamma)$.

However, even more is true. The graph Jacobian $\operatorname{Jac}(\Gamma)$ does not depend on the choice of orientation of Γ. By [**BL78**, Corollary 6.2.8], graphic matroids have a unique reorientation class, hence any oriented matroid having underlying matroid $M(\Gamma)$ is in fact obtained from an orientation of Γ. Therefore, we can construct $\operatorname{Jac}(\overrightarrow{M}(\Gamma))$ directly from the matroid $M(\Gamma)$ by choosing any structure of an oriented matroid, completely bypassing graphs. The resulting object, which we denote $\operatorname{Jac}(M(\Gamma))$, is isomorphic to the Jacobian $\operatorname{Jac}(\Gamma)$ of any graph whose matroid

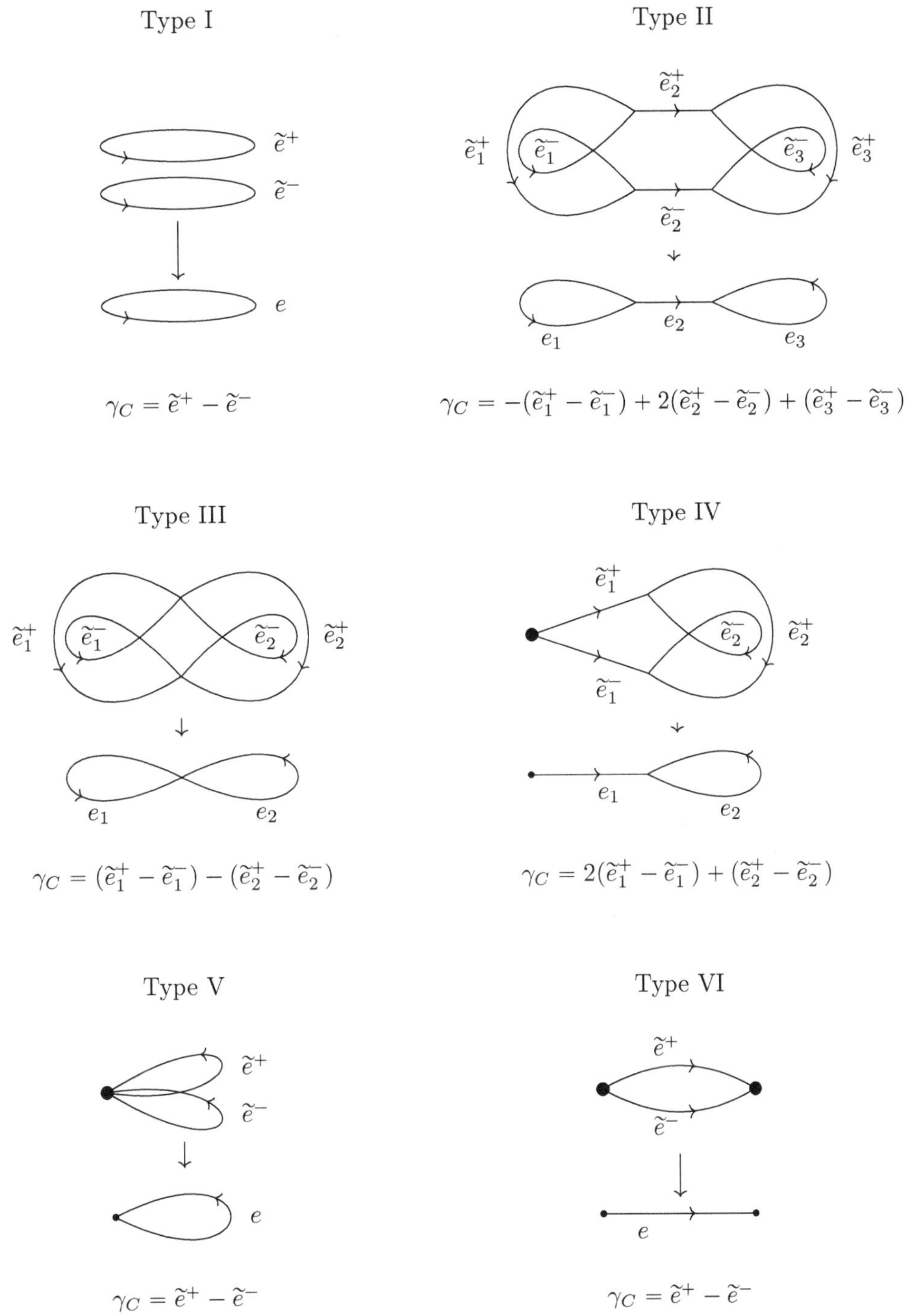

FIGURE 5. Circuits of the signed graphic matroid. Undilated vertices of valency 2 are not shown. Bold dots represent dilated vertices. The cycle γ_C is the fundamental cycle associated to C (see Definition 4.2).

is isomorphic (in a way that preserves edge lengths) to $M(\Gamma)$. We summarize the reconstruction procedure in the following diagram:

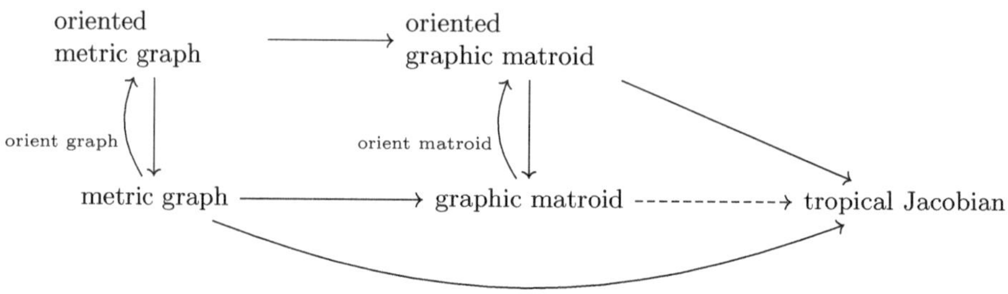

In this section, we describe an analogue of the fundamental cycle construction to determine the principalized Prym variety $\mathrm{Prym}_p(\widetilde{\Gamma}/\Gamma)$ of a double cover $\pi : \widetilde{\Gamma} \to \Gamma$ from the signed graphic matroid $M(\widetilde{\Gamma}/\Gamma)$ and its dual $M^*(\widetilde{\Gamma}/\Gamma)$:

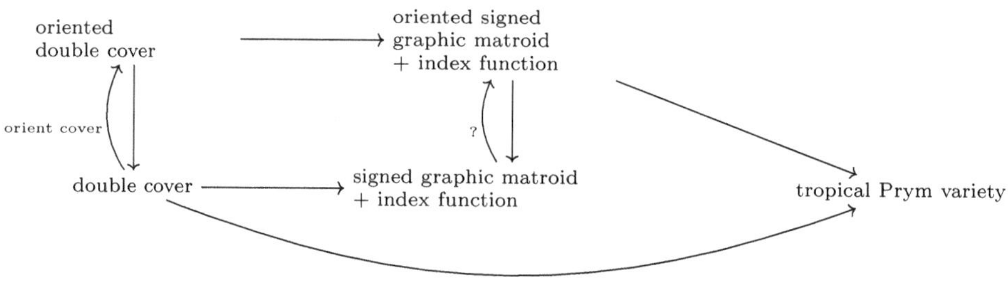

First, it is necessary to equip $M^*(\widetilde{\Gamma}/\Gamma)$ with an auxiliary index function, this is explained in Section 4.3. We then explain in Section 4.4 how an orientation of the double cover determines an orientation on $M(\widetilde{\Gamma}/\Gamma)$ and $M^*(\widetilde{\Gamma}/\Gamma)$. We then construct the Prym variety $\mathrm{Prym}_p(\widetilde{\Gamma}/\Gamma)$ from the oriented matroid $\overrightarrow{M}(\widetilde{\Gamma}/\Gamma)$, the index function, and the edge length function. Unlike the case of graphs, it is not true that all orientations of $M(\widetilde{\Gamma}/\Gamma)$ come from a choice of orientation on the double cover (for example, we show in Section 5.2 that $U_{2,4}$ is a signed graphic matroid, and [**GRS95**, Example 1] shows that $U_{2,4}$ has three reorientation classes). Hence we are not presently able to fully reconstruct $\mathrm{Prym}_p(\widetilde{\Gamma}/\Gamma)$ solely from the matroid $M(\widetilde{\Gamma}/\Gamma)$, we still need to retain the reorientation class that is determined by the double cover. In addition, we are not currently able to recover the induced polarization and reconstruct the non-principally polarized $\mathrm{Prym}(\widetilde{\Gamma}/\Gamma)$.

4.2. The fundamental cycle construction. In this section, we describe an analogue of the fundamental cycle construction for double covers. We first observe that the dilated edges of G do not contribute to the matroid $M(\widetilde{G}/G)$, and we show that they are not seen by the Prym variety either.

LEMMA 4.1. *Let $\pi : \widetilde{\Gamma} \to \Gamma$ be a double cover of metric graphs and let $\pi_{\mathrm{ef}} : \widetilde{\Gamma}_{\mathrm{ef}} \to \Gamma_{\mathrm{ef}}$ be the edge-free double cover obtained by contracting all dilated edges of Γ (see Example 2.2). Then the signed graphic matroids and the Prym varieties of the two double covers are isomorphic:*

$$M(\widetilde{\Gamma}/\Gamma) \cong M(\widetilde{\Gamma}_{\mathrm{ef}}/\Gamma_{\mathrm{ef}}), \quad \mathrm{Prym}(\widetilde{\Gamma}/\Gamma) \cong \mathrm{Prym}(\widetilde{\Gamma}_{\mathrm{ef}}/\Gamma_{\mathrm{ef}}), \quad \mathrm{Prym}_p(\widetilde{\Gamma}/\Gamma) \cong \mathrm{Prym}_p(\widetilde{\Gamma}_{\mathrm{ef}}/\Gamma_{\mathrm{ef}}).$$

PROOF. We already established that $M(\widetilde{\Gamma}/\Gamma) \cong M(\widetilde{\Gamma}_{\mathrm{ef}}/\Gamma_{\mathrm{ef}})$ in the last part of the proof of Proposition 3.9. To prove that the Pryms are isomorphic, we consider what happens when a single dilated edge e is contracted. Let $\pi_e : \widetilde{\Gamma}_e \to \Gamma_e$ be the double cover obtained by contracting e. There are natural surjective contraction maps $H_1(\widetilde{\Gamma}, \mathbb{Z}) \to H_1(\widetilde{\Gamma}_e, \mathbb{Z})$ and $H_1(\Gamma, \mathbb{Z}) \to H_1(\Gamma_e, \mathbb{Z})$, given by setting the corresponding edges to zero. If γ is a cycle in $\operatorname{Ker}\big(H_1(\widetilde{\Gamma}, \mathbb{Z}) \to H_1(\Gamma, \mathbb{Z})\big)$, then e does not occur in γ, hence γ corresponds to an element of $\operatorname{Ker}\big(H_1(\widetilde{\Gamma}_e, \mathbb{Z}) \to H_1(\Gamma_e, \mathbb{Z})\big)$, and therefore the two kernels are identified by the contraction maps. We similarly identify $(\operatorname{Coker}\pi^*)^{\mathrm{tf}}$ and $(\operatorname{Coker}\pi_e^*)^{\mathrm{tf}}$, and it is clear that the pairings and polarizations agree. Hence we have $\operatorname{Prym}(\widetilde{\Gamma}/\Gamma) \cong \operatorname{Prym}(\widetilde{\Gamma}_e/\Gamma_e)$ and therefore $\operatorname{Prym}_p(\widetilde{\Gamma}/\Gamma) \cong \operatorname{Prym}_p(\widetilde{\Gamma}_e/\Gamma_e)$ as well. □

By Lemma 4.1 it is possible to restrict attention to edge-free double covers, in other words replace each connected component of the dilation subgraph by a single dilated vertex. We now recall the notation for $\operatorname{Ker} p_*$ of a double cover $p : \widetilde{G} \to G$ introduced in Sections 2.2 and 2.7. Recall that we chose a labeling $p^{-1}(v) = \{\widetilde{v}^{\pm}\}$ of the undilated vertices $v \in V(G)\backslash V(G_{\mathrm{dil}})$. We then label $p^{-1}(h) = \{\widetilde{h}^{\pm}\}$ the preimages of the undilated half-edges, and require $r(\widetilde{h}^{\pm}) = \widetilde{r(h)}^{\pm}$ if an undilated half-edge is rooted at an undilated vertex. The involution on $\widetilde{G}$ is given by a parity assignment $\sigma : E_{\mathrm{ud}}(G) \to \{\pm 1\}$ via $\iota(\widetilde{h}^{\pm}) = \widetilde{\iota(h)}^{\pm\sigma(e)}$, where $\sigma(h) = \sigma\big(\iota(h)\big) = \sigma(e)$ for an edge $e = \big\{h, \iota(h)\big\} \in E(G)$, and we can further relabel to assume that $\sigma(e) = +1$ if e has at least one dilated root vertex. Choosing consistent orientations for $\widetilde{G}$ and G determines a labeling $p^{-1}(e) = \{\widetilde{e}^+, \widetilde{e}^-\}$ for undilated edges $e \in E_{\mathrm{ud}}(G)$, and we represent elements $\gamma \in \operatorname{Ker} p_*$ as chains on the target graph (see Equation (4)):

$$\gamma = \sum_{e \in E(G)} \gamma(e) \cdot (\widetilde{e}^+ - \widetilde{e}^-) = \sum_{e \in E(G)} \gamma(e) \cdot e \in C_1(G, \mathbb{Z}).$$

A chain $\gamma \in C_1(G, \mathbb{Z})$ represents an element of $\operatorname{Ker} p_*$ if the above expression is closed as a chain on $\widetilde{G}$. This is most conveniently stated in the language of oriented matroids, which we introduce in Section 4.4.

We observed in Proposition 3.10 that a set $F \subset E(G)$ is dependent for the signed graphic matroid $M(\widetilde{G}/G)$ if and only if the induced map $p[F]_*$ on homology has nontrivial kernel. If F is a circuit, in other words a minimal dependent set, then the kernel is one-dimensional and has a unique generator up to sign.

DEFINITION 4.2. Let $p : \widetilde{G} \to G$ be an oriented double cover, let C be a circuit of the signed graphic matroid $M(\widetilde{G}/G)$, and let $p[C] : p^{-1}(G[C]) \to G[C]$ be the restriction of p. The *fundamental cycle* γ_C *associated to* C

$$\gamma_C = \sum_{e \in E(G)} \gamma_C(e) \cdot (\widetilde{e}^+ - \widetilde{e}^-) = \sum_{e \in E(G)} \gamma_C(e) \cdot e$$

is the unique (up to sign) generator of $\operatorname{Ker} p[C]_*$.

Figure 5 shows the six circuit types and gives formulas for the fundamental cycles. Specifically, define the cycles $\gamma'_C \in \operatorname{Ker} p_*$ by the formulas

$$\gamma'_C(e) = \begin{cases} \pm 2, & \text{if } e \in C \text{ is a bridge of } G[C], \\ \pm 1, & \text{if } e \in C \text{ is not a bridge of } G[C], \\ 0, & \text{if } e \notin C, \end{cases} \tag{8}$$

where the signs are chosen in such a way that γ'_C is closed. Then it is easy to see that

$$\gamma_C = \frac{\gamma'_C}{\gcd\left\{\left|\gamma'_C(e)\right| : e \in C\right\}} = \begin{cases} \gamma'_C, & \text{if } C \text{ is of types I-V,} \\ \gamma'_C/2, & \text{if } C \text{ is of type VI.} \end{cases} \tag{9}$$

4.3. The index function. We now describe the additional decorations of the signed cographic matroid of a double cover that are needed to determine the coefficients of the cycles γ_C, and hence reconstruct $\operatorname{Ker} p_*$ and the Prym variety.

Equation (8) tells us that we need a way to detect whether an edge e of a circuit is a bridge or not. This information is contained in the matroid $M(G)$, but not in $M(\widetilde{G}/G)$. Instead, we observe that Equation (7) for the volume of the Prym (in contrast to Equation (6) for the Jacobian) contains an extra quantity, namely the index of an ogod, which must therefore play a role in the reconstruction. We now extend this function from the ogods, which are the bases of $M^*(\widetilde{G}/G)$, to all independent sets of $M^*(\widetilde{G}/G)$.

Definition 4.3. For a double cover $p : \widetilde{G} \to G$, we endow the matroid $M^*(\widetilde{G}/G)$ with the *index function* defined as

$$\operatorname{ind} : \mathcal{I}\big(M^*(\widetilde{G}/G)\big) \longrightarrow \mathbb{Z}_{>0}, \qquad F \longmapsto \text{number of connected components of } G\backslash F.$$

We emphasize that the index function is defined graph-theoretically and cannot be computed from the matroid $M^*(\widetilde{G}/G)$. In particular, when counting the connected components of $G\backslash F$, isolated dilated vertices are taken into account, so Zaslavsky's definition of $M^*(\widetilde{G}/G)$ cannot be used to determine the index. We collect some basic properties of the index function in the following lemma, which we leave without proof.

Lemma 4.4. *Let $p : \widetilde{G} \to G$ be a double cover of graphs, let $M^* = M^*(\widetilde{G}/G)$ be the signed cographic matroid, and let* ind *be the index function. Then the following hold.*

1. *If $F \in \mathcal{I}(M^*)$ and $F' \subset F$, then $\operatorname{ind}(F') \leqslant \operatorname{ind}(F)$.*
2. *For every $F \in \mathcal{I}(M^*)$ we have $\operatorname{ind}(F) \leqslant \operatorname{rk} M^* + 1$. If F is a basis then we also have $d(\widetilde{G}/G) \leqslant \operatorname{ind}(F)$.*
3. *If $B, B' \in \mathcal{B}(M^*)$ differ by only one element, then $\left|\operatorname{ind}(B') - \operatorname{ind}(B)\right| \leqslant 1$.*

Let C be a circuit of the signed graphic matroid $M(\widetilde{G}/G)$. We now show how to determine the absolute values of the coefficients of the fundamental cycle $\gamma_C \in \operatorname{Ker} p_*$. Let $e \in C$ be an edge. The set $C\backslash\{e\}$ is independent in $M(\widetilde{G}/G)$ and is contained in a basis $E(G)\backslash F$, where $F \subset E(G)$ is an ogod. Each connected component of $G\backslash F$ contains a unique undilated cycle or a dilated component, but not both. Hence if $e \in C$ is a bridge, then the two connected components of $C\backslash\{e\}$ lie in different connected components of $G\backslash F$. In other words, e is a bridge of C if and only if $G\backslash F$ has one more connected component than $G\backslash\big(F\backslash\{e\}\big)$. Therefore, we have proved the following.

Proposition 4.5. *Let $p : \widetilde{G} \to G$ be a double cover and let C be a circuit of the signed graphic matroid $M(\widetilde{G}/G)$. Then the absolute values of the coefficients in*

Equation (8) *can be computed as*

$$|\gamma'_C(e)| = \begin{cases} 2^{\mathrm{ind}(F)-\mathrm{ind}(F\setminus\{e\})}, & \text{if } e \in C, \\ 0, & \text{else,} \end{cases}$$

where F is any ogod such that $C\setminus\{e\} \subset E(G)\setminus F$.

4.4. Oriented matroids. To reconstruct the signs of the coefficients $\gamma_C(e)$ of the fundamental cycles, it is convenient to use the language of oriented matroids. We recall the definitions (a standard reference is [**BLVS**$^+$**99**]).

Given a finite set $E = \{e_1, \ldots, e_n\}$, a *signed subset* $\vec{S}$ of E is a subset of $E^+ \sqcup E^-$, where $E^+ = \{e_1^+, \ldots, e_n^+\}$ and $E^- = \{e_1^-, \ldots, e_n^-\}$ are copies of E. For a signed subset $\vec{S}$ of E define

- the *opposite set* $-\vec{S} = \{e^+ : e^- \in \vec{S}\} \cup \{e^- : e^+ \in \vec{S}\}$,
- the *underlying set* $S = \{e : e^+ \in \vec{S} \text{ or } e^- \in \vec{S}\}$, and
- the *positive (resp. negative) part* $S^+ = \{e : e^+ \in \vec{S}\}$ (resp. $S^- = \{e : e^- \in \vec{S}\}$).

It is understood that the underlying sets of S^+ and S^- are disjoint. Therefore, for $e \in E$ we write

$$\vec{S}(e) = \begin{cases} +1, & \text{if } e^+ \in \vec{S}, \\ -1, & \text{if } e^- \in \vec{S}, \\ 0, & \text{otherwise.} \end{cases}$$

For our purposes, it is convenient to define an *oriented matroid* $\vec{M}$ on the ground set E in terms of its *oriented circuits* $\mathcal{C}(\vec{M})$, a family of signed subsets of E satisfying the following axioms:

(1) $\varnothing \notin \mathcal{C}(\vec{M})$.
(2) If $\vec{S} \in \mathcal{C}(\vec{M})$ then $-\vec{S} \in \mathcal{C}(\vec{M})$.
(3) For $\vec{S}, \vec{T} \in \mathcal{C}(\vec{M})$ with $S \subset T$ we have either $\vec{S} = \vec{T}$ or $\vec{S} = -\vec{T}$.
(4) For all $\vec{S}, \vec{T} \in \mathcal{C}(\vec{M})$ with $\vec{S} \neq \pm\vec{T}$ and $e \in S^+ \cap T^-$ there exists $\vec{U} \in \mathcal{C}(\vec{M})$ such that $U^+ \subset (S^+ \cup T^+)\setminus\{e\}$ and $U^- \subset (S^- \cup T^-)\setminus\{e\}$.

For any oriented matroid $\vec{M}$ on E, there is an underlying matroid M on E whose circuits are the underlying sets C of the oriented circuits $\vec{C} \in \mathcal{C}(\vec{M})$. Furthermore, for every circuit C of M there are precisely two circuits $\vec{C}_1, \vec{C}_2 \in \mathcal{C}(\vec{M})$ with underlying set C and $\vec{C}_1 = -\vec{C}_2$. A matroid M is called *orientable* if there is an oriented matroid $\vec{M}$ having underlying matroid M.

In [**Zas91**], Zaslavsky described how to orient the signed graphic matroid of a signed graph, and here we translate his construction into the language of double covers. To simplify exposition, we first describe how to orient the graphic matroid of an ordinary graph.

EXAMPLE 4.6. Let $G = (V(G), H(G), r, \iota)$ be a graph. An orientation on G is defined by a function $o : H(G) \to \{\pm 1\}$ on the half-edges satisfying the property

$$o(h)o(\iota(h)) = -1 \tag{10}$$

for any edge $e = \{h, h'\} \in E(G)$. We view a half-edge $h \in H(G)$ rooted at $v = r(h)$ as pointing towards v if $o(h) = +1$ and away from v if $o(h) = -1$. With respect to this choice of orientation, we can define the simplicial chain complex of G as

$$d:\mathbb{Z}^{E(G)}\longrightarrow\mathbb{Z}^{V(G)},\qquad e\longmapsto\sum_{h\in e}o(h)\cdot r(h),$$

and the simplicial homology group as

$$H_1(G,\mathbb{Z})=\operatorname{Ker}d=\left\{\sum_{e\in E(G)}\gamma(e)\cdot e\in\mathbb{Z}^{E(G)}:\sum_{h\in T_vG}o(h)\gamma(h)=0\text{ for all }v\in V(G)\right\},$$

where $\gamma(h)=\gamma(e)$ for $h\in e$. Each circuit C of $M(G)$ supports a primitive cycle $\delta_C\in H_1(G,\mathbb{Z})$, unique up to sign, such that $\delta_C(e)=\pm1$ for all $e\in C$ and 0 otherwise. The circuits $\vec{C}=C^+\sqcup C^-$ of the *oriented graphic matroid* $\vec{M}(G)$ of the oriented graph G have the form

$$C^+=\{e\in C:\delta_C(e)=+1\}\qquad\text{and}\qquad C^-=\{e\in C:\delta_C(e)=-1\},$$

where C ranges over the circuits of $M(G)$ and δ_C is one of the two primitive cycles supported on C. It is elementary to verify that $\vec{M}(G)$ is an oriented matroid on the ground set $E(G)$, having underlying matroid $M(G)$. The cycles δ_C can be recovered from $\vec{M}(G)$ up to sign and span $H_1(G,\mathbb{Z})$. Hence the oriented matroid $\vec{M}(G)$ determines, after choosing a length function $\ell:E(G)\to\mathbb{R}_{>0}$, the tropical Jacobian.

We now consider a double cover $p:\widetilde{G}\to G$. Choose an orientation for G, in other words choose a function $o:H(G)\to\{\pm1\}$ satisfying Equation (10). This induces an orientation on $\widetilde{G}$, and in terms of the labeling introduced in Section 4.2, any element $\gamma\in\operatorname{Ker}p_*$ has the form

$$\gamma=\sum_{e\in E_{\mathrm{ud}}(G)}\gamma(e)\cdot(\widetilde{e}^+-\widetilde{e}^-).$$

Denoting $\gamma(h)=\gamma(e)$ when $h\in e$, it is then elementary to verify that

(11)

$$\operatorname{Ker}p_*=\left\{\sum_{e\in E_{\mathrm{ud}}(G)}\gamma(e)\cdot(\widetilde{e}^+-\widetilde{e}^-):\sum_{h\in T_vG}\tau(h)\gamma(h)=0\text{ for all }v\in V(G)\backslash V(G_{\mathrm{dil}})\right\}$$

where the function $\tau:H_{\mathrm{ud}}(G)\to\{\pm1\}$ is defined by

$$\tau(h)=\begin{cases}\sigma(e), & \text{if }o(h)=1\text{ and }h\in e,\\ -1, & \text{if }o(h)=-1.\end{cases}$$

The function τ satisfies

$$\tau(h)\tau\big(\iota(h)\big)=-\sigma(e)$$

and defines an *orientation of the signed graph* (G,σ) (see Equation (2.2) in [**Zas91**]). This in turn defines the structure of an oriented matroid $\vec{M}(\widetilde{G}/G)$ on $M(\widetilde{G}/G)$, and hence an induced orientation on $M^*(\widetilde{G}/G)$, as follows (see Theorem 3.3 in [**Zas91**]). Given an orientation $\tau:H_{\mathrm{ud}}(G)\to\{\pm1\}$, Equation (11) determines which 1-chains supported on $E_{\mathrm{ud}}(G)$ represent elements of $\operatorname{Ker}p_*$. In particular, given a circuit $C\in\mathcal{C}\big(M(\widetilde{G}/G)\big)$, we can now determine the signs of the coefficients of the fundamental cycle γ_C, the absolute values having been determined in Section 4.3:

$$(12)\qquad \gamma_C=\sum_{e\in E_{\mathrm{ud}}(G)}\vec{C}(e)\big|\gamma_C(e)\big|\cdot e=\sum_{e\in E_{\mathrm{ud}}(G)}\vec{C}(e)\big|\gamma_C(e)\big|\cdot(\widetilde{e}^+-\widetilde{e}^-).$$

The functions $\overrightarrow{C}(e)$ then represent the two oriented circuits of $\overrightarrow{M}(\widetilde{G}/G)$ lying over C, for the two possible generators γ_C of $\operatorname{Ker} p[C]_*$.

4.5. Reconstructing the Prym variety from the matroid. We now explain how to recover the Prym variety of a double cover $\pi : \widetilde{\Gamma} \to \Gamma$ from the signed graphic matroid $M(\widetilde{\Gamma}/\Gamma)$ equipped with the index function $\operatorname{ind} : \mathcal{I}(M^*(\widetilde{\Gamma}/\Gamma)) \to \mathbb{Z}_{>0}$, orientation $\overrightarrow{M}(\widetilde{\Gamma}/\Gamma)$, and edge length function $\ell : E(\Gamma) \to \mathbb{R}_{>0}$. The circuits $C \in \mathcal{C}(M(\widetilde{\Gamma}/\Gamma))$ determine fundamental cycles $\gamma_C \in \operatorname{Ker} \pi_*$. Specifically, the index determines the $|\gamma_C(e)|$ by Equation (9) and Proposition 4.5, while the orientation determines the signs by Equation (12). This data is sufficient to reconstruct $\operatorname{Ker} \pi_*$.

PROPOSITION 4.7. *The lattice* $\operatorname{Ker} \pi_* \subset H_1(\widetilde{\Gamma}, \mathbb{Z})$ *of a double cover* $\pi : \widetilde{\Gamma} \to \Gamma$ *is spanned by the fundamental cycles* γ_C*, where* C *ranges over the circuits of* $M(\widetilde{\Gamma}/\Gamma)$.

PROOF. By definition, $\gamma_C \in \operatorname{Ker} \pi_*$ for all circuits C of $M(\widetilde{\Gamma}/\Gamma)$. Conversely, let $\pi : \widetilde{\Gamma} \to \Gamma$ be a double cover. If the dilation index $d(\widetilde{\Gamma}/\Gamma) = 1$, then in fact $\operatorname{Ker} \pi_*$ has a basis consisting of fundamental cycles γ_C obtained by choosing an ogod $F \subset \mathcal{B}(M^*(\widetilde{\Gamma}/\Gamma))$ of minimal index; we prove this in Proposition 4.14. Hence we assume that $d(\widetilde{\Gamma}/\Gamma) > 1$. Let $\gamma \in \operatorname{Ker} \pi_*$. Choose an edge $e_1 \in E_{\mathrm{ud}}(\Gamma)$ lying on a type VI circuit C_1 (for example, choose any path connecting distinct connected components of Γ_{dil}). Then $\gamma_{C_1}(e_1) = \pm 1$, and we choose γ_{C_1} so that $\gamma_{C_1}(e_1) = 1$. We then have $\gamma = \gamma(e_1)\gamma_{C_1} + \gamma_1$, where $\gamma_1 \in \operatorname{Ker} \pi_*$ and is supported on $\Gamma \backslash \{e_1\}$.

We now remove e_1 from Γ and proceed by induction to obtain $\gamma = \sum \gamma(e_i)\gamma_{C_i} + \gamma'$, where each γ_{C_i} is a type VI circuit and $\gamma' \in \operatorname{Ker} \pi_*$ is supported on the disconnected graph $\Gamma' = \Gamma \backslash \{e_1, \ldots, e_k\}$, which has no type VI circuits. It follows that each connected component of Γ' has connected dilation subgraph and hence dilation index one, so the restriction of γ' to each connected component lies in the span of the fundamental circuits by Proposition 4.14. This completes the proof. □

For the reconstruction of $\operatorname{Prym}_p(\widetilde{\Gamma}/\Gamma)$ it is convenient to collect all the necessary matroidal data in a definition.

DEFINITION 4.8. Consider the following data:

- A signed graphic matroid M on a ground set E, together with an orientation $\overrightarrow{M}$ induced by an oriented double cover representing M.
- An index function $\operatorname{ind} : \mathcal{I}(M^*) \to \mathbb{Z}_{>0}$ induced from the oriented double cover representing $\overrightarrow{M}$.
- An edge length function ℓ on E.

We define a pptav $(\Lambda, \Lambda, [\cdot, \cdot])$, which depends on all of this data but which we simply denote $\operatorname{Prym}(M)$. For each circuit $C \in \mathcal{C}(M)$, let $\gamma_C \in \mathbb{Z}^E$ be the fundamental cycle given by Equation (12). We let Λ be the lattice spanned in $\mathbb{Z}^E$ by the γ_C for C ranging over the oriented circuits of M, and define the pairing to be

$$[\gamma_{C_1}, \gamma_{C_2}] = \sum_{e \in E} 2\gamma_{C_1}(e)\gamma_{C_2}(e)\ell(e). \tag{13}$$

Finally, the polarization $\xi : \Lambda \to \Lambda$ is the identity map.

We now state the main result of our paper.

THEOREM 4.9. *Let $\pi : \widetilde{\Gamma} \to \Gamma$ be a double cover of metric graphs. Choose an orientation for π, and let $M = M(\widetilde{\Gamma}/\Gamma)$ be the signed graphic matroid of π with induced orientation $\overrightarrow{M}$. Let $\mathrm{ind} : \mathcal{I}(M^*) \to \mathbb{Z}_{>0}$ be the index function and let $\ell : E(\Gamma) \to \mathbb{R}_{>0}$ be the edge length function. Then* $\mathrm{Prym}(M)$ *is a pptav and*

$$\mathrm{Prym}_p(\widetilde{\Gamma}/\Gamma) \cong \mathrm{Prym}(M)$$

as pptavs. In particular, $\mathrm{Prym}(M)$ *only depends on the reorientation class of $\overrightarrow{M}$ and not on $\overrightarrow{M}$ itself.*

PROOF. Recall that the Prym variety $\mathrm{Prym}(\widetilde{\Gamma}/\Gamma)$ is the polarized tropical abelian variety defined by the lattices $K = (\mathrm{Coker}\,\pi^*)^{\mathrm{tf}}$, $K' = \mathrm{Ker}\,\pi_*$, and the pairing $[\cdot,\cdot]$ and polarization $\xi : K' \to K$ induced by the integration pairing and polarization on $\widetilde{\Gamma}$. The principalization $\mathrm{Prym}_p(\widetilde{\Gamma}/\Gamma)$ is obtained by replacing K by the image $\mathrm{Im}\,\xi$. By Proposition 4.7, $K' = \mathrm{Im}\,\xi = \Lambda$ is the lattice spanned by the fundamental cycles γ_C corresponding to the oriented circuits of M. To verify that (13) is the pairing on $\mathrm{Prym}_p(\widetilde{\Gamma}/\Gamma)$ given in Equation (5), we simply note that

$$[\widetilde{e}_1^+ - \widetilde{e}_1^-, \widetilde{e}_2^+ - \widetilde{e}_2^-] = \begin{cases} 2\ell(e_1), & \text{if } e_1 = e_2, \\ 0, & \text{if } e_1 \neq e_2 \end{cases}$$

for any two undilated edges $e_1, e_2 \in E_{\mathrm{ud}}(\Gamma)$. In particular, $\mathrm{Prym}(M)$ is a pptav and does not change under reorientation of M, which is equivalent to reorienting the cover $\widetilde{\Gamma} \to \Gamma$. □

REMARK 4.10. We note that we are not currently able to reconstruct, directly from the matroid, the Prym variety $\mathrm{Prym}(\widetilde{\Gamma}/\Gamma)$, except in the following two cases. Recall that the induced polarization on $\mathrm{Prym}(\widetilde{\Gamma}/\Gamma)$ has type $(1,\ldots,1,2,\ldots,2)$, where the number of 1s is equal to $d(\widetilde{\Gamma}/\Gamma) - 1$. If $d(\widetilde{\Gamma}/\Gamma) = 1$ (in other words, if $\pi : \widetilde{\Gamma} \to \Gamma$ is free or has connected dilation subgraph), then the induced polarization is twice a principal polarization, and we can reconstruct K by doubling the lattice $K' = \mathrm{Ker}\,\pi_*$. On the other hand, if $d(\widetilde{\Gamma}/\Gamma) = g(\widetilde{\Gamma}) - g(\Gamma) + 1$ (equivalently, if Γ is a tree, as in Example 4.12), then the induced polarization on $\mathrm{Prym}(\widetilde{\Gamma}/\Gamma)$ is already principal and $\mathrm{Prym}_p(\widetilde{\Gamma}/\Gamma) = \mathrm{Prym}(\widetilde{\Gamma}/\Gamma)$. We also note that the dilation index $d(\widetilde{\Gamma}/\Gamma)$ (and hence the polarization type) can in fact be read off the indexed matroid $M(\widetilde{\Gamma}/\Gamma)$ as the minimal index of an ogod $F \in \mathcal{B}(M^*(\widetilde{\Gamma}/\Gamma))$ (see Lemma 4.4 and Lemma 4.13).

The following is an immediate consequence of Theorem 4.9.

COROLLARY 4.11. *Let $\pi_i : \widetilde{\Gamma}_i \to \Gamma_i$ for $i = 1, 2$ be double covers. Let $(M_i, \overrightarrow{M}_i, \mathrm{ind}_i, \ell_i)$ be the associated matroidal data packages, i.e. $M_i = M(\widetilde{\Gamma}_i/\Gamma_i)$ with oriented matroid structure $\overrightarrow{M}_i$ induced by the choice of some orientation on π_i, ind_i is the index function on M_i^* and ℓ_i the edge-length function on M_i. If there is an isomorphism of the matroidal data packages, i.e. a bijection $E_{\mathrm{ud}}(\Gamma_1) \to E_{\mathrm{ud}}(\Gamma_2)$ commuting with the ℓ_i which induces a bijection $\mathcal{I}(M_1^*) \to \mathcal{I}(M_2^*)$ commuting with the ind_i and which induces an isomorphism between $\overrightarrow{M}_1$ and a reorientation of $\overrightarrow{M}_2$ then $\mathrm{Prym}_p(\widetilde{\Gamma}_1/\Gamma_1) \cong \mathrm{Prym}_p(\widetilde{\Gamma}_2/\Gamma_2)$ as pptavs.*

A natural question is to what extent a double cover can be recovered from the associated matroid (cf. Problem 5.1 in [**Zas82**]). We give an example, and explore this question from a matroid-theoretic perspective in Section 5.

EXAMPLE 4.12. Consider a double cover $\pi : \widetilde{\Gamma} \to \Gamma$ where Γ is a tree consisting of n undilated edges all of length 1, and such that every vertex is dilated. We claim that all such covers have the same Prym variety. Indeed, since all vertices are dilated, removing edges never relatively disconnects π. Therefore the signed cographic matroid $M^*(\widetilde{\Gamma}/\Gamma) = U_{n,n}$ is uniform, and the index function is $\operatorname{ind}(F) = |F| + 1$, neither of which depend on the tree Γ. Moreover, since $U_{n,n}$ is in fact a graphic matroid, it is uniquely orientable up to reorientation. This shows that up to reorientation the matroidal package of π depends only on n, and we conclude from Corollary 4.11 that any two double covers π_1 and π_2 of the aforementioned form have the same Prym variety.

4.6. Effective reconstruction and a basis of fundamental cycles. In Theorem 4.9 we reconstruct the principalized tropical Prym variety from the matroidal data of the double cover, by considering the fundamental cycles corresponding to all circuits of $M(\widetilde{\Gamma}/\Gamma)$. This procedure is not computationally effective. In the case of graphs, to reconstruct the Jacobian, it is sufficient to use the circuits associated to a single spanning tree, obtained by adding each of the complementary edges one by one. We now explore whether an analogous construction produces the Prym variety. In Lemma 4.4, we noted that the index function on a basis of $M^*(\widetilde{G}/G)$ is bounded from below by the dilation index $d(\widetilde{G}/G)$. We now show that this bound is sharp.

LEMMA 4.13. *Let $p : \widetilde{G} \to G$ be a double cover with $d(\widetilde{G}/G) = 1$ (so p is either free or has connected dilation subgraph). Then there exists an ogod $F \subset E(G)$ of index one.*

PROOF. If p is free, we pick an odd cycle on G, having connected preimage in $\widetilde{G}$. We then construct F by iteratively removing edges of G without disconnecting G or breaking the chosen cycle. If p is dilated with connected dilation subgraph, we assume for simplicity that p is edge-free with a unique dilated vertex. Then F is the set of edges in the complement of any spanning tree. □

We now show that a basis for $\operatorname{Ker} p_*$ can be constructed by starting with an ogod of index one. This gives an exact analogue for Pryms of the fundamental cycle construction for Jacobians.

PROPOSITION 4.14. *Let $p : \widetilde{G} \to G$ be a double cover, and suppose that $F = \{e_1, \ldots, e_h\} \subset E(G)$ is an ogod of index one. For each $i = 1, \ldots, h$, let C_i denote the unique circuit of $M(\widetilde{G}/G)$ in $G\backslash F \cup \{e_i\}$. Then the fundamental cycles γ_{C_i} are a basis for* $\operatorname{Ker} p_*$.

PROOF. By Lemma 4.4 the index function is monotonous, so $\operatorname{ind}(F) = 1$ implies that $\operatorname{ind}(F\backslash\{e_i\}) = 1$ for all $i = 1, \ldots, h$. Hence Proposition 4.5 shows that $\gamma_{C_i}(e_i) = \pm 1$, and we choose signs so that $\gamma_{C_i}(e_i) = 1$. In addition, $e_j \notin C_i$ and therefore $\gamma_{C_i}(e_j) = 0$ for $j \neq i$.

Now let $\gamma \in \operatorname{Ker} p_*$. Since F is an ogod, the map $p^{-1}(G\backslash F) \to G\backslash F$ is injective on H_1. Hence $\gamma = 0$ if $\gamma(e_i) = 0$ for all i. It follows that

$$\gamma = \gamma(e_1)\gamma_{C_1} + \cdots + \gamma(e_h)\gamma_{C_h},$$

and therefore the γ_{C_i} are a basis. □

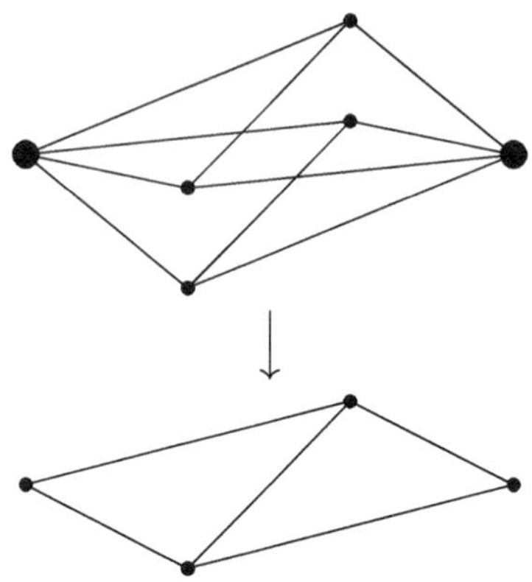

FIGURE 6. Double cover with no ogod producing a basis for $\operatorname{Ker} p_*$

The argument above shows that the fundamental cycles γ_{C_i} associated to an ogod F of index $\operatorname{ind}(F) \geqslant 2$ are linearly independent and span a finite index sublattice of $\operatorname{Ker} p_*$. The index of this lattice in $\operatorname{Ker} p_*$ does not seem to be directly related to $\operatorname{ind}(F)$. For example, the unique ogod of Example 4.12 has index n but gives a basis for $\operatorname{Ker} p_*$. On the other hand, Figure 6 gives an example of a double cover with dilation index two for which no ogod gives a basis for $\operatorname{Ker} p_*$. Hence the fundamental cycle construction using a single ogod cannot generally be used when the dilation subgraph is disconnected.

5. Invariance of $\operatorname{Prym}_p(\widetilde{\Gamma}/\Gamma)$ under the simplification of $M^*(\widetilde{\Gamma}/\Gamma)$

The papers [**CV10**] and [**BMV11**] considered the tropical Torelli problem, in other words whether and to what extent a metric graph Γ can be reconstructed from its Jacobian $\operatorname{Jac}(\Gamma)$. The authors reconstruct $\operatorname{Jac}(\Gamma)$ from the graphic matroid $M(\Gamma)$ and show that the Jacobian does not change when $M(\Gamma)$ is replaced by its simplification, and that conversely the simplification class of the matroid determines the Jacobian.

In this section, we construct the simplification of the signed cographic matroid $M^*(\widetilde{\Gamma}/\Gamma)$ of a double cover of metric graphs $\pi : \widetilde{\Gamma} \to \Gamma$, and show that $\operatorname{Prym}_p(\widetilde{\Gamma}/\Gamma)$ does not change under simplification. We then give an example of two double covers having non-isomorphic simple matroids but isomorphic Pryms, contrary to the case of Jacobians. Hence the fibers of the tropical Prym–Torelli map are not fully described by the simplification procedure.

5.1. Simplification of the signed cographic matroid. We recall that the simplification M_{sim} of a matroid M is given by deleting all 1-circuits, and then iteratively deleting elements in 2-circuits until there are no more 2-circuits. For the cographic matroid $M^*(\Gamma)$, the simplification $M^*_{\text{sim}}(\Gamma)$ is obtained by contracting all bridges in Γ, and an edge e_1 in each pair $\{e_1, e_2\}$ of separating edges. It is shown in [**CV10**] that if we also set the new length of e_2 to be $\ell(e_1) + \ell(e_2)$, then $\operatorname{Jac}(\Gamma)$ does not change.

We now adapt this procedure to the signed cographic matroid $M^*(\widetilde{\Gamma}/\Gamma)$ of a double cover $\pi : \widetilde{\Gamma} \to \Gamma$ of metric graphs. A circuit of $M^*(\widetilde{\Gamma}/\Gamma)$ is a minimal set of undilated edges F of Γ such that the restricted cover $\pi^{-1}(\Gamma \backslash F) \to \Gamma \backslash F$ is not relatively connected. The types of the 1- and 2-circuits of $M^*(\widetilde{\Gamma}/\Gamma)$ are shown on Figures 3 and 4. We label the edges of a 2-circuit with multiplicities as follows.

DEFINITION 5.1. Let f_i be an edge lying in a 2-circuit $F = \{f_1, f_2\}$ of the signed cographic matroid $M^*(\widetilde{\Gamma}/\Gamma)$. The *multiplicity of f_i* is

$$\mathrm{mult}(f_i) = 2^{\mathrm{ind}(\{f_i\})-\mathrm{ind}(\varnothing)} = 2^{\mathrm{ind}(\{f_i\})-1} \in \{1, 2\}.$$

We note that $\mathrm{mult}(f) = 2$ if and only if f is a bridge edge of Γ. We now define a non-canonical simplification process for double covers of metric graphs.

DEFINITION 5.2. Let $\pi : \widetilde{\Gamma} \to \Gamma$ be a double cover of metric graphs. The *simplification* $\pi_{\mathrm{sim}} : \widetilde{\Gamma}_{\mathrm{sim}} \to \Gamma_{\mathrm{sim}}$ of π is defined as follows. First, we contract all edges of Γ that are 1-circuits of $M^*(\widetilde{\Gamma}/\Gamma)$ (see Figure 3). Then, for every 2-circuit $\{f_1, f_2\}$ of $M^*(\widetilde{\Gamma}/\Gamma)$ (see Figure 4), we contract one of the edges, say f_1, and redefine the length of the other edge to be $\ell(f_2) + \left(\frac{\mathrm{mult}(f_1)}{\mathrm{mult}(f_2)}\right)^2 \ell(f_1)$. The contracted edge can be chosen arbitrarily except for a 2-circuit $\{f_1, f_2\}$ with $\mathrm{mult}(f_1) = 2$ and $\mathrm{mult}(f_2) = 1$ (lower left diagram on Figure 4), for which we always contract the edge f_1 having multiplicity 2.

By construction, $M^*(\widetilde{\Gamma}_{\mathrm{sim}}/\Gamma_{\mathrm{sim}})$ is the simplification of $M^*(\widetilde{\Gamma}/\Gamma)$. We note that we do not contract a loop whose preimage is a pair of parallel edges, except for the 2-circuit on the top right of Figure 4. This type of circuit only occurs for free double covers, and the resulting contraction is a double cover with a single dilated vertex. Therefore, simplification does not change the dilation index of a double cover. We now show that the Prym variety does not change under simplification.

THEOREM 5.3. *Let $\pi : \widetilde{\Gamma} \to \Gamma$ be a double cover of metric graphs and let $\pi_{\mathrm{sim}} : \widetilde{\Gamma}_{\mathrm{sim}} \to \Gamma_{\mathrm{sim}}$ be a simplification of π. Then* $\mathrm{Prym}_p(\widetilde{\Gamma}/\Gamma) \cong \mathrm{Prym}_p(\widetilde{\Gamma}_{\mathrm{sim}}/\Gamma_{\mathrm{sim}})$.

PROOF. Choose an orientation for the double cover π. This induces oriented matroid structures on $M(\widetilde{\Gamma}/\Gamma)$ and $M^*(\widetilde{\Gamma}/\Gamma)$. By [**BLVS$^+$99**, Theorem 3.4.3] for every oriented circuit $\vec{C} \in \mathcal{C}(\vec{M}(\widetilde{\Gamma}/\Gamma))$ and every oriented circuit $\vec{D} \in \mathcal{C}(\vec{M}^*(\widetilde{\Gamma}/\Gamma))$, either $C \cap D = \varnothing$ or there are distinct edges $e, f \in C \cap D$ such that $\vec{C}(e)\vec{D}(e) = -\vec{C}(f)\vec{D}(f)$. This implies that if $D = \{f\}$ is a 1-circuit of $M^*(\widetilde{\Gamma}/\Gamma)$, then f does not lie in any circuit of $M(\widetilde{\Gamma}/\Gamma)$. But the Prym variety is generated by the fundamental cycles of the circuits of $M(\widetilde{\Gamma}/\Gamma)$, and thus contracting f does not change the Prym variety.

Now let $D = \{f_1, f_2\}$ be a 2-circuit of $M^*(\widetilde{\Gamma}/\Gamma)$ and let C_1, C_2 be any circuits of $M(\widetilde{\Gamma}/\Gamma)$, possibly $C_1 = C_2$. We show that the pairing (5)

$$[\gamma_{C_1}, \gamma_{C_2}] = \sum_{e \in E(\Gamma)} 2\gamma_{C_1}(e)\gamma_{C_2}(e)\ell(e) = \sum_{e \in C_1 \cap C_2} 2|\gamma_{C_1}(e)||\gamma_{C_2}(e)|\vec{C}_1(e)\vec{C}_2(e)\ell(e)$$

does not change under the suggested contraction of f_1. If $C_1 \cap C_2 \cap D = \varnothing$ then neither f_1 nor f_2 contribute to $[\gamma_{C_1}, \gamma_{C_2}]$, so we assume without loss of generality that $f_1 \in C_1 \cap C_2$, and hence $f_2 \in C_1 \cap C_2$ as well by the aforementioned theorem. Still without loss of generality we may assume that $\vec{C}_1(f_1) = \vec{C}_2(f_1)$ (if not, simply replace $\vec{C}_1$ with $-\vec{C}_1$) and hence $\vec{C}_1(f_1)\vec{C}_2(f_1) = 1$. It follows that

$$-\vec{C}_1(f_2)\vec{D}(f_2) = \vec{C}_1(f_1)\vec{D}(f_1) = \vec{C}_2(f_1)\vec{D}(f_1) = -\vec{C}_2(f_2)\vec{D}(f_2),$$

and hence $\vec{C}_1(f_2) = \vec{C}_2(f_2)$ and $\vec{C}_1(f_2)\vec{C}_2(f_2) = 1$ as well. Therefore, the combined contribution from f_1 and f_2 to $[\gamma_{C_1}, \gamma_{C_2}]$ is equal to

$$2|\gamma_{C_1}(f_1)||\gamma_{C_2}(f_1)|\ell(f_1) + 2|\gamma_{C_1}(f_2)||\gamma_{C_2}(f_2)|\ell(f_2). \tag{14}$$

We now contract f_1 and rescale the length of f_2 to $\ell'(f_2) = \ell(f_2)+\left(\frac{\mathrm{mult}(f_1)}{\mathrm{mult}(f_2)}\right)^2\ell(f_1)$. Denote $C_1' = C_1\backslash\{f_1\}$ and $C_2' = C_2\backslash\{f_1\}$ the contractions of C_1 and C_2, respectively. Orientations are preserved under contraction, hence $\vec{C}'_1(f_2)\vec{C}'_2(f_2) = \vec{C}_1(f_2)\vec{C}_2(f_2) = 1$ and the contribution from f_2 to $[\gamma_{C_1'}, \gamma_{C_2'}]$ is equal to

$$\begin{aligned} &2|\gamma_{C_1'}(f_2)||\gamma_{C_2'}(f_2)|\ell'(f_2) \\ &= 2|\gamma_{C_1'}(f_2)||\gamma_{C_2'}(f_2)|\left(\frac{\mathrm{mult}(f_1)}{\mathrm{mult}(f_2)}\right)^2\ell(f_1) + 2|\gamma_{C_1'}(f_2)||\gamma_{C_2'}(f_2)|\ell(f_2). \end{aligned} \tag{15}$$

To show that the right hand side of (15) is equal to (14), it is enough to check for any circuit $C \in M(\widetilde{\Gamma}/\Gamma)$ containing $D = \{f_1, f_2\}$ and contracting to $C' = C\backslash\{f_1\}$, we have

$$|\gamma_C(f_2)| = |\gamma_{C'}(f_2)| \qquad \text{and} \qquad |\gamma_C(f_1)| = |\gamma_{C'}(f_2)|\frac{\mathrm{mult}(f_1)}{\mathrm{mult}(f_2)}.$$

This can be verified on a case-by-case basis by comparing types of 2-circuits of $M^*(\widetilde{\Gamma}/\Gamma)$ (Figure 4) and circuits of $M(\widetilde{\Gamma}/\Gamma)$ (Figure 5). Hence the Prym variety does not change when f_1 is contracted. Proceeding in this manner, we see that $\mathrm{Prym}_p(\widetilde{\Gamma}_{\mathrm{sim}}/\Gamma_{\mathrm{sim}})$ is isomorphic to $\mathrm{Prym}_p(\widetilde{\Gamma}/\Gamma)$. □

5.2. The Prym does not determine the simplified matroid. We now show that, unlike the case of graphs, double covers with non-isomorphic simple matroids may still have isomorphic Prym varieties.

Consider the double cover $\pi : \widetilde{\Gamma} \to \Gamma$ depicted on Figure 7 on the left, with edges $E(\Gamma) = \{e_1, e_2, e_3, e_4\}$ of lengths $x_i = \ell(e_i)$. The signed graphic matroid $M(\widetilde{\Gamma}/\Gamma) \cong U_{2,4}$ is the uniform matroid of rank 2 on 4 elements and its dual $M^*(\widetilde{\Gamma}/\Gamma) = U_{2,4}^* = U_{2,4}$ is simple. To compute the Prym variety, we apply Proposition 4.14 to the index-1 ogod $\{e_3, e_4\}$. We obtain the following basis for $\mathrm{Ker}\,\pi_*$:

$$\gamma_1 = e_1 + e_2 + e_3, \qquad \gamma_2 = e_1 + 2e_2 + e_4$$

Thus the pairing $[\cdot,\cdot]$ on $\mathrm{Prym}_p(\widetilde{\Gamma}/\Gamma)$ has Gram matrix

$$\begin{pmatrix} x_1 + x_2 + x_3 & x_1 + 2x_2 \\ x_1 + 2x_2 & x_1 + 4x_2 + x_4 \end{pmatrix}.$$

Now we study the double cover $\sigma : \widetilde{\Delta} \to \Delta$ on the right of Figure 7 with edges $E(\Delta) = \{f_1, f_2, f_3\}$ of lengths $\ell(f_j) = y_j$. This time the signed graphic matroid is $M(\widetilde{\Delta}/\Delta) = U_{1,3}$ and again the dual matroid $M^*(\widetilde{\Delta}/\Delta) = U_{1,3}^* = U_{2,3}$ is simple, but obviously $U_{2,3} \not\cong U_{2,4}$. The ogod $\{f_2, f_3\}$ produces the basis

$$\delta_1 = f_1 + f_2, \qquad \delta_2 = 2f_1 + f_3$$

for $\mathrm{Ker}\,\sigma_*$, and the Gram matrix of the pairing is

$$\begin{pmatrix} y_1 + y_2 & 2y_1 \\ 2y_1 & 4y_1 + y_3 \end{pmatrix}.$$

It is elementary to verify that the two Gram matrices are equal, and hence $\mathrm{Prym}_p(\widetilde{\Gamma}/\Gamma) \cong \mathrm{Prym}_p(\widetilde{\Delta}/\Delta)$, for appropriate positive values of x_i and y_i.

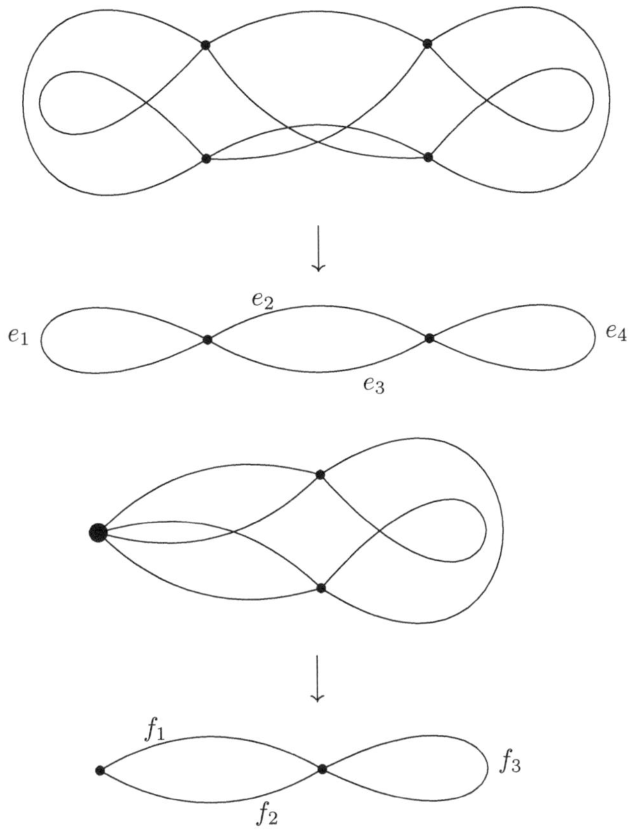

FIGURE 7. Double covers with isomorphic Pryms and non-isomorphic simple matroids.

Acknowledgments

The authors would like to thank Kevin Kühn, Yoav Len, Dhruv Ranganathan, Victoria Schleis, Benjamin Schröter, and Martin Ulirsch for insightful discussions. The first author gratefully acknowledges support from the SFB-TRR 195 "Symbolic Tools in Mathematics and their Application" of the German Research Foundation (DFG).

References

[ABKS14] Y. An, M. Baker, G. Kuperberg, and F. Shokrieh, *Canonical representatives for divisor classes on tropical curves and the matrix-tree theorem*, Forum Math. Sigma **2** (2014), Paper No. e24, 25, DOI 10.1017/fms.2014.25. MR3264262

[BD87] A. Beauville and O. Debarre, *Sur le problème de Schottky pour les variétés de Prym* (French), Ann. Scuola Norm. Sup. Pisa Cl. Sci. (4) **14** (1987), no. 4, 613–623 (1988). MR963492

[BF11] M. Baker and X. Faber, *Metric properties of the tropical Abel-Jacobi map*, J. Algebraic Combin. **33** (2011), no. 3, 349–381, DOI 10.1007/s10801-010-0247-3. MR2772537

[BL78] R. G. Bland and M. Las Vergnas, *Orientability of matroids*, J. Combinatorial Theory Ser. B **24** (1978), no. 1, 94–123, DOI 10.1016/0095-8956(78)90080-1. MR485461

[BL04] C. Birkenhake and H. Lange, *Complex abelian varieties*, 2nd ed., Grundlehren der mathematischen Wissenschaften [Fundamental Principles of Mathematical Sciences], vol. 302, Springer-Verlag, Berlin, 2004, DOI 10.1007/978-3-662-06307-1. MR2062673

[BLVS+99] A. Björner, M. Las Vergnas, B. Sturmfels, N. White, and G. M. Ziegler, *Oriented matroids*, 2nd ed., Encyclopedia of Mathematics and its Applications, vol. 46, Cambridge University Press, Cambridge, 1999, DOI 10.1017/CBO9780511586507. MR1744046

[BMV11] S. Brannetti, M. Melo, and F. Viviani, *On the tropical Torelli map*, Adv. Math. **226** (2011), no. 3, 2546–2586, DOI 10.1016/j.aim.2010.09.011. MR2739784

[CV10] L. Caporaso and F. Viviani, *Torelli theorem for graphs and tropical curves*, Duke Math. J. **153** (2010), no. 1, 129–171, DOI 10.1215/00127094-2010-022. MR2641941

[Fay06] J. D. Fay, *Theta functions on Riemann surfaces*, Lecture Notes in Mathematics, Vol. 352, Springer-Verlag, Berlin-New York, 1973. MR335789

[FRSS18] T. Foster, J. Rabinoff, F. Shokrieh, and A. Soto, *Non-Archimedean and tropical theta functions*, Math. Ann. **372** (2018), no. 3-4, 891–914, DOI 10.1007/s00208-018-1646-3. MR3880286

[Ger82] L. Gerritzen, *Die Jacobi-Abbildung über dem Raum der Mumfordkurven* (German), Math. Ann. **261** (1982), no. 1, 81–100, DOI 10.1007/BF01456412. MR675209

[GK10] S. Grushevsky and I. Krichever, *Integrable discrete Schrödinger equations and a characterization of Prym varieties by a pair of quadrisecants*, Duke Math. J. **152** (2010), no. 2, 317–371, DOI 10.1215/00127094-2010-014. MR2656091

[GRS95] I. M. Gelfand, G. L. Rybnikov, and D. A. Stone, *Projective orientations of matroids*, Adv. Math. **113** (1995), no. 1, 118–150, DOI 10.1006/aima.1995.1036. MR1332809

[Gun82] R. C. Gunning, *Some curves in abelian varieties*, Invent. Math. **66** (1982), no. 3, 377–389, DOI 10.1007/BF01389218. MR662597

[GZ23] A. Ghosh and D. Zakharov, *The Prym Variety of a Dilated Double Cover of Metric Graphs*, Ann. Comb. **29** (2025), no. 2, 439–460, DOI 10.1007/s00026-024-00707-0. MR4913020

[JL18] D. Jensen and Y. Len, *Tropicalization of theta characteristics, double covers, and Prym varieties*, Selecta Math. (N.S.) **24** (2018), no. 2, 1391–1410, DOI 10.1007/s00029-017-0379-6. MR3782424

[Kri06] I. Krichever, *Integrable linear equations and the Riemann-Schottky problem*, Algebraic geometry and number theory, Progr. Math., vol. 253, Birkhäuser Boston, Boston, MA, 2006, pp. 497–514, DOI 10.1007/978-0-8176-4532-8_8. MR2263198

[Kri10] I. Krichever, *Characterizing Jacobians via trisecants of the Kummer variety*, Ann. of Math. (2) **172** (2010), no. 1, 485–516, DOI 10.4007/annals.2010.172.485. MR2680424

[LU21] Y. Len and M. Ulirsch, *Skeletons of Prym varieties and Brill-Noether theory*, Algebra Number Theory **15** (2021), no. 3, 785–820, DOI 10.2140/ant.2021.15.785. MR4261102

[LZ22] Y. Len and D. Zakharov, *Kirchhoff's theorem for Prym varieties*, Forum Math. Sigma **10** (2022), Paper No. e11, 54, DOI 10.1017/fms.2021.75. MR4382460

[Mum74] D. Mumford, *Prym varieties. I*, Contributions to analysis (a collection of papers dedicated to Lipman Bers), Academic Press, New York-London, 1974, pp. 325–350. MR379510

[MZ08] G. Mikhalkin and I. Zharkov, *Tropical curves, their Jacobians and theta functions*, Curves and abelian varieties, Contemp. Math., vol. 465, Amer. Math. Soc., Providence, RI, 2008, pp. 203–230, DOI 10.1090/conm/465/09104. MR2457739

[Oxl11] J. Oxley, *Matroid theory*, 2nd ed., Oxford Graduate Texts in Mathematics, vol. 21, Oxford University Press, Oxford, 2011, DOI 10.1093/acprof:oso/9780198566946.001.0001. MR2849819

[RZ22] Felix Röhrle and Dmitry Zakharov. The tropical n-gonal construction. Preprint, `arXiv:2210.02267`, 2022.

[Wel83] G. E. Welters, *On flexes of the Kummer variety (note on a theorem of R. C. Gunning)*, Nederl. Akad. Wetensch. Indag. Math. **45** (1983), no. 4, 501–520. MR731833

[Wel84] G. E. Welters, *A criterion for Jacobi varieties*, Ann. of Math. (2) **120** (1984), no. 3, 497–504, DOI 10.2307/1971084. MR769160

[Zas82] T. Zaslavsky, *Signed graphs*, Discrete Appl. Math. **4** (1982), no. 1, 47–74, DOI 10.1016/0166-218X(82)90033-6. MR676405

[Zas83] T. Zaslavsky, *Erratum: "Signed graphs"*, Discrete Appl. Math. **5** (1983), no. 2, 248, DOI 10.1016/0166-218X(83)90047-1. MR683518

[Zas91] T. Zaslavsky, *Orientation of signed graphs*, European J. Combin. **12** (1991), no. 4, 361–375, DOI 10.1016/S0195-6698(13)80118-7. MR1120422

Universität Tübingen, Fachbereich Mathematik, Auf der Morgenstelle 10, 72076 Tübingen, Germany
Email address: roehrle@math.uni-tuebingen.de

Department of Mathematics, Central Michigan University, Mount Pleasant, Michigan 48859
Email address: dvzakharov@gmail.com

Contemporary Mathematics
Volume **823**, 2025
https://doi.org/10.1090/conm/823/16491

A Lax form for quantum q-$P_{\rm VI}$ equation

Yasuhiko Yamada

ABSTRACT. A quantization of Jimbo-Sakai's q-$P_{\rm VI}$ equation and its Lax form were constructed by Hasegawa many years ago. We will give another Lax form for the quantum q-$P_{\rm VI}$ equation. The new Lax form has a similar expression with the commutative case and shows its geometric structure clearly. Utilizing this, we derive the Lax forms also for the degenerate cases.

1. Introduction

The Lax formulation is a powerful method to study nonlinear systems as the compatibility condition for linear ones. Sometimes it brings beautiful results if it fits suitable geometric picture. The most remarkable example is the construction of the algebrogeometric solutions for nonlinear equations by I. M. Krichever [**11**].

The differential Painlevé equations also admit Lax formulation. In fact, the sixth Painlevé equation $P_{\rm VI}$ first arises in this form [**6**] as an iso-monodromic deformation equation. Its q-difference analog q-$P_{\rm VI}$ is obtained by Jimbo-Sakai [**9**] also as the compatibility for a pair of linear q-difference equations. Due to the recent developments in discrete Painlevé equations, the geometric structure of the differential and discrete Painlevé equations becomes very clear [**14**] and such a geometric study has been applied to various problems: special solutions, Lax formulations, identification of the equations etc.[**5**][**10**]. In this note, we will add one more application i.e. quantization.

The quantization of differential and discrete Painlevé equations is a natural problem. It is important in view of the AGT relation [**1**] and its q analog [**3**]. In differential case, the quantization of Painlevé equations has been studied in various way, in particular with the relation to conformal filed theory. For the discrete case, however, the study of their quantization is rather limited (see [**2**][**4**][**12**][**13**] and references therein). One concrete example is the quantization of q-$P_{\rm VI}$ which we call qq-$P_{\rm VI}$ for short. It was first formulated by Hasegawa [**7**] as a translation of certain representation of the extended affine Weyl group of type $D_5^{(1)}$ acting on non-commutative variables. This construction reflect the algebraic background of the system and the resulting equation has almost the same form as the commutative case.

Hasegawa also gave a Lax formulation for qq-$P_{\rm VI}$ based on the quantum group $U_q(A_1^{(1)})$ in [**8**]. However, since the construction is rather heavy, it is difficult to

2020 *Mathematics Subject Classification.* Primary 39A06; 33E05.

This work is supported in part by Grants-in-Aid for Scientific Research (Kakenhi); 22H01116.

see the underlying geometric structure. The aim of this note is to present another Lax form for qq-P_{VI} equation. The new Lax form has almost the same form as the commutative case [**10**] and hence its geometric structure is transparent. Thanks to this advantage, we can obtain the Lax forms for degenerate cases following the same limiting procedure as the commutative case.

The paper is organized as follows. In §2, we formulate a main Theorem which gives a new Lax form for qq-P_{VI} equation. In §3, we prove the Theorem, namely we show that the compatibility of the Lax pair is equivalent to the qq-P_{VI} equation. Since the direct computation becomes very complicated, we show a method to reduce the task. In §4, we derive the Lax forms and evolution equations for degenerate cases. In Appendix A, we recall a formulation of the qq-P_{VI} equation based on the affine Weyl group symmetry.

2. The main Theorem

Since we will consider quantum and q-deformed Painlevé equations, we need two base parameters. In this note, we use parameter q to specify the non-commutativity and parameter p for the base of multiplicative shift. Let $\mathcal{K}$ be a skew field over $\mathbb{C}$ of rational functions of the parameters (or variables) $\mathrm{q}, \mathrm{p}, \kappa_1, \kappa_2, \mathrm{v}_1, \ldots, \mathrm{v}_8$ and $\mathbf{x}, \mathbf{y}$. Where κ_i, v_i are central elements with a constraint

$$\kappa_1^2\kappa_2^2 = \mathrm{p}\,\mathrm{q}\,\mathrm{v}_1\,\mathrm{v}_2\cdots\mathrm{v}_8, \tag{2.1}$$

and the variables $\mathbf{x}, \mathbf{y}$ are subject to the q-commutation relation

$$\mathbf{x}\mathbf{y} = \mathrm{q}\,\mathbf{y}\mathbf{x}. \tag{2.2}$$

For simplicity, we will use the following notation:

$$\begin{aligned} V_{12}(z) &= (z - \frac{1}{\mathrm{v}_1})(z - \frac{1}{\mathrm{v}_2}), \quad V_{34}(z) = (z - \mathrm{v}_3)(z - \mathrm{v}_4),\\ V_{56}(z) &= (z - \frac{\mathrm{v}_5}{\kappa_2})(z - \frac{\mathrm{v}_6}{\kappa_2}), \quad V_{78}(z) = (z - \frac{\kappa_1}{\mathrm{v}_7})(z - \frac{\kappa_1}{\mathrm{v}_8}). \end{aligned} \tag{2.3}$$

DEFINITION 2.1. Define qq-P_{VI} equation as the following algebra automorphism T on the skew field $\mathcal{K} = \mathbb{C}(\mathrm{q}, \mathrm{p}, \kappa_1, \kappa_2, v_1, \ldots, v_8, \mathbf{x}, \mathbf{y})$:

$$T := \{\kappa_1 \to \mathrm{p}^{-1}\kappa_1, \kappa_2 \to \mathrm{p}\,\kappa_2, \mathrm{v}_i \to \mathrm{v}_i, \mathbf{x} \to \overline{\mathbf{x}}, \mathbf{y} \to \overline{\mathbf{y}}\}, \tag{2.4}$$

where

$$\overline{\mathbf{x}} = \frac{\mathrm{q}\,\mathrm{v}_3\,\mathrm{v}_4}{\mathbf{x}}\frac{V_{56}(\mathbf{y})}{V_{12}(\mathbf{y})} = \frac{\mathrm{q}\,\mathrm{v}_3\,\mathrm{v}_4}{\mathbf{x}}\frac{(\mathbf{y} - \frac{\mathrm{v}_5}{\kappa_2})(\mathbf{y} - \frac{\mathrm{v}_6}{\kappa_2})}{(\mathbf{y} - \frac{1}{\mathrm{v}_1})(\mathbf{y} - \frac{1}{\mathrm{v}_2})}, \tag{2.5}$$

$$\overline{\mathbf{y}} = \frac{1}{\mathrm{v}_1\,\mathrm{v}_2\,\mathbf{y}}\frac{\overline{V_{78}}(\overline{\mathbf{x}})}{V_{34}(\overline{\mathbf{x}})} = \frac{1}{\mathrm{v}_1\,\mathrm{v}_2\,\mathbf{y}}\frac{(\overline{\mathbf{x}} - \frac{\kappa_1}{\mathrm{p}\,\mathrm{v}_7})(\overline{\mathbf{x}} - \frac{\kappa_1}{\mathrm{p}\,\mathrm{v}_8})}{(\overline{\mathbf{x}} - \mathrm{v}_3)(\overline{\mathbf{x}} - \mathrm{v}_4)}. \tag{2.6}$$

For $x \in \mathcal{K}$, its up shift and down shift are denoted as $T(x) = \overline{x}$ and $T^{-1}(x) = \underline{x}$. For instance, as the down shift of (2.6) we have

$$\mathbf{y} = \frac{1}{\mathrm{v}_1\,\mathrm{v}_2\,\underline{\mathbf{y}}}\frac{V_{78}(\mathbf{x})}{V_{34}(\mathbf{x})}. \tag{2.7}$$

Note that from (2.5) (2.6) we have

$$\overline{\mathbf{x}}\mathbf{y} = \mathrm{q}^{-1}\mathbf{y}\overline{\mathbf{x}}, \quad \mathbf{x}\underline{\mathbf{y}} = \mathrm{q}^{-1}\underline{\mathbf{y}}\mathbf{x}, \quad \overline{\mathbf{x}}\overline{\mathbf{y}} = \mathrm{q}\,\overline{\mathbf{y}}\overline{\mathbf{x}}. \tag{2.8}$$

The equations (2.5) (2.6) are essentially the same as the equation obtained by Hasegawa [**7**] from a representation of the affine Weyl group of type $D_5^{(1)}$. In Appendix A, we give a formulation based on the affine Weyl group in terms of geometric parameters. In the commutative case $\mathrm{q}=1$, the qq-$P_{\rm VI}$ equation reduces to the q-$P_{\rm VI}$ equation introduced by Jimbo-Sakai [**9**].

The following is the main result of this note.

THEOREM 2.2. *The qq-$P_{\rm VI}$ equation* (2.5) (2.6) *is the compatibility condition of the following linear equations for $\psi(z)$:*

$$\begin{aligned} L_1(z) &:= \{\frac{z\,\mathrm{v}_1\,\mathrm{v}_2\,V_{12}(\mathbf{y})}{\mathrm{p}\,\mathbf{y}} - \frac{\mathrm{q}\,\mathrm{v}_1\,\mathrm{v}_2\,\mathrm{v}_3\,\mathrm{v}_4}{\mathbf{x}}\frac{V_{56}(\mathbf{y})}{\mathbf{y}}\}\psi(z) \\ &+\frac{V_{78}(z)}{\mathrm{p}(\mathbf{x}-z)}\{\frac{1}{\mathbf{y}}\psi(z)-\psi(\mathrm{p}\,z)\} + \frac{\mathrm{v}_1\,\mathrm{v}_2\,V_{34}(\frac{z}{\mathrm{p}})}{\frac{\mathbf{x}}{\mathrm{q}}-\frac{z}{\mathrm{p}}}\{\mathbf{y}\psi(z)-\psi(\tfrac{z}{\mathrm{p}})\}=0, \end{aligned} \tag{2.9}$$

$$L_2(z) := (1-\frac{\mathbf{x}\,\mathrm{p}}{z})\overline{\psi}(\frac{z}{\mathrm{p}}) - \{\frac{1}{\mathbf{y}}\psi(\frac{z}{\mathrm{p}})-\psi(z)\}=0. \tag{2.10}$$

The proof is given in the next section.

REMARK 2.3. The equation $L_1(z)$ is symmetric under the following transformation

$$\mathbf{x}\leftrightarrow z, \quad \mathbf{y}\leftrightarrow T_z^{-1}, \quad \mathrm{q}\leftrightarrow \mathrm{p}, \tag{2.11}$$

where $T_z : z \mapsto \mathrm{p}\,z$ is the p-shift operator for z.

3. Proof of the Theorem

The logic of the proof is the same as the commutative case [**10**]. Namely, from the equations $L_1(z)=0$, $L_2(z)=0$ and their p-shifts in z, we derive linear relations $L_3(z)=0, L_4(z)=0$ and $L_5(z)=0$ step by step. The last one $L_5(z)=0$ gives a p-difference equation among $\overline{\psi}(\mathrm{p}\,z)$, $\overline{\psi}(z)$ and $\overline{\psi}(\mathrm{p}^{-1}\,z)$. Then what we should show is that the equation $L_5(z)=0$ is equivalent to the up-shift of $L_1(z)=0$. In practice, we will use the relation (2.5) at each step to obtain a concise expression.

First, from $L_2(\mathrm{p}\,z)=0$ and $L_2(z)=0$ we have

$$\frac{1}{\mathbf{y}}\psi(z)-\psi(\mathrm{p}\,z) = (1-\frac{\mathbf{x}}{z})\overline{\psi}(z), \quad \psi(\frac{z}{\mathrm{p}})-\mathbf{y}\psi(z) = (1-\frac{\mathbf{x}\,\mathrm{p}}{\mathrm{q}\,z})\mathbf{y}\overline{\psi}(\frac{z}{\mathrm{p}}). \tag{3.1}$$

Substituting these into $L_1(z)=0$, we obtain

$$L_3(z) := (\frac{z}{\mathrm{p}}-\overline{\mathbf{x}})\frac{\mathrm{v}_1\,\mathrm{v}_2\,V_{12}(\mathbf{y})}{\mathbf{y}}\psi(z) + \frac{\mathrm{v}_1\,\mathrm{v}_2\,\mathrm{p}\,V_{34}(\frac{z}{\mathrm{p}})}{z}\mathbf{y}\overline{\psi}(\frac{z}{\mathrm{p}}) - \frac{V_{78}(z)}{\mathrm{p}\,z}\overline{\psi}(z)=0, \tag{3.2}$$

where the coefficient of $\psi(z)$ is rewritten using (2.5):

$$\begin{aligned} \frac{z\,\mathrm{v}_1\,\mathrm{v}_2\,V_{12}(\mathbf{y})}{\mathrm{p}\,\mathbf{y}} - \frac{\mathrm{q}\,\mathrm{v}_1\,\mathrm{v}_2\,\mathrm{v}_3\,\mathrm{v}_4}{\mathbf{x}}\frac{V_{56}(\mathbf{y})}{\mathbf{y}} &= \frac{z\,\mathrm{v}_1\,\mathrm{v}_2\,V_{12}(\mathbf{y})}{\mathrm{p}\,\mathbf{y}} - \overline{\mathbf{x}}\frac{\mathrm{v}_1\,\mathrm{v}_2\,V_{12}(\mathbf{y})}{\mathbf{y}} \\ &= (\frac{z}{\mathrm{p}}-\overline{\mathbf{x}})\frac{\mathrm{v}_1\,\mathrm{v}_2\,V_{12}(\mathbf{y})}{\mathbf{y}}. \end{aligned} \tag{3.3}$$

Next we consider the following combination:

$$L_4(z) := L_3(z) - (\frac{z}{\mathrm{p}}-\overline{\mathbf{x}})\frac{\mathrm{v}_1\,\mathrm{v}_2\,V_{12}(\mathbf{y})}{\mathbf{y}}L_2(z)=0. \tag{3.4}$$

Explicitly, it is given as follows

$$
\begin{aligned}
L_4(z) &= (\frac{z}{\mathrm{p}} - \overline{\mathbf{x}})\frac{\mathrm{v}_1\,\mathrm{v}_2\,V_{12}(\mathbf{y})}{\mathbf{y}}\psi(z) + \frac{\mathrm{v}_1\,\mathrm{v}_2\,\mathrm{p}\,V_{34}(\frac{z}{\mathrm{p}})}{z}\mathbf{y}\overline{\psi}(\tfrac{z}{\mathrm{p}}) - \frac{V_{78}(z)}{\mathrm{p}\,z}\overline{\psi}(z) \\
&\quad - (\frac{z}{\mathrm{p}} - \overline{\mathbf{x}})\frac{\mathrm{v}_1\,\mathrm{v}_2\,V_{12}(\mathbf{y})}{\mathbf{y}}\Big[(1 - \frac{\mathbf{x}\,\mathrm{p}}{z})\overline{\psi}(\tfrac{z}{\mathrm{p}}) - \frac{1}{\mathbf{y}}\psi(\tfrac{z}{\mathrm{p}}) + \psi(z)\Big] \\
&= \frac{\mathrm{v}_1\,\mathrm{v}_2\,\mathrm{p}\,V_{34}(\frac{z}{\mathrm{p}})}{z}\mathbf{y}\overline{\psi}(\tfrac{z}{\mathrm{p}}) - \frac{V_{78}(z)}{\mathrm{p}\,z}\overline{\psi}(z) \\
&\quad - (\frac{z}{\mathrm{p}} - \overline{\mathbf{x}})\frac{\mathrm{v}_1\,\mathrm{v}_2\,V_{12}(\mathbf{y})}{\mathbf{y}}\Big[(1 - \frac{\mathbf{x}\,\mathrm{p}}{z})\overline{\psi}(\tfrac{z}{\mathrm{p}}) - \frac{1}{\mathbf{y}}\psi(\tfrac{z}{\mathrm{p}})\Big] = 0.
\end{aligned} \tag{3.5}
$$

Finally we compute the following combination:

$$
L_5(z) := \frac{1}{\frac{z}{\mathrm{p}} - \frac{\overline{\mathbf{x}}}{\mathrm{q}}}\frac{1}{\mathbf{y}}L_3(z) - \frac{1}{z - \overline{\mathbf{x}}}L_4(\mathrm{p}\,z) = 0. \tag{3.6}
$$

For the coefficient of $\psi(z)$ in $L_5(z)$, we have

$$
\frac{1}{\frac{z}{\mathrm{p}} - \frac{\overline{\mathbf{x}}}{\mathrm{q}}}(\frac{z}{\mathrm{p}} - \frac{\overline{\mathbf{x}}}{\mathrm{q}})\frac{1}{\mathbf{y}}\frac{\mathrm{v}_1\,\mathrm{v}_2\,V_{12}(\mathbf{y})}{\mathbf{y}} - \frac{1}{z - \overline{\mathbf{x}}}(z - \overline{\mathbf{x}})\frac{\mathrm{v}_1\,\mathrm{v}_2\,V_{12}(\mathbf{y})}{\mathbf{y}}\frac{1}{\mathbf{y}} = 0. \tag{3.7}
$$

The coefficient of $\overline{\psi}(\mathrm{p}\,z)$ in $L_5(z)$ is

$$
\frac{1}{z - \overline{\mathbf{x}}}\frac{V_{78}(\mathrm{p}\,z)}{z\,\mathrm{p}^2} = \frac{\overline{V_{78}}(z)}{z(z - \overline{\mathbf{x}})}. \tag{3.8}
$$

The coefficient of $\overline{\psi}(\frac{z}{\mathrm{p}})$ is

$$
\frac{1}{\frac{z}{\mathrm{p}} - \frac{\overline{\mathbf{x}}}{\mathrm{q}}}\frac{1}{\mathbf{y}}\frac{\mathrm{v}_1\,\mathrm{v}_2\,\mathrm{p}\,V_{34}(\frac{z}{\mathrm{p}})}{z}\mathbf{y} = \frac{\mathrm{v}_1\,\mathrm{v}_2\,V_{34}(\frac{z}{\mathrm{p}})}{\frac{z}{\mathrm{p}}(\frac{z}{\mathrm{p}} - \frac{\overline{\mathbf{x}}}{\mathrm{q}})}. \tag{3.9}
$$

Finally, the coefficient $A(z)$ of $\overline{\psi}(z)$ is given as

(3.10)

$$
\begin{aligned}
A(z) &= \frac{1}{\frac{z}{\mathrm{p}} - \frac{\overline{\mathbf{x}}}{\mathrm{q}}}\frac{1}{\mathbf{y}}\{-\frac{V_{78}(z)}{\mathrm{p}\,z}\} - \frac{1}{z - \overline{\mathbf{x}}}\{\frac{\mathrm{v}_1\,\mathrm{v}_2\,V_{34}(z)}{z}\mathbf{y} - (z - \overline{\mathbf{x}})\frac{\mathrm{v}_1\,\mathrm{v}_2\,V_{12}(\mathbf{y})}{\mathbf{y}}(1 - \frac{\mathbf{x}}{z})\} \\
&= -\frac{1}{\frac{z}{\mathrm{p}} - \frac{\overline{\mathbf{x}}}{\mathrm{q}}}\frac{V_{78}(z)}{\mathrm{p}\,z\mathbf{y}} - \frac{\mathrm{v}_1\,\mathrm{v}_2\,V_{34}(z)}{(z - \overline{\mathbf{x}})z}\mathbf{y} + \frac{\mathrm{v}_1\,\mathrm{v}_2\,V_{12}(\mathbf{y})}{\mathbf{y}} - \frac{\mathrm{v}_1\,\mathrm{v}_2\,V_{12}(\mathbf{y})}{\mathbf{y}}\frac{\mathbf{x}}{z} \\
&= -\frac{1}{\frac{z}{\mathrm{p}} - \frac{\overline{\mathbf{x}}}{\mathrm{q}}}\frac{V_{78}(z)}{\mathrm{p}\,z\mathbf{y}} - \frac{\mathrm{v}_1\,\mathrm{v}_2\,V_{34}(z)}{(z - \overline{\mathbf{x}})z}\mathbf{y} + \frac{\mathrm{v}_1\,\mathrm{v}_2\,V_{12}(\mathbf{y})}{\mathbf{y}} - \mathrm{v}_1\,\mathrm{v}_2\,\frac{\mathrm{q}^2\,\mathrm{v}_3\,\mathrm{v}_4}{\overline{\mathbf{x}}}\frac{V_{56}(\frac{\mathbf{y}}{\mathrm{q}})}{z\mathbf{y}},
\end{aligned}
$$

where we used eq.(2.5): $V_{12}(\mathbf{y})\mathbf{x} = \frac{\mathrm{q}\,\mathrm{v}_3\,\mathrm{v}_4}{\overline{\mathbf{x}}}V_{56}(\frac{\mathbf{y}}{\mathrm{q}})$ in the last line. Hence we have

$$
L_5(z) = A(z)\overline{\psi}(z) + \frac{\overline{V_{78}}(z)}{z(z - \overline{\mathbf{x}})}\overline{\psi}(\mathrm{p}\,z) + \frac{\mathrm{v}_1\,\mathrm{v}_2\,V_{34}(\frac{z}{\mathrm{p}})}{\frac{z}{\mathrm{p}}(\frac{z}{\mathrm{p}} - \frac{\overline{\mathbf{x}}}{\mathrm{q}})}\overline{\psi}(\tfrac{z}{\mathrm{p}}) = 0. \tag{3.11}
$$

The equation (3.11) is the three term difference equation for $\overline{\psi}(z)$. The compatibility we want to show is

$$
L_5(z) \propto T(L_1(z)) = L_1(z)\Big|_{\kappa_1 \to \kappa_1/\,\mathrm{p},\kappa_2 \to \mathrm{p}\,\kappa_2,\mathbf{x}\to\overline{\mathbf{x}},\mathbf{y}\to\overline{\mathbf{y}},\psi(z)\to\overline{\psi}(z)}. \tag{3.12}
$$

Recalling (2.9)

$$L_1(z) = A_0(z)\psi(z) + \frac{V_{78}(z)}{\mathrm{p}(z-\mathbf{x})}\psi(\mathrm{p}\,z) + \frac{\mathrm{v}_1\,\mathrm{v}_2\,V_{34}(\frac{z}{\mathrm{p}})}{\frac{z}{\mathrm{p}}-\frac{\mathbf{x}}{\mathrm{q}}}\psi(\frac{z}{\mathrm{p}}),$$

$$(3.13)\quad A_0 = \frac{z\,\mathrm{v}_1\,\mathrm{v}_2\,V_{12}(\mathbf{y})}{\mathrm{p}\,\mathbf{y}} - \frac{\mathrm{q}\,\mathrm{v}_1\,\mathrm{v}_2\,\mathrm{v}_3\,\mathrm{v}_4}{\mathbf{x}}\frac{V_{56}(\mathbf{y})}{\mathbf{y}} + \frac{V_{78}(z)}{\mathrm{p}(\mathbf{x}-z)}\frac{1}{\mathbf{y}} + \frac{\mathrm{v}_1\,\mathrm{v}_2\,V_{34}(\frac{z}{\mathrm{p}})}{\frac{\mathbf{x}}{\mathrm{q}}-\frac{z}{\mathrm{p}}}\mathbf{y},$$

it is easy to see $L_5(z) = \frac{\mathrm{p}}{z}T(L_1(z))$ for the coefficients of $\overline{\psi}(\frac{z}{\mathrm{p}})$ and $\overline{\psi}(\mathrm{p}\,z)$. In order to compare the coefficient of $\overline{\psi}(z)$, we need to rewrite $A(z)$ in (3.10) further.

To do this efficiently, we note that $A(z)$ is a sum of the terms of the form (rational function in $z, \overline{\mathbf{x}}$) $\times$ (rational function in $\mathbf{y}$). In particular this expression is independent of $\mathbf{x}$, and z appears only through the first factor. Hence, by fixing the operator ordering in this way, one can treat $A(z)$ as a rational function in z with commutative parameter $\overline{\mathbf{x}}, \mathbf{y}$. Using this method, we have

$$\begin{aligned}
\lim_{z\to\infty} A(z) &= \frac{-1}{\mathbf{y}} + \frac{(\mathrm{v}_1\,\mathbf{y}-1)(\mathrm{v}_2\,\mathbf{y}-1)}{\mathbf{y}} - \mathrm{v}_1\,\mathrm{v}_2\,\mathbf{y} = -(\mathrm{v}_1+\mathrm{v}_2),\\
\mathrm{Res}_{z=\overline{\mathbf{x}}}\, A(z) &= -\frac{\mathrm{v}_1\,\mathrm{v}_2\,V_{34}(\overline{\mathbf{x}})}{\overline{\mathbf{x}}}\mathbf{y},\\
\mathrm{Res}_{z=\frac{\mathrm{p}}{\mathrm{q}}\overline{\mathbf{x}}}\, A(z) &= -\frac{V_{78}(\frac{\mathrm{p}}{\mathrm{q}}\overline{\mathbf{x}})}{\frac{\mathrm{p}}{\mathrm{q}}\overline{\mathbf{x}}}\frac{1}{\mathbf{y}},\\
\mathrm{Res}_{z=0}\, A(z) &= \frac{\mathrm{q}}{\overline{\mathbf{x}}}\frac{1}{\mathrm{p}\,\mathbf{y}}\frac{\kappa_1^2}{\mathrm{v}_7\,\mathrm{v}_8} + \frac{\mathrm{v}_1\,\mathrm{v}_2}{\overline{\mathbf{x}}}\mathrm{v}_3\,\mathrm{v}_4\,\mathbf{y} - \mathrm{v}_1\,\mathrm{v}_2\frac{\mathrm{q}^2\,\mathrm{v}_3\,\mathrm{v}_4}{\overline{\mathbf{x}}}\frac{V_{56}(\frac{\mathbf{y}}{\mathrm{q}})}{\mathbf{y}}\\
&= \frac{\mathrm{q}}{\overline{\mathbf{x}}}\frac{1}{\mathrm{p}\,\mathbf{y}}\frac{\kappa_1^2}{\mathrm{v}_7\,\mathrm{v}_8} + \frac{\mathrm{v}_1\,\mathrm{v}_2}{\overline{\mathbf{x}}}\mathrm{v}_3\,\mathrm{v}_4\,\mathbf{y} - \mathrm{v}_1\,\mathrm{v}_2\frac{\mathrm{q}^2\,\mathrm{v}_3\,\mathrm{v}_4}{\overline{\mathbf{x}}}\frac{1}{\mathbf{y}}\Big(\frac{\mathbf{y}^2}{\mathrm{q}^2} - \frac{\mathrm{v}_5+\mathrm{v}_6}{\kappa_2}\frac{\mathbf{y}}{\mathrm{q}} + \frac{\mathrm{v}_5\,\mathrm{v}_6}{\kappa_2^2}\Big)\\
&= \frac{\mathrm{q}\,\mathrm{v}_1\,\mathrm{v}_2\,\mathrm{v}_3\,\mathrm{v}_4(\mathrm{v}_5+\mathrm{v}_6)}{\kappa_2\overline{\mathbf{x}}} = \frac{\kappa_1^2(\frac{\kappa_2}{\mathrm{v}_5}+\frac{\kappa_2}{\mathrm{v}_6})}{\mathrm{p}\,\mathrm{v}_7\,\mathrm{v}_8\,\overline{\mathbf{x}}}.
\end{aligned}$$

Then by the partial fraction expansion, we obtain

$$(3.14)\qquad A(z) = \frac{\kappa_1^2(\frac{\kappa_2}{\mathrm{v}_5}+\frac{\kappa_2}{\mathrm{v}_6})}{\mathrm{p}\,\mathrm{v}_7\,\mathrm{v}_8\,z\overline{\mathbf{x}}} - (\mathrm{v}_1+\mathrm{v}_2) - \frac{\mathrm{v}_1\,\mathrm{v}_2\,V_{34}(\overline{\mathbf{x}})}{\overline{\mathbf{x}}(z-\overline{\mathbf{x}})}\mathbf{y} - \frac{V_{78}(\frac{\mathrm{p}}{\mathrm{q}}\overline{\mathbf{x}})}{\frac{\mathrm{p}}{\mathrm{q}}\overline{\mathbf{x}}(z-\frac{\mathrm{p}}{\mathrm{q}}\overline{\mathbf{x}})}\frac{1}{\mathbf{y}}.$$

Of course, this expression holds when $\overline{\mathbf{x}}, \mathbf{y}$ are non-commutative variables as well.

Similarly, for the coefficient $A_0(z)$ in (3.13), we have

$$(3.15)\qquad \frac{\mathrm{p}}{z}A_0(z) = \frac{\kappa_1^2(\frac{\kappa_2}{\mathrm{v}_5}+\frac{\kappa_2}{\mathrm{v}_6})}{\mathrm{v}_7\,\mathrm{v}_8\,z\mathbf{x}} - (\mathrm{v}_1+\mathrm{v}_2) - \frac{V_{78}(\mathbf{x})}{\mathbf{x}(z-\mathbf{x})}\frac{1}{\mathbf{y}} - \frac{\mathrm{p}\,\mathrm{v}_1\,\mathrm{v}_2\,V_{34}(\frac{\mathbf{x}}{\mathrm{q}})}{\frac{\mathbf{x}}{\mathrm{q}}(z-\frac{\mathrm{p}}{\mathrm{q}}\mathbf{x})}\mathbf{y},$$

and its up-shift is given by

$$(3.16)\quad \frac{\mathrm{p}}{z}T(A_0(z)) = \frac{\kappa_1^2(\frac{\kappa_2}{\mathrm{v}_5}+\frac{\kappa_2}{\mathrm{v}_6})}{\mathrm{p}\,\mathrm{v}_7\,\mathrm{v}_8\,z\overline{\mathbf{x}}} - (\mathrm{v}_1+\mathrm{v}_2) - \frac{\overline{V_{78}}(\overline{\mathbf{x}})}{\overline{\mathbf{x}}(z-\overline{\mathbf{x}})}\frac{1}{\overline{\mathbf{y}}} - \frac{\mathrm{p}\,\mathrm{v}_1\,\mathrm{v}_2\,V_{34}(\frac{\overline{\mathbf{x}}}{\mathrm{q}})}{\frac{\overline{\mathbf{x}}}{\mathrm{q}}(z-\frac{\mathrm{p}}{\mathrm{q}}\overline{\mathbf{x}})}\overline{\mathbf{y}}.$$

Thanks to the equation (2.6) for $\overline{\mathbf{y}}$:

$$(3.17)\qquad \mathrm{v}_1\,\mathrm{v}_2\,V_{34}(\overline{\mathbf{x}})\mathbf{y} = \overline{V_{78}}(\overline{\mathbf{x}})\frac{1}{\overline{\mathbf{y}}},\quad V_{78}(\frac{\mathrm{p}}{\mathrm{q}}\overline{\mathbf{x}})\frac{1}{\mathbf{y}} = \mathrm{p}^2\,\mathrm{v}_1\,\mathrm{v}_2\,V_{34}(\frac{\overline{\mathbf{x}}}{\mathrm{q}})\overline{\mathbf{y}},$$

we have $A(z) = \frac{\mathrm{p}}{z}T(A_0(z))$ as desired. This completes the proof of the Theorem. □

4. Degenerate cases

Since the obtained Lax form (2.9) (2.10) for qq-P_{VI} has the same structure with the commutative case, we can study the degenerate cases

$$D_5^{(1)} \to A_4^{(1)} \to (A_2+A_1)^{(1)} \to (A_1+A_1)^{(1)} \begin{array}{l} \to \tilde{A}_1^{(1)} \to A_0^{(1)} \\ \searrow A_1^{(1)} \end{array} \tag{4.1}$$

by the same limiting procedure as in [**10**]. In terms of the Newton polygons for the corresponding quantum curves [**13**] the degeneration is given by the following diagram.

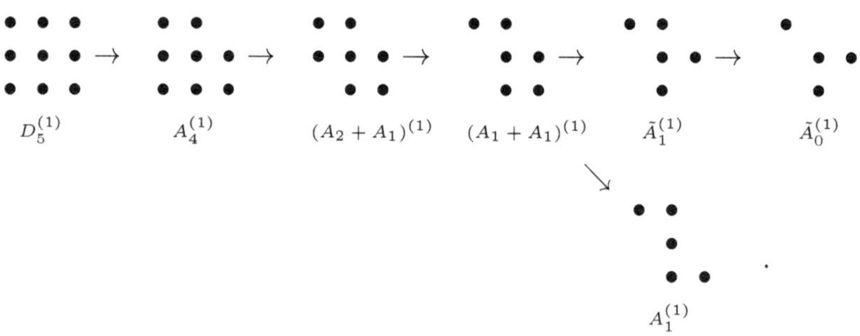

We give L_1 and the compatibility conditions for each degenerate case. The operator L_2 is omitted since it is the same as (2.10) for each case. In what follows $T_z : z \mapsto \mathrm{p}\, z$ denotes the p-shift operator for z.

• $A_4^{(1)}$ case: This is obtained from the $D_5^{(1)}$ case (i.e. qq-P_{VI}) by a limit $\epsilon \to 0$ under the rescaling $\mathrm{v}_2 \to \mathrm{v}_2\,\epsilon, \mathrm{v}_3 \to \mathrm{v}_3\,\epsilon^{-1}$. Then the L_1 operator is given by

$$L_1 = \Big\{-\frac{\mathrm{p\,q\,v_1\,v_2\,v_3}(\mathbf{x}-\mathrm{q\,v_4})}{\mathbf{x}(\mathbf{x}\,\mathrm{p}-z\,\mathrm{q})}\mathbf{y}+\frac{\mathrm{p\,q\,v_1\,v_2\,v_3\,v_4}(\mathrm{v_5}+\mathrm{v_6})}{\mathbf{x}z\kappa_2}-\mathrm{v_1}+\frac{(\mathbf{x}\,\mathrm{v_7}-\kappa_1)(\mathbf{x}\,\mathrm{v_8}-\kappa_1)}{\mathbf{x}(\mathbf{x}-z)\,\mathrm{v_7\,v_8}}\frac{1}{\mathbf{y}}\Big\}$$
$$+\frac{\mathrm{p\,q\,v_1\,v_2\,v_3}(z-\mathrm{p\,v_4})}{z(\mathbf{x}\,\mathrm{p}-z\,\mathrm{q})}T_z^{-1}-\frac{(z\,\mathrm{v_7}-\kappa_1)(z\,\mathrm{v_8}-\kappa_1)}{(\mathbf{x}-z)z\,\mathrm{v_7\,v_8}}T_z.$$

The compatibility of L_1, L_2 gives the following evolution equation

$$\mathbf{x}\overline{\mathbf{x}} = -\,\mathrm{q\,v_2\,v_3\,v_4}\frac{(\mathbf{y}-\frac{\mathrm{v_5}}{\kappa_2})(\mathbf{y}-\frac{\mathrm{v_6}}{\kappa_2})}{\mathbf{y}-\frac{1}{\mathrm{v_1}}}, \quad \mathbf{y}\overline{\mathbf{y}} = -\frac{1}{\mathrm{v_1\,v_2\,v_3}}\frac{(\overline{\mathbf{x}}-\frac{\kappa_1}{\mathrm{p\,v_7}})(\overline{\mathbf{x}}-\frac{\kappa_1}{\mathrm{p\,v_8}})}{\overline{\mathbf{x}}-\mathrm{v_4}}. \tag{4.2}$$

• $(A_2+A_1)^{(1)}$ case: This is obtained from the $A_4^{(1)}$ case by a limit $\epsilon \to 0$ under the rescaling $\mathrm{v}_6 \to \mathrm{v}_6\,\epsilon, \mathrm{v}_7 \to \mathrm{v}_7\,\epsilon^{-1}$. Then we have

$$L_1 = \Big\{-\frac{\mathrm{p\,q\,v_1\,v_2\,v_3}(\mathbf{x}-\mathrm{q\,v_4})}{\mathbf{x}(\mathbf{x}\,\mathrm{p}-z\,\mathrm{q})}\mathbf{y}+\frac{\kappa_2\kappa_1^2}{\mathbf{x}z\,\mathrm{v_6\,v_7\,v_8}}-\mathrm{v_1}+\frac{\mathbf{x}\,\mathrm{v_8}-\kappa_1}{(\mathbf{x}-z)\,\mathrm{v_8}}\frac{1}{\mathbf{y}}\Big\}+\frac{\mathrm{p\,q\,v_1\,v_2\,v_3}(z-\mathrm{p\,v_4})}{z(\mathbf{x}\,\mathrm{p}-z\,\mathrm{q})}\psi T_z^{-1}$$
$$-\frac{(z\,\mathrm{v_8}-\kappa_1)}{(\mathbf{x}-z)\,\mathrm{v_8}}T_z.$$

The compatibility condition is

$$\mathbf{x}\overline{\mathbf{x}} = -\,\mathrm{q\,v_2\,v_3\,v_4}\frac{\mathbf{y}(\mathbf{y}-\frac{\mathrm{v_5}}{\kappa_2})}{\mathbf{y}-\frac{1}{\mathrm{v_1}}}, \quad \mathbf{y}\overline{\mathbf{y}} = -\frac{1}{\mathrm{v_1\,v_2\,v_3}}\frac{\overline{\mathbf{x}}(\overline{\mathbf{x}}-\frac{\kappa_1}{\mathrm{p\,v_8}})}{\overline{\mathbf{x}}-\mathrm{v_4}}. \tag{4.3}$$

• $(A_1+A_1)^{(1)}$ case: This is obtained from the $(A_2+A_1)^{(1)}$ case by a limit $\epsilon\to 0$ under the rescaling $\mathrm{v}_5\to\mathrm{v}_5\,\epsilon, \mathrm{v}_6\to\mathrm{v}_6\,\epsilon^{-1}$. Then we have

$$L_1=\{-\frac{\mathrm{p\,q\,v_1\,v_2\,v_3}(\mathbf{x}-\mathrm{q\,v_4})}{\mathbf{x}(\mathbf{x}\,\mathrm{p}-z\,\mathrm{q})}\mathbf{y}-\mathrm{v}_1+\frac{\mathbf{x}\,\mathrm{v}_8-\kappa_1}{(\mathbf{x}-z)\,\mathrm{v}_8}\frac{1}{\mathbf{y}}\}+\frac{\mathrm{p\,q\,v_1\,v_2\,v_3}(z-\mathrm{p\,v_4})}{z(\mathbf{x}\,\mathrm{p}-z\,\mathrm{q})}T_z^{-1}-\frac{(z\,\mathrm{v}_8-\kappa_1)}{(\mathbf{x}-z)\,\mathrm{v}_8}T_z.$$

The compatibility condition is

$$\mathbf{x}\overline{\mathbf{x}}=-\mathrm{q\,v_2\,v_3\,v_4}\frac{\mathbf{y}^2}{\mathbf{y}-\frac{1}{\mathrm{v}_1}},\quad \mathbf{y}\overline{\mathbf{y}}=-\frac{1}{\mathrm{v_1\,v_2\,v_3}}\frac{\overline{\mathbf{x}}(\overline{\mathbf{x}}-\frac{\kappa_1}{\mathrm{p\,v_8}})}{\overline{\mathbf{x}}-\mathrm{v}_4}. \tag{4.4}$$

• $A_1^{(1)}$ case: This is obtained from the $(A_1+A_1)^{(1)}$ case by a limit $\epsilon\to 0$ under the rescaling $\mathrm{v}_1\to\mathrm{v}_1\,\epsilon, \mathrm{v}_2\to\mathrm{v}_2\,\epsilon^{-1}$. Then we have

$$L_1=\{-\frac{\mathrm{p\,q\,v_1\,v_2\,v_3}(\mathbf{x}-\mathrm{q\,v_4})}{\mathbf{x}(\mathbf{x}\,\mathrm{p}-z\,\mathrm{q})}\mathbf{y}+\frac{\mathbf{x}\,\mathrm{v}_8-\kappa_1}{(\mathbf{x}-z)\,\mathrm{v}_8}\frac{1}{\mathbf{y}}\}\psi(z)+\frac{\mathrm{p\,q\,v_1\,v_2\,v_3}(z-\mathrm{p\,v_4})}{z(\mathbf{x}\,\mathrm{p}-z\,\mathrm{q})}T_z^{-1}-\frac{(z\,\mathrm{v}_8-\kappa_1)}{(\mathbf{x}-z)\,\mathrm{v}_8}T_z.$$

The compatibility condition is

$$\mathbf{x}\overline{\mathbf{x}}=\mathrm{q\,v_1\,v_2\,v_3\,v_4}\,\mathbf{y}^2,\quad \mathbf{y}\overline{\mathbf{y}}=-\frac{1}{\mathrm{v_1\,v_2\,v_3}}\frac{\overline{\mathbf{x}}(\overline{\mathbf{x}}-\frac{\kappa_1}{\mathrm{p\,v_8}})}{\overline{\mathbf{x}}-\mathrm{v}_4}. \tag{4.5}$$

• $\tilde{A}_1^{(1)}$ case: This is obtained from the $(A_1+A_1)^{(1)}$ case by a limit $\epsilon\to 0$ under the rescaling $\mathrm{v}_8\to\mathrm{v}_8\,\epsilon, \mathrm{v}_1\to\mathrm{v}_1\,\epsilon^{-1}$ and $L_1\to\mathrm{v}_8\,L_1$. Then we have

$$L_1=\{-\frac{\mathrm{p\,q\,v_1\,v_2\,v_3\,v_8}(\mathbf{x}-\mathrm{q\,v_4})}{\mathbf{x}(\mathbf{x}\,\mathrm{p}-z\,\mathrm{q})}\mathbf{y}-\mathrm{v_1\,v_8}-\frac{\kappa_1}{(\mathbf{x}-z)}\frac{1}{\mathbf{y}}\}+\frac{\mathrm{p\,q\,v_1\,v_2\,v_3\,v_8}(z-\mathrm{p\,v_4})}{z(\mathbf{x}\,\mathrm{p}-z\,\mathrm{q})}T_z^{-1}+\frac{\kappa_1}{\mathbf{x}-z}T_z.$$

The compatibility condition is

$$\mathbf{x}\overline{\mathbf{x}}=-\mathrm{q\,v_2\,v_3\,v_4}\,\mathbf{y},\quad \mathbf{y}\overline{\mathbf{y}}=\frac{\kappa_1}{\mathrm{p\,v_1\,v_2\,v_3\,v_8}}\frac{\overline{\mathbf{x}}}{\overline{\mathbf{x}}-\mathrm{v}_4}. \tag{4.6}$$

• $A_0^{(1)}$ case: This is obtained from the $\tilde{A}_1^{(1)}$ case by a limit $\epsilon\to 0$ under the rescaling $\mathrm{v}_3\to\mathrm{v}_3\,\epsilon, \mathrm{v}_4\to\mathrm{v}_4\,\epsilon^{-1}$.

$$L_1=\{\frac{\mathrm{q}\,\kappa_1^2\kappa_2^2}{\mathbf{x}(\mathbf{x}\,\mathrm{p}-z\,\mathrm{q})\,\mathrm{v_5\,v_6\,v_7}}\mathbf{y}-\mathrm{v_1\,v_8}-\frac{\kappa_1}{(\mathbf{x}-z)}\frac{1}{\mathbf{y}}\}-\frac{\mathrm{p}\,\kappa_1^2\kappa_2^2}{z(\mathbf{x}\,\mathrm{p}-z\,\mathrm{q})\,\mathrm{v_5\,v_6\,v_7}}T_z^{-1}+\frac{\kappa_1}{\mathbf{x}-z}T_z.$$

The compatibility condition is

$$\mathbf{x}\overline{\mathbf{x}}=-\mathrm{q\,v_2\,v_3\,v_4}\,\mathbf{y},\quad \mathbf{y}\overline{\mathbf{y}}=-\frac{\kappa_1}{\mathrm{p\,v_1\,v_2\,v_3\,v_4\,v_8}}\overline{\mathbf{x}}. \tag{4.7}$$

In summary we obtained the Lax pair for qq-$P_{\rm VI}$ and its degenerations. The generalization to $E_6^{(1)}, E_7^{(1)}, E_8^{(1)}$ and elliptic case is an interesting future problem.

Acknowledgments

This work is supported in part by Grants-in-Aid for Scientific Research (Kakenhi); 22H01116. The author thanks the collaborators of the related works [**2**], [**10**],[**13**] for valuable discussions. He also thank M. Bershtein, A. Dzhamay, P. Gavrylenko, I. Krichever and A. Marshakov, for organizing the nice workshop "Isomonodromic Deformations, Painlevé Equations, and Integrable Systems", Columbia University (online), June 27-July 1, 2022. Last but not least, the author is very grateful to the referee for the careful reading of the paper and for the valuable comments.

Appendix A. qq-$P_{\rm VI}$ equation from the Weyl group

For completeness, we will recall a formulation of qq-$P_{\rm VI}$ based on the affine Weyl group representation. The following is a modification of the result by Hasegawa [**7**] using the geometric notation in [**10**][**13**].

LEMMA A.1. *On the skew field* $\mathcal{K} = \mathbb{C}(\mathsf{p}, \mathsf{q}, e_1, \dots, e_8, h_1, h_2, \mathsf{f}, \mathsf{g})$ *with* $\mathsf{fg} = \mathsf{q}\,\mathsf{gf}$, *define the algebra automorphisms* $s_0, \dots, s_5, \pi$ *by*

$$
\begin{aligned}
&s_0 = \{e_7 \leftrightarrow e_8\},\\
&s_1 = \{e_3 \leftrightarrow e_4\},\\
&s_2 = \{e_3 \to \frac{h_1}{e_7},\ e_7 \to \frac{h_1}{e_3},\ h_2 \to \frac{h_1 h_2}{e_3 e_7},\ \mathsf{g} \to \frac{\mathsf{f} - e_3}{\mathsf{f} - \frac{h_1}{e_7}}\mathsf{g}\},\\
&s_3 = \{e_1 \to \frac{h_2}{e_5},\ e_5 \to \frac{h_2}{e_1},\ h_1 \to \frac{h_1 h_2}{e_1 e_5},\ \mathsf{f} \to \mathsf{f}\frac{\mathsf{g} - \frac{1}{e_1}}{\mathsf{g} - \frac{e_5}{h_2}}\},\\
&s_4 = \{e_1 \leftrightarrow e_2\},\\
&s_5 = \{e_5 \leftrightarrow e_6\},\\
&\pi = \{e_i \to e_{\sigma(i)},\ {\scriptstyle \sigma=(1,5)(2,6)(3,7)(4,8)},\ \mathsf{f} \to \mathsf{q}^{-1} h_1 \mathsf{f}^{-1},\ \mathsf{g} \to \mathsf{q}^{-1} h_2^{-1} \mathsf{g}^{-1}\}.
\end{aligned}
\tag{A.1}
$$

Then the actions $s_0, \dots, s_5$ *generate the affine Weyl group* $W(D_5^{(1)})$ *corresponding to the following Dynkin diagram*

$$
\begin{array}{ccccccc}
 & & s_1 & & s_4 & & \\
 & & | & & | & & \\
s_0 & - & s_2 & - & s_3 & - & s_5\ ,
\end{array}
\tag{A.2}
$$

and π *gives an extension* $\tilde{W}(D_5^{(1)})$ *by the relation* $\pi^2 = 1$ *and* $\pi s_{\{0,1,2,3,4,5\}} = s_{\{1,0,2,3,5,4\}}\pi$.

The parameters e_i, h_i can be identified with the standard basis of Picard lattice (written in multiplicative notation) associated with the eight points blow-up of $\mathbb{P}^1 \times \mathbb{P}^1$. The proof of the Lemma is obtained by direct check of the Weyl group relations. As noted by Hasegawa, all the relations except for $\pi s_{\{0,1,2,3,4,5\}} = s_{\{1,0,2,3,5,4\}}\pi$ hold without assuming the q-commutation relation $\mathsf{fg} = \mathsf{q}\,\mathsf{gf}$.

The following composition

$$
T = \pi s_3 s_4 s_5 s_3 s_2 s_1 s_0 s_2,
\tag{A.3}
$$

is a translation of $\tilde{W}(D_5^{(1)})$, and generates the qq-$P_{\rm VI}$ equation. In fact, one can derive the explicit actions of the translation T on variables f, g as

$$
\begin{aligned}
T(\mathsf{f}) &= \frac{h_1 h_2^2}{\mathsf{q} e_1 e_2 e_5 e_6} \frac{(\mathsf{g} - \frac{e_5}{h_2})(\mathsf{g} - \frac{e_6}{h_2})}{(\mathsf{g} - \frac{1}{e_1})(\mathsf{g} - \frac{1}{e_2})} \mathsf{f}^{-1},\\
T^{-1}(\mathsf{g}) &= \frac{e_3 e_4 e_7 e_8}{\mathsf{q} h_1^2 h_2} \mathsf{g}^{-1} \frac{(\mathsf{f} - \frac{h_1}{e_7})(\mathsf{f} - \frac{h_1}{e_8})}{(\mathsf{f} - e_3)(\mathsf{f} - e_4)},
\end{aligned}
\tag{A.4}
$$

and the actions on the parameters are
(A.5)

$$
\begin{aligned}
&T^{-1}(h_1, h_2) = s^2 (d_1 d_3 h_1, \frac{1}{d_2 d_4} h_2),\\
&T^{-1}(e_1, \dots, e_8) = s(d_1 e_1, d_1 e_2, \frac{1}{d_4} e_3, \frac{1}{d_4} e_4, d_3 e_5, d_3 e_6, \frac{1}{d_2} e_7, \frac{1}{d_2} e_8),\\
&s = \frac{h_1 h_2}{e_3 e_4 e_7 e_8}, \quad d_1 = \frac{h_1}{e_1 e_2}, \quad d_2 = \frac{h_2}{e_3 e_4}, \quad d_3 = \frac{h_1}{e_5 e_6}, \quad d_4 = \frac{h_2}{e_7 e_8}, \quad d_1 d_2 d_3 d_4 = \mathsf{q}\mathsf{p}.
\end{aligned}
$$

Note that the parameters $d_1, \dots, d_4$ corresponds to the four components of anti-canonical divisor of the surface and are almost frozen under $\tilde{W}(D_5^{(1)})$. The equations (A.4) together with (A.5) are equivalent to qq-P_{VI} (2.5)(2.6) by suitable rescaling and gauge transformation depending on the discrete time.

References

[1] L. F. Alday, D. Gaiotto, and Y. Tachikawa, *Liouville correlation functions from four-dimensional gauge theories*, Lett. Math. Phys. **91** (2010), no. 2, 167–197, DOI 10.1007/s11005-010-0369-5. MR2586871

[2] H. Awata, K. Hasegawa, H. Kanno, R. Ohkawa, S. Shakirov, J. Shiraishi, and Y. Yamada, *Non-stationary difference equation and affine Laumon space: quantization of discrete Painlevé equation*, SIGMA Symmetry Integrability Geom. Methods Appl. **19** (2023), Paper No. 089, 47, DOI 10.3842/SIGMA.2023.089. MR4665306

[3] H. Awata and Y. Yamada, *Five-dimensional AGT conjecture and the deformed Virasoro algebra*, J. High Energy Phys. **1** (2010), 125, 11, DOI 10.1007/JHEP01(2010)125. MR2660773

[4] M. Bershtein, P. Gavrylenko, and A. Marshakov, *Cluster integrable systems, q-Painlevé equations and their quantization*, J. High Energy Phys. **2** (2018), 077, front matter+33, DOI 10.1007/jhep02(2018)077. MR3789593

[5] A. Dzhamay and T. Takenawa, *On some applications of Sakai's geometric theory of discrete Painlevé equations*, SIGMA Symmetry Integrability Geom. Methods Appl. **14** (2018), Paper No. 075, 20, DOI 10.3842/SIGMA.2018.075. MR3830210

[6] R. Fuchs, "Sur quelques équations différentielles linéaires du second ordre", C. R. Acad. Sci. (Paris) **141** (1905) 555-558.

[7] K. Hasegawa, *Quantizing the Bäcklund transformations of Painlevé equations and the quantum discrete Painlevé VI equation*, Exploring new structures and natural constructions in mathematical physics, Adv. Stud. Pure Math., vol. 61, Math. Soc. Japan, Tokyo, 2011, pp. 275–288, DOI 10.2969/aspm/06110275. MR2867149

[8] K. Hasegawa, *Quantizing the discrete Painlevé VI equation: the Lax formalism*, Lett. Math. Phys. **103** (2013), no. 8, 865–879, DOI 10.1007/s11005-013-0620-y. MR3063949

[9] M. Jimbo and H. Sakai, *A q-analog of the sixth Painlevé equation*, Lett. Math. Phys. **38** (1996), no. 2, 145–154, DOI 10.1007/BF00398316. MR1403067

[10] K. Kajiwara, M. Noumi, and Y. Yamada, *Geometric aspects of Painlevé equations*, J. Phys. A **50** (2017), no. 7, 073001, 164, DOI 10.1088/1751-8121/50/7/073001. MR3609039

[11] I. M. Krichever, *Integration of nonlinear equations by the methods of algebraic geometry* (Russian), Funkcional. Anal. i Priložen. **11** (1977), no. 1, 15–31, 96. MR494262

[12] G. Kuroki, *Quantum groups and quantization of Weyl group symmetries of Painlevé systems*, Exploring new structures and natural constructions in mathematical physics, Adv. Stud. Pure Math., vol. 61, Math. Soc. Japan, Tokyo, 2011, pp. 289–325, DOI 10.2969/aspm/06110289. MR2867150

[13] S. Moriyama and Y. Yamada, *Quantum representation of affine Weyl groups and associated quantum curves*, SIGMA Symmetry Integrability Geom. Methods Appl. **17** (2021), Paper No. 076, 24, DOI 10.3842/SIGMA.2021.076. MR4298802

[14] H. Sakai, *Rational surfaces associated with affine root systems and geometry of the Painlevé equations*, Comm. Math. Phys. **220** (2001), no. 1, 165–229, DOI 10.1007/s002200100446. MR1882403

Department of Mathematics, Kobe University
Current address: Rokko, Kobe, 657-8501, JAPAN
Email address: yamaday@math.kobe-u.ac.jp

Contemporary Mathematics
Volume **823**, 2025
https://doi.org/10.1090/conm/823/16492

Elliptic Calogero-Moser system, crossed and folded instantons, and bilinear identities

Andrei Grekov and Nikita Nekrasov

ABSTRACT. Affine analogues of the $Q, \tilde{Q}$-functions are constructed using *folded instantons* partition functions. They are shown to be the solutions of the quantum spectral curve of the N-body elliptic Calogero-Moser (eCM) system, the quantum Krichever curve. $Q, \tilde{Q}$ also solve the elliptic analogue of the quantum Wronskian equation. In the companion paper we present the quantum analogue of Krichever's Lax operator for eCM. A connection to crossed instantons on Taub-Nut spaces, and opers on a punctured torus is pointed out.

1. Elliptic Calogero-Moser system

Let

$$\tau = \tau_1 + \mathrm{i}\tau_2 \in \mathbb{C},\ \tau_2 > 0,\ \mathfrak{q} = e^{2\pi \mathrm{i}\tau}\ . \tag{1.1}$$

Let

$$E_\tau = \mathbb{C}/\mathbb{Z} + \tau\mathbb{Z} \approx \mathbb{C}^\times/\mathfrak{q}^{\mathbb{Z}} \tag{1.2}$$

be the corresponding elliptic curve.

Fix an integer $N \geq 1$, and $n, \hbar \in \mathbb{C}$. Quantum elliptic Calogero-Moser system (eCM) can be defined in many ways. At simplified level it is a collection of commuting differential operators $\hat{H}_1, \hat{H}_2, \ldots, \hat{H}_N$ in N variables $z_1, \ldots, z_N$,

$$\begin{aligned} &\hat{H}_1 = \sum_{i=1}^{N} \frac{\partial}{\partial z_i}, \\ &\hat{H}_2 = -\frac{1}{2}\sum_{i=1}^{N} \frac{\partial^2}{\partial z_i^2} + n(n-1)\sum_{i<j} \wp(z_i - z_j|\tau) \\ &\ldots \end{aligned} \tag{1.3}$$

invariant under the action of the double affine Weyl group

$$(z_i)_{i=1}^N \mapsto \left(z_{\sigma(i)} + a_i + \tau b_i\right)_{i=1}^N,\ \sigma \in S(N),\ a_k, b_k \in \mathbb{Z} \tag{1.4}$$

The elliptic Calogero-Moser system admits several interesting degenerations. One such limit, $\tau \to \mathrm{i}\infty$, produces the trigonometric/hyperbolic CM system. Another limit is more subtle: let $z_k = y_k + \tau k/N$, $k = 1, \ldots, N$, $\Lambda^{2N} = \mathfrak{q}\left(-(\hbar n)^2\right)^N$. Send

$\tau \to \mathrm{i}\infty, n \to \infty$ while keeping $\mathbf{y} = (y_k)_{k=1}^N$ and $\Lambda, \hbar$ finite. The result is the N-body periodic Toda chain [**22**]:

$$\begin{aligned} \hat{H}_1 &= \sum_{i=1}^N \frac{\partial}{\partial y_i}, \\ \hat{H}_2 &= -\frac{1}{2}\sum_{i=1}^N \frac{\partial^2}{\partial y_i^2} + (\Lambda/\hbar)^2 \sum_{i=1}^N e^{y_i - y_{i+1}}, \\ &\dots \\ y_{N+1} &\equiv y_1 \end{aligned} \tag{1.5}$$

1.1. Spectral curve of eCM. In the classical limit $\hbar \to 0$, $n \to \infty$, $\hbar n = \nu$ fixed, the operators (1.3) go over to the Poisson-commuting functions

$$H_1(p,z), \dots, H_N(p,z) \text{ on } \mathcal{P}_{\mathbb{C}} = \left(T^* E_\tau\right)^N / S(N),$$

or on its smooth resolution

$$\tilde{\mathcal{P}}_{\mathbb{C}} = Hilb^{[N]}(S), \tag{1.6}$$

the Hilbert scheme of N points on the complex symplectic surface S, which is an affine bundle over the cotangent bundle to E_τ (which can be analytically identified with the algebraic torus $\mathbb{C}^\times \times \mathbb{C}^\times$, or, non-canonically, mapped to T^*E_τ, isomorphically outside a fiber over one point in E_τ). In [**28**] the functions H_k are constructed with the help of the Lax operator

$$\mathbf{L}(u) = \|L_{ij}(u)\|_{i,j=1}^N, \ L_{ij}(u) = p_i \delta_{ij} + \nu(1-\delta_{ij})\frac{\theta(z_i - z_j + u|\tau)\theta'(0|\tau)}{\theta(z_i - z_j|\tau)\theta(u|\tau)} \tag{1.7}$$

where

$$\theta(u|\tau) = \sum_{l \in \mathbb{Z}} (-1)^l e^{2\pi \mathrm{i} l u} \mathfrak{q}^{\frac{l(l-1)}{2}} \tag{1.8}$$

obeys

$$\theta(u+1|\tau) = \theta(u|\tau) = -e^{2\pi \mathrm{i} u}\theta(u+\tau|\tau) \tag{1.9}$$

One defines the *spectral curve* $\mathcal{C} \subset T^*\left(E_\tau \backslash \{u=0\}\right)$ via:

$$R(\lambda, u) \equiv \mathrm{Det}\left(\lambda \cdot \mathrm{Id}_N - \mathbf{L}(u)\right) = 0 \tag{1.10}$$

As a function of λ, R is a degree N polynomial, with coefficients meromorphic double-periodic functions of u, with poles only at $u = 0$.

An innocent looking change of variables (used in [**28**])

$$\lambda = x - \nu \frac{\theta'(u|\tau)}{\theta(u|\tau)} \tag{1.11}$$

makes (1.10) into

$$\mathcal{R}(x,u) = \mathrm{Det}\left(x \cdot \mathrm{Id}_N - \tilde{\mathbf{L}}(u)\right) = 0 \tag{1.12}$$

with $\tilde{L}(u)$ have a first order pole at $u \in \mathbb{Z} + \tau\mathbb{Z}$ with rank one residue. Therefore, the function $\mathcal{R}(x,u)$ has, at fixed x only a first order pole at $u = 0$. At fixed u it is still a degree N polynomial in x. However, it is not a double-periodic function of u at fixed x, rather:

$$\mathcal{R}(x - 2\pi\mathrm{i}\nu, u+\tau) = \mathcal{R}(x, u+1) = \mathcal{R}(x,u), \tag{1.13}$$

Thus $\mathcal{R} = 0$ is an equation for a curve $\mathcal{C}$ in the total space S_ν of affine bundle over T^*E_τ, again with the fiber over $u = 0$ removed. Note that S_ν has a rich set of holomorphic functions, e.g. any Laurent polynomial in

$$U_1 = \exp\left(\frac{x}{\nu}\right),\ U_2 = \exp\left(2\pi\mathrm{i}u + \tau\frac{x}{\nu}\right) \tag{1.14}$$

Thus, for $\nu \neq 0$, outside the fiber over $u = 0$ the surface S_ν is isomorphic to the algebraic torus $\mathbb{C}^\times \times \mathbb{C}^\times$. The curve $\mathcal{C}$, however, is not algebraic, only analytic, in the (U_1, U_2) coordinate system.

2. Quantum Krichever curve

In a very naive way the equation $\mathcal{R}(x, u) = 0$ can be quantized, for example, by making x, u the generators $\hat{x}, \hat{u}$ of a non-commutative algebra, obeying

$$[\hat{u}, \hat{x}] = \hbar \cdot 1 \tag{2.1}$$

where $\hbar \in \mathbb{C}$ is a parameter. It is not obvious how to order $\hat{x}$ and $\hat{u}$ in the complicated function such as $\mathcal{R}(x, u)$, or, say, $R(x, u)$. However, $\mathcal{R}(x, u)$ is much better suited for quantization than $R(x, u)$ for the following reason. As we pointed out earlier, the operator $\tilde{\mathbf{L}}(u)$ has a first order pole at $u = 0$ (modulo the lattice $\mathbb{Z} + \mathbb{Z}\tau$) with the rank one residue. Thus,

$$\mathcal{R}(x, u) = \sum_{i=0}^{N} x^{N-i} f_i(u) \tag{2.2}$$

with the coefficients $f_i(u)$ having only a first order pole at $u \in \mathbb{Z}{+}\mathbb{Z}\tau$. Of course, the shifted periodicity (1.13) means that $f_i(u)$'s are not double-periodic meromorphic functions of u (with the exception of $f_0(u) \equiv 1$). Now, let us multiply $\mathcal{R}(x, u)$ by $\theta(u|\tau)$, and define the Fourier coefficients $\tilde{\mathfrak{R}}_l(x)$ by:

$$\mathfrak{R}(x, u) = \theta(u|\tau)\mathcal{R}(x, u) = \sum_{l\in\mathbb{Z}} (-1)^l e^{2\pi\mathrm{i}lu} \mathfrak{q}^{\frac{l(l-1)}{2}} \tilde{\mathfrak{R}}_l(x) \tag{2.3}$$

It is clear from the definition and (1.13) that $\mathfrak{R}(x, u)$ is an entire function of x, u, degree N polynomial in x at fixed u, obeying

$$\mathfrak{R}(x, u+1) = -e^{2\pi\mathrm{i}u}\,\mathfrak{R}(x - 2\pi\mathrm{i}\nu, u + \tau) = \mathfrak{R}(x, u)\ . \tag{2.4}$$

Thus $\tilde{\mathfrak{R}}_l(x)$ are all degree N (monic) polynomials in x, obeying

$$\tilde{\mathfrak{R}}_l(x - 2\pi\mathrm{i}\nu) = \tilde{\mathfrak{R}}_{l+1}(x) \tag{2.5}$$

In other words

$$\tilde{\mathfrak{R}}_l(x) = \mathcal{Y}(\mathsf{w} - \nu l) \tag{2.6}$$

where $x = 2\pi\mathrm{i}\mathsf{w}$, and $\mathcal{Y} = \tilde{\mathfrak{R}}_0$ is a degree N monic polynomial. Now we quantize (2.3) by replacing u in the exponent by

$$\hat{u} = \hbar\frac{\partial}{\partial x} = \frac{\hbar}{2\pi\mathrm{i}}\frac{\partial}{\partial \mathsf{w}} \tag{2.7}$$

We can place $e^{2\pi\mathrm{i}l\hat{u}}$ to the right of $\tilde{\mathfrak{R}}_l(x)$, or we can place it to the left, amounting to a renormalization $\nu \to \nu \pm \hbar$. We can also imagine a dual quantization, where $e^{2\pi\mathrm{i}l\hat{u}}$ is to the right of $\tilde{\mathfrak{R}}_l(x)$ but acts on the left. We shall see both versions realized below.

The quantum version of the spectral curve can, therefore, be represented as an infinite order difference operator $\hat{\mathfrak{R}}$ with polynomial coefficients, acting on functions of one variable w, or its dual. The corresponding solutions $\mathbf{\Psi}$, $\mathbf{\Psi}^{\vee}$ are the functions of w, obeying:

$$\text{(2.8)} \quad \begin{aligned} \hat{\mathfrak{R}}\mathbf{\Psi}(\mathsf{w}) &= \sum_{l\in\mathbb{Z}}(-1)^l \mathfrak{q}^{\frac{l(l-1)}{2}}\mathcal{Y}(\mathsf{w}-\nu l)\mathbf{\Psi}(\mathsf{w}+\hbar l) = 0 \\ \hat{\mathfrak{R}}^*\mathbf{\Psi}^{\vee}(\mathsf{w}) &= \sum_{l\in\mathbb{Z}}(-1)^l \mathfrak{q}^{\frac{l(l-1)}{2}}\mathcal{Y}(\mathsf{w}-\nu l-\hbar l)\mathbf{\Psi}^{\vee}(\mathsf{w}-\hbar l) = 0 \end{aligned}$$

The quantum version of the spectral curve can, equivalently, be represented in u-variable, with

$$\widehat{\mathsf{w}} = \frac{\hbar}{2\pi\mathrm{i}}\frac{\partial}{\partial u} \tag{2.9}$$

as the N'th order meromorphic differential operator on E_τ, with a regular singularity at $u = 0$, a PGL_N-oper. The corresponding Fourier transforms

$$\text{(2.10)} \quad \begin{aligned} \chi_\alpha(u) &= \sum_{\mathsf{w}\in\alpha+\hbar\mathbb{Z}} e^{\frac{2\pi\mathrm{i}\mathsf{w}}{\hbar}u}\,\mathbf{\Psi}(\mathsf{w}) \\ \chi_\alpha^{\vee}(u) &= \sum_{\mathsf{w}\in\alpha+\hbar\mathbb{Z}} e^{-\frac{2\pi\mathrm{i}\mathsf{w}}{\hbar}u}\,\mathbf{\Psi}^{\vee}(\mathsf{w}) \end{aligned}$$

where α is chosen so that the series on the right hand side of (2.10) converge are the local solutions of the oper. They should play a rôle in the quantum separation of variables for eCM and its spectral dual $Y(\widehat{\mathfrak{gl}}(1))$-based spin chain, analogously to the genus zero ($\mathfrak{sl}_{r+1}$ spin chain) case studied in [**23**]. Also, the construction of [**12**] seems to generalize to genus one case, giving rise to the L^2, or $z, \bar{z}$-version of the elliptic Calogero-Moser system, where the wave-function of the N-body system is an L^2-function on $E_\tau^N/S(N)$, which is a common eigenvector of the Hamiltonians of the conventional, meromorphic, eCM and its complex conjugate. The parameters α and the coefficients of the polynomial $\mathcal{Y}$ would have to be fixed by a requirement of the $PGL_N(\mathbb{R})$-holonomy of the oper. The $\mathcal{N} = 2^*$ counterpart of this construction seems to be placing the theory on extremely elongated four dimensional ellipsoid with a surface defect filling up a two dimensional equatorial section thereof.

We shall return to this picture elsewhere.

3. $T - Q$-equation

Functional Bethe ansatz [**41**] (see [**11, 14, 42, 44**] for the reviews of the foundational work) approach to $\mathfrak{sl}_2$-spin chains or quantum periodic N-body Toda chain uses the following $T - Q$-equation [**3, 40**]:

$$P_+(\mathsf{w}-2\hbar)\mathbf{Q}(\mathsf{w}+\hbar) + \mathfrak{q}P_-(\mathsf{w}+2\hbar)\mathbf{Q}(\mathsf{w}-\hbar) = (1+\mathfrak{q})\mathbf{T}(\mathsf{w})\mathbf{Q}(\mathsf{w}) \tag{3.1}$$

with degree N monic polynomial $\mathbf{T}(\mathsf{w})$ and entire function $\mathbf{Q}(\mathsf{w})$. The Eq. (3.1) describes the eigenvalues $\mathbf{T}(\mathsf{w})$ of the transfer matrix $\hat{T}(\mathsf{w})$. There is a *spectral dual* Lax operator $\mathbf{L}(z)$ of a particular case of $\mathfrak{sl}_N$ Gaudin-Garnier model, whose eigenvalues belong to the w line. The equation (3.1) can be derived from the

Yangian version of the q-characters of $\mathfrak{sl}_2$ [**13**], from the $\varepsilon_2 \to 0$ limit of the qq-characters of the A_1-type $\mathcal{N}=2$ superconformal gauge theory ($= N_f = 2N_c$ super-QCD) [**32**], it can also be viewed as a quantum version of the Seiberg-Witten curve of the A_1 type gauge theory (for the topological recursion approach to quantum curves of that type see [**7**], and for the extension to the local CY geometries [**9**]). In this paper we define (and solve) an $\widehat{\mathfrak{gl}(1)}$-analogue of (3.1).

4. Partitions, characters, and contents

In what follows Λ denotes the set of all partitions, i.e. non-increasing sequences

$$\Lambda = \left\{ \lambda \,|\, \lambda = \left(\lambda_1 \geq \lambda_2 \ldots \geq \lambda_{\ell(\lambda)} > 0 = \lambda_{\ell(\lambda)+1} \right) \right\} \tag{4.1}$$

The size and length maps

$$|\cdot|, \ell : \Lambda \longrightarrow \mathbb{Z}_{\geq 0}, \qquad |\lambda| \equiv \lambda_1 + \ldots + \lambda_{\ell(\lambda)}, \tag{4.2}$$

and the "transposition" map: $\lambda \mapsto \lambda^t$, $\ell(\lambda^t) = \lambda_1$, $|\lambda^t| = |\lambda|$, best described by viewing partitions as finite subsets $\lambda \subset \mathbb{Z}_{>0} \times \mathbb{Z}_{>0}$:

$$\lambda = \left\{ \square = (i,j) \,|\, i,j \geq 1 \;\; 1 \leq j \leq \lambda_i, \; 1 \leq i \leq \lambda_j^t \right\} \tag{4.3}$$

Define the *contents* of $\square$:

(4.4)

$$\xi_\square = \hbar\left(i - j + n(j-1)\right), \; \upsilon_\square = \hbar\left(n(j-i) + 1 - j\right), \; \zeta_\square = \hbar\left(i - 1 - n(j-1)\right)$$

Define the *boundary sets* $\Gamma_\lambda^+, \Gamma_\lambda^-$ by:

$$\sum_{(i,j)\in\Gamma_\lambda^+} \langle i-1, j-1 \rangle - \sum_{(i,j)\in\Gamma_\lambda^-} \langle i,j \rangle = \sum_{i=1}^{\ell(\lambda)+1} \langle i-1, \lambda_i \rangle - \sum_{i=1}^{\ell(\lambda)} \langle i, \lambda_i \rangle \tag{4.5}$$

where $\langle i,j \rangle$ is any linear function $\langle i,j \rangle = (i-1) + f(j-1)$, for $f \in \mathbb{C}\backslash\mathbb{Q}$.

Finally, define the character

$$\chi_{ab}(\lambda) = \sum_{\square\in\lambda} q_a^{i-1} q_b^{j-1} \tag{4.6}$$

for $a, b \in \{1,2,3,4\}$

5. Q-functions, $\mathsf{X},\mathsf{Y},\mathsf{Z}$-observables, and $\mathcal{X},\mathcal{Y},\mathcal{Z}$-characters

Supersymmetric gauge theories produce [**32**] (see Appendix **A** for some details) entire functions $Q(\mathsf{w})$, which, in the limit $\varepsilon_2 \to 0, \varepsilon_1 = \hbar$ of [**31**] solve *Bethe equations*

$$\frac{Q(\mathsf{w}+\hbar)Q(\mathsf{w}-\hbar n)Q(\mathsf{w}+\hbar(n-1))}{Q(\mathsf{w}-\hbar)Q(\mathsf{w}+\hbar n)Q(\mathsf{w}+\hbar(1-n))} = -\mathfrak{q}, \; \forall \mathsf{w} \in Q^{-1}(0) \; . \tag{5.1}$$

Define

$$\mathsf{X}(\mathsf{w}) = \frac{Q(\mathsf{w})}{Q(\mathsf{w}+\hbar n)}, \; \mathsf{Y}(\mathsf{w}) = \frac{Q(\mathsf{w})}{Q(\mathsf{w}-\hbar)}, \; \mathsf{Z}(\mathsf{w}) = \frac{Q(\mathsf{w})}{Q(\mathsf{w}+\hbar(1-n))} \, . \tag{5.2}$$

Then $U(N)$ $\mathcal{N}=2^*$ theory with adjoint hypermultiplet of mass $\hbar n$ on Ω-deformed $\mathbf{R}^2_\hbar \times \mathbb{R}^{1,1}$ is characterized by the property that

$$\mathsf{Y}(\mathsf{w}) \sim \mathsf{w}^N + \ldots \text{ at } \mathsf{w} \to \infty \tag{5.3}$$

(it is called the Y-observable in [**32**]). Define the following triplet of observables built out of $\mathsf{X}, \mathsf{Y}, \mathsf{Z}$, respectively:

$$\begin{aligned}
\mathcal{X}(\mathsf{w}) &= \sum_\lambda \mathfrak{q}^{|\lambda|} \frac{\prod_{\square \in \Gamma^+_\lambda} \mathsf{X}(\mathsf{w} - \hbar n + \xi_\square)}{\prod_{\square \in \Gamma^-_\lambda} \mathsf{X}(\mathsf{w} + \xi_\square)}, \\
\mathcal{Y}(\mathsf{w}) &= \sum_\lambda \mathfrak{q}^{|\lambda|} \frac{\prod_{\square \in \Gamma^+_\lambda} \mathsf{Y}(\mathsf{w} + \hbar + \upsilon_\square)}{\prod_{\square \in \Gamma^-_\lambda} \mathsf{Y}(\mathsf{w} + \upsilon_\square)}, \\
\mathcal{Z}(\mathsf{w}) &= \sum_\lambda \mathfrak{q}^{|\lambda|} \frac{\prod_{\square \in \Gamma^+_\lambda} \mathsf{Z}(\mathsf{w} + \hbar(n-1) + \zeta_\square)}{\prod_{\square \in \Gamma^-_\lambda} \mathsf{Z}(\mathsf{w} + \zeta_\square)}
\end{aligned} \tag{5.4}$$

Then $\mathcal{Y}(\mathsf{w})$ is a degree N polynomial in w, as (5.1) guarantees the absence of poles, and the asymptotics (5.3) fixes the rest (it is a particular, $\hat{A}_0$, case of the main theorem of [**32**]). In fact, $\mathcal{Y}(\mathsf{w})$ is the $\varepsilon_2 \to 0$ limit of the the fundamental qq-character of $\hat{A}_0$-theory, introduced in [**32**]. It can be related to the q-character of [**13**] for the $\widehat{\mathfrak{gl}}(1)$-algebra. Our main *new* result is the set of three

6. Bilinear equations

$$\begin{aligned}
\mathcal{Y} \star \mathcal{X}(\mathsf{w}) &\equiv \sum_{l \in \mathbb{Z}} (-1)^l \mathfrak{q}^{\frac{l(l+1)}{2}} \mathcal{Y}(\mathsf{w} - \hbar n(l+1)) \mathcal{X}(\mathsf{w} - \hbar l) = 0 \\
\mathcal{Y} \star \mathcal{Z}(\mathsf{w}) &\equiv \sum_{l \in \mathbb{Z}} (-1)^l \mathfrak{q}^{\frac{l(l+1)}{2}} \mathcal{Y}(\mathsf{w} + \hbar(n-1)(l+1)) \mathcal{Z}(\mathsf{w} - \hbar l) = 0 \\
\mathcal{X} \star \mathcal{Z}(\mathsf{w}) &\equiv \sum_{l \in \mathbb{Z}} (-1)^l \mathfrak{q}^{\frac{l(l+1)}{2}} \mathcal{X}(\mathsf{w} + \hbar(n-1)(l+1)) \mathcal{Z}(\mathsf{w} + \hbar n l) = 0
\end{aligned} \tag{6.1}$$

A quick comparison with (2.8) reveals $\mathcal{X}$ and $\mathcal{Z}$ are the dual solutions of the quantum spectral curve (1.12). Having these solutions is a step towards the quantum separation of variables [**43**] for the elliptic CM system.

7. Discussion: gauge theory and integrable system

Gauge theory approach to elliptic Calogero-Moser system was started in [**18**]. It was shown there that the phase space and the Hamiltonians of the many-body system can be obtained by Hamiltonian reduction from an infinite-dimensional phase space, which is the space of two-dimensional gauge fields and adjoint-valued one-forms (mathematically called Higgs fields, although they were introduced and studied by N. Hitchin in [**21**], the construction of [**18**] being a degenerate version [**30**]). In this realization the classical and quantum evolution of the many-body system is identified with that of some $2+1$ dimensional gauge theory. The proposal [**10**] identified the family of spectral curves found in [**18**] (which thanks to [**28**] are the spectral curves of elliptic Calogero-Moser system) with the family of Seiberg-Witten curves of $\mathcal{N}=2^*$ theory with gauge group $SU(N)$ (an earlier paper [**17**] pointed out a connection between the Seiberg-Witten proposal for the geometry of vacua of $\mathcal{N}=2$ theories and integrable systems). The analysis [**29**] allows to extract the low-instanton number non-perturbative corrections to the low-energy effective action from the careful expansion of the periods of a degenerating family of curves. The bottom up approach [**31**], starting with supersymmetric gauge theory, computing the effects of the non-perturbative dynamics on the geometry of its moduli space of vacua, produced a two-parameter generalization of the Seiberg-Witten geometry. It was pointed out that these generalizations might lead to the quantization of the integrable system, as well as other deformations. In [**33, 34**] important steps were made in showing that indeed, for $\mathcal{N} = 2^*$ theory (among others) the gauge theory computation produces the spectral curve of the elliptic Calogero-Moser system.

Our present achievement is to find, within gauge theory, the observables whose expectation values play the role of the *single-particle wave-functions* for the quantum version of the spectral curve. We build them using the technique of *folded instantons*, introduced in [**32**]. Again, we stress that our results are derived, not conjectured. In this way we made independent checks of many string duality based claims e.g. [**1, 2, 6, 15, 24**]. The proper embedding of folded instanton observables in gauge theory remains an interesting open problem (see [**35**] for string theory considerations).

Likewise, the meaning of the sums over l in (6.1) begs for a gauge theory explanation. Our conjecture is that just like the $\mathcal{Y}$-observable can be interpreted as a partition function of an auxiliary four dimensional theory living on a copy of $\mathbf{R}^4$ transverse to the physical spacetime $\mathbf{R}^4$ in some ambient eight-dimensional special holonomy geometry, the sum in $\mathfrak{R}(x, u)$ is the partition function of a similar $U(1)$ theory placed in the Taub-Nut background. The sum over $l \in \mathbb{Z}$ is the sum over the fluxes of the abelian gauge field, which is supported by the L^2-normalizable self-dual harmonic two-form. One needs to find a Taub-Nut $\times\mathbb{R}^4$ analogue of the moduli space of crossed instantons. The ADHM construction for instantons on Taub-Nut spaces has been beautifully generalized in [**8**].

Of course, the relation between (2.8) and the quantum spectral curve of elliptic Calogero-Moser system is not very direct. We simply match the families of curves, and their non-commutative versions. However we can claim the match of the coordinate systems, the coordinates $z_1, \ldots, z_N$ of the particles having gauge-theoretic significance. We know from [**32**] the regular surface defect of $\mathcal{N} = 2^*$ theory obeys eCM Schrödinger equation, with $z_i - z_{i+1}$ being the complex Kähler moduli of the surface defect. As we said in the beginning, we are discussing a simpler version of eCM, defined in complex-analytic terms, so there are separate "in" and "out" wave-functions, related by $\hbar \mapsto -\hbar$ transformations. A short form of (6.1) reads

$$(7.1)\quad 0 = \left(\sum_{l\in\mathbb{Z}} (-1)^l \mathfrak{q}^{\frac{l(l+1)}{2}} \mathcal{Y}(\mathsf{w} - \hbar n(l+1)) e^{-\hbar l \partial_{\mathsf{w}}}\right) \Psi,$$

$$0 = \left(\sum_{l\in\mathbb{Z}} (-1)^l \mathfrak{q}^{\frac{l(l+1)}{2}} e^{\hbar l \partial_{\mathsf{w}}} \mathcal{Y}(\mathsf{w} - \hbar n(l+1))\right) \Psi^\vee,$$

where $\Psi(\mathsf{w}) = \mathcal{X}(\mathsf{w})$, $\Psi^\vee(\mathsf{w}) = \mathcal{Z}(\mathsf{w} + \hbar(1 - n))$. The quantum separation of variables which would express the regular surface defect as an integral transform of a product of the single-particle wave-functions, which should be the $\Psi \sim \mathcal{X}$ or $\Psi^\vee \sim \mathcal{Z}$-observables.

The solutions $\mathcal{X}$ and $\mathcal{Z}$ are the $\hat{A}_0$ analogues of the two independent (over the quasiconstants, i.e. $\hbar$-periodic functions) solutions of the $T-Q$ equations. The third bilinear equation $\mathcal{X} \star \mathcal{Z} = 0$ is the analogue of the very important *quantum Wronskian relation* (see the Appendix **B**).

In the forthcoming publication [**20**] we shall establish another detail in the gauge theory-integrable system package dictionary: the Lax operators as gauge theory observables. Our main equations (6.1) will get a matrix-vector dress explaining the origin of so many beautiful *zero-curvature representations* in the theory of integrable systems. The manipulation (2.3) leading from $\mathcal{R}$ to $\mathfrak{R}$ will also get a matrix version, generalizing that of [**19, 45**]. The story of the $\mathcal{N} = 2^*$ ($\hat{A}_0$) theory we discussed generalizes to the $\hat{A}_r$-type quiver theories, with (6.1) mapping, upon a Fourier transform, to PGL_N-opers on elliptic curve E_τ with $r+1$ regular singularities with minimal monodromies around the poles (another limit, $\varepsilon_1 + \varepsilon_2 \to 0$ of this theory was studied recently in [**5**]).

The quantum separation of variables for eCM is still around the corner...

Our equations should also play a rôle in the solution of the quantum intermediate long-wave equation, a $N \to \infty$ counterpart of eCM system, related to the six dimensional gauge theories and sigma models on instanton moduli spaces [**38**], as well as to the integrable structure of two dimensional conformal field theory [**4**], and quantum hydrodynamics [**27, 37**].

8. Appendix A. Derivation of the main bilinear equations

In this section we use the notations of [**32**]. Let $\varepsilon_1, \varepsilon_2, \varepsilon_3, \varepsilon_4 \in \mathbb{C}$ sum to zero,

$$\varepsilon_1 + \varepsilon_2 + \varepsilon_3 + \varepsilon_4 = 0 \tag{8.1}$$

and let $q_a = e^{\varepsilon_a}$, $P_a = 1 - q_a$, $P_{ab} = P_a P_b$, $a, b, = 1, \dots, 4$. The $\mathcal{Y} \star \mathcal{X}$ of (6.1) is the $\varepsilon_2 \to 0$ limit of the sum over l of

$$(-1)^l \mathfrak{q}^{\frac{l(l+1)}{2}} \times Z_{\varepsilon_1, \varepsilon_2 = 0, \varepsilon_3, \varepsilon_4}(N_{ab}), \tag{8.2}$$

the partition function of the gauge origami theory with

$$\varepsilon_1 = \hbar,\ \varepsilon_3 = -\hbar n,\ \varepsilon_4 = \hbar(n-1) \tag{8.3}$$

and the Chan-Paton characters

$$N_{12} = \sum_{\alpha=1}^{N} e^{a_\alpha},\ N_{34} = e^{\mathsf{w} + \varepsilon_3(l+1)},\ N_{14} = e^{\mathsf{w} - \varepsilon_1 l}, \tag{8.4}$$

the $\mathcal{Y} \star \mathcal{Z}$ corresponds to the origami with Chan-Paton characters

$$N_{12} = \sum_{\alpha=1}^{N} e^{a_\alpha},\ N_{34} = e^{\mathsf{w} + \varepsilon_4(l+1)},\ N_{13} = e^{\mathsf{w} - \varepsilon_1 l}, \tag{8.5}$$

finally $\mathcal{X} \star \mathcal{Z}$ corresponds to the origami with

$$N_{12} = \sum_{\alpha=1}^{N} e^{a_\alpha},\ N_{14} = e^{\mathsf{w} + \varepsilon_4(l+1)},\ N_{13} = e^{\mathsf{w} - \varepsilon_3 l}, \tag{8.6}$$

We prove the $\mathcal{Y} \star \mathcal{X} = 0$ equation from (6.1) by constructing a involution

$$\Upsilon : \Lambda \times \Lambda \times \mathbb{Z} \longrightarrow \Lambda \times \Lambda \times \mathbb{Z}, \tag{8.7}$$

sending (λ, μ, l) to $(\tilde{\lambda}, \tilde{\mu}, \tilde{l})$, such that:

$$\text{(8.8)} \qquad \begin{aligned} &1) \qquad |l-\tilde{l}| = 1 \\ &2) \qquad P_1 S_{34}(l,\lambda) + P_3 S_{14}(l,\mu) = P_1 S_{34}(\tilde{l},\tilde{\lambda}) + P_3 S_{14}(\tilde{l},\tilde{\mu}) \end{aligned}$$

the latter equation implying

$$\text{(8.9)} \qquad \frac{l(l+1)}{2} + |\lambda| + |\mu| = \frac{\tilde{l}(\tilde{l}+1)}{2} + |\tilde{\lambda}| + |\tilde{\mu}|$$

where

$$\text{(8.10)} \quad S_{34}(l,\lambda) = e^{\mathsf{w}} q_3^{l+1} \left(1 - P_{34}\chi_{34}(\lambda)\right), \qquad S_{14}(l,\mu) = e^{\mathsf{w}} q_1^{-l} \left(1 - P_{14}\chi_{14}(\mu)\right),$$

The involution acts by:
(8.11)

$$\Upsilon(\lambda,\mu,l) = \begin{cases} \left(\tilde{\lambda},\tilde{\mu},l-1\right) \quad , & \tilde{\lambda}_1 = \mu_1 + l,\ \tilde{\lambda}_{i+1} = \lambda_i,\ \mu_{i+1} = \tilde{\mu}_i,\ i \geq 1, \\ & \text{when } \mu_1 + l \geq \lambda_1, \\ \left(\hat{\lambda},\hat{\mu},l+1\right) \quad , & \hat{\mu}_1 = \lambda_1 - l - 1,\ \hat{\lambda}_i = \lambda_{i+1},\ \hat{\mu}_{i+1} = \mu_i,\ i \geq 1, \\ & \text{when } \mu_1 + l < \lambda_1 \,. \end{cases}$$

Let us check (8.8) in the case $\mu_1 + l \geq \lambda_1$, i.e. $\tilde{l} = l - 1$. In that case, the second equation in (8.8) reads

$$\text{(8.12)} \qquad \frac{1-q_4^l}{1-q_4} + q_3\chi_{34}(\lambda) + q_4^l \chi_{14}(\mu) = \chi_{34}(\tilde{\lambda}) + q_4^l q_1 \chi_{14}(\tilde{\mu})$$

Let us assume $l \geq 1$ (the case $l \leq 0$ is analogous) and expand the left hand side of (8.12) in q_1, q_4:

$$\text{(8.13)} \qquad \sum_{i=1}^{\ell(\lambda)} q_1^{-i} \sum_{j=1}^{\lambda_i} q_4^{j-i-1} + \sum_{j=1}^{\mu_1+l} q_4^{j-1} + \sum_{i=2}^{\ell(\mu)} q_1^{i-1} \sum_{j=l+1}^{\mu_i+l} q_4^{j-1}$$

while the right hand side of (8.12) expands, similarly, as

$$\text{(8.14)} \qquad \sum_{i=1}^{\ell(\tilde{\mu})} q_1^{i} \sum_{j=l+1}^{\tilde{\mu}_i+l} q_4^{j-1} + \sum_{i=1}^{\ell(\tilde{\lambda})} q_1^{1-i} \sum_{j=1}^{\tilde{\lambda}_i} q_4^{j-i}$$

Comparing the q_1^0, $q_1^{<0}, q_1^{\geq 0}$ terms, respectively, we get the precise match, given (8.11).

9. Appendix B. Limit to periodic Toda and relation to earlier work

The Toda limit, i.e.

$$\text{(9.1)} \qquad n \to \infty, \mathfrak{q} \to 0,$$

while keeping finite

$$\text{(9.2)} \qquad \Lambda^{2N} = \mathfrak{q}(-1)^N \left(\hbar n\right)^{2N}$$

is a bit delicate, as far as the $\mathcal{X}, \mathcal{Z}, \mathsf{X}, \mathsf{Z}$ functions are concerned. The Q-function has a smooth limit, with the Bethe equations (5.1) going over to

$$\frac{Q(\mathsf{w}+\hbar)}{Q(\mathsf{w}-\hbar)} = -\Lambda^{2N}, \ \forall \mathsf{w} \in Q^{-1}(0) \ . \tag{9.3}$$

More precisely, this is the limit of $Q(\mathsf{w})$ at fixed w. On the other hand, Q with the shifted arguments have somewhat singular limits, which we need to account for:

$$Q(\mathsf{w}-\hbar nl) \sim \ (nl)^{\frac{N}{2}} \, e^{Nnl} \, (-\hbar nl)^{\frac{N\mathsf{w}}{\hbar}-Nnl} \times \left(1+O(n^{-1})\right) \tag{9.4}$$

Specifically, we shall need (9.4) for $l=-1,1$ and a limit of the product

$$\mathsf{q}^{\frac{\mathsf{w}}{\hbar}+p} Q(\mathsf{w}+\hbar(p-n))Q(\mathsf{w}+\hbar(n-1+p)) \sim \frac{(-1)^{N\left(n+\frac{1}{2}\right)}}{\hbar^N}\Lambda^{2N\left(\frac{\mathsf{w}}{\hbar}+p\right)} \tag{9.5}$$

The series for $\mathcal{X}(\mathsf{w}), \mathcal{Z}(\mathsf{w})$, in our limit, receive contributions only from partitions $\lambda = (1^p)$, for $p=0,1,\ldots$, giving:

$$\begin{aligned} \mathcal{X}(\mathsf{w}) &\sim \tilde{Q}(\mathsf{w}-\hbar), \\ \mathcal{Z}(\mathsf{w}) &\sim \tilde{Q}(\mathsf{w}-\hbar), \end{aligned} \tag{9.6}$$

where $\sim$ means up to a factor $e^{c_1(n)\mathsf{w}+c_0(n)}$,

$$\tilde{Q}(\mathsf{w}) = \Lambda^{2N\frac{\mathsf{w}}{\hbar}}\, Q(\mathsf{w}) \sum_{p=0}^{\infty} \frac{\Lambda^{2Np}}{Q(\mathsf{w}+\hbar p)Q(\mathsf{w}+\hbar(p+1))} \ . \tag{9.7}$$

The $\hbar n$-shifted $\mathcal{X}(\mathsf{w}), \mathcal{Z}(\mathsf{w})$ functions have Q-function as its principal asymptotics:

$$\mathcal{X}(\mathsf{w}+\hbar n) \sim Q(\mathsf{w}), \ \ \mathcal{Z}(\mathsf{w}+\hbar(1-n)) \sim Q(\mathsf{w}) \tag{9.8}$$

Finally, the series for $\mathcal{Y}(\mathsf{w})$ retains only two terms in the limit (9.1):

$$\mathcal{Y}(\mathsf{w}) = \mathsf{Y}(\mathsf{w}+\hbar) + \frac{\Lambda^{2N}}{\mathsf{Y}(\mathsf{w})} \tag{9.9}$$

Either by recalling the relation between $Q(\mathsf{w})$ and $\mathsf{Y}(\mathsf{w})$, or by taking the limit of (6.1) we obtain the *T-Q equation* (cf. [**23**,**32**,**39**]), obeyed both by $Q(\mathsf{w})$ and $\tilde{Q}(\mathsf{w})$:

$$\begin{aligned} Q(\mathsf{w}+\hbar) + \Lambda^{2N} Q(\mathsf{w}-\hbar) &= \mathcal{Y}(\mathsf{w})Q(\mathsf{w}) \\ \tilde{Q}(\mathsf{w}+\hbar) + \Lambda^{2N} \tilde{Q}(\mathsf{w}-\hbar) &= \mathcal{Y}(\mathsf{w})\tilde{Q}(\mathsf{w}) \end{aligned} \tag{9.10}$$

in complete agreement with (9.3). Note that (9.10) can also be obtained by a limit of (3.1).

The Toda limit (9.1) of (6.1) gives (9.10) and the Wronskian relation (cf. [**26**])

$$\tilde{Q}(\mathsf{w})Q(\mathsf{w}+\hbar) - Q(\mathsf{w})\tilde{Q}(\mathsf{w}+\hbar) = \Lambda^{\frac{2N\mathsf{w}}{\hbar}} \tag{9.11}$$

Both Q and $\tilde{Q}$ functions prominently feature (in disguise) in the separation of variables for the quantum Toda system (the Ref. [**25**] used determinant representation of Gaudin-Pasquier [**16**], which can be related to our Q, $\tilde{Q}$'s).

We leave the verification to the reader as an exercise in application of (9.6), (9.5), (9.4).

Acknowledgments

We enormously benefited from discussions and collaboration with I. Krichever in 1995-2022. We also thank S. Cherkis and S. Grushevsky for discussions. Research was partly supported by NSF PHY Award 2310279. Any opinions expressed are solely our own and do not represent the views of the National Science Foundation.

References

[1] L. F. Alday, D. Gaiotto, and Y. Tachikawa, *Liouville correlation functions from four-dimensional gauge theories*, Lett. Math. Phys. **91** (2010), no. 2, 167–197, DOI 10.1007/s11005-010-0369-5. MR2586871

[2] M. Aganagic, R. Dijkgraaf, A. Klemm, M. Mariño, and C. Vafa, *Topological strings and integrable hierarchies*, Comm. Math. Phys. **261** (2006), no. 2, 451–516, DOI 10.1007/s00220-005-1448-9. MR2191887

[3] R. J. Baxter, *Generalized ferroelectric model on a square lattice*, Studies in Appl. Math. **50** (1971), 51–69, DOI 10.1002/sapm197150151. MR307646

[4] V. V. Bazhanov, S. L. Lukyanov, and A. B. Zamolodchikov, *Integrable structure of conformal field theory, quantum KdV theory and thermodynamic Bethe ansatz*, Comm. Math. Phys. **177** (1996), no. 2, 381–398. MR1384140

[5] G. Bonelli, F. Del Monte, P. Gavrylenko, and A. Tanzini, *Circular quiver gauge theories, isomonodromic deformations and W_N fermions on the torus*, Lett. Math. Phys. **111** (2021), no. 3, Paper No. 83, 38 pp., DOI 10.1007/s11005-020-01343-4. MR4277314

[6] M. Bullimore, H.-C. Kim, and P. Koroteev, *Defects and quantum Seiberg-Witten geometry*, J. High Energy Phys. **5** (2015), 95 pp., DOI 10.1007/JHEP05(2015)095. MR3358908

[7] L. Chekhov, B. Eynard, and S. Ribault, *Seiberg-Witten equations and non-commutative spectral curves in Liouville theory*, J. Math. Phys. **54** (2013), no. 2, 022306, 21, DOI 10.1063/1.4792241. MR3076379

[8] S. A. Cherkis, *Moduli spaces of instantons on the Taub-NUT space*, Comm. Math. Phys. **290** (2009), no. 2, 719–736, DOI 10.1007/s00220-009-0863-8. MR2525636

[9] I. Coman, E. Pomoni, and J. Teschner, *From quantum curves to topological string partition functions*, Comm. Math. Phys. **399** (2023), no. 3, 1501–1548, DOI 10.1007/s00220-022-04579-4. MR4580528

[10] R. Donagi and E. Witten, *Supersymmetric Yang-Mills theory and integrable systems*, Nuclear Phys. B **460** (1996), no. 2, 299–334, DOI 10.1016/0550-3213(95)00609-5. MR1377167

[11] V. G. Drinfel′d, *A new realization of Yangians and of quantum affine algebras* (Russian), Dokl. Akad. Nauk SSSR **296** (1987), no. 1, 13–17; English transl., Soviet Math. Dokl. **36** (1988), no. 2, 212–216. MR0914215

[12] P. Etingof, E. Frenkel, and D. Kazhdan, *Hecke operators and analytic Langlands correspondence for curves over local fields*, Duke Math. J. **172** (2023), no. 11, 2015–2071, DOI 10.1215/00127094-2022-0068. MR4627247

[13] E. Frenkel and N. Reshetikhin, *The q-characters of representations of quantum affine algebras and deformations of $\mathcal{W}$-algebras*, Recent developments in quantum affine algebras and related topics (Raleigh, NC, 1998), Contemp. Math., vol. 248, Amer. Math. Soc., Providence, RI, 1999, pp. 163–205, DOI 10.1090/conm/248/03823. MR1745260

[14] L. A. Tahtadžjan and L. D. Faddeev, *The quantum method for the inverse problem and the XYZ Heisenberg model* (Russian), Uspekhi Mat. Nauk **34** (1979), no. 5 (209), 13–63, 256 pp. MR0562799

[15] D. Gaiotto, *$N = 2$ dualities*, J. High Energy Phys. **8** (2012), 34 pp., DOI 10.1007/JHEP08(2012)034. MR3006961

[16] V. Pasquier and M. Gaudin, *The periodic Toda chain and a matrix generalization of the Bessel function recursion relations*, J. Phys. A **25** (1992), no. 20, 5243–5252. MR1192958

[17] A. Gorsky, I. M. Krichever, A. Marshakov, A. Mironov, and A. Morozov, *Integrability and Seiberg-Witten exact solution*, Phys. Lett. B **355** (1995), no. 3-4, 466–474, DOI 10.1016/0370-2693(95)00723-X. MR1345598

[18] A. Gorsky and N. Nekrasov, *Elliptic Calogero-Moser system from two dimensional current algebra*, `arXiv:hep-th/9401021` [hep-th].

[19] A. Grekov and A. Zotov, *Characteristic determinant and Manakov triple for the double elliptic integrable system*, SciPost Phys. **10** (2021), no. 3, Paper No. 055, 34 pp., DOI 10.21468/scipostphys.10.3.055. MR4237749

[20] A. Grekov, I. Krichever, N. Nekrasov, *Lax operators from enumerative geometry*, to appear

[21] N. Hitchin, *Stable bundles and integrable systems*, Duke Math. J. **54** (1987), no. 1, 91–114, DOI 10.1215/S0012-7094-87-05408-1. MR885778

[22] M. C. Gutzwiller, *The quantum mechanical Toda lattice. II*, Ann. Physics **133** (1981), no. 2, 304–331, DOI 10.1016/0003-4916(81)90253-0. MR626693

[23] S. Jeong and N. Nekrasov, *Opers, surface defects, and Yang-Yang functional*, Adv. Theor. Math. Phys. **24** (2020), no. 7, 1789–1916, DOI 10.4310/ATMP.2020.v24.n7.a4. MR4313236

[24] S. Katz, A. Klemm, and C. Vafa, *Geometric engineering of quantum field theories*, Nuclear Phys. B **497** (1997), no. 1-2, 173–195, DOI 10.1016/S0550-3213(97)00282-4. MR1467889

[25] S. Kharchev and D. Lebedev, *Integral representations for the eigenfunctions of quantum open and periodic Toda chains from the QISM formalism*, J. Phys. A **34** (2001), no. 11, 2247–2258, DOI 10.1088/0305-4470/34/11/317. MR1831292

[26] K. K. Kozlowski and J. Teschner, *TBA for the Toda chain*, doi:10.1142/9789814324373_0011

[27] A. V. Litvinov, *On spectrum of ILW hierarchy in conformal field theory*, J. High Energy Phys. **11** (2013), 13 pp., DOI 10.1007/JHEP11(2013)155. MR3132164

[28] I. M. Krichever, *Elliptic solutions of the Kadomcev-Petviašvili equations, and integrable systems of particles* (Russian), Funktsional. Anal. i Prilozhen. **14** (1980), no. 4, 45–54, 95. MR595728

[29] E. D'Hoker, I. M. Krichever, and D. H. Phong, *The effective prepotential of $N = 2$ supersymmetric* $\mathrm{SU}(N_c)$ *gauge theories*, Nuclear Phys. B **489** (1997), no. 1-2, 179–210, DOI 10.1016/S0550-3213(97)00035-7. MR1443801

[30] N. Nekrasov, *Holomorphic bundles and many-body systems*, Comm. Math. Phys. **180** (1996), no. 3, 587–603. MR1408519

[31] N. A. Nekrasov, *Seiberg-Witten prepotential from instanton counting*, Adv. Theor. Math. Phys. **7** (2003), no. 5, 831–864, DOI 10.4310/atmp.2003.v7.n5.a4. MR2045303

[32] N. Nekrasov, *BPS/CFT correspondence III: Gauge origami partition function and qq-characters*, Comm. Math. Phys. **358** (2018), no. 3, 863–894, DOI 10.1007/s00220-017-3057-9. MR3778345

[33] N. A. Nekrasov and A. Okounkov, *Seiberg-Witten theory and random partitions*, The unity of mathematics, Progr. Math., vol. 244, Birkhäuser Boston, Boston, MA, 2006, pp. 525–596, DOI 10.1007/0-8176-4467-9_15. MR2181816

[34] N. Nekrasov and V. Pestun, *Seiberg-Witten geometry of four-dimensional $\mathcal{N} = 2$ quiver gauge theories*, SIGMA Symmetry Integrability Geom. Methods Appl. **19** (2023), Paper No. 047, 141, DOI 10.3842/SIGMA.2023.047. MR4614418

[35] N. Nekrasov and N. S. Prabhakar, *Spiked Instantons from Intersecting D-branes*, Nucl. Phys. B **914**, 257-300 (2017)

[36] N. Nekrasov and A. Tsymbaliuk, *Surface defects in gauge theory and KZ equation*, Lett. Math. Phys. **112** (2022), no. 2, Paper No. 28, 53 pp., DOI 10.1007/s11005-022-01511-8. MR4400676

[37] A. Okounkov, *Deep waters of quantum cohomology*, `https://scgp.stonybrook.edu/video_portal/video.php?id=524`

[38] A. Okounkov and R. Pandharipande, *Quantum cohomology of the Hilbert scheme of points in the plane*, Invent. Math. **179** (2010), no. 3, 523–557, DOI 10.1007/s00222-009-0223-5. MR2587340

[39] G. Poghosyan and R. Poghossian, *VEV of Baxter's Q-operator in $N = 2$ gauge theory and the BPZ differential equation*, J. High Energy Phys. **11** (2016), DOI 10.1007/JHEP11(2016)058. MR3584491

[40] N. Yu. Reshetikhin, *A method of functional equations in the theory of exactly solvable quantum systems*, Lett. Math. Phys. **7** (1983), no. 3, 205–213, DOI 10.1007/BF00400435. MR706209

[41] E. K. Sklyanin, *Quantum inverse scattering method. Selected topics*, Quantum group and quantum integrable systems, Nankai Lectures Math. Phys., World Sci. Publ., River Edge, NJ, 1992, pp. 63–97. MR1239668

[42] E. K. Sklyanin, *The quantum Toda chain*, Nonlinear equations in classical and quantum field theory (Meudon/Paris, 1983/1984), Lecture Notes in Phys., vol. 226, Springer, Berlin, 1985, pp. 196–233, DOI 10.1007/3-540-15213-X_80. MR802110

[43] E. K. Sklyanin, *Separation of variables—new trends*, Progr. Theoret. Phys. Suppl. **118** (1995), 35–60, DOI 10.1143/PTPS.118.35. Quantum field theory, integrable models and beyond (Kyoto, 1994). MR1356514

[44] L. A. Takhtajan, *Lectures on quantum groups*, Introduction to quantum group and integrable massive models of quantum field theory (Nankai, 1989), Nankai Lectures Math. Phys., World Sci. Publ., River Edge, NJ, 1990, pp. 69–197. MR1145243

[45] M. Vasilyev and A. Zotov, *On factorized Lax pairs for classical many-body integrable systems*, Rev. Math. Phys. **31** (2019), no. 6, 1930002, 45, DOI 10.1142/s0129055x19300024. MR3974680

[46] E. Witten, *Integrable lattice models from gauge theory*, Adv. Theor. Math. Phys. **21** (2017), no. 7, 1819–1843, DOI 10.4310/ATMP.2017.v21.n7.a10. MR3780281

Simons Center for Geometry and Physics^{n}, Yang Institute for Theoretical Physics^{g,n}, Stony Brook University, Stony Brook New York 11794-3636

Email address: andrei.grekov@stonybrook.edu

Email address: nnekrasov@scgp.stonybrook.edu

Contemporary Mathematics
Volume **823**, 2025
https://doi.org/10.1090/conm/823/16493

Hypergeometric integrals, hook formulas and Whittaker vectors

G. Felder, A. Smirnov, V. Tarasov, and A. Varchenko

In memory of Igor Krichever (1950–2022)

ABSTRACT. The classical hook-length formula in combinatorics relates the number of standard Young tableaux with given shape to the product of hook-lengths of the Young diagram. A generalization to skew Young diagrams is due to Naruse, who obtains it as a limit of a multivariate generalization arising from equivariant Schubert calculus. We show that these combinatorial objects appear in a relation between multidimensional hypergeometric integrals motivated by 3D mirror symmetry and in the diagonalization of the action of the centre of $U(\mathfrak{gl}_n)$ on the space of Whittaker vectors in the tensor product of dual Verma modules with fundamental modules.

CONTENTS

2020 *Mathematics Subject Classification.* Primary 33C70; Secondary 14N10, 32G34.

Key words and phrases. Singular vectors, Young diagrams, excited diagrams, hooks, master function, hypergeometric integrals, Whittaker vectors.

The first author was supported in part by the SNSF under grants 196892, 205607.

The second author was Supported in part by the NSF under grant DMS - 2054527 and by the RSF under grant 19-11-00062.

The third author was Supported in part by the Simons Foundation under grants 430235, 852996.

The fourth author was supported in part by the NSF under grant DMS-1954266.

1. Introduction

We determine the coefficient of proportionality between two multidimensional hypergeometric integrals. One of them is a solution of the dynamical difference equations associated with a Young diagram and the other is the vertex integral associated with the Young diagram. The coefficient of proportionality is the inverse of the product of weighted hooks of the Young diagram. The same coefficient appears in the problem of diagonalizing the action of the center of the universal enveloping algebra of $\mathfrak{gl}_n$ on the space of Whittaker vectors in the tensor product of dual Verma modules with fundamental modules..

The standard basis (u_λ) of a fundamental $\mathfrak{gl}_n$-module U_r is labeled by the Young diagrams λ inscribed in an $(n-r)\times r$-rectangle. One assigns to a basis vector u_λ a system of dynamical difference equations

(1.1)
$$I(z_1,\dots,z_i+\kappa,\dots,z_{n-1},\kappa)=a_i(z_1,\dots,z_{n-1},\kappa)\,I(z_1,\dots,z_{n-1},\kappa),\quad i=1,\dots,n-1,$$

where $I(z_1,\dots,z_{n-1},\kappa)$ is an unknown scalar function and a_i are suitable coefficients defined in terms of the $\mathfrak{gl}_n$-action on U_r. The equations were introduced in [**TV**], and solutions were constructed in [**MV**]. A solution $I_\lambda(z,\kappa)$ to (1.1) is given by a hypergeometric integral of dimension equal to the number of boxes in λ.

We also introduce another hypergeometric integral $V_\lambda(z,\kappa)$ of the same dimension associated with λ. We show that it is proportional to $I_\lambda(z,\kappa)$ and determine the coefficient of proportionality between the two integrals in Theorem 3.1

$$V_\lambda(z,\kappa)=\frac{1}{\prod_{\square\in\lambda}h(\square)(z)}\,I_\lambda(z,\kappa),\tag{1.2}$$

where $h(\square)(z)$ is the hook-weight of a box $\square$ of λ, see the definition in (2.5).

Our motivation for considering (1.2) is the following. On the one hand, the enumerative geometry of quiver varieties is controlled by two important objects: the vertex function and the capping operator of a quiver variety [**O**]. These are the generating functions counting quasimaps to the quiver variety with nonsingular or relative boundary conditions.

On the other hand, with any quiver variety X one can associate a hypergeometric integral. The 3D-mirror symmetry predicts that this integral computes the vertex function of the mirror variety $X^!$. The capping operator of $X^!$ is obtained in the same way with additional insertion of the stable envelope functions to the integral [**AO**] (also known as weight functions in the theory of qKZ equations). For example, see [**SmV1**, **SmV2**] where the integral formulas of this type are discussed in the case of cotangent bundle over Grassmannian $X^!=T^*Gr(k,n)$.

Let X_λ be the zero-dimensional Nakajima quiver variety of type A associated to a Young diagram λ [**DS1**, **DS2**]. The hypergeometric integral assigned to X_λ is given by the function $V_\lambda(z,\kappa)$ and the integral for the capping operator is given by $I_\lambda(z,\kappa)$. Since the cohomology of X_λ are one-dimensional, it is expected from [**O**] that both integrals are proportional. Our formula (1.2) establishes the coefficient of proportionality explicitly.

Note, however, that in this case, the mirror $X^!_\lambda$ is a $2|\lambda|$-dimensional variety which can not be realized as a quiver variety. Thus, the vertex function for $X^!_\lambda$ is not defined by the methods of [**O**]. To clarify this point, we refer to the function $V_\lambda(z,\kappa)$ as the vertex integral, instead of the "vertex function of $X^!_\lambda$".

We note also that the coefficient in (1.2) has a geometric meaning: the mirror variety $X^!_\lambda$ is equipped with a torus action with unique fixed point. The denominator of the coefficient in (1.2) is the product over the half of the tangent weights at this point with parameters $z_1,\dots,z_n$ understood as the equivariant parameters of the torus [**DS2**].

To determine the coefficient of proportionality we consider the tensor product $M\otimes U_r$, where M is a Verma module and analyze singular weight vectors in $M\otimes U_r$ of the form

$$v(\lambda)=\sum_{\mu\leqslant\lambda} v_\mu\otimes u_\mu, \qquad \text{with } v_\mu\in M.$$

The collection of vectors (v_μ) is quite a nontrivial object. We simplify it by choosing a suitable linear function $\psi : M\to\mathbb{C}$ and considering instead the collection of numbers $(\psi(v_\mu))$. We develop simple recurrence relations and formulas for these numbers. We also show that

$$V_\lambda(z,\kappa)/I_\lambda(z,\kappa)=\psi(v_\emptyset)/\psi(v_\lambda).$$

Together with formulas for $\psi(v_\mu)$ this equation proves formula (1.2).

The numbers $\psi(v_\mu)$ are functions of the highest weight of M associated to the skew Young diagram λ/μ. We show that these numbers arise in the problem of diagonalizing the action of the center Z of the universal enveloping algebra on the space of Whittaker vectors of $M'\otimes U_r$ where M' is the dual of the Verma module M.

A Whittaker vector in a $\mathfrak{gl}_n$-module is a vector on which the nilpotent subalgebra of lower triangular matrices acts via a fixed regular character, see Section 3.4. The center Z acts on the space of Whittaker vectors $\mathrm{Wh}(M'\otimes U_r)$. A Whittaker vector $\beta\in M'\otimes U_r$ is uniquely determined by its contraction $\beta(v)\in U_r$ with the highest weight vector v of M. For generic highest weight of M, we show that a basis of eigenvectors is given by

$$\beta_\lambda(v)=\sum_{\mu\leqslant\lambda}\sum_{\nu\in E(\lambda/\mu)}\frac{1}{\prod_{\Box\in\lambda\smallsetminus\nu}h(\Box)(z)}\,u_\mu,$$

where λ runs over the set of Young diagrams fitting in an $(n-r)\times r$ rectangle and z is an affine function of the highest weight of M. The set $E(\lambda/\mu)$ is the set of Ikeda–Naruse excited diagrams, which are subsets of λ obtained from μ by moving boxes according to certain rules. The coefficient of u_μ is $\psi(v_\mu)$. In particular the coefficient of $u_\emptyset$ is the coefficient of proportionality in (1.2) (the coefficient of u_λ is normalized to be 1).

2. Singular vectors

2.1. Linear function ψ. Consider the complex Lie algebra $\mathfrak{gl}_n$ with standard generators e_{ij}, $i,j=1,\dots,n$, simple roots α_i, $i=1,\dots,n-1$, half-sum ρ of positive roots. Denote $e_i=e_{i,i+1}$, $f_i=e_{i+1,i}$, $h_i=e_{i,i}-e_{i+1,i+1}$ for $i=1,\dots,n-1$.

Let M be a $\mathfrak{gl}_n$ Verma module with highest weight vector v. Define a linear function $\psi : M \to \mathbb{C}$ as follows. Any vector $v' \in M$ can be written (in general non-uniquely) as a finite linear combination of the products of elements $f_1, \dots, f_{n-1}$ applied to v,

$$v' = \sum c_{i_m, i_{m-1}, \dots, i_1} f_{i_m} f_{i_{m-1}} \cdots f_{i_1} v,$$

where $1 \leqslant i_j \leqslant n-1$ and $c_{i_m, i_{m-1}, \dots, i_1} \in \mathbb{C}$. Set

$$\psi(v') = \sum c_{i_m, i_{m-1}, \dots, i_1}. \tag{2.1}$$

The function ψ is well-defined since it is zero on Serre's relations $f_i^2 f_{i\pm 1} - 2 f_i f_{i\pm 1} f_i + f_{i\pm 1} f_i^2 = 0$. It is in fact a Whittaker vector in the dual of M, as will be discussed in Section 3.4.

2.2. Fundamental representations. Let U_r, $r = 1, \dots, n-1$, be the r-th fundamental representation of $\mathfrak{gl}_n$. Its highest weight is $(1, \dots, 1, 0, \dots, 0)$ with r ones.

The U_1 is the vector representation $\mathbb{C}^n$ with standard basis u_i, $i = 1, \dots, n$. The U_r is the r-th exterior power $\wedge^r \mathbb{C}^n$ of the vector representation with standard basis

$$u_I := u_{i_1} \wedge u_{i_2} \wedge \dots \wedge u_{i_r}, \tag{2.2}$$

where $I = \{i_1 < i_2 < \dots < i_r\}$ is any r-element subset of $\{1, \dots, n\}$. Denote by $\mathcal{I}_r$ the set of such subsets.

The decomposition

$$U_r = \oplus_{I \in \mathcal{I}_k} \mathbb{C} u_I$$

is the weight decomposition. We have $e_{ii} u_I = u_I$ if $i \in I$ and $e_{ii} u_I = 0$ otherwise. Thus, the weight $w(u_I)$ of u_I is the n-vector whose i-th coordinate is 1 if $i \in I$ and is 0 otherwise. The vector $u_{I^{min}}$ with $I^{min} = \{1 < 2 < \dots < r\}$ is a highest weight vector.

2.3. Young diagrams. The set $\mathcal{I}_r$ is identified with the set of sequences of nonnegative integers

$$\{0 \leqslant \lambda_1 \leqslant \ldots \leqslant \lambda_r \leqslant n-r\}$$

by the formula $\{i_1 < i_2 < \dots < i_r\} \mapsto \{i_1 - 1 \leqslant i_2 - 2 \leqslant \ldots \leqslant i_r - r\}$. The set of such sequences is identified with the set of Young diagrams inscribed in the $(n-r) \times r$-rectangle R. Thus the set $\mathcal{I}_r$ is identified with the set of Young diagrams inscribed in the rectangle R.

For example, I^{min} corresponds to the empty Young diagram $\emptyset$, and the rectangle R corresponds to the subset $\{n-r+1 < n-r+2 < \dots < n\}$.

We conclude that the basis (u_I) of U_r is labeled by the Young diagrams. A vector u_I will also be denoted u_λ if I corresponds to a Young diagram λ. The weight $w(u_I)$ of u_I will also be denoted by $w(\lambda)$.

The set of Young diagrams is partially ordered with respect to inclusion of the diagrams. We write $\mu \leqslant \lambda$ if the Young diagram λ contains the Young diagram μ.

2.4. Singular vectors. Let M be the $\mathfrak{gl}_n$ Verma module with highest weight $t-\rho$ and highest weight vector v, where $t=(t_1,\dots,t_n)$.

Fix a Young diagram $\lambda \in \mathcal{I}_r$. Consider the vector subspace $\cap_{i=1}^{n-1} \operatorname{Ker} e_i$ of singular vectors in $M \otimes U_r$ of weight $w(\lambda)+t-\rho$. For generic t, this space is one-dimensional with a generator of the form

$$v(\lambda) := v \otimes u_\lambda + \sum_{\mu<\lambda} v_\mu \otimes u_\mu \tag{2.3}$$

for suitable vectors $v_\mu \in M$. Recall the linear function $\psi : M \to \mathbb{C}$. Define the following scalar functions $F_{\lambda/\mu}$ of t:

$$F_{\lambda/\mu} = \psi(v_\mu) \quad \text{for} \quad \mu<\lambda \quad \text{and} \quad F_{\lambda/\lambda}=1. \tag{2.4}$$

More precisely, each $F_{\lambda/\mu}$ is a function of $z=(z_1,\dots,z_{n-1})$ where $z_i = t_{i+1}-t_i$.

The main result of this paper is recurrence relations and a formula for functions $F_{\lambda/\mu}$.

2.5. Hooks. The $(n-r)\times r$-rectangle R lies in the positive quadrant in $\mathbb{R}^2$ and consists of unit boxes $\square_{i,j}$, $i=1,\dots,n-r$, $j=1,\dots,r$. The center of a box $\square_{i,j}$ has coordinates $\left(i-\frac{1}{2}, j-\frac{1}{2}\right)$. Every nonempty Young diagram $\lambda \in \mathcal{I}_r$ contains the corner box $\square_{1,1}$.

To every box $\square_{i,j}$ we assign one of $z_1,\dots,z_{n-1}$ by the rule:

$$z(\square_{i,j}) := z_{i-j+r}.$$

For example, $z(\square_{1,r})=z_1$, $z(\square_{1,1})=z_r$, $z(\square_{n-r,1})=z_{n-1}$, $z(\square_{n-r,r})=z_{n-r}$. We say that z_{i-j+r} is the z-label of a box $\square_{i,j}$.

Recall that a hook $H_\lambda(\square_{i,j})$ of a box $\square_{i,j}$ in a Young diagram λ is the set of all boxes $\square_{a,b}$ in λ such that $a=i$, $b \geqslant j$, or $a \geqslant i$, $b=j$. We define the hook-weight of a box $\square$ of λ by the formula

$$h(\square) = 1 + \sum_{\square' \in H_\lambda(\square)} z(\square'). \tag{2.5}$$

THEOREM 2.1. *We have*

$$F_{\lambda/\emptyset} = \frac{1}{\prod_{\square\in\lambda} h(\square)}. \tag{2.6}$$

For example, if $(r,n)=(2,4)$ and $\lambda=(2,1)$. Then λ consists of the three boxes $\square_{1,1}$, $\square_{1,2}$, $\square_{2,1}$, and

$$F_{\lambda/\emptyset} = \frac{1}{(z_1+z_2+z_3+1)(z_1+1)(z_3+1)}.$$

In Section 2.7 we give a formula for all coefficients $F_{\lambda/\mu}$. Theorem 2.1 follows from Theorem 2.5 below.

2.6. Recurrence relations. Let λ/μ be a skew-diagram. Let k_i be the number of boxes in λ/μ with z-label z_i, where $i=1,\dots,n-1$. We put $k_n=0$. Define the *z-content* of λ/μ by the formula

$$G_{\lambda/\mu} = \sum_{i=1}^{n-1} k_i(k_i-k_{i+1}+z_i).$$

For example, if $(r,n)=(2,4)$, $\lambda=(2,2)$, $\mu_1=\emptyset$, $\mu_2=(1)$, $\mu_3=(1,1)$, $\mu_4=(2)$, $\mu_5=(2,1)$, then

$$G_{\lambda/\mu_1}=z_1+2z_2+z_3+2,\qquad G_{\lambda/\mu_2}=z_1+z_2+z_3+1,$$
$$G_{\lambda/\mu_3}=z_1+z_2+1,\qquad G_{\lambda/\mu_4}=z_2+z_3+1,\qquad G_{\lambda/\mu_5}=z_2+1.$$

THEOREM 2.2. *The following recurrence relations hold:*

$$F_{\lambda/\mu}=\frac{1}{G_{\lambda/\mu}}\sum_{\mu'}F_{\lambda/\mu'},\tag{2.7}$$

where the sum is over all the Young diagrams μ' such that $\mu<\mu'\leqslant\lambda$ and the skew-diagram μ'/μ consists of one box.

For example, if $(r,n)=(2,4)$, $\lambda=(2,2)$, and μ_i are as before, then we have

(2.8)
$$\begin{aligned}F_{\lambda/\emptyset}&=\frac{1}{z_1+2z_2+z_3+2}F_{\lambda/\mu_2}=\frac{1}{(z_1+2z_2+z_3+2)(z_1+z_2+z_3+1)}[F_{\lambda/\mu_3}+F_{\lambda/\mu_4}]\\&=\frac{1}{(z_1+2z_2+z_3+2)(z_1+z_2+z_3+1)}\Big(\frac{1}{z_1+z_2+1}+\frac{1}{z_2+z_3+1}\Big)F_{\lambda/\mu_5}\\&=\frac{1}{(z_1+z_2+z_3+1)(z_1+z_2+1)(z_2+z_3+1)(z_2+1)}\end{aligned}$$

where in the last step we used $F_{\lambda/\lambda}=1$.

PROOF. Denote $\beta=t-\rho-\sum_{i=1}^{n-1}k_i\alpha_i$ where k_i are some nonnegative integers. Let $M[\beta]$ be the weight subspace of M of weight β.

LEMMA 2.3. *For any $v'\in M[\beta]$, we have*

$$\psi((e_1+\cdots+e_{n-1})v')=-\sum_{i=1}^{n-1}k_i(k_i-k_{i+1}+z_i)\,\psi(v').\tag{2.9}$$

PROOF. The proof is straightforward. It is enough to check formula (2.9) for $v'=f_{m_1}\cdots f_{m_k}v$ where $1\leqslant m_j\leqslant n-1$ and for any i the sequence $m_1,\dots,m_k$ has exactly k_i elements equal to i.

For example, for $\beta=t-\rho-2\alpha_1-\alpha_2$ and $v'=f_1f_1f_2v$ we have

$$\begin{aligned}\psi((e_1+e_2+e_3)v')&=\psi(h_1f_1f_2v+f_1h_1f_2v+f_1f_1h_2v)\\&=-\psi((z_1+2)f_1f_2v+z_1f_1f_2v+(z_2+1)f_1f_1v)=-(2z_1+z_2+3),\end{aligned}$$

while $\psi(v')=1$. □

To prove the theorem notice that the vector $v(\lambda)$ is singular and hence $\psi((e_1+\cdots+e_{n-1})v(\lambda))=0$. By Lemma 2.3, we also have

$$\psi((e_1+\cdots+e_{n-1})v(\lambda))=\sum_{\mu<\lambda}\Big(-G_{\lambda/\mu}F_{\lambda/\mu}+\sum_{\mu'}F_{\lambda/\mu'}\Big)u_\mu,\tag{2.10}$$

where the second sum is over all the Young diagrams μ' such that $\mu<\mu'\leqslant\lambda$ and the skew-diagram μ'/μ consists of one box. Since $(u_\mu)_{\mu\in\mathcal{I}_r}$ is a basis of U_r, the coefficient of each u_μ in (2.10) must be equal to zero. This proves the theorem. □

COROLLARY 2.4. *Let d be the number of boxes in λ/μ. Then*

$$F_{\lambda/\mu}=\sum_{\mu=\mu_1<\mu_2<\cdots<\mu_d<\lambda}\frac{1}{\prod_{i=1}^d G_{\lambda/\mu_i}}.\tag{2.11}$$

See an example in formula (2.8).

2.7. Excited diagrams. Let λ/μ be a skew-diagram and D a subset of the Young diagram λ. A box $\square_{i,j}$ of D is called *active* if the boxes $\square_{i+1,j}, \square_{i+1,j+1}, \square_{i,j+1}$ are all in $\lambda - D$. Let $b = \square_{i,j}$ be an active box of D, define D_b to be the set obtained by replacing $\square_{i,j}$ in D by $\square_{i+1,j+1}$. We call this replacement an *elementary excitation*. An *excited diagram* of λ/μ is a subset of boxes of λ obtained from the Young diagram μ after a sequence of elementary excitations on active boxes. Let $E(\lambda/\mu)$ be the set of excited diagrams of λ/μ, see this definition in [**IN**, **Na**, **MPP1**], also see [**MPP2**].

THEOREM 2.5. *We have*

$$F_{\lambda/\mu} = \frac{1}{\prod_{\square\in\lambda} h(\square)} \sum_{\nu\in E(\lambda/\mu)} \prod_{\square\in\nu} h(\square). \tag{2.12}$$

For example, in the notation of formula (2.8), the set $E(\lambda/\mu_2)$ consists of two elements $\{\square_{1,1}\}$ and $\{\square_{2,2}\}$. Then

$$F_{\lambda/\mu_2} = \frac{z_1 + 2z_2 + x_3 + 2}{(z_1 + z_2 + z_3 + 1)(z_1 + z_2 + 1)(z_2 + z_3 + 1)(z_2 + 1)},$$

where

$$h(\square_{1,1}) + h(\square_{2,2}) = z_1 + 2z_2 + x_3 + 2.$$

REMARK. The equality between (2.11) and (2.12) in the case $\mu = \emptyset$ is a generalization of the classical hook-length formula relating the number of standard Young tableaux of shape λ to the inverse product of hook-lengths. It converges to it in the limit where all z_i are equal and tend to infinity. In the same limit for general $\mu \subset \lambda$ we obtain Naruse's generalization for skew diagrams [**Na**], also see [**MPP2**].

2.8. Proof of Theorem 2.5. Theorem 2.5 follows from Corollary 2.4 and Naruse's formula in [**Na**] by a change of parameters $z_1, \dots, z_{n-1}$.

More precisely, let $\lambda \in \mathcal{I}_r$ be a nonempty Young diagram. We say that a box $\square_{i,j} \in \lambda$ is a *boundary box* if $\square_{i+1,j+1} \notin \lambda$. Let $\square_{i,j} \in \lambda$ be a boundary box. If $\square_{i,j+1} \notin \lambda$ and $\square_{i+1,j} \notin \lambda$, then $\square_{i,j}$ is an active boundary box according to the definition in Section 2.7 (with $D = \lambda$). If $\square_{i,j+1} \in \lambda$ and $\square_{i+1,j} \in \lambda$, then we say that $\square_{i,j}$ is a *corner boundary box*. If $\square_{i,j+1} \in \lambda$ and $\square_{i+1,j} \notin \lambda$, or if $\square_{i,j+1} \notin \lambda$ and $\square_{i+1,j} \in \lambda$, then we say that $\square_{i,j}$ is a *flat boundary box*.

If λ has a box with z-label z_i, then λ has an exactly one boundary box with z-label z_i. We define new parameters y_i by the following formulas. We define

$z_i = y_i - 1$, if the boundary box with label z_i is an active boundary box,

$z_i = y_i + 1$, if the boundary box with label z_i is a corner boundary box,

$z_i = y_i$, if the boundary box with label z_i is a flat boundary box.

To every box $\square_{i,j} \in \lambda$ we assign one of $y_1, \dots, y_{n-1}$ by the rule:

$$y(\square_{i,j}) := y_{i-j+r}.$$

We say that y_{i-j+r} is the y-label of a box $\square_{i,j}$.

LEMMA 2.6.

(i) *Let* $\square$ *be a box in* λ *with hook-weight* $h(\square)(z) = 1 + z_a + z_{a+1} + \cdots + z_b$ *for some* a, b. *Then*

$$h(\square)(z(y)) = y_a + y_{a+1} + \cdots + y_b. \tag{2.13}$$

(ii) *Let* $\mu < \lambda$ *and let*

$$G_{\lambda/\mu}(z) = \sum_{i=1}^{n-1} k_i(k_i - k_{i+1} + z_i)$$

be the z*-content of the skew-diagram* λ/μ. *Then*

$$G_{\lambda/\mu}(z(y)) = \sum_{i=1}^{n-1} k_i y_i. \tag{2.14}$$

For example, in the notation of formula (2.8), the change of variables for λ is

$$z_1 = y_1, \qquad z_2 = y_2 - 1, \qquad z_3 = y_3.$$

Then $G_{\lambda/\emptyset}(z) = z_1 + 2z_2 + z_3 + 2$ and $G_{\lambda/\emptyset}(z(y)) = y_1 + 2y_2 + y_3$. Similarly, $G_{\lambda/\mu_2}(z) = z_1 + z_2 + z_3 + 1$ and $G_{\lambda/\mu_2}(z(y)) = y_1 + y_2 + y_3$.

PROOF. The proof of the lemma is straightforward. For example, we prove part (i). Let $\square_{i,j}$ be a box in λ with hook-weight $h(\square_{i,j})(z) = 1 + z_a + z_{a+1} + \cdots + z_b$ for some a, b. The boxes $\square_{i,a}$ and $\square_{b,j}$ are boundary boxes of λ. Let us walk from the box $\square_{i,a}$ to the box $\square_{b,j}$ through the boundary boxes of λ. This walk consists of $b - a + 1$ boundary boxes with z-labels $z_a, z_{a+1}, \dots, z_b$. Let ℓ be the number of active boundary boxes in this walk. Then the walk has exactly $\ell - 1$ corner boundary boxes. Hence our change of variables transforms $h(\square)(z) = 1 + z_a + z_{a+1} + \cdots + z_b$ to $1 + y_a + y_{a+1} + \cdots + y_b - \ell + (\ell - 1) = y_a + y_{a+1} + \cdots + y_b$. Part (i) is proved. $\square$

Having Lemma 2.6 we rewrite Corollary 2.4 in terms of the variables y_i. Namely, define the y-hook-weight of a box $\square \in \lambda$ by the formula

$$\tilde{h}(\square) = \sum_{\square' \in H_\lambda(\square)} y(\square'),$$

and the y-content of a skew-diagram λ/μ by the formula

$$\tilde{G}_{\lambda/\mu} = \sum_{i=1}^{n-1} k_i y_i,$$

if k_i is the number of boxes in λ/μ with y-label y_i. Then

$$\tilde{h}(\square)(y) = h(\square)(z(y)), \qquad \tilde{G}_{\lambda/\mu}(y) = G_{\lambda/\mu}(z(y))$$

by Lemma 2.6. Formula (2.11) takes the form:

$$F_{\lambda/\mu}(z(y)) = \sum_{\mu=\mu_1<\mu_2<\cdots<\mu_d<\lambda} \frac{1}{\prod_{i=1}^{d} \tilde{G}_{\lambda/\mu_i}(y)}. \tag{2.15}$$

On the other hand, H. Naruse's formula [**Na**, page 13] states that

$$\sum_{\mu=\mu_1<\mu_2<\cdots<\mu_d<\lambda} \frac{1}{\prod_{i=1}^{d} \tilde{G}_{\lambda/\mu_i}(y)} = \frac{1}{\prod_{\square\in\lambda} \tilde{h}(\square)(y)} \sum_{\nu\in E(\lambda/\mu)} \prod_{\square\in\nu} \tilde{h}(\square)(y), \tag{2.16}$$

see also [**IN**, **MPP1**]. Hence,

$$F_{\lambda/\mu}(z(y)) = \frac{1}{\prod_{\square\in\lambda} h(\square)(z(y))} \sum_{\nu\in E(\lambda/\mu)} \prod_{\square\in\nu} h(\square)(z(y)), \tag{2.17}$$

and Theorem 2.5 is proved.

2.9. Change of variable and weight shift. The change of variables $y \mapsto z = z(y)$ can be understood in terms of weights as follows:

LEMMA 2.7. *Let* $\zeta\colon \mathbb{C}^n \to \mathbb{C}^{n-1}$ *be the linear map* $t \mapsto (t_2 - t_1, \dots, t_n - t_{n-1})$. *If* $z = \zeta(t)$ *then* $y = \zeta(t - w(\lambda))$.

PROOF. The weight corresponding to the Young diagram λ is $w(\lambda) = (\epsilon_1, \dots, \epsilon_n)$ where $\epsilon_i \in \{0, 1\}$ with $\epsilon_i = 1$ iff $i \in \{i_1 < \cdots < i_r\}$ where $i_k = \lambda_k + k$ $(k = 1, \dots, r)$, see Section 2.3. Then for each $i = 1, \dots, n-1$, we have

- $\epsilon_{i+1} - \epsilon_i = -1$ if $i = i_k$ for some $k \in \{1, \dots, r\}$ and $i_k < i_{k+1} - 1$, where we set $i_{r+1} = n + 1$,
- $\epsilon_{i+1} - \epsilon_i = 1$ if $i + 1 = i_k$ for some $k \in \{1, \dots, r\}$ and $i_{k-1} < i_k - 1$, where we set $i_0 = 0$,
- $\epsilon_{i+1} - \epsilon_i = 0$, otherwise.

The first alternative occurs iff $i = i_k$ and $\lambda_k < \lambda_{k+1}$ where we set $\lambda_{r+1} = n - r$. This is exactly the condition for the box with coordinates $(\lambda_k, r - k)$, which has z-label z_i, to be an active boundary box.

The second alternative occurs iff $i+1 = i_k$ and $\lambda_{k-1} < \lambda_k$ where we set $\lambda_0 = 0$. This is exactly the condition for the boundary box with coordinates $(\lambda_k, r - k + 1)$, which has z-label z_i, to be a corner boundary box. $\square$

3. Applications

3.1. Master function. Let λ be a Young diagram inscribed in the $(n-r) \times r$ rectangle R. Let λ have k_i boxes with z-label z_i for $i = 1, \dots, n-1$. Denote $k = k_1 + \cdots + k_{n-1}$.

Consider $\mathbb{C}^k$ with coordinates $x = (x_{i,j})$, $i = 1, \dots, n-1$, $j = 1, \dots, k_i$. Define the *master function* [1]

(3.1)

$$\Phi_\lambda(x, z) = \prod_{i=1}^{n-1} \prod_{j=1}^{k_i} x_{i,j}^{z_i+1} \prod_{j=1}^{k_r} (x_{r,j} - 1)^{-1} \prod_{i=1}^{n-1} \prod_{j<j'} (x_{i,j} - x_{i,j'})^2 \prod_{i=1}^{n-2} \prod_{j=1}^{k_i} \prod_{j'=1}^{k_{i+1}} (x_{i,j} - x_{i+1,j'})^{-1}.$$

The linear functions $x_{i,j}$, $x_{r,j} - 1$, $x_{i,j} - x_{i,j'}$, $x_{i,j} - x_{i+1,j'}$ appearing in the master function define an arrangement $\mathcal{C}$ of hyperplanes in $\mathbb{C}^k$.

The group $G = S_{k_1} \times \cdots \times S_{k_{n-1}}$ acts on $\mathbb{C}^k$ by permuting the coordinates $(x_{i,j})$ with the same first index i. The arrangement $\mathcal{C}$ and master function $\Phi_\lambda(x, z)$ are G-invariant.

For $\kappa \in \mathbb{C}^\times$, the multivalued function $\Phi_\lambda(x, z)^{1/\kappa}$ defines a rank one local system $\mathcal{L}_\kappa$ on the complement $X = \mathbb{C}^k \backslash \mathcal{C}$ to the arrangement. The group G acts on the homology $H_*(X; \mathcal{L}_\kappa)$ and cohomology $H^*(X; \mathcal{L}_\kappa)$. Let $H_k(X; \mathcal{L}_\kappa)^- \subset H_k(X; \mathcal{L}_\kappa)$ and $H^k(X; \mathcal{L}_\kappa)^- \subset H^k(X; \mathcal{L}_\kappa)$ be the isotypical components corresponding to the sign representation. It is known that for generic κ, we have $\dim H^k(X; \mathcal{L}_\kappa)^- = \dim H_k(X; \mathcal{L}_\kappa)^- = 1$ since the space $H^k(X; \mathcal{L}_\kappa)^-$ can be identified with the space

[1]The *superpotential* in the terminology of enumerative geometry.

of singular vectors in $M \otimes U_r$ of weight $w(\lambda) + t - \rho$, which is of dimension 1, see [**SV**].

3.2. Weight function of u_λ. Let u_λ be the basis vector of U_r corresponding to the diagram λ, see notations in Section 2.3. The vector u_λ is related to the highest weight vector $u_\emptyset$ by the formula

$$u_\lambda = f_{\ell_k} \dots f_{\ell_2} f_{\ell_1} u_\emptyset, \tag{3.2}$$

where $f_{\ell_k}, \dots, f_{\ell_2}, f_{\ell_1}$ is a certain (*admissible*) sequence of Cartan generators $f_1, \dots, f_{n-1}$ in which there are exactly k_i elements f_i for every $i = 1, \dots, n-1$. Let $\mathcal{F}_\lambda$ be the set of all such admissible sequences.

Let $f = \{f_{\ell_k}, \dots, f_{\ell_1}\}$ be an admissible sequence. Define the function $W_f^\circ(x)$,

$$W_f^\circ(x) = \frac{1}{(x_{a_k,b_k} - x_{a_{k-1},b_{k-1}}) \dots (x_{a_3,b_3} - x_{a_2,b_2})(x_{a_2,b_2} - x_{a_1,b_1})(x_{a_1,b_1} - 1)} \tag{3.3}$$

such that

(i) each variable $x_{i,j}$ is present in (3.3),
(ii) if $(x_{a_c,b_c} - x_{a_{c-1},b_{c-1}})$ is any of the factors, then $a_c = \ell_c$,
(iii) for any i and $1 \leqslant j < j' \leqslant k_i$, the variable $x_{i,j}$ appears in (3.3) on the right from the variable $x_{i,j'}$.

These properties determine the function $W_f^\circ(x)$ uniquely.

Define

$$W_\lambda(x) = \mathrm{Sym}_{x_{1,1},\dots,x_{1,k_1}} \dots \mathrm{Sym}_{x_{n-1,1},\dots,x_{n-1,k_{n-1}}} \Big[\sum_{f \in \mathcal{F}_\lambda} W_f^\circ(x) \Big],$$

where we use the notation $\mathrm{Sym}_{t_1,\dots,t_j} P(t_1, \dots, t_j) := \sum_{\sigma \in S_j} P(t_{\sigma(1)}, \dots, t_{\sigma(j)})$. The function $W_\lambda(x)$ is called the *weight function* of the vector $v \otimes u_\lambda$ in $M \otimes U_r$.

For example, if $(r, n) = (2, 4)$, $\lambda = (2, 2)$, then $x = (x_{1,1}, x_{2,1}, x_{2,2}, x_{3,1})$. There are two admissible sequences

$$u_\lambda = f_2 f_3 f_1 f_2 u_\emptyset = f_2 f_1 f_3 f_2 u_\emptyset,$$

and

$$\begin{aligned} W_\lambda(x) = & \frac{1}{(x_{2,2} - x_{3,1})(x_{3,1} - x_{1,1})(x_{1,1} - x_{2,1})(x_{2,1} - 1)} \\ & + \frac{1}{(x_{2,1} - x_{3,1})(x_{3,1} - x_{1,1})(x_{1,1} - x_{2,2})(x_{2,2} - 1)} \\ & + \frac{1}{(x_{2,2} - x_{1,1})(x_{1,1} - x_{3,1})(x_{3,1} - x_{2,1})(x_{2,1} - 1)} \\ & + \frac{1}{(x_{2,1} - x_{1,1})(x_{1,1} - x_{3,1})(x_{3,1} - x_{2,2})(x_{2,2} - 1)}. \end{aligned}$$

3.3. Two integrals. Let $\gamma \in H_k(X; \mathcal{L}_\kappa)^-$ be a generator. Let

$$\wedge_{i,j} dx_{i,j}$$

denote the wedge product in the lexicographic order of all the differentials $dx_{i,j}$. Define two functions

$$I_\lambda(z, \kappa) = \int_\gamma \Phi_\lambda(x, z)^{1/\kappa} W_\lambda(x) (\wedge_{i,j} dx_{i,j}), \quad V_\lambda(z, \kappa) = \int_\gamma \Phi_\lambda(x, z)^{1/\kappa} \frac{1}{\prod_{i,j} x_{i,j}} (\wedge_{i,j} dx_{i,j}).$$

Both function are multiplied by the same nonzero constant if we choose a different generator.

As shown in [**MV**], the first function is a *hypergeometric solution* of the dynamical difference equations associated with the weight subspace $U_r[w(\lambda)]$ of the $\mathfrak{gl}_n$-module U_r. The dynamical equations were introduced in [**TV**]. The (hypergeometric) solutions of the dynamical equations were constructed in [**MV**]. The dynamical equations is a system of difference equations of the form

$$I(z_1,\dots,z_i+\kappa,\dots,z_{n-1},\kappa)=a_i(z_1,\dots,z_{n-1},\kappa)I(z_1,\dots,z_{n-1},\kappa),\qquad i=1,\dots,n-1,$$

for suitable coefficients a_i defined in terms of the $\mathfrak{gl}_n$-action on U_r.

We call the second function $V_\lambda(z,\kappa)$ – the *vertex integral* associated with the weight subspace $U_r[w(\lambda)]$ of the $\mathfrak{gl}_n$-module U_r.

THEOREM 3.1. *We have*

$$V_\lambda(z,\kappa)=\frac{1}{\prod_{\Box\in\lambda}h(\Box)(z)}\,I_\lambda(z,\kappa). \tag{3.4}$$

The starting goal of this project was to find the coefficient of proportionality between the vertex integral $V_\lambda(z,\kappa)$ and the hypergeometric solution $I_\lambda(z,\kappa)$ which turned out to be the inverse of the product of the hook-weights of the boxes of the Young diagram λ.

PROOF. In [**SV**], given $M\otimes U_r$ and $\lambda\in\mathcal{I}_r$, a vector $\bar v(\lambda)$ is constructed,

$$\bar v(\lambda):=\bar v_\lambda\otimes u_\lambda+\sum_{\mu<\lambda}\bar v_\mu\otimes u_\mu,\qquad \bar v_\lambda,\bar v_\mu\in H^k(X;\mathcal{L}_\kappa)^-\otimes M.$$

Thus $\bar v(\lambda)\in H^k(X;\mathcal{L}_\kappa)^-\otimes M\otimes U_r$. The vector $\bar v(\lambda)$ has $\mathfrak{gl}_n$-weight $w(\lambda)+t-\rho$ and is singular with respect to the factors $M\otimes U_r$. The vector $\bar v(\lambda)$ is a cohomological version of the vector $v(\lambda)$ defined in (2.3) and studied in the previous sections.

The vector $\bar v_\lambda$ is represented by the differential form

$$\big(\Phi_\lambda(x,z)^{1/\kappa}W_\lambda(x)\big(\wedge_{i,j}\,dx_{i,j}\big)\big)\otimes v,$$

see [**SV**].

The vector $\bar v_\emptyset$ is represented by a differential form constructed as follows. A sequence $f_{\ell_k},\dots,f_{\ell_2},f_{\ell_1}$ is called *weakly admissible* if for $i=1,\dots,n-1$, the sequence contains exactly k_i elements f_i. Let $\mathcal{F}^\star_\lambda$ be the set of all weakly admissible sequences.

For example, if $(r,n)=(2,4)$ and $\lambda=(2,2)$, then $\mathcal{F}^\star_\lambda$ consists of 12 sequences: $\{f_2,f_2,f_1,f_3\}$, ..., $\{f_3,f_1,f_2,f_2\}$.

Let $f=\{f_{\ell_k},\dots,f_{\ell_1}\}$ be a weakly admissible sequence. Define the function $W^\star_f(x)$ by the formula

$$W^\star_f(x)=\frac{1}{(x_{a_k,b_k}-x_{a_{k-1},b_{k-1}})\dots(x_{a_3,b_3}-x_{a_2,b_2})(x_{a_2,b_2}-x_{a_1,b_1})x_{a_1,b_1}} \tag{3.5}$$

such that

(i) each variable $x_{i,j}$ is present in (3.5),
(ii) if $(x_{a_c,b_c}-x_{a_{c-1},b_{c-1}})$ is any of the factors, then $a_c=\ell_c$,
(ii') $(a_1,b_1)=(\ell_1,1)$,
(iii) for any i and $1\leqslant j<j'\leqslant k_i$, the variable $x_{i,j}$ appears in (3.3) on the right from the variable $x_{i,j_i'}$.

These properties determine the function $W_f^\star(x)$ uniquely.

Notice that the last factors in (3.3) and (3.5) are different.

Define the function $W_f(x)$ by the formula

$$W_f(x) = \mathrm{Sym}_{x_{1,1},\dots,x_{1,k_1}} \dots \mathrm{Sym}_{x_{n-1,1},\dots,x_{n-1,k_{n-1}}} \big[W_f^\star(x)\big]. \tag{3.6}$$

Then the vector $v_\emptyset$ is represented by the differential form

$$\sum_{f=\{f_{\ell_k},\dots,f_{\ell_1}\}\in\mathcal{F}_\lambda^\star} \big(\Phi_\lambda(x,z)^{1/\kappa} W_f(x)\big(\wedge_{i,j}\, dx_{i,j}\big)\big)\otimes f_{\ell_k}\dots f_{\ell_1} v,$$

see [**SV**].

Let $\gamma \in H_k(X;\mathcal{L}_\kappa)^-$ be a generator. The integral of $\bar v(\lambda)$ over γ is a scalar multiple the vector $v(\lambda)$,

$$\int_\gamma \bar v(\lambda) = c(z,\kappa)\, v(\lambda).$$

We apply the linear function $\psi : M \to \mathbb{C}$ to both sides of this equation and equate the coefficients of u_λ and $u_\emptyset$. Then

$$c(z,\kappa) = \int_\gamma \Phi(x,z)^{1/\kappa} W_\lambda(x)\big(\wedge_{i,j}\, dx_{i,j}\big),$$

$$c(z,\kappa)\, F_{\lambda/\emptyset}(z) = \int_\gamma \Phi(x,z)^{1/\kappa}\Big(\sum_{f\in\mathcal{F}_\lambda^\star} W_f(x)\Big)\big(\wedge_{i,j}\, dx_{i,j}\big).$$

Using the formula

$$\sum_{\sigma\in S_k} \frac{1}{(G_{\sigma(k)}-G_{\sigma(k-1)})(G_{\sigma(k-1)}-G_{\sigma(k-2)})\dots(G_{\sigma(2)}-G_{\sigma(1)})G_{\sigma(a)}} = \frac{1}{\prod_{j=1}^k G_j}$$

and the definition of $W_f(x)$ we conclude that

$$\sum_{f\in\mathcal{F}_\lambda^\star} W_f(x) = \frac{1}{\prod_{i,j} x_{i,j}}.$$

Hence

$$c(z,\kappa) = I_\lambda(z,\kappa), \qquad c(z,\kappa)\, F_{\lambda/\emptyset}(z) = V_\lambda(z,\kappa).$$

Now formula (2.6) implies Theorem 3.1. □

3.4. Whittaker vectors. Let $\mathfrak{n}^-$ be the maximal nilpotent subalgebra of $\mathfrak{gl}_n$ of lower triangular matrices. It is generated by $f_1,\dots,f_{n-1}$. Let $\eta\colon \mathfrak{n}^- \to \mathbb{C}$ be the character of the Lie algebra $\mathfrak{n}^-$ such that $\eta(f_i) = -1$ for all i.

A *Whittaker vector* in a $\mathfrak{gl}_n$-module V is a vector $u \in V$ so that $xu = \eta(x)u$ for all $x \in \mathfrak{n}^-$. This notion was introduced and studied by B. Kostant, [**Ko**]. The space of Whittaker vectors in V is denoted $\mathrm{Wh}(V)$. It is a module over the center Z of the universal enveloping algebra of $\mathfrak{gl}_n$. A Whittaker vector $u \neq 0$ such that $zu = \chi(z)u$ for all $z \in Z$ and some character $\chi\colon Z \to \mathbb{C}$ is said to have infinitesimal character χ.

For example let $M' = \mathrm{Hom}_{\mathbb{C}}(M,\mathbb{C})$ be the dual of a Verma module M. It is a $\mathfrak{gl}_n$-module for the action $(x\alpha)(m) = -\alpha(xm)$, $\alpha \in M$, $m \in M$, $x \in \mathfrak{gl}_n$. Central elements $z \in Z$ act on M as multiples $\chi_{M'}(z)$ of the identity, for some character $\chi_{M'}\colon Z \to \mathbb{C}$. The linear function $\psi \in M'$ of Section 2.1 is defined by the conditions $f_i\psi = -\psi$ and $\psi(v) = 1$ and is in particular a Whittaker vector. On the other hand, any Whittaker vector in M' is uniquely determined by its value on v since v generates M as a module over $U(\mathfrak{n}^-)$. Thus:

LEMMA 3.2. *The space of Whittaker vectors* $\mathrm{Wh}(M')$ *is one-dimensional, spanned by the Whittaker vector* ψ *of infinitesimal weight* $\chi_{M'}$.

More generally let us consider the problem of describing the Z-module of Whittaker vectors in the $\mathfrak{gl}_n$-module $M' \otimes U \cong \mathrm{Hom}_{\mathbb{C}}(M, U)$ for a Verma module M and a fundamental module U. By definition $\alpha\colon M \to U$ is a Whittaker vector if and only if

$$x\alpha(m) = \alpha(xm) + \eta(x)\alpha(m), \quad \forall m \in M, x \in \mathfrak{n}^-.$$

It follows that a Whittaker vector α is again uniquely determined by its value on the highest weight vector $v \in M$.

Let $M_{t-\rho}$ denote the Verma module of highest weight $t - \rho \in \mathbb{C}^n$. Let $\chi(t) = \chi_{M'_{t-\rho}}$ be the infinitesimal character of its dual.

PROPOSITION 3.3. *Let* $r \in \{1, \dots, n-1\}$ *and* $t \in \mathbb{C}^n$ *be generic and set* $z_i = t_{i+1} - t_i$, $(i = 1, \dots, n-1)$. *Then for each Young diagram* $\lambda \in \mathcal{I}_r$ *there is a unique Whittaker vector* $\alpha_{\lambda,t} \in \mathrm{Hom}_{\mathbb{C}}(M_{w(\lambda)+t-\rho}, U_r)$ *of infinitesimal character* $\chi(t)$ *such that*

$$\alpha_{\lambda,t}(v) = \sum_{\mu \leqslant \lambda} F_{\lambda/\mu}(z) u_\mu,$$

where

$$F_{\lambda/\mu}(z) = \frac{1}{\prod_{\Box \in \lambda} h(\Box)} \sum_{\nu \in E(\lambda/\mu)} \prod_{\Box \in \nu} h(\Box),$$

and $h(\Box) = 1 + \sum_{\Box' \in H_\lambda(\Box)} z(\Box')$.

PROOF. The morphism of $\mathfrak{gl}_n$-modules $M_{w(\lambda)+t-\rho} \to M_{t-\rho} \otimes U_r$ sending the highest weight vector to the singular vector $v(\lambda) = \sum_{\mu \leqslant \lambda} v_\mu \otimes u_\mu$, see Section 2.4, induces a morphism

$$M'_{t-\rho} \to \mathrm{Hom}_{\mathbb{C}}(M_{w(\lambda)+t-\rho}, U_r).$$

The morphism property implies that it sends the Whittaker vector ψ of infinitesimal character $\chi(t)$ to a Whittaker vector α with the same infinitesimal character. By definition $\alpha(v) = \sum_{\mu \leqslant \lambda} \psi(v_\mu) u_\mu$ and $F_{\lambda/\mu} = \psi(v_\mu)$ is given in Theorem 2.5. □

We thus obtain an explicit diagonalization of the action of the center Z on the space of Whittaker vectors in $M' \otimes U$ for a generic Verma module M and a fundamental module U:

THEOREM 3.4. *Let* $t \in \mathbb{C}^n$, $r \in \{1, \dots, n-1\}$ *and* $y_i = t_{i+1} - t_i$ $(i = 1, \dots, n-1)$. *Then*

$$W = \mathrm{Wh}(\mathrm{Hom}_{\mathbb{C}}(M_{t-\rho}, U_r))$$

has dimension $\dim(U_r)$. *For generic* t, W *decomposes into a direct sum* $W = \oplus_{\lambda \in \mathcal{I}_r} W_\lambda$ *of* Z*-invariant one-dimensional subspaces on which* Z *acts by the character* $\chi(t - w(\lambda))$. *The subspace* W_λ *is spanned by the Whittaker vector* $\beta_{\lambda,t}$, *such that* $\beta_{\lambda,t}(v) = \sum_{\mu \leqslant \lambda} \tilde{F}_{\lambda/\mu}(y) u_\mu$ *with*

$$\tilde{F}_{\lambda/\mu}(y) = \frac{1}{\prod_{\Box \in \lambda} \tilde{h}(\Box)} \sum_{\nu \in E(\lambda/\mu)} \prod_{\Box \in \nu} \tilde{h}(\Box),$$

and $\tilde{h}(\Box) = \sum_{\Box' \in H_\lambda(\Box)} y(\Box')$.

PROOF. By Proposition 3.3 the Whittaker vectors $\beta_{\lambda,t} = \alpha_{\lambda,t-w(\lambda)}$ belong to W and have infinitesimal character $\chi(t - w(\lambda))$. Since $\beta_\lambda(v) = u_\lambda$ plus a linear combination of u_μ with $\mu < \lambda$, these vectors are linearly independent. We need to show that they span W. Let β is a Whittaker vector. Since the vectors $\beta_\lambda(v)$ form a basis of U_r there exist coefficients $c_i \in \mathbb{C}$ so that $\gamma = \beta - \sum_{\lambda\in\mathcal{I}_r} c_\lambda\beta_\lambda$ is a Whittaker vector which vanishes on v. Since a Whittaker vector is uniquely determined by its value on v, γ must be zero.

Let $t' = t - w(\lambda)$ and $z_i' = t_{i+1}' - t_i'$, $(i = 1, \ldots, n-1)$. We need to compute the coeffcients $F_{\lambda/\mu}(z')$. By Lemma 2.7, $z' = z(y)$ defined in Section 2.8 and therefore $F_{\lambda/\mu}(z') = \tilde{F}_{\lambda/\mu}(y)$. □

Acknowledgments

The authors thank I. Pak for useful comments on references and notations and for clarifying that Theorem 2.2 can also be derived from formula (5.2) in [**IN**] using the change of variables discussed in Section 2.8, cf. formula (3) in [**Kon**]. The fourth author thanks FIM at ETH Zurich and IHES in Bures-sur-Yvette for hospitality in June-July 2023. The fourth author also thanks E. Mukhin for useful discussions.

References

[AO] M. Aganagic and A. Okounkov, *Quasimap counts and Bethe eigenfunctions*, Mosc. Math. J. **17** (2017), no. 4, 565–600, DOI 10.17323/1609-4514-2016-16-4-565-600. MR3734654

[DS1] H. Dinkins, A. Smirnov, *Quasimaps to zero-dimensional A_∞-quiver varieties*, International Mathematics Research Notices, Vol. 2022, No. 2, pp. 1123–1153. Advance Access Publication June 4, 2020, `https://doi.org/10.1093/imrn/rnaa129`.

[DS2] H. Dinkins and A. Smirnov, *Capped vertex with descendants for zero dimensional A_∞ quiver varieties*, Adv. Math. **401** (2022), Paper No. 108324, 30, DOI 10.1016/j.aim.2022.108324. MR4394686

[IN] T. Ikeda and H. Naruse, *Excited Young diagrams and equivariant Schubert calculus*, Trans. Amer. Math. Soc. **361** (2009), no. 10, 5193–5221, DOI 10.1090/S0002-9947-09-04879-X. MR2515809

[Kon] M. Konvalinka, *A bijective proof of the hook-length formula for skew shapes*, Sém. Lothar. Combin. **80B** (2018), Art. 13, 12. MR3940588

[Ko] B. Kostant, *On Whittaker vectors and representation theory*, Invent. Math. **48** (1978), no. 2, 101–184, DOI 10.1007/BF01390249. MR507800

[MO] D. Maulik and A. Okounkov, *Quantum groups and quantum cohomology* (English, with English and French summaries), Astérisque **408** (2019), ix+209, DOI 10.24033/ast. MR3951025

[MPP1] A. H. Morales, I. Pak, and G. Panova, *Hook formulas for skew shapes I. q-analogues and bijections*, J. Combin. Theory Ser. A **154** (2018), 350–405, DOI 10.1016/j.jcta.2017.09.002. MR3718070

[MPP2] A. Morales, I. Pak, G. Panova, *Hook formulas for skew shapes IV. Increasing tableaux and factorial Grothendieck polynomials*, `arXiv:2108.10140`, 1–26

[MV] Y. Markov and A. Varchenko, *Hypergeometric solutions of trigonometric KZ equations satisfy dynamical difference equations*, Adv. Math. **166** (2002), no. 1, 100–147, DOI 10.1006/aima.2001.2027. MR1882860

[Na] H. Naruse, *Schubert calculus and hook formula*, talk slides at 73rd Sem. Lothar. Combin., Strobl, Austria, 2014; available at `www.mat.univie.ac.at/~slc/wpapers/s73vortrag/naruse.pdf`

[O] A. Okounkov, *Lectures on K-theoretic computations in enumerative geometry*, Geometry of moduli spaces and representation theory, IAS/Park City Math. Ser., vol. 24, Amer. Math. Soc., Providence, RI, 2017, pp. 251–380, DOI 10.1090/pcms/024. MR3752463

[SV] V. V. Schechtman and A. N. Varchenko, *Arrangements of hyperplanes and Lie algebra homology*, Invent. Math. **106** (1991), no. 1, 139–194, DOI 10.1007/BF01243909. MR1123378

[SmV1] A. Smirnov and A. Varchenko, *The p-adic approximations of vertex functions via 3D mirror symmetry*, Lett. Math. Phys. **115** (2025), no. 3, Paper No. 56, DOI 10.1007/s11005-025-01944-x. MR4908958

[SmV2] A. Smirnov, A. Varchenko, *Polynomial superpotential for Grassmannian* $\mathrm{Gr}(k,n)$ *from a limit of vertex function*,Arnold Mathematical Journal, `https://doi.org/10.1007/s40598-024-00245-w`, 1–18.

[TV] V. Tarasov and A. Varchenko, *Difference equations compatible with trigonometric KZ differential equations*, Internat. Math. Res. Notices **15** (2000), 801–829, DOI 10.1155/S1073792800000441. MR1780748

DEPARTMENT OF MATHEMATICS, ETH ZURICH, 8092 ZURICH, SWITZERLAND
Email address: `giovanni.felder@math.ethz.ch`

DEPARTMENT OF MATHEMATICS, UNIVERSITY OF NORTH CAROLINA AT CHAPEL HILL, CHAPEL HILL, NORTH CAROLINA 27599-3250
Email address: `asmirnov@email.unc.edu`

DEPARTMENT OF MATHEMATICAL SCIENCES, INDIANA UNIVERSITY –PURDUE UNIVERSITY INDIANAPOLIS, 402 NORTH BLACKFORD ST, INDIANAPOLIS, INDIANA 46202-3216
Email address: `vtarasov@iupui.edu`

DEPARTMENT OF MATHEMATICS, UNIVERSITY OF NORTH CAROLINA AT CHAPEL HILL, CHAPEL HILL, NORTH CAROLINA 27599-3250
Email address: `anv@email.unc.edu`

Contemporary Mathematics
Volume **823**, 2025
https://doi.org/10.1090/conm/823/16494

Periodic and quasi-motivic pencils of flat connections

Pavel Etingof and Alexander Varchenko

To the memory of Igor Krichever

ABSTRACT. We introduce a new notion of a **periodic pencil of flat connections** on a smooth algebraic variety X. This is a family $\nabla(s_1, \ldots, s_n)$ of flat connections on a trivial vector bundle on X depending linearly on parameters $s_1, \ldots, s_n$ and generically invariant, up to isomorphism, under the shifts $s_i \mapsto s_i + 1$ for all i. If in addition ∇ has regular singularities, we call it a **quasi-motivic pencil**. We use tools from complex analysis to establish various remarkable properties of such pencils over $\mathbb{C}$. For example, we show that the monodromy of a quasi-motivic pencil is defined over the field of algebraic functions in $e^{2\pi i s_j}$, and that its singularities are constrained to an arrangement of hyperplanes with integer normal vectors. Then we show that many important examples of families of flat connections, such as Knizhnik-Zamolodchikov, Dunkl, and Casimir connections, are quasi-motivic and thus periodic pencils.

Besides being interesting in its own right, the periodic property of a pencil of flat connections turns out to be very useful in computing the eigenvalues of the p-curvature of its reduction to positive characteristic. This is done in the follow-up paper arXiv:2401.05652.

CONTENTS

2020 *Mathematics Subject Classification.* Primary 32G34, 81R12; Secondary 14D05, 34M35.
Key words and phrases. Flat connection, monodromy, Knizhnik-Zamolodchikov equations, Dunkl operators, Casimir connection.
The first author was partially supported by the NSF grant DMS-2001318.
The second author was partially supported by the NSF grant DMS-1954266.

1. Introduction

In this paper we introduce a new notion of a **periodic pencil of flat connections**. Namely, a **pencil of flat connections** is a family

$$\nabla(\mathbf{s}) = d - s_1 B_1 - \cdots - s_n B_n$$

of flat connections on the trivial rank N vector bundle over a smooth irreducible variety X over a field $\mathbf{k}$, where $B_j \in \Omega^1(X) \otimes \mathrm{Mat}_N(\mathbf{k})$ for $1 \le j \le n$. Such a pencil is said to be **periodic**[1] if it admits **shift operators**

$$A_j \in GL_N(\mathbf{k}(\mathbf{s})[X])$$

such that

$$\nabla(\mathbf{s} + \mathbf{e}_j) \circ A_j(\mathbf{s}) = A_j(\mathbf{s}) \circ \nabla(\mathbf{s}), \ 1 \le j \le n.$$

There are many interesting examples of periodic pencils, falling into three large classes: Knizhnik-Zamolodchikov (KZ) connections, Dunkl connections, and Casimir connections. Another important source of examples (often belonging to one of these classes) is equivariant quantum connections of conical symplectic resolutions of singularities with finitely many torus fixed points.

At the same time, periodic pencils enjoy many remarkable properties, especially when they have regular singularities. We call periodic pencils with regular singularities **quasi-motivic**. For example, we show that quasi-motivic pencils give rise to local systems defined over $\overline{\mathbb{Q}}$ in both de Rham and Betti realizations, and their singularities can only occur on an arrangement of hyperplanes.

In a follow-up paper [**EV2**], we describe the spectrum of the p-curvature of a periodic pencil in characteristic p. When applied to equivariant quantum connections for conical resolutions with finitely many torus fixed points, this has applications to symplectic geometry, since it has been shown in [**Lee**] that the p-curvature of such a connection coincides with the equivariant version of Fukaya's quantum Steenrod operation (at least up to a nilpotent correction that does not affect the spectrum).

The organization of the paper is as follows. In Section 2 we discuss preliminaries and auxiliary results. In Section 3 we discuss properties of periodic pencils of flat connections and of **pencils with periodic monodromy**, which is a weaker but more readily verifiable condition that makes sense over $\mathbb{C}$. This condition requires that the monodromy representation of $\nabla(\mathbf{s})$ up to isomorphism generically depend only on the exponentials $q_j := e^{2\pi i s_j}$. The main result in this section is Theorem 3.9, which states that the condition of periodic monodromy is equivalent to the monodromy of $\nabla(\mathbf{s})$ being defined over a finite Galois extension of the field $\mathbb{C}(q_1, \ldots, q_n)$ and stable under the Galois group of this extension. We also explain that equivariant quantum connections of conical symplectic resolutions with finitely many torus fixed points give rise to periodic pencils.

In Section 4 we introduce the notion of a **quasi-motivic pencil**, which is a pencil with periodic monodromy and regular singularities. We show in Theorem 4.1 that such a pencil is necessarily periodic. Then we proceed to study properties of quasi-motivic pencils, showing that their singularities in $\mathbf{s}$ can only occur on translates of hyperplanes in $\mathbb{C}^n$ defined over $\mathbb{Q}$. We also discuss **motivic** pencils, which are a large special class of quasi-motivic pencils arising as Gauss-Manin

[1] In fact, we introduce this definition in a larger generality of **families of flat connections**, where the dependence on s_j is allowed to be polynomial, but our main results only apply to pencils, where this dependence is linear.

connections. Finally, we discuss the notion of a **quasi-geometric** local system, which is a local system defined over $\overline{\mathbb{Q}}$ both in the de Rham and the Betti realization. It is a generalization of the notion of a geometric local system due to Deligne [**De1**]. We explain how quasi-motivic pencils give rise to quasi-geometric local systems, and then make a conjecture that local systems arising from braided fusion categories are quasi-geometric, discussing some evidence in support of this conjecture.

Finally, in Section 5 we review numerous examples of quasi-motivic (hence periodic) pencils, and discuss their irreducibility, unitarity and generic semisimplicity.

2. Preliminaries

2.1. Galois-stability. Let K be a field of characteristic 0. Let G be an affine algebraic group and Y an affine G-variety, both defined over K.

DEFINITION 2.1. A point $y \in Y(\overline{K})$ is said to be **Galois-stable** up to G-action if for any $\gamma \in \mathrm{Gal}(\overline{K}/K)$ there exists $g \in G(\overline{K})$ such that $\gamma(y) = gy$.

It is clear that whether or not a point y is Galois-stable up to G-action is determined by its $G(\overline{K})$-orbit, and that if $gy \in Y(K)$ for some $g \in G(\overline{K})$ then y is Galois-stable. The converse is false, however: e.g., take $K = \mathbb{R}$, $Y \subset \mathbb{A}^1$ defined by the equation $x^2 = -1$ (so $Y(K) = \emptyset$), $G = \{1, -1\}$ acting on x by multiplication, and $y = i$. So in general being Galois-stable up to G-action is a strictly weaker condition than being conjugate to a point defined over K.

EXAMPLE 2.2. Let A be a finitely generated K-algebra, N a positive integer, Y the variety of N-dimensional matrix representations of A, and $G = PGL_N$ acting on Y by conjugation. Then a representation $\eta : A \to \mathrm{Mat}_N(\overline{K})$ is Galois-stable up to G-action if and only if for any $\gamma \in \mathrm{Gal}(\overline{K}/K)$, the representations η and $\gamma(\eta)$ are isomorphic. In particular, this notion applies to representations of finitely generated groups, by taking A to be a group algebra. For brevity, from now on we will speak about Galois-stable representations, dropping the words "up to G-action".

Now let G, Y be defined over $\mathbb{C}$ and let $K := \mathbb{C}(\mathbf{q})$, where $\mathbf{q} := (q_1, \dots, q_n)$ is a collection of variables. Then for every $y \in Y(\overline{K})$ there exists a nonempty affine open subset $U \subset \mathbb{C}^n$ and a finite unramified Galois cover $p : \widetilde{U} \to U$ such that $y \in Y(\mathbb{C}[\widetilde{U}])$. Thus, given $\mathbf{a} := (a_1, \dots, a_n) \in U$ and $\widetilde{\mathbf{a}} \in p^{-1}(\mathbf{a})$, we can define the specialization $y|_{\widetilde{\mathbf{a}}} \in Y(\mathbb{C})$ of y at $\widetilde{\mathbf{a}}$. However, if η is Galois-stable up to G-action then the $G(\mathbb{C})$-orbit of $y|_{\widetilde{\mathbf{a}}}$ depends only on $\mathbf{a}$. So when we only care about the $G(\mathbb{C})$-orbit of this specialization, we will denote it by $y|_{\mathbf{a}}$.

2.2. Exponentially bounded holomorphic functions. Let V be a finite dimensional complex vector space, and $f : \mathbb{C}^n \to V$ a holomorphic function. We say that f is **exponentially bounded** if there exists $C > 0$ such that

$$\|f(\mathbf{s})\| = O(e^{C\|s\|}), \ \mathbf{s} \to \infty. \tag{2.1}$$

For $\mathbf{b} = (b_1, \dots, b_n) \in \mathbb{C}^n$ let $e^{\mathbf{b}} := (e^{b_1}, \dots, e^{b_n})$.

LEMMA 2.3. *Let $g, h : \mathbb{C}^n \to \mathbb{C}$ be exponentially bounded holomorphic functions such that $h \neq 0$. Suppose that the meromorphic function g/h is periodic under the lattice $\mathbb{Z}^n$. Then g/h is a rational function of $e^{2\pi i\mathbf{s}}$, $\mathbf{s} \in \mathbb{C}^n$.*

PROOF. We first prove the statement for $n = 1$. In this case by a well known theorem in complex analysis [**SS**, Chapter 5, Theorem 2.1], the number of zeros of g and h in the disk of radius R is $O(R)$ as $R \to \infty$. Thus so is the number of zeros and poles of g/h. But $g(s)/h(s) = f(e^{2\pi i s})$ where f is a meromorphic function on $\mathbb{C}^\times$. It follows that f has finitely many zeros and poles. Also, since g and h are exponentially bounded, f has at most power growth at zero and infinity. Thus f is rational, as claimed.

Now we prove the result in general. By the one-dimensional case, f is rational in each variable when the other variables are fixed. This implies that f is rational by the Hurwitz-Weierstrass theorem, [**BM**], 9.5, Theorem 5 (p.201). □

2.3. Generically periodic holomorphic maps up to G-action. Let $\mathbf{e}_1, \dots, \mathbf{e}_n$ be the standard basis of the standard n-dimensional space. Let G be an complex affine algebraic group and V a finite dimensional algebraic representation of G. Let $\rho : \mathbb{C}^n \to V$ be a holomorphic map.

DEFINITION 2.4. We say that ρ is **generically periodic** up to G-action if for each j the vectors $\rho(\mathbf{s} + \mathbf{e}_j)$ and $\rho(\mathbf{s})$ are G-conjugate for an analytically Zariski dense set of $\mathbf{s} \in \mathbb{C}^n$.

THEOREM 2.5. *Let $\rho : \mathbb{C}^n \to V$ be an exponentially bounded holomorphic map which is generically periodic up to G-action. Then there exists a nonempty affine open subset U of $(\mathbb{C}^\times)^n$, a finite unramified Galois cover $p : \widetilde{U} \to U$, and a Galois-stable up to G-action regular function $\eta : \widetilde{U} \to V$ such that $\rho(\mathbf{s})$ is $G(\mathbb{C})$-conjugate to $\eta|_{e^{2\pi i \mathbf{s}}}$ for $e^{2\pi i \mathbf{s}} \in U$. Moreover, η is unique up to action of $G(\overline{\mathbb{C}(\mathbf{q})})$.*

PROOF. Consider the closed subvarieties Y_i of V of vectors $y \in V$ with stabilizer $G_y \subset G$ of codimension $\le i$. Thus $Y_0 \subset Y_1 \subset \cdots \subset Y_{\dim G} = V$. Let d be the smallest integer such that $\rho : \mathbb{C}^n \to Y_d$.

Let $E \subset V$ be a generic affine subspace of codimension d (a *gauge fixing subspace*). Then for sufficiently generic $\mathbf{s}$ (namely, outside of a proper analytic subset $Z \subset \mathbb{C}^n$) the set $\Xi_{\mathbf{s}}$ of $v \in E$ which are G-conjugate to $\rho(\mathbf{s})$ has a finite cardinality m.

So we obtain a meromorphic map $R : \mathbb{C}^n \to \Sigma^m V$ (m-th symmetric power of V as a variety) with poles at Z given by $R(\mathbf{s}) = \Xi_{\mathbf{s}}$; i.e. for each regular function Ψ on $\Sigma^m V$ the function $f_\Psi := \Psi \circ R : \mathbb{C}^n \to \mathbb{C}$ is meromorphic with poles at Z.

Moreover, since ρ is generically periodic up to G-action, $R(\mathbf{s} + \mathbf{e}_j) = R(\mathbf{s})$ and therefore $f_\Psi(\mathbf{s} + \mathbf{e}_j) = f_\Psi(\mathbf{s})$ for an analytically Zariski dense set of $\mathbf{s}$. Thus $f_\Psi(\mathbf{s} + \mathbf{e}_j) = f_\Psi(\mathbf{s})$ as meromorphic functions on $\mathbb{C}^n$. So $f_\psi(\mathbf{s}) = F_\Psi(e^{2\pi i \mathbf{s}})$, where $F_\Psi : (\mathbb{C}^\times)^n \to \mathbb{C}$ is a meromorphic function.

Finally, by construction, f_Ψ can be written as g/h, where $g, h : \mathbb{C}^n \to \mathbb{C}$ are polynomials of $\rho(\mathbf{s})$, hence they are exponentially bounded. So it follows from Lemma 2.3 that F_Ψ is rational. Thus R descends to a rational map $\overline{R} : (\mathbb{C}^\times)^n \to \Sigma^m V$, which defines a point of $Q \in \Sigma^m V(\overline{\mathbb{C}(\mathbf{q})})$. Now any $\eta \in Q$ is a Galois-stable element of $\overline{\mathbb{C}(\mathbf{q})} \otimes V$, and in fact belongs to $\mathbb{C}[\widetilde{U}] \otimes V$ for some $U \subset (\mathbb{C}^\times)^n$ and $p : \widetilde{U} \to U$, such that η has the required property for $e^{2\pi i \mathbf{s}} \in U$.

It remains to show that η is unique up to G-action. Let η_1, η_2 be two such elements. We may assume that both of them are defined over $\mathbb{C}[\widetilde{U}]$ where $U \subset (\mathbb{C}^\times)^n$ is a nonempty affine open subset and $p : \widetilde{U} \to U$ is a finite unramified Galois cover. Let $\mathbf{a} \in U$, $U_{\mathbf{a}}$ be a small ball around $\mathbf{a}$, and U_1 be a connected component of $p^{-1}(U_{\mathbf{a}})$. There exists a connected component U_2 of $p^{-1}(U_{\mathbf{a}})$ such that for an analytically dense set of $\mathbf{b} \in U_{\mathbf{a}}$, $\eta_1|_{\mathbf{b}_1} = \eta_2|_{\mathbf{b}_2}$, where $\mathbf{b}_i$ are the preimages of $\mathbf{b}$ in U_i, $i = 1, 2$. So by Chevalley's elimination of quantifiers, there exists $g \in G(\overline{\mathbb{C}(\mathbf{q})})$ such that $g\eta_1 = \eta_2$. □

2.4. Algebraic independence of exponentials.

THEOREM 2.6. *Let $E : \mathbb{C}^n \to (\mathbb{C}^\times)^n$ be the exponential map given by*

$$E(s_1, \dots, s_n) := (e^{2\pi i s_1}, \dots, e^{2\pi i s_n}).$$

Suppose that $Y \subset \mathbb{C}^n$ is an irreducible algebraic hypersurface such that $E(Y)$ is also contained in an algebraic hypersurface. Then Y is a translate of a hyperplane defined over $\mathbb{Q}$. In other words, $E(Y)$ is an affine hypertorus in $(\mathbb{C}^\times)^n$.

PROOF. We may assume without loss of generality that Y contains 0, is smooth there, and is locally near 0 parametrized by the coordinates $s_1, \dots, s_{n-1}$. Then Y is defined near 0 by the equation

$$s_n = F(s_1, \dots, s_{n-1}),$$

where F is an algebraic function. Assume that Y is not a translate of a hyperplane defined over $\mathbb{Q}$. Then $s_1, \dots, s_n$ are analytic functions of $s_1, \dots, s_{n-1}$ linearly independent over $\mathbb{Q}$, and the Jacobi matrix $\frac{\partial(s_1,\dots,s_n)}{\partial(s_1,\dots,s_{n-1})}$ has rank $n-1$. Thus [**Ax**], Corollary 2, p.253 (for the complex field) implies that the transcendence degree of the field $\mathbb{C}(s_1, \dots, s_n, e^{2\pi i s_1}, \dots, e^{2\pi i s_n})$, is at least $2n-1$. So $e^{2\pi i s_1}, \dots, e^{2\pi i s_n}$ must be algebraically independent over $\mathbb{C}(s_1, \dots, s_n)$ (as it has transcendence degree $n-1$), hence they are algebraically independent over $\mathbb{C}$. □

2.5. Geometric local systems. In this subsection we briefly review Deligne's theory of geometric local systems (which we do not use here, and recall solely for motivational purposes). Let X be a smooth irreducible complex variety. The following definition is due to Deligne (see [**De1**]).

DEFINITION 2.7. A Betti $\overline{\mathbb{Q}}$-local system ρ on X is said to be **geometric** if there exists a non-empty Zariski open subset $X^\circ \subset X$ and a smooth proper surjective morphism $\pi : Y \to X^\circ$ such that $\rho|_{X^\circ}$ is a direct summand in the monodromy representation of the Gauss-Manin connection on $H^i(\pi^{-1}(\mathbf{x}), \overline{\mathbb{Q}})$, $\mathbf{x} \in X^\circ$, for some i.

REMARK 2.8.
(1) If ∇ has finite monodromy group G then it is geometric. Indeed, ∇ is a direct summand in a multiple of $\pi_*\nabla_{\rm triv}$, where $\nabla_{\rm triv}$ is the trivial connection on the trivial line bundle on a Galois cover $\pi : \widetilde{X} \to X$ with Galois group G, which implies the statement.
(2) By the Decomposition Theorem, any geometric system ρ is semisimple.
(3) The category of geometric systems is invariant under pullbacks, external products (hence tensor products), and duals. Thus for every X, we have the semisimple $\overline{\mathbb{Q}}$-linear Tannakian category $\mathcal{G}_{\rm B}(X)$ of geometric Betti local systems on X, which is a tensor subcategory of the Tannakian category $\mathcal{L}_{\rm B}(X, \overline{\mathbb{Q}})$ of all Betti $\overline{\mathbb{Q}}$-local systems on X. Similarly, the extension of scalars $\mathcal{G}_{\rm B}(X)_{\mathbb{C}}$ is a tensor subcategory of the Tannakian category $\mathcal{L}_{\rm B}(X, \mathbb{C})$ of all Betti $\mathbb{C}$-local systems on X.

(4) The same definition can be used in the de Rham realization (for $\mathcal{O}$-coherent D-modules with regular singularities). Thus for every X, we also have the semisimple $\mathbb{C}$-linear Tannakian category $\mathcal{G}_{\mathrm{dR}}(X)$ of geometric de Rham local systems on X, which is a tensor subcategory of the Tannakian category $\mathcal{L}_{\mathrm{dR}}(X)$ of all de Rham local systems on X with regular singularities. Moreover, if X is defined over $\overline{\mathbb{Q}}$ then we have the full category $\mathcal{L}_{\mathrm{dR}}(X,\overline{\mathbb{Q}}) \subset \mathcal{L}_{\mathrm{dR}}(X)$ of all de Rham local systems on X with regular singularities defined over $\overline{\mathbb{Q}}$.

(5) The Riemann-Hilbert correspondence (taking monodromy of a regular flat connection) sets up a canonical tensor equivalence $\mathcal{L}_{\mathrm{dR}}(X) \cong \mathcal{L}_{\mathrm{B}}(X,\mathbb{C})$ [**De2**], which in general (for X defined over $\overline{\mathbb{Q}}$) **does not** map the full subcategory $\mathcal{L}_{\mathrm{dR}}(X,\overline{\mathbb{Q}})$ into $\mathcal{L}_{\mathrm{B}}(X,\overline{\mathbb{Q}})_{\mathbb{C}}$. In other words, the monodromy of a holonomic system of linear differential equations defined over $\overline{\mathbb{Q}}$ need not be defined over $\overline{\mathbb{Q}}$. However, the Riemann-Hilbert correspondence **does** restrict to an equivalence between the subcategories $\mathcal{G}_{\mathrm{dR}}(X)$ and $\mathcal{G}_{\mathrm{B}}(X)_{\mathbb{C}}$, i.e., the monodromy of a direct summand of a Gauss-Manin connection is always defined over $\overline{\mathbb{Q}}$. In particular, every object of $\mathcal{G}_{\mathrm{dR}}(X)$ can be found in the de Rham cohomology of $\pi^{-1}(\mathbf{x})$ where $\pi : Y \to X^\circ$ is defined over $\overline{\mathbb{Q}}$, hence is itself defined over $\overline{\mathbb{Q}}$. This striking property follows from the fact that monodromy of a geometric de Rham local system can be computed in the Betti realization by following the deformation of twisted cycles in $Y_{\mathbf{x}}$ as $\mathbf{x}$ varies along X°.

3. Periodic families and pencils of flat connections

3.1. Polynomial families and pencils of flat connections. Let $\mathbf{k}$ be an algebraically closed field, X a smooth irreducible algebraic variety over $\mathbf{k}$, and V a finite dimensional $\mathbf{k}$-vector space.

DEFINITION 3.1. An n-**parameter (polynomial) family of flat connections** on X with values in V is a family of flat connections

$$\nabla(\mathbf{s}) = d - B(\mathbf{s}),$$

$\mathbf{s} := (s_1, \dots, s_n)$, on the trivial vector bundle $X \times V \to X$, where $B \in \Omega^1(X) \otimes \mathrm{End}\, V[\mathbf{s}]$ (i.e., $dB - [B,B] = 0$, where d is the de Rham differential on X and $[,]$ is the supercommutator in the graded algebra $\Omega^\bullet(X)\otimes\mathrm{End}\, V$). Such a family is called a **pencil** if B is linear homogeneous in $\mathbf{s}$, i.e. $B = \sum_{j=1}^n s_j B_j$, $B_j \in \Omega^1(X)\otimes\mathrm{End}\, V$.

Thus a pencil of flat connections is determined by a collection

$$B_j = \sum_{i=1}^r B_{ij} dx_i \in \Omega^1(X) \otimes \mathrm{End}\, V,\ 1 \le j \le n$$

of 1-forms on X with values in $\mathrm{End}\, V$ such that

$$dB_j = [B_j, B_k] = 0,\ 1 \le j,k \le n. \tag{3.1}$$

Of course, if $\dim X = 1$ then the flatness conditions (3.1) are vacuous, and B_j can be arbitrary. Yet this will already be an interesting case.

REMARK 3.2. These definitions extend mutatis mutandis to the case when V is a possibly nontrivial vector bundle on X. In this case the trivial connection d on the trivial bundle $X \times V \to X$ should be replaced by a fixed flat connection

$\nabla(0)$ on the bundle V. In particular, when ∇ is (affine) linear in $\mathbf{s}$, we have $\nabla(\mathbf{s}) = \nabla(0) - s_1B_1 - \cdots - s_nB_n$, where $B_i \in \Omega^1(X, \operatorname{End} V)$. In this case we will say that ∇ is an **affine pencil of flat connections**. Note that if V is trivial, we have $\nabla(0) = d - B_0$, so $\nabla(\mathbf{s}) = d - B$, where $B = B_0 + s_1B_1 + \cdots + s_nB_n$, which is an inhomogeneous (affine) linear function in $\mathbf{s}$, which justifies the terminology. Note that the notion of an affine pencil (unlike that of an ordinary pencil) is stable under gauge transformations, so it is the only meaningful notion when the bundle V is not necessarily trivial and there is no canonical reference connection $\nabla(0)$.

We will, however, mostly restrict ourselves to the case when the bundle V is trivial, since this is so in all the examples we consider.

3.2. Jumping loci of families of flat connections. For a flat connection ∇ on a vector bundle V on X, let $\Gamma(\nabla)$ be the space of flat global sections of ∇ in $\Gamma(X, V)$.

Now let $\nabla(\mathbf{s})$, $\mathbf{s} \in \mathbb{C}^n$ be a family of flat connections on the trivial bundle on X with fiber V. For $d \in \mathbb{N}$ let $Q_d(\nabla)$ be the set of $\mathbf{s} \in \mathbb{C}^n$ for which $\dim \Gamma(\nabla(\mathbf{s})) \geq d$.

LEMMA 3.3. *$Q_d(\nabla)$ is a countable union of Zariski closed subsets*[2] *of $\mathbb{C}^n$.*

PROOF. Fix a basis $\{f_i, i \geq 1\}$ of $\mathbb{C}[X]$, and for $N \leq \infty$ let $\mathbb{C}[X]_N$ be the span of f_i with $i \leq N$. For a flat connection ∇ on the trivial bundle on X with fiber V, let $\Gamma(\nabla)_N$ be the space of flat sections of ∇ in $\mathbb{C}[X]_N \otimes V$. In particular, $\Gamma(\nabla)_\infty = \Gamma(\nabla)$.

Now let $\nabla(\mathbf{s})$ be our family. For $d \in \mathbb{N}$ let $Q_{d,N}(\nabla)$ be the set of $\mathbf{s} \in \mathbb{C}^n$ for which $\dim \Gamma(\nabla(\mathbf{s}))_N \geq d$. Since $\Gamma(\nabla(\mathbf{s}))_N$ is defined inside $\mathbb{C}[X]_N \otimes V$ by a finite system of homogeneous linear equations with coefficients in $\mathbb{C}[\mathbf{s}]$, the subsets $Q_{d,N}(\nabla) \subset \mathbb{C}^n$ are Zariski closed. Thus $Q_d(\nabla) = Q_{d,\infty}(\nabla) = \bigcup_{N \in \mathbb{N}} Q_{d,N}(\nabla)$ is a countable union of Zariski closed subsets. □

It is clear that $Q_0(\nabla) = \mathbb{C}^n$, $Q_{d+1}(\nabla) \subset Q_d(\nabla)$, and $Q_{\dim V+1}(\nabla) = \emptyset$, so let $d = d(\nabla)$ be the largest integer in $[0, \dim V]$ for which $Q_d(\nabla) = \mathbb{C}^n$, i.e., the smallest possible value of $\dim \Gamma(\nabla(\mathbf{s}))$. The proper subset $J(\nabla) := Q_{d+1}(\nabla)$ of $\mathbb{C}^n$ is then called the **jumping locus** of ∇. Lemma 3.3 implies

COROLLARY 3.4. *The jumping locus $J(\nabla)$ is a countable union of proper Zariski closed subsets of $\mathbb{C}^n$.*

If a countable union of Zariski closed subsets of $\mathbb{C}^n$ is the whole $\mathbb{C}^n$ then one of them must be the whole $\mathbb{C}^n$, so there exists $N = N(\nabla) \in \mathbb{N}$ such $Q_{d,N}(\nabla) = \mathbb{C}^n$. Thus for any $\mathbf{s} \notin J(\nabla)$ we have $\Gamma(\nabla(\mathbf{s}))_N = \Gamma(\nabla(\mathbf{s}))$. Hence for any point $\mathbf{x} \in X$ the map $J(\nabla)^c \to \operatorname{Gr}(d, V)$ from the complement of $J(\nabla)$ to the Grassmannian of d-dimensional subspaces in V given by $\mathbf{s} \mapsto \Gamma(\nabla(\mathbf{s}))|_\mathbf{x}$ is rational, i.e., is the restriction of a rational map $\gamma(\nabla) : \mathbb{C}^n \to \operatorname{Gr}(d, V)$. Note that, as any rational map, this map is regular outside a Zariski closed subset $T(\nabla) \subset \mathbb{C}^n$ of codimension ≥ 2.

3.3. Periodic families and pencils of flat connections.

DEFINITION 3.5. A family ∇ is said to be **periodic** if there exist **shift operators**

$$A_j \in GL(V)(\mathbf{k}(\mathbf{s})[X]),\ 1 \leq j \leq n$$

---

[2]Note that $Q_d(\nabla)$ need not be Zariski closed. For example, if $\nabla(s) = d - \frac{s}{x}dx$ on $\mathbb{C}^\times$ then $Q_1(\nabla) = \mathbb{Z}$.

such that

$$\nabla(\mathbf{s}+\mathbf{e}_j)\circ A_j(\mathbf{s}) = A_j(\mathbf{s})\circ\nabla(\mathbf{s}),\ 1\le j\le n.$$

REMARK 3.6.

(1) Note that if $\nabla(\mathbf{s})$ has generically trivial endomorphism algebra (i.e., trivial over $\mathbb{C}(\mathbf{s})$) then A_j are unique up to multiplication by a scalar rational function of $\mathbf{s}$. In this case there exist scalar rational functions $\xi_{jk}(\mathbf{s})$ such that
$$A_k(\mathbf{s}+\mathbf{e}_j)A_j(\mathbf{s}) = \xi_{jk}(\mathbf{s})A_j(\mathbf{s}+\mathbf{e}_k)A_k(\mathbf{s}).$$
(2) As before, these definitions extend straightforwardly to the case when V is a possibly nontrivial vector bundle on X.

EXAMPLE 3.7. Let $\mathbf{k}=\mathbb{C}$, $X=\mathbb{C}^\times$, $V=\mathbb{C}^2$, and $\nabla(s) = d - B(s,x)$, where

$$B(s,x) := \frac{1}{x}\begin{pmatrix} s & 1\\ 0 & 0\end{pmatrix}.$$

This family is actually an affine pencil, since the dependence on s is affine linear. The equation for flat sections of $\nabla(s)$ is

$$(\tfrac{d}{dx} - B(s,x))F(x) = 0,$$

so it has a fundamental solution $F_s(x) := \begin{pmatrix} x^s & -\frac{1}{s}\\ 0 & 1\end{pmatrix}$. Thus $F_s(x)^{-1} = \begin{pmatrix} x^{-s} & \frac{x^{-s}}{s}\\ 0 & 1\end{pmatrix}$, so this family is periodic with

$$A(s,x) = F_{s+1}(x)F_s(x)^{-1} = \begin{pmatrix} x & \frac{x}{s}-\frac{1}{s+1}\\ 0 & 1\end{pmatrix}.$$

3.4. Families and pencils with periodic monodromy. Let $\nabla=\nabla(\mathbf{s})$ be a family of flat connections on a complex variety $X = X(\mathbb{C})$ with values in V parametrized by $\mathbb{C}^n$. Let $\mathbf{x}_0\in X$, and for $\mathbf{s}\in\mathbb{C}^n$ let

$$\rho_{\mathbf{s}} : \pi_1(X,\mathbf{x}_0)\to GL(V)$$

be the monodromy representation of $\nabla(\mathbf{s})$.

DEFINITION 3.8. We say that ∇ has **periodic monodromy** if for every $1\le j\le n$, the set S_j of $\mathbf{s}$ for which the representation $\rho_{\mathbf{s}}$ is isomorphic to $\rho_{\mathbf{s}+\mathbf{e}_j}$ is not contained in a countable union of proper analytic subsets of $\mathbb{C}^n$.

In particular, every periodic family has periodic monodromy.

Let $K := \mathbb{C}(\mathbf{q})$, where $\mathbf{q} = (q_1,\dots,q_n)$.

THEOREM 3.9. *A* ***pencil*** ∇ *of flat connections on* X *with fiber* V *has periodic monodromy if and only if there exist a nonempty affine open subset* $U\subset(\mathbb{C}^\times)^n$*, a finite unramified Galois cover* $p:\widetilde{U}\to U$*, and a Galois-stable representation* $\eta:\pi_1(X,\mathbf{x}_0)\to GL(V)(\mathbb{C}[\widetilde{U}])$ *such that the monodromy representation* $\rho_{\mathbf{s}}$ *of* $\nabla(\mathbf{s})$ *is isomorphic to* $\eta|_{e^{2\pi i\mathbf{s}}}$ *(hence also to* $\rho_{\mathbf{s}+\mathbf{e}_j}$ *for all* j*) when* $e^{2\pi i\mathbf{s}}\in U$*. Moreover, in this case* η *is unique up to action of* $GL(V)(\overline{K})$*.*

PROOF. The "if" direction is obvious, so we only need to prove the "only if" direction. It is well known that $\pi_1(X,\mathbf{x}_0)$ is finitely generated. Thus $\rho_{\mathbf{s}}$ may be viewed as a holomorphic family of representations (Example 2.2).

Since ∇ is a pencil (i.e., linear in $\mathbf{s}$), for any $g\in\pi_1(X,\mathbf{x}_0)$

$$\|\rho_{\mathbf{s}}(g)\| = O(e^{C_g\|\mathbf{s}\|}),\ \|\mathbf{s}\|\to\infty$$

for some $C_g > 0$, i.e., $\rho_{\mathbf{s}}(g)$ is exponentially bounded for each g. Since ∇ also has periodic monodromy, the result follows from Theorem 2.5. □

Let $\mathcal{U} := E^{-1}(U) \subset \mathbb{C}^n$ and $\pi : \widetilde{\mathcal{U}} \to \mathcal{U}$ be the pullback of the cover $p : \widetilde{U} \to U$ under the map $E : \mathcal{U} \to U$. Denote the corresponding map $\widetilde{\mathcal{U}} \to \widetilde{U}$ by $\widetilde{E}$. Thus we have a Cartesian square of unramified analytic covers

$$\begin{array}{ccc} \widetilde{\mathcal{U}} & \xrightarrow{\widetilde{E}} & \widetilde{U} \\ {\scriptstyle\pi}\downarrow & & \downarrow{\scriptstyle p} \\ \mathcal{U} & \xrightarrow{E} & U \end{array}$$

PROPOSITION 3.10.

(i) *An isomorphism* $\phi : \rho_{\pi(\widetilde{\mathbf{s}})} \to \eta|_{\widetilde{E}(\widetilde{\mathbf{s}})}$ *can be realized by a meromorphic function* $\phi(\widetilde{\mathbf{s}})$ *on* $\widetilde{\mathcal{U}}$ *with values in* $GL(V)$.

(ii) *Suppose* $\mathrm{End}_{\overline{K}}\, \eta = \overline{K}$, *and let* $U^\circ \subset U$ *be the (nonempty, Zariski open) set of* $\mathbf{a} \in U$ *for which* $\mathrm{End}\, \eta|_{\mathbf{a}} = \mathbb{C}$. *Let* $\phi_*(\widetilde{\mathbf{s}})$ *be the image of* $\phi(\widetilde{\mathbf{s}})$ *in* $PGL(V)$. *Then the meromorphic function* ϕ_* *on* $\widetilde{\mathcal{U}}$ *is holomorphic on the open subset* $\widetilde{\mathcal{U}}^\circ := (E \circ \pi)^{-1}(U^\circ) \subset \widetilde{\mathcal{U}}$.

PROOF.

(i) For $\phi \in \mathrm{End}\, V$, being a morphism of representations $\rho_{\pi(\widetilde{\mathbf{s}})} \to \eta|_{\widetilde{E}(\widetilde{\mathbf{s}})}$ is a system of linear equations with coefficients in the algebra R generated by some holomorphic functions $h_1, \ldots, h_m$ on $\widetilde{\mathcal{U}}$. Moreover, for an analytically dense set of $\widetilde{\mathbf{s}} \in \widetilde{\mathcal{U}}$, this system has an invertible solution. Thus be Chevalley's elimination of quantifiers, there is an invertible solution ϕ of this system over $\mathrm{Frac}(R)$.

(ii) This follows from the removable singularity theorem.

□

REMARK 3.11. Theorem 3.9 trivially extends to the case when $\nabla(\mathbf{s})$ is an affine pencil of flat connections on a possibly nontrivial vector bundle V on X (cf. Remark 3.2).

3.5. Quantum connections. An important class of periodic families of connections is **equivariant quantum connections** $\nabla(\mathbf{s})$ of conical symplectic resolutions $\widetilde{Y} \to Y$ (see [**BMO**], [**MO**] and references therein). These connections are known explicitly in many cases, such as cotangent bundles of partial flag varieties T^*G/P, resolutions of Slodowy slices of nilpotent orbits of a simple Lie algebra, quiver varieties, etc. The parameters s_j of such families are the equivariant parameters of the torus $\mathbf{T}$ acting on Y, the base is $X := H^2(\widetilde{Y}, \mathbb{C}^\times) \setminus D$ where D is a union of subtori, and the fiber is $H^*(\widetilde{Y}, \mathbb{C})$. It is known that such families admit **geometric shift operators** $A_j(\mathbf{s})$ which can be defined using enumerative geometry, even though they may be difficult to compute explicitly (especially for copies of $\mathbb{G}_m \subset \mathbf{T}$ that dilate the symplectic form). This implies that such families are indeed periodic.

Also in many examples the quantum connection $\nabla(\mathbf{s})$ is a pencil, i.e., linear in the equivariant parameters s_i. Most of the "trigonometric" examples of Section 5 (trigonometric KZ connections of type A, trigonometric Casimir connections, trigonometric Dunkl connections) have this property. In general, this happens, for example, when the torus $\mathbf{T}$ acting on X has finitely many fixed points (see e.g. [**Lee**] and references therein).

4. Quasi-motivic families and pencils of flat connections

4.1. Quasi-motivic families. It is often hard to prove directly that a given family ∇ of flat connections is periodic, as the operators A_j may be difficult to compute. However, if $\mathbf{k} = \mathbb{C}$ and ∇ has regular singularities, there is an approach to checking this property using the monodromy of ∇. This approach is based on the following theorem.

THEOREM 4.1. *A family ∇ of flat connections with regular singularities on a complex variety X has periodic monodromy if and only if it is periodic.*

PROOF. Any periodic family has periodic monodromy, so we just need to prove the "only if" direction.

Consider the family of flat connections $\nabla^{(j)}(\mathbf{s}) := \nabla(\mathbf{s}+\mathbf{e}_j) \otimes \nabla(\mathbf{s})^*$. Thus $\nabla^{(j)}(\mathbf{s})$ has monodromy representation $\rho_{\mathbf{s}+\mathbf{e}_j} \otimes \rho_{\mathbf{s}}^*$.

By assumption, the periodicity condition for index j holds for $\mathbf{s}$ belonging to some subset $S_j \subset \mathbb{C}^n$ which is not contained in a countable union of proper Zariski closed subsets. For a fixed j and $\mathbf{s} \in S_j$ fix an isomorphism $\psi_{\mathbf{s}} : \rho_{\mathbf{s}} \cong \rho_{\mathbf{s}+\mathbf{e}_j}$. There exists a flat holomorphic section $A_j(\mathbf{s})$ of $\nabla^{(j)}(\mathbf{s})$ such that $A_j(\mathbf{s}, \mathbf{x}_0) = \psi_{\mathbf{s}}$; in particular, $A_j(\mathbf{s}, \mathbf{x}) \in GL(V)$ for all $\mathbf{x} \in X$.

Fix a smooth compactification $\overline{X}$ of X such that $\overline{X} \setminus X$ is a normal crossing divisor (it exists by Hironaka's theorem). Since $\nabla^{(j)}(\mathbf{s})$ has regular singularities, the functions $A_j(\mathbf{s}, \mathbf{x})$ have at worst power growth near $\overline{X} \setminus X$. Hence they are meromorphic functions on $\overline{X}$. Thus $A_j(\mathbf{s}, \mathbf{x})$ are rational functions, i.e., $A_j(\mathbf{s}) \in GL(V)(\mathbb{C}[X])$.

Fix a basis $\{f_i, i \in \mathbb{Z}_{\geq 1}\}$ of $\mathbb{C}[X]$. For $M \in \mathbb{N}$ let $S_j(M) \subset S_j$ be the set of $\mathbf{s} \in S$ for which

$$A_j(\mathbf{s}, \mathbf{x}) = \sum_{l=1}^{M} A_{jl}(\mathbf{s}) f_l(\mathbf{x}).$$

for suitable matrices $A_{jl}(\mathbf{s})$. It is clear that $S_j(M) \subset S_j(M+1)$ and $\cup_M S_j(M) = S_j$. Since S_j is not contained in a countable union of proper Zariski closed subsets, there exists M for which $S_j(M)$ is Zariski dense in $\mathbb{C}^n$. The property that $A_j(\mathbf{s}, \mathbf{x})$ is a flat section of $\nabla^{(j)}(\mathbf{s})$ is then a system of homogeneous linear equations on the matrices $A_{jl}(\mathbf{s})$, $l = 1, \dots, M$, over $\mathbb{C}(\mathbf{s})$. By Chevalley's elimination of quantifiers, since this system has an invertible solution for $\mathbf{s} \in S_j(M)$, it must have one over $\mathbb{C}(\mathbf{s})$, where $\mathbf{s}$ is a collection of variables, as claimed. □

REMARK 4.2. If ∇ has irregular singularities, then it can have periodic monodromy without being periodic, for example, $\nabla(s) = d - s$ on $X = \mathbb{A}^1$ (the monodromy is trivial for all s since X is simply connected). We expect, however, that Theorem 4.1 can be extended to the irregular case (at least when $\dim X = 1$) if the condition of periodic monodromy is replaced by "periodic monodromy and Stokes data". We hope that this can be done using the techniques of [**Ka**].

Theorem 4.1 motivates the following definition.

DEFINITION 4.3. We say that a family ∇ of flat connections over an algebraically closed field $\mathbf{k}$ of characteristic zero is **quasi-motivic** if

(i) $\nabla(\mathbf{s})$ has regular singularities for all $\mathbf{s} \in \mathbf{k}^n$, and
(ii) ∇ is periodic.

By Theorem 4.1, if $\mathbf{k} = \mathbb{C}$ then ∇ is quasi-motivic if and only if it has regular singularities and periodic monodromy.

4.2. Jumping loci of quasi-motivic pencils. In the next few subsections we investigate various kinds of singularities of quasi-motivic pencils. Since we are using the Riemann-Hilbert correspondence, we will work over $\mathbf{k} = \mathbb{C}$, but as our statements are purely algebraic, they hold, by standard abstract nonsense, over any algebraically closed field $\mathbf{k}$ of characteristic zero.

Let ∇ be a quasi-motivic pencil. Let $Z := U^c \subset (\mathbb{C}^\times)^n$ be the complement of the set U from Theorem 3.9, a divisor in $(\mathbb{C}^\times)^n$. Let $Z_{\text{ht}} \subset Z$ be the union of the components of Z which are (affine) hypertori. Then $\mathcal{Z} := E^{-1}(Z_{\text{ht}}) \subset \mathbb{C}^n$ is an arrangement of translates of finitely many hyperplanes defined over $\mathbb{Q}$, with equations

$$a_{i1}s_1 + \cdots + a_{in}s_n = c_i + N, \; 1 \le i \le p,$$

where $a_{ij} \in \mathbb{Z}$ (coprime for each i), $N \in \mathbb{Z}$, $c_i \in \mathbb{C}$.

THEOREM 4.4. *Every codimension* 1 *irreducible Zariski closed subset H of the jumping locus $J(\nabla)$ is a hyperplane contained in $\mathcal{Z}$.*

PROOF. Let $\rho_{\mathbf{s}}$ be the monodromy representation of $\pi_1(X, \mathbf{x}_0)$ defined by $\nabla(\mathbf{s})$. Since $\nabla(\mathbf{s})$ has regular singularities, $\Gamma(\nabla(\mathbf{s})) \cong \mathrm{Hom}_{\pi_1(X,\mathbf{x}_0)}(\mathbb{C}, \rho_{\mathbf{s}})$, thus $J(\nabla)$ is the set of $\mathbf{s}$ for which $\dim \mathrm{Hom}_{\pi_1(X,\mathbf{x}_0)}(\mathbb{C}, \rho_{\mathbf{s}}) > d(\nabla)$. Since ∇ has periodic monodromy, by Theorem 3.9, $E(H)$ is contained in Z. So Theorem 2.6 implies that $E(H)$ is contained in Z_{ht}, as desired. $\square$

EXAMPLE 4.5. Since $\Gamma(\nabla \otimes \nabla^*) = \mathrm{End}(\nabla)$ is the endomorphism algebra of ∇, Theorem 4.4 applies to the jumping locus $J(\nabla \otimes \nabla^*)$ for the dimension of $\mathrm{End}(\nabla)$. Thus every codimension 1 irreducible Zariski closed subset H of the set of $\mathbf{s} \in \mathbb{C}^n$ where $\mathrm{End}(\nabla(\mathbf{s}))$ exceeds its generic value is a hyperplane contained in $\mathcal{Z}$.

4.3. Poles of shift operators for quasi-motivic pencils. Let us choose the shift operators A_j for ∇. Let $\mathcal{B}_j$ be the set of exceptional values of $\mathbf{s}$ where $\nabla(\mathbf{s})$ is not isomorphic to $\nabla(\mathbf{s} + \mathbf{e}_j)$. It is clear that $\mathcal{B}_j$ is a contained in the pole divisor $\mathcal{Z}_j \subset \mathbb{C}^n$ of the projection A_{j*} of A_j to $PGL(V)$. Thus every codimension 1 irreducible component of the Zariski closure $\overline{\mathcal{B}}_j$ is a component of $\mathcal{Z}_j$.

THEOREM 4.6.

(i) *All codimension* 1 *irreducible components of $\overline{\mathcal{B}}_j$ are contained in $\mathcal{Z}$.*
(ii) *If ∇ has generically trivial endomorphism algebra (i.e., trivial over $\mathbb{C}(\mathbf{s})$) then $\mathcal{Z}_j$ are contained in $\mathcal{Z}$.*

PROOF.

(1) Since ∇ has regular singularities, by the Riemann-Hilbert correspondence [**De2**] we have $\mathcal{B}_j \subset E^{-1}(Z)$. Hence $\overline{\mathcal{B}_j} \subset E^{-1}(Z)$, since $E^{-1}(Z)$ is closed in the Euclidean topology, and for a constructible set the Zariski and Euclidean closures are the same. Now, by Theorem 2.6, every component of $\mathcal{Z}_j$ that is contained in $E^{-1}(Z)$ is, in fact, contained in $E^{-1}(Z_{\text{ht}})$. This implies (i).

(2) Let $\widetilde{\mathcal{U}}$ be as in Proposition 3.10. Since ∇ has generically trivial endomorphism algebra, by Subsection 3.2 this is so outside a countable union of hypersurfaces, so $\mathrm{End}_{\overline{K}}\, \eta = \overline{K}$. Thus, in view of Theorem 4.4, by Theorem 3.9 and Proposition 3.10(ii) we have an isomorphism $\widehat{A}_j(\widetilde{\mathbf{s}}) : \rho_{\mathbf{s}} \to \rho_{\mathbf{s}+\mathbf{e}_i}$ which depends holomorphically on $\widetilde{\mathbf{s}} \in (E \circ \pi)^{-1}(U)$, where $\mathbf{s} = \pi(\widetilde{\mathbf{s}})$. Let $\widehat{A}_{j*}(\widetilde{\mathbf{s}})$ be the image of $\widehat{A}_j(\widetilde{\mathbf{s}})$ in $PGL(V)$. Since the endomorphism algebra of $\nabla(\mathbf{s})$ is trivial when $\widetilde{\mathbf{s}} \in \widetilde{\mathcal{U}}$, we have $\widehat{A}_{j*}(\widetilde{\mathbf{s}}) = A_{j*}(\mathbf{s})$, $\widetilde{\mathbf{s}} \in \widetilde{\mathcal{U}}$. It follows that A_{j*} is regular on $E^{-1}(U)$, so $\mathcal{Z}_j \subset E^{-1}(Z)$. Thus by Theorem 2.6, $\mathcal{Z}_j \subset E^{-1}(Z_{\mathrm{ht}}) = \mathcal{Z}$.

□

4.4. Endomorphism algebras of quasi-motivic pencils. For a finite dimensional $GL(V)$-module W, let ∇_W be the flat connection on X with fiber W associated to ∇. It is clear that if ∇ has regular singularities then so does ∇_W, and if $\nabla(\mathbf{s})$ has periodic monodromy, is periodic, or is quasi-motivic, then so is $\nabla_W(\mathbf{s})$

Let $d_W = d_W(\nabla) := d(\nabla_W)$. Let $Y \subset \mathrm{Gr}(d_W, W)$ be the Zariski closure of the image of the rational map $\gamma = \gamma(\nabla_W)$ defined in Subsection 3.2, an irreducible closed subvariety.

THEOREM 4.7. *The action of $GL(V)$ on Y has a (unique) dense orbit Y°.*

PROOF. By Rosenlicht's theorem in invariant theory [**PV**, 2.3], rational invariants separate generic orbits. Thus it suffices to show that $\mathbb{C}(Y)^{GL(V)} = \mathbb{C}$. Let F be a $GL(V)$-invariant rational function on Y. Then $F \circ \gamma$ is a rational function of $\mathbf{s} \in \mathbb{C}^n$. On the other hand, since ∇ has regular singularities, we have $\gamma(\mathbf{s}) = W^{\pi_1(X,\mathbf{x}_0)}$, where $\pi_1(X, \mathbf{x}_0)$ acts on W through the monodromy representation $\rho_{\mathbf{s}}$. Thus by Theorem 3.9, $F \circ \gamma$ is also rational in $e^{2\pi i \mathbf{s}}$. This implies the statement, since a function simultaneously rational in $\mathbf{s}$ and $e^{2\pi i \mathbf{s}}$ must be constant. □

Let $\mathcal{U}(\nabla, W)$ be the Zariski open set of $\mathbf{s} \in \mathbb{C}^n$ such that $\gamma(\mathbf{s})$ is defined and belongs to Y°. Thus for any two elements $\mathbf{s}, \mathbf{s}' \in \mathcal{U}(\nabla, W)$, the corresponding subspaces $\gamma(\mathbf{s})$ and $\gamma(\mathbf{s}')$ are conjugate under the action of $GL(V)$.

COROLLARY 4.8. *Every codimension 1 component H of the complement $\mathcal{U}(\nabla, W)^c$ is a hyperplane contained in $\mathcal{Z}$.*

PROOF. Since the membership of $\mathbf{s}$ in $\mathcal{U}(\nabla, W)$ depends only on $\rho_{\mathbf{s}}$, $E(H)$ is contained in Z. So by Theorem 2.6 it is contained in Z_{ht} and the statement follows. □

Now consider the special case $W = V \otimes V^*$. In this case $\gamma(\mathbf{s})$, when defined, is the **reduced endomorphism algebra** $\mathrm{End}_{\mathrm{red}}(\nabla(\mathbf{s}))$. If $\mathbf{s} \notin J(\nabla)$, it coincides with the usual endomorphism algebra $\mathrm{End}(\nabla(\mathbf{s}))$, otherwise it is a subalgebra in $\mathrm{End}(\nabla(\mathbf{s}))$ obtained as the limit of the endomorphism algebra from the generic locus.

COROLLARY 4.9. *For values of $\mathbf{s}$ outside finitely many hyperplanes contained in $\mathcal{Z}$ and a Zariski closed subset of $\mathbb{C}^n$ of codimension ≥ 2, the reduced endomorphism algebra $\mathrm{End}_{\mathrm{red}}(\nabla(\mathbf{s}))$ together with its action on V does not depend on $\mathbf{s}$ up to an isomorphism.*

PROOF. It suffices to apply Corollary 4.8 to $W = V \otimes V^*$. □

Let us regard $\mathrm{Aut}_{\mathrm{red}}(\nabla(\mathbf{s})) = \mathrm{End}_{\mathrm{red}}(\nabla(\mathbf{s}))^\times$ as a group scheme over $\mathcal{U}(\nabla, V \otimes V^*)$. We can then project the shift operators $A_j(\mathbf{s})$ to rational sections $A_{j\bullet}(\mathbf{s})$ of the bundle over $\mathcal{U}(\nabla, V \otimes V^*)$ with fiber $GL(V)/\mathrm{Aut}_{\mathrm{red}}(\nabla(\mathbf{s}))$. It is clear that $A_{j\bullet}(\mathbf{s})$ does not depend on the choice of A_j. Let $\mathcal{Z}_{j\bullet}$ be the pole divisor of $A_{j\bullet}$. We obtain the following generalization of Theorem 4.6(ii).

COROLLARY 4.10. *We have* $\mathcal{Z}_{j\bullet} \subset \mathcal{Z}$.

PROOF. Using the same argument as in the proof of Theorem 4.6(ii), one establishes that $E(\mathcal{Z}_{j\bullet}) \subset Z$. Thus by Theorem 2.6, $E(\mathcal{Z}_{j\bullet}) \subset Z_{\mathrm{ht}}$, so the result follows. □

4.5. Semisimplicity loci of quasi-motivic pencils. Another similar result is the following theorem, which is proved analogously to the above results by applying Theorem 2.6. Let $\mathrm{Semis}(\nabla) \subset \mathbb{C}^n$ be the subset of points $\mathbf{s}$ where $\nabla(\mathbf{s})$ is semisimple. This means that in some basis $\nabla(\mathbf{s})$ is block-diagonal with blocks of some sizes $N_1, \ldots, N_p$ (where $N_1 + \cdots + N_p = \dim V$), but it is not block-triangular with $p+1$ blocks in any basis. So similarly to Subsection 3.2, one shows that the complement $\mathrm{Semis}(\nabla)^c$ is a countable union of locally closed subsets of $\mathbb{C}^n$.

THEOREM 4.11. *Suppose that* ∇ *is generically semisimple (i.e., semisimple over* $\mathbb{C}(\mathbf{s})$*). Then the closure of every codimension* 1 *irreducible locally closed subset of* $\mathrm{Semis}(\nabla)^c$ *is a hyperplane contained in* $E^{-1}(T)$*, where* T *is a union of affine hypertori in* $(\mathbb{C}^\times)^n$.

REMARK 4.12. We will see in Section 5 that many natural examples of quasi-motivic pencils are generically semisimple.

REMARK 4.13. In general, the kind of result similar to Theorem 4.11 holds (with the same proof) for any property of ∇ that is algebraic in both de Rham and Betti realizations and preserved by the Riemann-Hilbert correspondence.

4.6. Motivic families. An important class of quasi-motivic families is **motivic families**. To define them, let us introduce some notation. Let Y be a smooth complex variety and $\Phi_1, \ldots, \Phi_n$ non-vanishing regular functions on Y. Denote by $\mathcal{L}(\Phi_1, \ldots, \Phi_n, \mathbf{s})$ the de Rham local system on Y generated by the multivalued function $\prod_{j=1}^n \Phi_j^{s_j}$.

Let $\nabla(\mathbf{s})$ be a family of flat connections on an irreducible affine complex variety X.

DEFINITION 4.14. We say that ∇ is **motivic** if there exists an irreducible variety Y, a smooth morphism $\pi : Y \to X$, and non-vanishing regular functions $\Phi_1, \ldots, \Phi_n$ on Y such that on some dense open set $X^\circ \subset X$, the connection $\nabla(\mathbf{s})$ is isomorphic to the Gauss-Manin connection $\pi_* \mathcal{L}(\Phi_1, \ldots, \Phi_n, \mathbf{s})$ on $H^i(\pi^{-1}(\mathbf{x}), \mathcal{L}(\Phi_1, \ldots, \Phi_n, \mathbf{s}))$ for Zariski generic $\mathbf{s}$ and some i.

For example, if π is affine, then in informal terms this means that the local flat sections $F(\mathbf{x})$ of $\nabla(\mathbf{s})$ can be obtained by integrating the function $\prod_{j=1}^n \Phi_j^{s_j}(\mathbf{y})$ against a regular differential form ω on Y over a twisted singular cycle $C_{\mathbf{x}}$ in the fiber $Y_{\mathbf{x}} = \pi^{-1}(\mathbf{x})$ of (real) dimension $\dim Y_{\mathbf{x}} - i$, varying continuously with $\mathbf{x}$:

$$F(\mathbf{x}) = \int_{C_{\mathbf{x}}} \prod_{j=1}^n \Phi_j^{s_j} \omega.$$

Let $K = \mathbb{Q}(\mathbf{q})$. We say that a quasi-motivic family ∇ on a complex variety X has **algebraic monodromy** if there exists a representation $\eta : \pi_1(X, \mathbf{x}_0) \to GL_N(\overline{K})$ such that the monodromy representation $\rho_{\mathbf{s}}$ of $\nabla(\mathbf{s})$ is isomorphic to $\eta|_{\mathbf{q}=e^{2\pi i \mathbf{s}}}$ for Zariski generic $e^{2\pi i \mathbf{s}} \in (\mathbb{C}^\times)^n$, and **rational monodromy** if η can be chosen to land in $GL_N(K)$. In light of Theorem 3.9, algebraic monodromy is just the requirement that the numerical coefficients of η, which are a priori complex numbers, actually belong to $\overline{\mathbb{Q}}$.

PROPOSITION 4.15. *Every motivic family ∇ is quasi-motivic and has rational monodromy.*

PROOF. This follows since a Gauss-Manin connection always has regular singularities and its monodromy can be computed in the Betti realization. □

Hence every motivic family is periodic by Theorem 4.1. In fact, this is also easy to see directly, as multiplication by Φ_j defines an isomorphism

$$a_k(\mathbf{s}) : \mathcal{L}(\Phi_1, \ldots, \Phi_n, \mathbf{s}) \cong \mathcal{L}(\Phi_1, \ldots, \Phi_n, \mathbf{s} + \mathbf{e}_j)$$

which gives rise to an isomorphism $A_k(\mathbf{s}) = \pi_*(a_k(\mathbf{s}))$ of the corresponding direct images.

REMARK 4.16. Our results about periodic, quasi-motivic and motivic families of flat connections extend in a straightforward way to **rational** families, i.e. those with coefficients in $\mathbb{C}(\mathbf{s})$ rather than in $\mathbb{C}[\mathbf{s}]$.

REMARK 4.17. We will see below that many quasi-motivic families are in fact motivic, or, more generally, can be realized as subquotients of motivic families (at least for generic $\mathbf{s}$). In fact, we do not know an example of a generically irreducible quasi-motivic family which does not have this property (possibly after shifting $\mathbf{s}$), although there are many examples of such families with no known motivic realization.

4.7. A speculative digression: quasi-geometric connections.

4.7.1. *Quasi-geometric connections.*

DEFINITION 4.18. Let X be a smooth irreducible complex algebraic variety defined over $\overline{\mathbb{Q}}$. Let ∇ be a flat connection on a vector bundle V on X with regular singularities. We say that ∇ is **quasi-geometric** if the monodromy representation of the corresponding flat holomorphic connection $\nabla_{\mathbb{C}}$ on X viewed as a complex manifold is also defined over $\overline{\mathbb{Q}}$.

In other words, a quasi-geometric connection is a flat connection with regular singularities defined over $\overline{\mathbb{Q}}$ both in the de Rham and the Betti realization.

Note that if $X^\circ \subset X$ is a non-empty Zariski open subset and $\mathbf{x}_0 \in X^\circ$ then the map $\pi_1(X^\circ, \mathbf{x}_0) \to \pi_1(X, \mathbf{x}_0)$ is surjective, so ∇ is quasi-geometric on X if and only if it is quasi-geometric on X°.

EXAMPLE 4.19.

(1) A geometric connection (Subsection 2.5) is quasi-geometric.
(2) The category of quasi-geometric connections is invariant under pullbacks, external products (hence tensor products), and duals. Thus for every X, we have a $\overline{\mathbb{Q}}$-linear Tannakian category $\mathcal{QG}(X)$ of quasi-geometric connections on X.

(3) If $\pi : Y \to X$ is a smooth morphism and ∇ is a quasi-geometric connection on Y then for all i the i-th cohomology of $\pi_*\nabla$ is quasi-geometric on its smooth locus $X^\circ \subset X$, since the direct image can be computed over $\overline{\mathbb{Q}}$ both in the de Rham and the Betti realization.
(4) A simple composition factor of a quasi-geometric connection is quasi-geometric.
(5) An irreducible **rigid** connection (i.e., one determined up to finitely many choices by conjugacy classes of its local monodromies) is quasi-geometric, since it is deformation-theoretically rigid both in the de Rham and the Betti realization. Note that for $X = \mathbb{P}^1 \setminus S$ where $S \subset \overline{\mathbb{Q}}$ is a finite set, N. Katz showed that rigid connections are geometric by using the method of middle convolution [**K**].

The relevance of the notion of a quasi-geometric connection to this paper is exhibited by the following proposition, whose proof is straightforward.

PROPOSITION 4.20. *If ∇ is a quasi-motivic family of flat connections on X defined over $\overline{\mathbb{Q}}$ with algebraic monodromy, then for $\mathbf{s} \in \mathbb{Q}^n$, $\nabla(\mathbf{s})$ is quasi-geometric.*

In particular, this means that the examples of Section 5 are quasi-geometric (we will see that they have algebraic monodromy).

An important property of quasi-geometric connections is quasiunipotent monodromy at infinity.

PROPOSITION 4.21. *Let $\overline{X}$ be a smooth compactification of X such that $\overline{X} \setminus X$ is a normal crossing divisor. If ∇ is a quasi-geometric connection on X, then its monodromy around every component D of $\overline{X} \setminus X$ is quasiunipotent.*

PROOF. Let us extend the underlying vector bundle of ∇ over D so that ∇ has a first order pole there (this is possible since ∇ has regular singularities, [**De2**]). Then the eigenvalues of the monodromy of ∇ around D are of the form $e^{2\pi i s}$, where s runs through the eigenvalues of the residue of ∇ at D. Thus both s and $e^{2\pi i s}$ are algebraic numbers. Hence it follows from the theorem of Gelfond and Schneider in transcendental number theory [**Ge**] that $s \in \mathbb{Q}$, i.e., $e^{2\pi i s}$ is a root of unity, as desired. □

REMARK 4.22. Since every geometric connection is quasi-geometric, Proposition 4.21 is a generalization of the Local Monodromy Theorem in Hodge theory: monodromy at infinity of a Gauss-Manin connection is quasiunipotent. In fact, our proof of Proposition 4.21 is the same as Brieskorn's proof of this theorem given at the end of [**De2**].

REMARK 4.23. There are many examples of irreducible quasi-geometric connections without a known geometric construction, even on $X = \mathbb{P}^1 \setminus S$. E.g., we can take the Dunkl connection 5.2.1 for exceptional Coxeter groups at a rational parameter c, restricted to a generic line in $\mathfrak{h}$ (not passing through 0). However, we do not know an example of an irreducible quasi-geometric connection which is provably non-geometric. For instance, Proposition 4.21 implies that a 1-dimensional connection on $\mathbb{P}^1 \setminus S$ which is quasi-geometric is actually geometric.

Let ∇ be a connection on $X := \mathbb{P}^1 \setminus S$ with regular singularities defined over $\overline{\mathbb{Q}}$ whose monodromies around points of S are quasiunipotent (i.e., the eigenvalues of the residues are rational). In this case it is usually difficult to show that ∇ is not quasi-geometric, because this entails establishing transcendence of certain numbers (such as traces of monodromy operators). Yet we expect that the quasi-geometric property is quite rare. Namely, we make the following conjecture.

CONJECTURE 4.24. *Fix a quasiunipotent Riemann symbol P for S (i.e., the list of eigenvalues of monodromies around points of S). Then there are finitely many quasi-geometric connections on $\mathbb{P}^1 \setminus S$ with Riemann symbol P.*

In particular, this would mean that already for rank 2 and $|S| = 4$, most connections with a given P are not quasi-geometric, since such connections are parametrized by two parameters in $\overline{\mathbb{Q}}$.

REMARK 4.25. Conjecture 4.24 holds for geometric connections by a result of Deligne [**De1**], which provides motivation and supporting evidence for this conjecture.

4.7.2. *Quasi-geometric connections and braided fusion categories.* In spite of likely being rare among general regular connections with quasiunipotent local monodromies, quasi-geometric connections are ubiquitous in different areas of mathematics. In particular, we would like to propose the following conjecture, stating that connections arising from braided fusion categories are always quasi-geometric.

Let $\mathcal{C}$ be a **braided fusion category** over $\mathbb{C}$ [**EGNO**, Definitions 4.1.1, 8.1.1]. By the Ocneanu rigidity theorem, $\mathcal{C}$ is defined over $\overline{\mathbb{Q}}$ [**EGNO**, Corollary 9.1.8]. Thus for any objects $X_1, \ldots, X_n, Y \in \mathcal{C}$, the space $\mathrm{Hom}(Y, X_1 \otimes \cdots \otimes X_n)$ carries an action of the pure braid group PB_n, which defines a Betti local system $\mathcal{L}_{X_1,\ldots,X_n,Y}$ on the configuration space $X_n := \mathbb{C}^n \setminus$ diagonals defined over $\overline{\mathbb{Q}}$. By the Riemann-Hilbert correspondence [**De2**], there is a unique up to isomorphism flat connection $\nabla_{X_1,\ldots,X_n,Y}$ on X_n with regular singularities whose monodromy is $\mathcal{L}_{X_1,\ldots,X_n,Y}$.

CONJECTURE 4.26. *The connection $\nabla_{X_1,\ldots,X_n,Y}$ is defined over $\overline{\mathbb{Q}}$. In other words, it is a quasi-geometric connection.*

If $\mathcal{C}$ is a modular category [**EGNO**, Subsection 8.14] then Conjecture 4.26 can be strengthened to say that the flat connections with regular singularities on the moduli space of curves of any genus g with n punctures attached to mapping class group representations on spaces of conformal blocks of the category $\mathcal{C}$ are defined over $\overline{\mathbb{Q}}$ (equivalently, are quasi-geometric). In genus 0 this is actually equivalent to Conjecture 4.26, since the category $\mathcal{C}$ can be assumed modular by pivotalization [**EGNO**, p. 180] and taking the Drinfeld center [**EGNO**, Subsection 7.13].

A possible approach to proof of Conjecture 4.26 would be to establish a de Rham analog of Ocneanu rigidity, using the de Rham description of the notion of a modular fusion category and modular functor in terms of the Teichmüller tower, due to Deligne and Beilinson-Feigin-Mazur (see [**BK**], Chapter 6 and references therein).

Some evidence for Conjecture 4.26 is provided by the following proposition.

PROPOSITION 4.27. *If $\mathcal{C}$ is the category of representations of a strongly rational vertex algebra $\mathcal{V}$ (in the sense of* [**M**]*, p.1) defined over $\overline{\mathbb{Q}}$ then Conjecture* 4.26 *holds for $\mathcal{C}$ (in the generalized form for any genus).*

PROOF. By Huang's theorem [**H**], $\mathcal{C}$ is a modular category. Since $\mathcal{V}$ is defined over $\overline{\mathbb{Q}}$, so are the KZ type connections on genus 0 conformal blocks for $\mathcal{V}$. On the other hand, the monodromy of these connections is also defined over $\overline{\mathbb{Q}}$ by Ocneanu rigidity [**EGNO**, Corollary 9.1.8], as already mentioned above. This proves the proposition. □

Note that it is conjectured (see e.g. [**GJ**], p.2) that any modular category $\mathcal{C}$ is the category of representations of a (strongly) rational vertex algebra $\mathcal{V}$. If so, and $\mathcal{V}$ is defined over $\overline{\mathbb{Q}}$, then Conjecture 4.26 would follow from Proposition 4.27.

We do not know if any rational vertex algebra is defined over $\overline{\mathbb{Q}}$ (although this seems to be true in all known examples). But we hope that this can be proved (maybe under mild assumptions) if $\mathcal{V}$ is deformation-theoretically rigid (has no non-trivial formal deformations), which could presumably be deduced from vanishing of a suitable second cohomology group. We do not know, however, how to prove such rigidity.

Also, in all examples of vertex algebras we know, the KZ type connections on genus 0 conformal blocks are Gauss-Manin connections (i.e., their local flat sections have integral representations). For instance, if $\mathcal{V}$ is the simple vertex algebra attached to the affine Kac-Moody algebra $\widehat{\mathfrak{g}}$ at level k then these connections are literally the KZ connections, whose Gauss-Manin realization (i.e., integral form of solutions) is given in [**SV**]. In such cases, one may say that the quasi-geometric nature of the corresponding flat connections is explained by the fact that they are Gauss-Manin connections (i.e., geometric).

Finally, we note that there are concrete examples of modular fusion categories (coming from subfactors) for which the flat KZ-type connections attached via the Riemann-Hilbert correspondence to their braid group representations are not known explicitly (the problem of computing a regular connection with given monodromy is, in general, highly transcendental, even on $X = \mathbb{P}^1 \setminus S$). In particular, they are not known to be Gauss-Manin connections, not even quasi-geometric ones, so Conjecture 4.26 is open for them. The simplest such example is perhaps the Drinfeld center of the Haagerup fusion category, see [**GI**]. This category is not known to be the representation category of a rational vertex algebra.

REMARK 4.28. If a braided fusion category $\mathcal{C}$ is unitary [**Tu**] then the representations of the pure braid group PB_n on its multiplicity spaces are also unitary, thus semisimple. Hence they are also semisimple if $\mathcal{C}$ is Galois conjugate to a unitary category. We do not know, however, whether representations of PB_n coming from braided fusion categories must be semisimple in general. This is true in the abelian case $n = 2$ by the Anderson-Moore-Vafa theorem [**EGNO**, 8.18], but nothing is known for $n > 2$.

4.8. Pseudo-pencils. Even though the pencil condition on a family of flat connections $\nabla(\mathbf{s})$ (i.e., linearity in $\mathbf{s}$) is satisfied in many interesting examples listed in the next section, it is still quite restrictive and unfortunately rules out many other important examples, for instance most motivic families. On the other hand, we have used this condition only in one place - to argue that the monodromy representation of $\nabla(\mathbf{s})$ is exponentially bounded. It turns out that the pencil condition can be relaxed to include many more examples while retaining this property. Namely, this is accomplished by the following notion of a **pseudo-pencil**, which we first define in the 1-parameter case.

DEFINITION 4.29. Let $\nabla(s)$ be a 1-parameter polynomial family of flat connections on a smooth irreducible variety X. We say that $\nabla(s)$ is a **pseudo-pencil** if after a gauge transformation over $\mathbf{k}(s)(X)$, $\nabla(s)$ takes the form $d - sB(s)$ where B is regular at $s = \infty$.

It is clear that any pencil of flat connections is a pseudo-pencil. On the other hand, pseudo-pencils are much more general. For instance, using the results of Mochizuki [**Mo**], one can prove the following proposition (see e.g. [**Sa**], Theorem 2 and Example 3):

PROPOSITION 4.30. [3] *Let $\pi : Y \to X$ be a smooth morphism between smooth complex varieties of relative dimension m and Φ a nonvanishing regular function on Y such that the map $(\pi, \Phi) : Y \to X \times \mathbb{G}_m$ is proper. Suppose that the critical points of Φ on a generic fiber $\pi^{-1}(\mathbf{x})$ are isolated. Then the motivic family $H^m(\pi_*\mathcal{L}(\Phi, s))$ (over some open subset of X) is a pseudo-pencil.*

There are also other similar situations where one can conclude that $H^m(\pi_*\mathcal{L}(\Phi, s))$ is a pseudopencil, see e.g. [**Sa**], Example 4.

Yet it is clear that the monodromy of any pseudo-pencil (for $\mathbf{k} = \mathbb{C}$) is exponentially bounded (indeed, we may compute the monodromy in the gauge where $\nabla(s) \sim d - sB(s)$, where it is obviously exponentially bounded, but this condition does not depend on the gauge).

This can be generalized to the multiparameter case as follows.

DEFINITION 4.31. Let $\nabla(\mathbf{s})$, $\mathbf{s} = (s_1, \ldots, s_n)$ be an n-parameter polynomial family of flat connections on X. We say that $\nabla(\mathbf{s})$ is a **pseudo-pencil** if it is so with respect to each s_i when the other coordinates s_j, $j \neq i$ are fixed.

Again, it is easy to see that any pencil is a pseudo-pencil, but pseudo-pencils are much more general. Yet the above results can be generalized to them.

To make this generalization, note that, as shown by the proof of Lemma 2.3, the exponential boundedness assumption in this Lemma and hence in Theorem 2.5 can be relaxed to the assumption of **coordinatewise exponential boundedness**: we say that $f : \mathbb{C}^n \to V$ is coordinatewise exponentially bounded for all i, $f(\mathbf{s})$ is exponentially bounded with respect to s_i when s_j for $j \neq i$ are fixed.

It remains to note that the monodromy of any pseudo-pencil is coordinatewise exponentially bounded, which enables the desired generalizations.

Finally, we remark that this discussion extends straightforwardly to the case when $\nabla(\mathbf{s})$ is a rational (rather than polynomial) function of $\mathbf{s}$.

5. Examples of periodic pencils

In this section we will review numerous examples of pencils $\nabla(\mathbf{s})$ of flat connections arising in representation theory and mathematical physics. All of them will be shown to be periodic, possibly after rescaling $\mathbf{s}$ by an integer factor. In most cases, this is done by showing that these pencils are quasi-motivic.

---

[3] We thank V. Vologodsky for explanations regarding this proposition

5.1. KZ connections.

5.1.1. *KZ connections for finite dimensional simple Lie algebras.* Let $\mathfrak{g}$ be a finite dimensional simple Lie algebra over $\mathbb{C}$ with triangular decomposition $\mathfrak{g} = \mathfrak{n}_- \oplus \mathfrak{h} \oplus \mathfrak{n}_+$ and root system $R \subset \mathfrak{h}^*$, and let $\{h_i\}$ be the Chevalley basis of $\mathfrak{h}$. Let $\Omega \in (S^2\mathfrak{g})^{\mathfrak{g}}$ be the Casimir tensor of $\mathfrak{g}$ corresponding to the invariant inner product $(,)$ on $\mathfrak{g}^*$ under which short roots of $\mathfrak{g}$ have squared length 2:

$$\Omega = \sum_{i=1}^{\mathrm{rank}(\mathfrak{g})} h_i \otimes h_i^* + \sum_{\alpha \in R} e_\alpha \otimes e_{-\alpha},$$

where h_i^* is the dual basis of $\mathfrak{h}$ under $(,)$ and e_α are suitably normalized root elements of $\mathfrak{g}$. Let $V_1, \ldots, V_r$ be representations of $\mathfrak{g}$ from category $\mathcal{O}$ with rational weights. Then for any rational weight $\mu \in \mathfrak{h}^*$ the weight subspace $(V_1 \otimes \cdots \otimes V_r)[\mu]$ is finite dimensional. The KZ connection on $X = \mathbb{C}^r \setminus$ diagonals with values in $(V_1 \otimes \cdots \otimes V_r)[\mu]$ has the form

$$\nabla(\hbar) := d - \hbar \sum_{i=1}^{r} \left(\sum_{j \neq i} \frac{\Omega^{ij}}{x_i - x_j} \right) dx_i, \tag{5.1}$$

where Ω^{ij} is Ω acting in the i-th and j-th factor. It is easy to see that this connection has regular singularities.

By the Drinfeld-Kohno theorem [**Dr**], [**EK**][4], the monodromy of $\nabla(\hbar)$ for all $\hbar$ except possibly a countable set is given by R-matrices of the quantum group $U_q(\mathfrak{g})$, where $q = e^{\pi i \hbar}$. So, given that the weights of V_j are rational, the monodromy depends on $e^{\pi i \hbar / T}$, where T is the common denominator of the weights of V_j.

Variants of this construction include restricting the KZ connection to singular vectors (i.e., $\mathfrak{n}_+$-invariants in $(V_1 \otimes \cdots \otimes V_r)[\mu]$) or, for $\mu = 0$, $\mathfrak{g}$-invariants.

This example can be generalized in several directions described in the following subsections, which gives many more examples of pencils of regular flat connections that turn out to be quasi-motivic.

5.1.2. *KZ connections for Kac-Moody algebras.* In 5.1.1, the Lie algebra $\mathfrak{g}$ may be replaced with a symmetrizable Kac-Moody algebra $\mathfrak{g}(A)$ for a generalized Cartan matrix A, or, more generally, with the Lie algebra $\widetilde{\mathfrak{g}}(A)$ without Serre relations attached to a rational matrix A (see [**V**]). In this case, the relevant version of the Drinfeld-Kohno theorem is proved in [**V**] (for Verma modules and integrable modules) and [**EK**] (in general).

5.1.3. *Multiparameter KZ connections.* We may consider a multiparameter version of the pencil 5.1.1. For simplicity consider the case when $V_i = M(\lambda_i)$ are Verma modules with highest weights $\lambda_i \in \mathfrak{h}^*$, and $\mu = \sum_{i=1}^r \lambda_i - \beta$, where $\beta \in Q_+$. In this case, we have a natural identification

$$(V_1 \otimes \cdots \otimes V_r)[\mu] \cong \bigoplus_{\beta_1, \ldots, \beta_r : \sum_{i=1}^r \beta_i = \beta} \bigotimes_{i=1}^{r} U(\mathfrak{n}_-)[\beta_i].$$

[4]At the end of [**Dr**], Drinfeld claims this theorem for finite dimensional representations of $\mathfrak{g}$, but the proof applies to the more general case of category $\mathcal{O}$ using the result of [**EK**] that the Etingof-Kazhdan functorial quantization of a simple Lie algebra with its standard quasitriangular structure is isomorphic to the usual Drinfeld-Jimbo quantization (for formal $\hbar$).

Realizing $M(\lambda)\otimes M(\mu)$ as $U(\mathfrak{n}_-)\otimes U(\mathfrak{n}_-)$, we can write the operator Ω on $M(\lambda)\otimes M(\mu)$ as

$$\Omega=(\lambda,\mu)+Q+\sum_k \lambda(h_k)P_k^{12}+\sum_k \mu(h_k)P_k^{21},$$

where Q, P_k do not depend on λ,μ and have rational coefficients. It follows that the KZ connection can be written as

$$\nabla=d-\hbar\sum_{i=1}^r\left(\sum_{j\neq i}\frac{(\lambda_i,\lambda_j)+Q^{ij}+\sum_k\lambda_i(h_k)(P_k^{ij}+P_k^{ji})}{x_i-x_j}\right)dx_i.$$

Define the modified KZ connection by conjugating ∇ by the multivalued analytic function $\psi_0=\prod_{i<j}(x_i-x_j)^{\hbar(\lambda_i,\lambda_j)}$:

$$\nabla_*=d-\hbar\sum_{i=1}^r\left(\sum_{j\neq i}\frac{Q^{ij}+\sum_k\lambda_i(h_k)(P_k^{ij}+P_k^{ji})}{x_i-x_j}\right)dx_i.$$

Now we can define $\mathbf{s}$ to be the vector with coordinates $s_0:=\hbar$ and $s_{jk}:=\hbar\lambda_j(h_k)$, and we get a pencil of flat connections $\nabla_*(\mathbf{s})$. Moreover, the Drinfeld-Kohno theorem [**Dr**], [**EK**] implies that the monodromy of this pencil depends on $q=e^{\pi i\hbar}$ and $e^{\pi i s_{jk}}=q^{\lambda_j(h_k)}$.

EXAMPLE 5.1. Let $\mathfrak{g}=\mathfrak{sl}_2$. In this case we have $M(\lambda)=\mathbb{C}[f]v_\lambda\cong\mathbb{C}[z]$. Thus, setting $\partial=\partial_z$, we have $\Omega=\frac12 h\otimes h+e\otimes f+f\otimes e$, so by direct computation we obtain

$$\Omega|_{M(\lambda)\otimes M(\mu)}=\tfrac12\lambda\mu+\lambda(\partial\otimes z-1\otimes z\partial)+\mu(z\otimes\partial-z\partial\otimes 1)-z\partial^2\otimes 1-1\otimes z\partial^2.$$

Thus

$$P=\partial\otimes z-1\otimes z\partial,\ Q=-z\partial^2\otimes 1-1\otimes z\partial^2.$$

EXAMPLE 5.2. Consider the multiparameter KZ pencil for $\mathfrak{g}=\mathfrak{sl}_2$ in the vector space $(M(\lambda_1)\otimes\cdots\otimes M(\lambda_r))[\sum_{j=1}^r\lambda_j-2]\cong\mathbb{C}^r$. Let $s_j:=\hbar\lambda_j$. Then from Example 5.1 we get that the KZ equations for a flat section $I(\mathbf{s},\mathbf{x})=(I_1(\mathbf{s},\mathbf{x}),\dots,I_r(\mathbf{s},\mathbf{x}))$ of $\nabla(\mathbf{s})$ (conjugated by ψ_0) have the form

$$\frac{\partial I_j}{\partial x_i}=s_i\frac{I_i-I_j}{x_i-x_j},\ j\neq i;\quad \frac{\partial I_i}{\partial x_i}=-\sum_{j\neq i}s_j\frac{I_i-I_j}{x_i-x_j}. \tag{5.2}$$

(In this special case the KZ connection does not explicitly depend on $\hbar$ but only depends on s_j, because $Q=0$). We have a $\nabla(\mathbf{s})$-invariant decomposition $\mathbb{C}^r=\mathbb{C}\mathbf{1}\oplus V_{\mathbf{s}}$, where $\mathbf{1}=(1,\dots,1)$ and $V_{\mathbf{s}}$ is the orthogonal complement to $\mathbf{s}$. Thus the shift operators A_k should decompose accordingly as

$$A_k(\mathbf{s},\mathbf{x})=A_k^0(\mathbf{s},\mathbf{x})\oplus\xi_k(\mathbf{s})(\mathbf{1}\otimes\mathbf{s}^T),$$

where ξ_k are scalar rational functions and

$$A_k^0(\mathbf{s},\mathbf{x})\mathbf{1}=0,\ A_k^0(\mathbf{s},\mathbf{x})(V_{\mathbf{s}})\subset V_{\mathbf{s}+\mathbf{e}_k}.$$

Then A_k^0 is uniquely determined up to scaling by a rational function of $\mathbf{s}$.

To compute A_k^0, note that by definition $A_k^0(\mathbf{s},\mathbf{x})$ expresses flat sections of $\nabla(\mathbf{s}+\mathbf{e}_k)$ via flat sections of $\nabla(\mathbf{s})$. Now, flat sections of $\nabla(\mathbf{s})$ are solutions of (5.2), so they have the form

$$I_i(\mathbf{x},\mathbf{s})=\int\Phi(t,\mathbf{s},\mathbf{x})\frac{dt}{t-x_i},$$

where

$$\Phi(t, \mathbf{s}, \mathbf{x}) := \prod_{j=1}^{r} (t - x_j)^{-s_j}$$

is the master function, and integration is over a suitable twisted cycle (see e.g. [**EFK**], 4.3). Thus for $i \neq k$, we get by a direct computation

$$I_i(\mathbf{s} + \mathbf{e}_k, \mathbf{x}) = \int \Phi(t, \mathbf{s}, \mathbf{x}) \frac{1}{(t - x_k)(t - x_i)} dt = \frac{I_i(\mathbf{s}, \mathbf{x}) - I_k(\mathbf{s}, \mathbf{x})}{x_i - x_k}.$$

and

$$I_k(\mathbf{s} + \mathbf{e}_k, \mathbf{x}) = \int \Phi(t, \mathbf{s}, \mathbf{x}) \frac{1}{(t - x_k)^2} dt = -\sum_{i \neq k} \frac{s_i}{s_k + 1} \frac{I_i(\mathbf{s}, \mathbf{x}) - I_k(\mathbf{s}, \mathbf{x})}{x_i - x_k}.$$

This yields the following formula for A_k^0:

$$A_k^0(\mathbf{s}, \mathbf{x}) = \sum_{i \neq k} \frac{(s_k + 1)(E_{ii} - E_{ik}) + s_i(E_{kk} - E_{ki})}{x_i - x_k},$$

where E_{ij} are the elementary matrices.

REMARK 5.3. The discussion of this subsection applies mutatis mutandis to the setting of 5.1.2. Moreover, one may consider the case of a generic complex matrix A and Lie algebra $\widetilde{\mathfrak{g}}(A)$ without Serre relations. In this case the entries a_{ij} of A multiplied by $\hbar$ should become parameters of the pencil (which can be done since the KZ connection is linear in a_{ij}).

REMARK 5.4. The integral formulas for solutions of KZ equations given in [**SV**], [**V**] imply that the pencils in 5.1.1, 5.1.2,5.1.3 are actually motivic (not just quasi-motivic).

5.1.4. *KZ connections in the Deligne category.* For $t \in \mathbb{C}$ let $\mathrm{Rep}(GL_t)$ be the (abelian) Deligne category [**EGNO**, Section 9.12], and let $V = [1, 0]$ be the generating object (the "vector representation"). Consider the space $\mathrm{Hom}(\mathbf{1}, V^{\otimes m} \otimes V^{*\otimes m}) = \mathrm{End}(V^{\otimes m})$. By the Schur-Weyl duality, this space is naturally identified with $\mathbb{C}S_m$. The Casimir operator Ω acts on $V \otimes V$ and $V^* \otimes V^*$ by the flip and on $V \otimes V^*$ by $-\mathrm{coev} \circ \mathrm{ev}$, where $\mathrm{ev} : V \otimes V^* \to \mathbf{1}$, $\mathrm{coev} : \mathbf{1} \to V \otimes V^*$ are the evaluation and coevaluation maps. Thus the operators Ω^{ij}, $1 \leq i < j \leq 2m$ on $\mathrm{Hom}(\mathbf{1}, V^{\otimes m} \otimes V^{*\otimes m}) = \mathbb{C}S_m$ act as follows: for $\sigma \in S_m$,

$$\Omega^{ij}\sigma = \begin{cases} (i, j) \circ \sigma \text{ if } i, j \leq m; \\ \sigma \circ (i - m, j - m) \text{ if } i, j > m; \\ -t\sigma \text{ if } \sigma(j - m) = i, \ i \leq m < j; \\ -(i, \sigma(j - m)) \circ \sigma = -\sigma \circ (\sigma^{-1}(i), j - m) \text{ if } \sigma(j - m) \neq i, \ i \leq m < j, \end{cases}$$

where (i, j) is the transposition of i and j. In other words, $\Omega^{ij} = \Omega_0^{ij} + t\Omega_1^{ij}$, where $\Omega_0^{ij}, \Omega_1^{ij}$ are independent on t.

We may now set $s_0 = \hbar$, $s_1 = \hbar t$ and consider the pencil of flat connections $\nabla_{KZ,m}(\mathbf{s})$ with fiber $\mathbb{C}S_m$ given by (5.1). Note that if $t = n \geq m$ is a positive integer then this connection coincides with the usual KZ connection for $\mathfrak{gl}_n$ on $(V^{\otimes m} \otimes V^{*\otimes m})^{\mathfrak{gl}_n}$. Thus, repeating the arguments of [**Dr**] in the Deligne category (which can be done since they are based on diagrammatic tensor calculus), we can interpolate the Drinfeld-Kohno theorem to non-integer t. This implies that the monodromy of $\nabla_{KZ,m}$ is given by the braiding operators of the tensor category $\mathrm{Rep}_q(GL_t)$, also called the oriented skein category (see e.g. [**B**]). These operators are rational functions in $q := e^{\pi i s_0}$ and $a := q^t = e^{\pi i s_1}$.

5.1.5. *KZ connections for Lie superalgebras.* The results of 5.1.1 can be generalized to the case when $\mathfrak{g}$ is a Lie superalgebra of type A-G, and $V_1, \dots, V_r$ are finite dimensional $\mathfrak{g}$-modules. In this case the relevant version of the Drinfeld-Kohno theorem is proved in [**G**]: again it turns out that the monodromy representation depends on $q := e^{\pi i \hbar}$.

This can be generalized in several directions:

(1) $V_1, \dots, V_r$ can be taken from category $\mathcal{O}$ for some Borel subalgebra of $\mathfrak{g}$ (with rational weights), as in 5.1.1. Note that for Lie superalgebras there are several types of Borel subalgebras up to conjugation.
(2) The setup can be generalized to Kac-Moody Lie superalgebras, as in 5.1.2 (for basic results on their structure and representation theory we refer the reader to [**S**]).
(3) $V_1, \dots, V_r$ can be taken to be Verma modules with complex highest weights for some Borel subalgebra, and their highest weights multiplied by $\hbar$ can be made parameters of a multiparameter pencil, as in 5.1.3.

REMARK 5.5. To be more precise, the Lie superalgebras of type A-G contain a unique continuous family $D(2,1,\alpha)$, $\alpha \in \mathbb{C}$. So the above discussion applies literally when $\alpha \in \mathbb{Q}$. For more general $\alpha \in \mathbb{C}$, we should make it a parameter of the pencil, similarly to Remark 5.3 (as the entries a_{ij} of the Cartan matrix of $D(2,1,\alpha)$ depend linearly on α).

5.1.6. *Trigonomeric KZ connections.* Let $\mathfrak{g}$ be a simple Lie algebra as in 5.1.1 and $\mathbf{r} \in \mathfrak{g} \otimes \mathfrak{g}$ a quasitriangular structure (classical r-matrix) such that $\mathbf{r} + \mathbf{r}^{21} = \Omega$. Such structures were classified by Belavin and Drinfeld, [**BD1**] and are labeled (up to abelian twists) by Belavin-Drinfeld triples. Given $\mathbf{r}$, we may define the affine r-matrix (with spectral parameter)

$$\widehat{\mathbf{r}}(z) := \frac{\mathbf{r}^{21} z + \mathbf{r}}{z-1}.$$

We may assume that the centralizer of $\mathbf{r}$ in $\mathfrak{g}$ is a Lie subalgebra $\mathfrak{h}_{\mathbf{r}} \subset \mathfrak{h}$ (this can always be achieved by conjugation). Let $s \in \mathfrak{h}_{\mathbf{r}}$, let $V_1, \dots, V_r$ be finite dimensional $\mathfrak{g}$-modules, and consider the connection on the trivial bundle with fiber $(V_1 \otimes \cdots \otimes V_r)[\mu]$, $\mu \in \mathfrak{h}_{\mathbf{r}}^*$ on $\mathbb{C}^r$ with first order poles at $x_i = 0$ and $x_i = x_j$, given by

$$\nabla(\hbar, s) := d - \sum_{i=1}^{r} \left(s^{(i)} + \hbar \sum_{j \neq i} \widehat{\mathbf{r}}(x_i/x_j) \right) \frac{dx_i}{x_i}. \tag{5.3}$$

This connection is flat and called **the trigonometric KZ connection**; it is a pencil with parameters $\mathbf{s} = (\hbar, s) \in \mathbb{C} \oplus \mathfrak{h}_{\mathbf{r}}$ (i.e., $\hbar = s_0, s = (s_1, \dots, s_n)$). The monodromy of the trigonometric KZ connection is computed in [**EG**] in terms of the Etingof-Kazhdan quantization of the quasitriangular Lie bialgebra $(\mathfrak{g}, \mathbf{r})$, using its relationship to the usual KZ connection. This quantization is described explicitly in [**ESS**], and this description implies that the monodromy of the trigonometric KZ connection depends on $q = e^{\pi i \hbar}$ and q^s.

REMARK 5.6.

(1) In the case when $\mathbf{r}$ is the standard classical r-matrix of $\mathfrak{g}$, the trigonometric KZ connection can be easily obtained from the rational one, as explained for example in [**EFK**], Subsection 3.8, or [**MV**]. In particular, the integral representations of solutions of the KZ equations from [**SV**] give integral representations of solutions of the trigonometric KZ equations. Hence the trigonometric KZ connection for the standard r-matrix is motivic.
(2) Moreover, in this case the shift operators $A_j(\mathbf{s})$ in the parameters s_j, $j \geq 1$ are the dynamical difference operators introduced in [**TV**] and studied further in [**EV1**] (the lattice part of the dynamical Weyl group for the affine Lie algebra $\widehat{\mathfrak{g}}$). However, the shift operator $A_0(\mathbf{s})$ in $s_0 = \hbar$ is more complicated, and its explicit expression is not known in general.

As before, this can be generalized to category $\mathcal{O}$ modules and to Lie superalgebras.

Another generalization arises if the trigonometric r-matrix $\widehat{\mathbf{r}}(z)$ corresponding to $\mathbf{r}$ is replaced by an arbitrary trigonometric r-matrix $\widehat{\mathbf{r}}(z)$ from [**BD2**] (i.e., corresponding to a Belavin-Drinfeld triple for a possibly twisted affine Lie algebra). This defines a pencil of flat connections (5.3) (where $s \in \mathfrak{h}$ preserves $\widehat{\mathbf{r}}$) whose monodromy can be computed using the trace interpretation of flat sections given in [**ES**], and shown to depend on $q = e^{\pi i \hbar}$ and q^s.

5.1.7. *Elliptic KZ connections.* Let $\tau \in \mathbb{C}_+$ and $E = \mathbb{C}^\times / e^{2\pi i \tau \mathbb{Z}}$ be an elliptic curve. Let $\mathfrak{g} = \mathfrak{sl}_n$ and $\mathbf{r}(z, \tau)$ be the Belavin elliptic r-matrix associated to some integer $1 \leq k \leq n-1$ coprime to n [**BD2**]; it has simple poles at points of order n on E. Let $V_1, \dots, V_r$ be finite dimensional representations of $\mathfrak{sl}_n$. The **elliptic KZ connection** is the connection on the trivial bundle on E^r with fiber $V_1 \otimes \cdots \otimes V_r$ and first order poles at $x_i = \zeta x_j$, where $\zeta \in E$, $\zeta^n = 1$, given by the formula

$$\nabla(\hbar) = d - \hbar \sum_{i=1}^{r} \sum_{j \neq i} \mathbf{r}(x_i/x_j, \tau) \frac{dx_i}{x_i}.$$

This is a pencil of flat connections on its regular locus $X \subset E^r$. The monodromy of this pencil was computed in [**E1**] and shown to depend on $q = e^{\pi i \hbar}$.

5.2. Dunkl connections.

5.2.1. *Dunkl connections.* Let W be an irreducible finite Coxeter group with reflection representation $\mathfrak{h}$, $S \subset W$ the set of reflections, and for any $w \in S$ let $\alpha_w \in \mathfrak{h}^* \setminus 0$ be such that $w\alpha_w = -\alpha_w$. Let $\mathfrak{h}_{\text{reg}} \subset \mathfrak{h}$ be the set of $\mathbf{x} \in \mathfrak{h}$ such that $\alpha_w(\mathbf{x}) \neq 0$ for all $w \in S$. Let V be a finite dimensional representation of W. Consider the pencil of connections

$$\nabla(c) = d - c \sum_{w \in S} \frac{d\alpha_w}{\alpha_w}(w-1)$$

on the trivial bundle with fiber V over $\mathfrak{h}_{\text{reg}}$. It was proved by Dunkl that $\nabla(c)$ is flat, and it is called the **Dunkl connection**. It is not hard to show that the Dunkl connection has regular singularities and its monodromy is given by the representation V_q of the **Hecke algebra** $H_q(W)$ attached to V, where $q = e^{2\pi i c}$ (see [**GGOR**]). This representation is determined by q, or in rare cases by $q^{\frac{1}{2}}$, due to presence of Opdam's KZ twists (see e.g. [**BC**], 7.2).

This can be generalized in several directions. First of all, if W is of type I_{2m} (even dihedral type), B_m, $m \geq 3$, or F_4, then S falls into two conjugacy classes S_1, S_2, so we can consider the 2-parameter pencil of Dunkl connections

$$\nabla(c_1, c_2) = d - c_1 \sum_{w \in S_1} \frac{d\alpha_w}{\alpha_w}(w-1) - c_2 \sum_{w \in S_2} \frac{d\alpha_w}{\alpha_w}(w-1).$$

The above discussion extends verbatim to this case.

Secondly, we can generalize this story to the case when W is a complex reflection group. In this case we should fix a conjugation-invariant function $c : S \to \mathbb{C}$, and the Dunkl connection (introduced in this generality in [**BMR**, **DO**]) has the form

$$\nabla_c = d - \sum_{w \in S} c(w) \frac{d\alpha_w}{\alpha_w}(w-1).$$

Let $\overline{S}$ be the set reflection hyperplanes, and pick representatives $E_1, \ldots, E_m$ of W-orbits in $\overline{S}$. For each $1 \leq k \leq m$, the stabilizer of a generic point in E_k is $\mathbb{Z}/\ell_k$, generated by an element w_k which has nontrivial eigenvalue $e^{\frac{2\pi i}{\ell_k}}$ on $\mathfrak{h}$. Let $c_{jk} := c(w_k^j)$ where $1 \leq j \leq \ell_k - 1$. Then $c = (c_{jk})$. Let us make the following linear transformation of these coordinates:

$$s_{lk} := \sum_{j=1}^{\ell_k - 1} e^{\frac{2\pi i j l}{\ell_k}} c_{jk}, \; 0 \leq l \leq \ell_k - 1.$$

Thus for each k we have $\sum_{l=0}^{\ell_k-1} s_{lk} = 0$. Let $\mathbf{s} = (s_{lk})$ (subject to these conditions) and $\nabla(\mathbf{s}) := \nabla_c$. It is shown in [**GGOR**] that the monodromy of $\nabla(\mathbf{s})$ factors through the **cyclotomic Hecke algebra** $H_{\mathbf{q}}(W)$ where $\mathbf{q} = e^{2\pi i \mathbf{s}}$, and it is now known that, as originally conjectured in [**BMR**], this algebra is a flat deformation of the group algebra $\mathbb{C}W$, i.e. has dimension $|W|$ (see [**E2**] and references therein).

5.2.2. *Thrigonometric Dunkl (Dunkl-Cherednik) connections.* Now let W be the Weyl group of an irreducible root system $R \subset \mathfrak{h}^*$ of rank r with a polarization $R = R_+ \cup R_-$. For $\alpha \in R_+$ let $s_\alpha \in W$ be the reflection corresponding to α. Let $Q \subset \mathfrak{h}^*$ be the root lattice of R spanned by R over $\mathbb{Z}$ and $P^\vee \subset \mathfrak{h}$ be the dual lattice of coweights. Let $H = \mathfrak{h}/2\pi i P^\vee$ be the maximal torus of the corresponding adjoint simple complex Lie group G. For $\alpha \in Q$ denote by e^α the character of H corresponding to α. Let H_{reg} be the set of $g \in H$ such that $e^\alpha(g) \neq 1$ for any $\alpha \in R_+$.

Following Drinfeld and Lusztig, define the **degenerate affine Hecke algebra** $\mathcal{H}(W)$ to be the algebra generated by W and $\operatorname{Sym}\mathfrak{h}$ with defining commutation relations

$$s_\alpha h - s_\alpha(h) s_\alpha = \alpha(h), \; h \in \mathfrak{h}$$

for simple roots α. It is well known that the multiplication map $\mathbb{C}W \otimes \operatorname{Sym}\mathfrak{h} \to \mathcal{H}(W)$ in either order is an isomorphism of vector spaces (the PBW theorem for $\mathcal{H}(W)$, see e.g. [**L**]). Let V be a finite dimensional representation of $\mathcal{H}(W)$. Consider the pencil of connections on the trivial bundle over H_{reg} with fiber V given by

$$\nabla(c) = d - c\left(\sum_{i=1}^{r} \omega_i^\vee d\alpha_i + \sum_{\alpha \in R_+} \frac{e^\alpha d\alpha}{1 - e^\alpha}(s_\alpha - 1)\right),$$

where α_i are the simple roots and $\omega_i^\vee$ the fundamental coweights (here s_α and $\omega_i^\vee$ act on V as elements of $\mathcal{H}(W)$). In other words, using $x_i := e^{\alpha_i}$ as coordinates on H, we obtain the following formula for the covariant derivatives:

$$\nabla_i(c) = \partial_i - \frac{c}{x_i}\left(\omega_i^\vee + \sum_{\alpha\in R_+}\alpha(\omega_i^\vee)\frac{\prod_{j=1}^r x_j^{\alpha(\omega_j^\vee)}}{1-\prod_{j=1}^r x_j^{\alpha(\omega_j^\vee)}}(s_\alpha - 1)\right).$$

This connection is flat, has regular singularities, and is called the **Dunkl-Cherednik connection** (also known as trigonometric Dunkl connection, or the affine KZ connection for R); it was introduced in [**Ch**]. Cherednik showed that the monodromy of this connection is given by the representation V_q of the **affine Hecke algebra** $\mathcal{H}_q(W)$ corresponding to V via Lusztig's map, [**L**], where $q = e^{2\pi i c}$.

This can be upgraded to a multiparameter version. To this end, define the family of **induced representations** V_λ of $\mathcal{H}(W)$ for a weight λ. Namely, given $\lambda \in \mathfrak{h}^*$, let $V_\lambda := \mathcal{H}(W) \otimes_{\mathrm{Sym}\,\mathfrak{h}} \mathbb{C}_\lambda$, where $\mathfrak{h}$ acts in $\mathbb{C}_\lambda$ by $h \mapsto \lambda(h)$. The PBW theorem for $\mathcal{H}(W)$ yields an isomorphism $V_\lambda \cong \mathbb{C}W$ as a $\mathbb{C}W$-module. Moreover, the commutation relations of $\mathcal{H}(W)$ imply that for any $h \in \mathfrak{h}$, the action of h in V_λ is given by

$$h \circ w = \lambda(w^{-1}h)w + \xi_w(h)$$

where $\xi_w(h)$ comprises the lower length terms independent of λ.

Thus the pencil $\nabla(c)$ on V_λ may be regarded as a multiprameter pencil $\nabla(\mathbf{s})$ with parameter $\mathbf{s} := (c, c\lambda)$. Cherednik's monodromy theorem implies that the monodromy of $\nabla(\mathbf{s})$ depends of $e^{2\pi i \mathbf{s}}$.

As in the rational case, this can be generalized to unequal parameters for different W-orbits of roots. Let us present this generalization in the multiparameter form. Namely, let R^l be the W-orbits of R, and c_l the corresponding parameters. In the case of several parameters the degenerate affine Hecke algebra $\mathcal{H}_{\mathbf{c}}(W)$ depends on $\mathbf{c}$; namely it is the algebra generated by W and $\mathrm{Sym}\,\mathfrak{h}$ with defining commutation relations

$$s_\alpha h - s_\alpha(h)s_\alpha = c_l\alpha(h),\ h \in \mathfrak{h}$$

for simple roots $\alpha \in R_l$. Let $V_{\mathbf{c},\lambda}$ be the induced representation $\mathcal{H}_{\mathbf{c}} \otimes_{\mathrm{Sym}\,\mathfrak{h}} \mathbb{C}_\lambda$. Note that for any $t \neq 0$ we have a natural isomorphism $\mathcal{H}_{\mathbf{c}}(W) \cong \mathcal{H}_{t\mathbf{c}}(W)$ which preserves $w \in W$ and maps $h \in \mathfrak{h}$ to th, and that under this isomorphism $V_{\mathbf{c},\lambda}$ maps to $V_{t\mathbf{c},t\lambda}$. Thus in the above single-parameter case for $c \neq 0$, the representation $V_{c,\lambda}$ of $\mathcal{H}_c(W)$ corresponds to the representation $V_{\lambda/c}$ of $\mathcal{H}(W)$. As before, $V_{\mathbf{c},\lambda}$ is naturally identified with $\mathbb{C}W$ as a $\mathbb{C}W$-module with

$$h \circ w = \lambda(w^{-1}h)w + \sum_l c_l\xi_{l,w}(h), \tag{5.4}$$

where $\xi_{l,w} : \mathfrak{h} \to \mathbb{C}W$ is a linear map (landing in the span of elements of length $< \ell(w)$). Then the Dunkl-Cherednik connection has the form

$$\nabla(\mathbf{c}) = d - \sum_{i=1}^r \omega_i^\vee d\alpha_i - \sum_l c_l \sum_{\alpha\in R_+^l}\frac{e^\alpha d\alpha}{1-e^\alpha}(s_\alpha - 1).$$

As before, $\nabla(\mathbf{c})$ is flat and has regular singularities, and by Cherednik's theorem its monodromy on $V_{\mathbf{c},\lambda}$ depends on $e^{2\pi i\mathbf{c}}$ and $e^{2\pi i\lambda}$. Also in view of (5.4), $\nabla(\mathbf{c})|_{V_{\mathbf{c},\lambda}}$ is a pencil $\nabla(\mathbf{s})$ with parameter $\mathbf{s} = (\mathbf{c}, \lambda)$.

REMARK 5.7. In type A ($W = S_n$) the (trigonometric) Dunkl connections are special cases of the (trigonometric) KZ connections for $\mathfrak{g} = \mathfrak{sl}_n$. Namely, we should take $V_1, \ldots, V_n$ to be the vector representation $\mathbb{C}^n$ and consider the connection on $(V_1 \otimes \cdots \otimes V_n)[0]$.

5.3. Casimir connections.

5.3.1. *Rational Casimir connections.* Consider the setting of 5.1.1. Let $\mathfrak{h}_{\rm reg}$ be the regular part of the Cartan subalgebra $\mathfrak{h} \subset \mathfrak{g}$ as in 5.2.1. Let V be a finite dimensional $\mathfrak{g}$-module. The **Casimir connection** (also called the **dynamical connection**) is the connection on the trivial bundle on $\mathfrak{h}_{\rm reg}$ with fiber V given by

$$\nabla(\hbar) = d - \hbar \sum_{\alpha \in R_+} \frac{e_{-\alpha}e_\alpha + e_{-\alpha}e_\alpha}{2} \frac{d\alpha}{\alpha},$$

where e_α are the root elements of $\mathfrak{g}$ defined in 5.1.1. This connection is flat and has regular singularities. It was first defined by De Concini around 1995 (unpublished), and then was studied in the papers [**FMTV**], [**TL1**], [**MTL**] and many subsequent ones. In particular, Toledano-Laredo proved [**TL2**], [**TL3**] that the monodromy of $\nabla(\hbar)$ is given by the quantum Weyl group action on the representation V_q of the quantum group $U_q(\mathfrak{g})$ corresponding to V, where $q = e^{\pi i \hbar}$ (as had been conjectured independently by Toledano-Laredo and C. De Concini and proved for $\mathfrak{g} = \mathfrak{sl}_n$ in [**TL1**]). Since ∇ is $\mathfrak{h}$-invariant, the same is clearly true on every weight subspace $V[\mu]$.

This can be generalized in various ways. First of all, one can take V to be a module from category $\mathcal{O}$ with rational weights and restrict $\nabla(\hbar)$ to a weight subspace $V[\mu]$. The monodromy of ∇ is then still given by the (pure) quantum Weyl group [**ATL**].

Also we can consider the multiparameter version. To this end take $V = M(\lambda)$, the Verma module over $\mathfrak{g}$ with highest weight λ, and take $\mu = \lambda - \beta$ for fixed $\beta \in Q_+$. Then $V[\mu] \cong U(\mathfrak{n}_-)[-\beta]$, and in this realization the operators $e_{-\alpha}e_\alpha$ depend linearly (inhomogeneously) on λ. Thus $\nabla(\hbar)|_{M(\lambda)[\lambda-\beta]}$ can be viewed as a multiparameter pencil with poarameter $\mathbf{s} = (\hbar, \hbar\lambda)$. By [**ATL**], the monodromy of $\nabla(\mathbf{s})$ is given by the pure quantum Weyl group, so depends on $e^{2\pi i \mathbf{s}}$.

Finally, this whole story extends to the case when $\mathfrak{g}$ is a symmetrizable Kac-Moody algebra, since this is the generality of the monodromy result of [**ATL**].

5.3.2. *Trigonometric Casimir connections.* The trigonometric version of the Casimir connection was introduced in [**TL4**]. To define it, keep the setting of the previous subsection and the notation of 5.2.2. Let $Y(\mathfrak{g})$ be Drinfeld's **Yangian** of $\mathfrak{g}$ in which the Planck constant $\hbar$ is set to 1, and let $J : \mathfrak{g} \to Y(\mathfrak{g})$, be the $\mathfrak{g}$-invariant map defining the Drinfeld generators of $Y(\mathfrak{g})$ [**TL4**, Section 3]. Let V be a finite dimensional irreducible representation of $Y(\mathfrak{g})$ on which $J(\omega_i^\vee)$ acts with rational eigenvalues (i.e., its Drinfeld polynomial has rational roots). The **trigonometric Casimir connection** is the connection on the trivial bundle on $H_{\rm reg}$ with fiber V given by

$$\nabla(\hbar) = d - \hbar \left(\sum_{\alpha \in R_+} \frac{1}{4} \frac{e^\alpha + 1}{e^\alpha - 1} (e_\alpha e_{-\alpha} + e_{-\alpha}e_\alpha) d\alpha - \sum_{i=1}^{\mathrm{rank}(\mathfrak{g})} J(\omega_i^\vee) d\alpha_i \right).$$

It is shown in [**TL4**] that this connection is flat, and it clearly has regular singularities. Note that ∇ falls into a direct sum of connections on $\mathfrak{h}$-weight spaces of V.

Also ∇ is a pencil, and it arises as the equivariant quantum connection for quiver varieties of finite type [**MO**], hence is periodic and quasi-motivic.

Also for $\mathfrak{g} = \mathfrak{sl}_2$ it is shown in [**GTL**] that the monodromy of $\nabla(\hbar)$ is given by the action of the quantum Weyl group on the corresponding representation V_q of the quantum affine algebra $U_q(\widehat{\mathfrak{g}})$, and the proof for general $\mathfrak{g}$ is forthcoming in a joint work of Gautam and Toledano-Laredo. This also will imply the quasi-motivic and periodic properties of $\nabla(\hbar)$.

There is also a multiparameter version of this pencil. To describe it, denote by $V(z)$ the shift of the representation V by $z \in \mathbb{C}$, which is V as a $\mathfrak{g}$-module with

$$J(a)|_{V(z)} = (J(a) + za)|_V.$$

Let $V_1, \ldots, V_n$ be irreducible finite dimensional $Y(\mathfrak{g})$-modules whose Drinfeld polynomials have rational roots, and let $V := V_1(z_1) \otimes \cdots \otimes V_n(z_n)$ (tensor product using the Hopf structure on $Y(\mathfrak{g})$). Recall that

$$\Delta(J(h)) = J(h) \otimes 1 + 1 \otimes J(h) + \frac{1}{2}[h \otimes 1, \Omega].$$

Thus $V = V_1 \otimes \cdots \otimes V_n$ as a $\mathfrak{g}$-module, and

$$J(h)|_V = \sum_{j=1}^{n} z_j (J(h)|_{V_j})^{(j)} + J_0(h)|_{V_1 \otimes \cdots \otimes V_n},$$

where $J_0(h) \in U(\mathfrak{g})^{\otimes n}$ does not depend on z_j. Thus the pencil $\nabla(\hbar)|_V$ can be viewed as a multiparameter pencil $\nabla(\mathbf{s})$ with parameter $\mathbf{s} = (\hbar, \hbar\mathbf{z})$, where $\mathbf{z} := (z_1, \ldots, z_n)$. This pencil arises as the equivariant quantum connection for quiver varieties of finite type [**MO**], so it is quasi-motivic and periodic. This will also follow from the forthcoming results of Gautam and Toledano-Laredo on the monodromy of $\nabla(\mathbf{s})$.

REMARK 5.8.

(1) If $\mathfrak{g} = \mathfrak{sl}_n$ then the (trigonometric) Casimir connection is equivalent to the (trigonometric) KZ connection for $\mathfrak{gl}_m$, via the Howe duality between $\mathfrak{gl}_n$ and $\mathfrak{gl}_m$, see [**GTL**], [**TL1**]. As explained in these papers, this duality can be used to compute the monodromy of the trigonometric Casimir connection in type A.

(2) Let $V = \mathfrak{g}$ be the adjoint representation of $\mathfrak{g}$. Then the Casimir connection on $V[0]$ coincides with the Dunkl connection for the Weyl group W of $\mathfrak{g}$ in its reflection representation $\mathfrak{h}$. Similarly, in the trigonometric case consider the adjoint representation V of the Yangian $Y(\mathfrak{g})$ (evaluated at 0), which is $\mathfrak{g}$ for type A and $\mathfrak{g} \oplus \mathbb{C}$ otherwise as a $\mathfrak{g}$-module. Then the trigonometric Casimir connection in $V[0]$ coincides with the trigonometric Dunkl connection on the reflection representation of the degenerate affine Hecke algebra $\mathcal{H}(W)$, which is $\mathfrak{h}$ in type A and $\mathfrak{h} \oplus \mathbb{C}$ otherwise.

5.4. Irreducibility. In many of the cases described above, the connection $\nabla(\mathbf{s})$ is irreducible for generic $\mathbf{s}$, so that Theorem 4.6(ii) applies. For example, it follows from Theorem 1.2 in [**MTL**] and a result of the first author (included in [**MTL**] as Theorem 1.8) that the Casimir connection $\nabla(\hbar, \hbar\mu)$ on the weight subspace $M(\mu)[\mu - \beta]$ in the Verma module $M(\mu)$ is irreducible for Zariski generic $(\hbar, \mu)$. Another example where the connection is generically irreducible is the Dunkl connection on an irreducible W-module V (this follows easily by the same argument as in the proof of [**MTL**], Theorem 1.2).

5.5. Unitarity and generic semisimplicity. In many of the above examples, the monodromy representation of the connection is unitary, hence semisimple for small real values of $\mathbf{s}$. Thus such pencils are generically semisimple.

For example, consider the KZ connection (5.1) in $\mathrm{Hom}_{\mathfrak{g}}(V_0, V_1 \otimes \cdots \otimes V_r)$ where V_i are finite dimensional representations of $\mathfrak{g}$. As noted above, the Drinfeld-Kohno theorem identifies the monodromy of the KZ connection with the representation of the pure braid group PB_r by R-matrices of the quantum group $U_q(\mathfrak{g})$, where $q = e^{\pi i \hbar}$. If $\hbar$ is real (i.e., $|q| = 1$), this quantum group has a real structure, and its category of representations is a Hermitian tensor category (when q is not a root of 1). This means that multiplicity spaces of tensor products of representations of $U_q(\mathfrak{g})$ carry a nondegenerate Hermitian inner product. Moreover, if $\hbar = 0$, this is the usual positive inner product on multiplicity spaces for the corresponding compact Lie group. Thus by deformation argument, when the representations V_i are fixed and $\hbar \in \mathbb{R}$ is small, then the monodromy representation of PB_r on $\mathrm{Hom}_{\mathfrak{g}}(V_0, V_1 \otimes \cdots \otimes V_r)$ is unitary. The same argument can be used for the multiparameter KZ connection considered in 5.1.3.

Similarly, as noted above, the monodromy of the Casimir connection discussed in 5.3.1 is given by the quantum Weyl group operators of $U_q(\mathfrak{g})$ on the q-deformation V_q of a finite-dimensional $\mathfrak{g}$-module V. This module carries a contravariant Hermitian form which is preserved by this action and is positive definite if $|q| = 1$ and q is close to 1. Thus in this case the monodromy representation is also unitary.

Finally, the monodromy of the rational Dunkl connection discussed in 5.2.1 is given by representations of the Hecke algebra $H_q(W)$, where $q = e^{2\pi i c}$, which are unitary for small real c.

Using this approach, one can establish generic semisimplicity for many other examples considered above. We omit the details.

Acknowledgments

This paper belongs to the area of quantum integrable systems which took shape in the work of our friend and colleague Igor Krichever (1950-2022). Igor influenced us in many ways throughout our lives, and we dedicate this paper to his memory.

We thank Jae Hee Lee and Vadim Vologodsky for useful discussions.

References

[ATL] A. Appel and V. Toledano Laredo, *Pure braid group actions on category $\mathcal{O}$ modules*, Pure Appl. Math. Q. **20** (2024), no. 1, 29–79, DOI 10.4310/pamq.2024.v20.n1.a3. MR4734869

[Ax] J. Ax, *On Schanuel's conjectures*, Ann. of Math. (2) **93** (1971), 252–268, DOI 10.2307/1970774. MR277482

[BC] Y. Berest and O. Chalykh, *Quasi-invariants of complex reflection groups*, Compos. Math. **147** (2011), no. 3, 965–1002, DOI 10.1112/S0010437X10005063. MR2801407

[BK] B. Bakalov and A. Kirillov Jr., *Lectures on tensor categories and modular functors*, University Lecture Series, vol. 21, American Mathematical Society, Providence, RI, 2001, DOI 10.1090/ulect/021. MR1797619

[BD1] A. A. Belavin and V. G. Drinfel′d, *Triangle equations and simple Lie algebras*, Mathematical physics reviews, Vol. 4, Soviet Sci. Rev. Sect. C: Math. Phys. Rev., vol. 4, Harwood Academic Publ., Chur, 1984, pp. 93–165. Translated from the Russian. MR768939

[BD2] A. A. Belavin and V. G. Drinfel′d, *Solutions of the classical Yang-Baxter equation for simple Lie algebras* (Russian), Funktsional. Anal. i Prilozhen. **16** (1982), no. 3, 1–29, 96. MR674005

[BM] S. Bochner and W. T. Martin, *Several Complex Variables*, Princeton Mathematical Series, vol. 10, Princeton University Press, Princeton, NJ, 1948. MR27863

[BMO] A. Braverman, D. Maulik, and A. Okounkov, *Quantum cohomology of the Springer resolution*, Adv. Math. **227** (2011), no. 1, 421–458, DOI 10.1016/j.aim.2011.01.021. MR2782198

[BMR] M. Broué, G. Malle, and R. Rouquier, *Complex reflection groups, braid groups, Hecke algebras*, J. Reine Angew. Math. **500** (1998), 127–190. MR1637497

[B] J. Brundan and C. Stroppel, *Highest weight categories arising from Khovanov's diagram algebra III: category $\mathcal{O}$*, Represent. Theory **15** (2011), 170–243, DOI 10.1090/S1088-4165-2011-00389-7. MR2781018

[Ch] I. Cherednik, *A unification of Knizhnik-Zamolodchikov and Dunkl operators via affine Hecke algebras*, Invent. Math. **106** (1991), no. 2, 411–431, DOI 10.1007/BF01243918. MR1128220

[De1] P. Deligne, *Un théorème de finitude pour la monodromie* (French), Discrete groups in geometry and analysis (New Haven, Conn., 1984), Progr. Math., vol. 67, Birkhäuser Boston, Boston, MA, 1987, pp. 1–19, DOI 10.1007/978-1-4899-6664-3_1. MR900821

[De2] P. Deligne, *Équations différentielles à points singuliers réguliers* (French), Lecture Notes in Mathematics, Vol. 163, Springer-Verlag, Berlin-New York, 1970. MR417174

[Dr] V. G. Drinfel′d, *Quasi-Hopf algebras* (Russian), Algebra i Analiz **1** (1989), no. 6, 114–148; English transl., Leningrad Math. J. **1** (1990), no. 6, 1419–1457. MR1047964

[DO] C. F. Dunkl and E. M. Opdam, *Dunkl operators for complex reflection groups*, Proc. London Math. Soc. (3) **86** (2003), no. 1, 70–108, DOI 10.1112/S0024611502013825. MR1971464

[E1] P. I. Etingof, *Representations of affine Lie algebras, elliptic r-matrix systems, and special functions*, Comm. Math. Phys. **159** (1994), no. 3, 471–502. MR1259404

[E2] P. Etingof, *Proof of the Broué-Malle-Rouquier conjecture in characteristic zero (after I. Losev and I. Marin—G. Pfeiffer)*, Arnold Math. J. **3** (2017), no. 3, 445–449, DOI 10.1007/s40598-017-0069-7. MR3692497

[EFK] P. I. Etingof, I. B. Frenkel, and A. A. Kirillov Jr., *Lectures on representation theory and Knizhnik-Zamolodchikov equations*, Mathematical Surveys and Monographs, vol. 58, American Mathematical Society, Providence, RI, 1998, DOI 10.1090/surv/058. MR1629472

[EG] P. Etingof and N. Geer, *Monodromy of trigonometric KZ equations*, Int. Math. Res. Not. IMRN **24** (2007), Art. ID rnm123, 15 pp., DOI 10.1093/imrn/rnm123. MR2377010

[EGNO] P. Etingof, S. Gelaki, D. Nikshych, and V. Ostrik, *Tensor categories*, Mathematical Surveys and Monographs, vol. 205, American Mathematical Society, Providence, RI, 2015, DOI 10.1090/surv/205. MR3242743

[EK] P. Etingof and D. Kazhdan, *Quantization of Lie bialgebras. VI. Quantization of generalized Kac-Moody algebras*, Transform. Groups **13** (2008), no. 3-4, 527–539, DOI 10.1007/s00031-008-9029-6. MR2452604

[ES] P. Etingof and O. Schiffmann, *Twisted traces of intertwiners for Kac-Moody algebras and classical dynamical R-matrices corresponding to generalized Belavin-Drinfeld triples*, Math. Res. Lett. **6** (1999), no. 5-6, 593–612, DOI 10.4310/MRL.1999.v6.n6.a1. MR1739218

[ESS] P. Etingof, T. Schedler, and O. Schiffmann, *Explicit quantization of dynamical r-matrices for finite dimensional semisimple Lie algebras*, J. Amer. Math. Soc. **13** (2000), no. 3, 595–609, DOI 10.1090/S0894-0347-00-00333-7. MR1758755

[EV1] P. Etingof and A. Varchenko, *Dynamical Weyl groups and applications*, Adv. Math. **167** (2002), no. 1, 74–127, DOI 10.1006/aima.2001.2034. MR1901247

[EV2] P. Etingof, A. Varchenko, Periodic pencils of flat connections and their p-curvature, `arXiv:2401.05652`.

[FMTV] G. Felder, Y. Markov, V. Tarasov, and A. Varchenko, *Differential equations compatible with KZ equations*, Math. Phys. Anal. Geom. **3** (2000), no. 2, 139–177, DOI 10.1023/A:1009862302234. MR1797943

[GJ] T. Gannon and C. Jones, *Vanishing of categorical obstructions for permutation orbifolds*, Comm. Math. Phys. **369** (2019), no. 1, 245–259, DOI 10.1007/s00220-019-03288-9. MR3959559

[GTL] S. Gautam and V. Toledano Laredo, *Monodromy of the trigonometric Casimir connection for $\mathfrak{sl}_2$*, Noncommutative birational geometry, representations and combinatorics,

Contemp. Math., vol. 592, Amer. Math. Soc., Providence, RI, 2013, pp. 137–176, DOI 10.1090/conm/592/11864. MR3087943

[G] N. Geer, *Etingof-Kazhdan quantization of Lie superbialgebras*, Adv. Math. **207** (2006), no. 1, 1–38, DOI 10.1016/j.aim.2005.11.005. MR2264064

[Ge] A. O. Gel′fond, *Transcendental and algebraic numbers*, Dover Publications, Inc., New York, 1960. Translated from the first Russian edition by Leo F. Boron. MR111736

[GGOR] V. Ginzburg, N. Guay, E. Opdam, and R. Rouquier, *On the category $\mathcal{O}$ for rational Cherednik algebras*, Invent. Math. **154** (2003), no. 3, 617–651, DOI 10.1007/s00222-003-0313-8. MR2018786

[GI] P. Grossman and M. Izumi, *Drinfeld centers of fusion categories arising from generalized Haagerup subfactors*, Quantum Topol. **13** (2022), no. 4, 593–668, DOI 10.4171/qt/167. MR4540757

[H] Y.-Z. Huang, *Rigidity and modularity of vertex tensor categories*, Commun. Contemp. Math. **10** (2008), no. suppl. 1, 871–911, DOI 10.1142/S0219199708003083. MR2468370

[K] N. M. Katz, *Rigid local systems*, Annals of Mathematics Studies, vol. 139, Princeton University Press, Princeton, NJ, 1996, DOI 10.1515/9781400882595. MR1366651

[Ka] M. Kashiwara, *Riemann-Hilbert correspondence for irregular holonomic $\mathcal{D}$-modules*, Jpn. J. Math. **11** (2016), no. 1, 113–149, DOI 10.1007/s11537-016-1564-7. MR3510681

[Lee] J. H. Lee, Quantum Steenrod operations of symplectic resolutions, arXiv:2312.02100, 2023.

[L] G. Lusztig, *Affine Hecke algebras and their graded version*, J. Amer. Math. Soc. **2** (1989), no. 3, 599–635, DOI 10.2307/1990945. MR991016

[MO] D. Maulik and A. Okounkov, *Quantum groups and quantum cohomology* (English, with English and French summaries), Astérisque **408** (2019), ix+209, DOI 10.24033/ast. MR3951025

[M] R. McRae, On rationality for C_2-cofinite vertex operator algebras, arXiv:2108.01898, 2021.

[MTL] J. J. Millson and V. Toledano Laredo, *Casimir operators and monodromy representations of generalised braid groups*, Transform. Groups **10** (2005), no. 2, 217–254, DOI 10.1007/s00031-005-1008-6. MR2195601

[MV] Y. Markov and A. Varchenko, *Hypergeometric solutions of trigonometric KZ equations satisfy dynamical difference equations*, Adv. Math. **166** (2002), no. 1, 100–147, DOI 10.1006/aima.2001.2027. MR1882860

[Mo] T. Mochizuki, *Wild harmonic bundles and wild pure twistor D-modules* (English, with English and French summaries), Astérisque **340** (2011), x+607. MR2919903

[PV] V. Popov, E. Vinberg, Invariant theory, Encyclopaedia of Mathematical Sciences, Volume 55, Springer, 1994.

[Sa] C. Sabbah, https://perso.pages.math.cnrs.fr/users/claude.sabbah/livres/sabbah_notredame-improvement2310.pdf

[SV] V. V. Schechtman and A. N. Varchenko, *Arrangements of hyperplanes and Lie algebra homology*, Invent. Math. **106** (1991), no. 1, 139–194, DOI 10.1007/BF01243909. MR1123378

[S] V. Serganova, *Structure and representation theory of Kac-Moody superalgebras*, Highlights in Lie algebraic methods, Progr. Math., vol. 295, Birkhäuser/Springer, New York, 2012, pp. 65–102, DOI 10.1007/978-0-8176-8274-3_3. MR2866847

[SS] E. M. Stein and R. Shakarchi, *Complex analysis*, Princeton Lectures in Analysis, vol. 2, Princeton University Press, Princeton, NJ, 2003. MR1976398

[TL1] V. Toledano Laredo, *A Kohno-Drinfeld theorem for quantum Weyl groups*, Duke Math. J. **112** (2002), no. 3, 421–451, DOI 10.1215/S0012-9074-02-11232-0. MR1896470

[TL2] V. Toledano Laredo, *Quasi-Coxeter algebras, Dynkin diagram cohomology, and quantum Weyl groups*, Int. Math. Res. Pap. IMRP, posted on 2008, Art. ID rpn009, 167, DOI 10.1093/imrp/rpn009. MR2470574

[TL3] V. Toledano-Laredo, Quasi-Coxeter quasitriangular quasibialgebras and the Casimir connection, arXiv:1601.04076, 2016.

[TL4] V. Toledano Laredo, *The trigonometric Casimir connection of a simple Lie algebra*, J. Algebra **329** (2011), 286–327, DOI 10.1016/j.jalgebra.2010.05.025. MR2769327

[TV] V. Tarasov and A. Varchenko, *Difference equations compatible with trigonometric KZ differential equations*, Internat. Math. Res. Notices **15** (2000), 801–829, DOI 10.1155/S1073792800000441. MR1780748

[Tu] V. G. Turaev, *Quantum invariants of knots and 3-manifolds*, De Gruyter Studies in Mathematics, vol. 18, Walter de Gruyter & Co., Berlin, 1994. MR1292673

[V] A. Varchenko, *Multidimensional hypergeometric functions and representation theory of Lie algebras and quantum groups*, Advanced Series in Mathematical Physics, vol. 21, World Scientific Publishing Co., Inc., River Edge, NJ, 1995, DOI 10.1142/2467. MR1384760

Department of Mathematics, MIT, Cambridge, Massachussetts 02139
Email address: `etingof@math.mit.edu`

Department of Mathematics, University of North Carolina at Chapel Hill, CB# 3250 Phillips Hall Chapel Hill, North Carolina 27599
Email address: `anv@email.unc.edu`

Contemporary Mathematics
Volume **823**, 2025
https://doi.org/10.1090/conm/823/16495

Hecke algebras for the 1st congruence subgroup and bundles on $\mathbb{P}^1$ I: the case of finite field

Alexander Braverman and David Kazhdan

ABSTRACT. Let G be a split reductive group over a finite field $\mathbf{k}$ In this note we study the space V of finitely supported functions on the set $\mathrm{Bun}_G(\mathbb{P}^1)_{0,\infty}(\mathbf{k})$ of isomorphism classes of G-bundles on the projective line $\mathbb{P}^1$ endowed with a trivialization at 0 and ∞. We show that V is naturally isomorphic to the regular bimodule over the Hecke algebra A of the group $G(\mathbf{k}((t)))$ with respect to the first congruence subgroup. As a byproduct we show that Hecke operators at points different from 0 and ∞ generate the "stable center" of A. We provide an expression of the character of the lifting of an irreducible cuspidal representation of $\mathrm{GL}(N,\mathbf{k})$ to $\mathrm{GL}(N,\mathbf{k}')$ where $\mathbf{k}'$ is a finite extension of $\mathbf{k}$ in terms of these generators.

In a subsequent publication we plan to develop analogous constructions in the case when $\mathbf{k}$ is replaced by a local non-archimedian field.

1. Introduction and setup

1.1. Let $\mathbf{k}$ be the finite field $\mathbb{F}_q$ with a fixed algebraic closure $\overline{\mathbf{k}}$. For every $i \geq 1$ let $\mathbf{k}_i$ denote the degree i extension of $\mathbf{k}$ inside $\overline{\mathbf{k}}$; we have the norm map $\mathrm{Norm}_{i,j} : \mathbf{k}_i^\times \to \mathbf{k}_j^\times$ for every $i \geq j$. In particular the groups $\mathbf{k}_i^\times$ form a projective system. Let also $\mathbb{K} = \mathbf{k}((t)), \mathbb{O} = \mathbf{k}[[t]]$.

Let G be a split connected reductive group over $\mathbf{k}$ with a Borel subgroup B and a maximal torus T.

Denote by K_1 the first congruence subgroup of $G(\mathbb{O})$ (consisting of elements equal to 1 modulo t). We denote by A the Hecke algebra of $G(\mathbb{K})$ with respect to K_1 with coefficients in an algebraically closed field F of characteristic 0; this is an associative algebra containing the group algebra of $G(\mathbf{k})$ as a subalgebra. We denote by $G^\vee$ the Langlands dual group of G; unless otherwise stated we shall think about it as a group over F.

Let $\mathcal{Z}(A)$ denote the center of A. It follows from [**6**] that $\mathcal{Z}(A)$ contains the algebra $F[Z(G^\vee,\mathbf{k})]^{G^\vee}$ of $G^\vee$-invariant functions on the scheme $Z_{G^\vee,\mathbf{k}}$ consisting of pairs (s,u) where

(1) $s \in G^\vee$
(2) $u : \lim\limits_{\leftarrow} \mathbf{k}_i^\times \to G^\vee$

subject to the relation $sus^{-1} = u^q$ (it is easy to see that $Z_{G^\vee,\mathbf{k}}$ is indeed a scheme of finite type over F; this follows from the fact that there exists some i depending only on G such that for any pair (s, u) as above the homomorphism u factorizes through $\mathbf{k}_i^\times$).

One of the purposes of this note is to give an alternative characterization of the homomorphism $F[Z(G^\vee, \mathbf{k})]^{G^\vee} \to \mathcal{Z}(A)$ (in the future we shall call the image of this homomorphism *the stable center of* A and we shall denote it by $\mathcal{Z}_{st}(A)$). For this we are going to use some global constructions discussed in the next subsection.

1.2. Bundles on $\mathbb{P}^1$ and $\mathcal{Z}(A)$. Let $\mathrm{Bun}_G(\mathbb{P}^1)_{0,\infty}$ denote the moduli stack of bundles on $\mathbb{P}^1$ (over $\mathbf{k}$) with trivializations at 0 and ∞. Let V denote the space of F-valued functions on $\mathrm{Bun}_G(\mathbb{P}^1)_{0,\infty}(\mathbf{k})$ with finite support. The space V has two natural commuting actions of the algebra A (by Hecke operators at 0 and ∞). Our first statement is that V is essentially the regular bimodule over A. More precisely, we claim that there exists an anti-involution $\iota : A \to A$ on A such that if we use it to turn V into an A-bimodule – keeping the (left) action of A at 0 intact and making the action of A at ∞ into a right action by using ι, then we have

THEOREM 1.1. *There is a natural isomorphism $V \simeq A$ of A-bimodules.*

COROLLARY 1.2. *We have $\mathcal{Z}(A) = \mathrm{End}_{A\otimes A}(V)$.*

We do not know any simple description of ι but it will satisfy the following properties:

(1) It preserves the subalgebra $F[G(\mathbf{k})] \subset A$, and on this subalgebra it is induced by the map $g \mapsto g^{-1}$.
(2) Let e_{fin} be the idempotent $\frac{1}{\#G(\mathbf{k})} \sum_{g\in G(\mathbf{k})} g \in F[G(\mathbf{k})] \subset A$. Then the induced anti-involution on $e_{fin}Ae_{fin}$ is equal to identity (note that $e_{fin}Ae_{fin}$ is the spherical Hecke algebra of G which is commutative).

Let us also note that ι is not completely local in the sense that it depends on the choice of a generator of the maximal ideal of $\mathbf{k}[[t]]$.

We shall explain the strategy of the proof of Theorem 1.1 in Subsection 1.5.

1.3. Hecke operators at other points and stable center. Let D be a subset of $\overline{\mathbf{k}}^\times$ invariant under the Galois group of $\overline{\mathbf{k}}$ over $\mathbf{k}$. Let also $f \in F[G^\vee]^{G^\vee}$. To this data one can in the standard way associate the Hecke operator $h_{D,f} \in \mathrm{End}_{A\otimes A}(V)$. According to Corollary 1.2 we can regard $h_{D,f}$ as an element of $\mathcal{Z}(A)$.

THEOREM 1.3.

(i) *We have $h_{D,f} \in \mathcal{Z}_{st}$. In particular, $h_{D,f}$ corresponds to a function $\phi_{D,f}$ on $Z(G^\vee, \mathbf{k})$.*
(ii) *Assume that D is the $\mathrm{Gal}(\overline{\mathbf{k}}/\mathbf{k})$-orbit of some $x \in \mathbf{k}_i^\times$ which does not belong to any $\mathbf{k}_j$ with $j < i$. Then*

(1) $$\phi_{D,f}(s, u) = f(s^i\, u(y))$$

where $y \in \mathbf{k}_n^\times$ for some $n \geq i$ is such that u factorizes through $\mathbf{k}_n^\times$, n is divisible by i and $\mathrm{Norm}_{n,i}(y) = x$ (in particular, we claim that the RHS of (1) does not depend on the choice of n and y).

It is easy to see that Theorem 1.3 implies the following

COROLLARY 1.4. *$\mathcal{Z}_{st}(A)$ is the subalgebra of $\mathcal{Z}(A)$ generated by all the $h_{D,f}$.*

1.4. Action on cuspidal representations of $G(\mathbf{k})$. It is clear that stack $\mathrm{Bun}_G(\mathbb{P}^1)_{0,\infty}$ is acted on by $G \times G$ (by changing the trivializations at 0 and ∞), hence the group $G(\mathbf{k}) \times G(\mathbf{k})$ acts on V. Let V_{cusp} denote the maximal cuspidal $G(\mathbf{k}) \times G(\mathbf{k})$-subrepresentation of V (it is easy to see that V_{cusp} is also the maximal cuspidal $G(\mathbf{k})$-subrepresentation of V with respect to either action of $G(\mathbf{k})$ – this follows immediately from Theorem 1.1). It is clear that V_{cusp} is invariant under the action of $A \otimes A$ and also under the operators $h_{D,f}$.

The following lemma is easy:

LEMMA 1.5. *Assume that G is semi-simple. Then $V_{\mathrm{cusp}} = F_{\mathrm{cusp}}(G(\mathbf{k}))$ (i.e. the cuspidal part of the regular representation of $G(\mathbf{k})$.*

In particular, the restriction of every $h_{D,f}$ to V_{cusp} defines a $G(\mathbf{k}) \times G(\mathbf{k})$-endomorphism of $F_{\mathrm{cusp}}(G(\mathbf{k}))$. Such a data is equivalent to a function $\eta_{D,f}$: $\mathrm{Irr}_{\mathrm{cusp}}(G(\mathbf{k})) \to F$ where $\mathrm{Irr}_{\mathrm{cusp}}(G(\mathbf{k}))$ denotes the set of isomorphism classes of irreducible cuspidal representations of $G(\mathbf{k})$. We will now describe this function explicitly in the case when $G = \mathrm{PGL}(N)$ and f is the trace function.

First of all, for every $a \geq 1$ let us denote by $\mathcal{R}_a$ the Grothedieck group of representations of $\mathrm{PGL}(N, \mathbf{k}_a)$; we shall write $\mathcal{R}$ instead of $\mathcal{R}_1$. We have a natural map $\mathrm{Lift}_a : \mathcal{R} \to \mathcal{R}_a$ defined as follows. It is known (that $\mathcal{R}_a$ is freely spanned by the classes of Deligne-Lusztig virtual representations $R(T,\theta)$ where T is a maximal torus in $\mathrm{PGL}(N)$ defined over $\mathbf{k}$ and and $\theta : T(\mathbf{k}) \to \mathbb{C}^\times$ (the representation $R(T,\theta)$ depends only on the pair (T,θ) up to conjugacy). Given $a \geq 1$ and a pair (T,θ) we get by extension of scalars a torus T_a in $\mathrm{PGL}(N, \mathbf{k}_a)$ endowed with the norm map $T_a(\mathbf{k}) \to T(\mathbf{k})$; composing this map with θ we get a character θ_a of $T_a(\mathbf{k})$. The map Lift_a is characterized by the condition that it sends the class of $R(T,\theta)$ to the class of $R(T_a, \theta_a)$.

We shall also assume that D is as in Theorem 1.3(2). Then we have the following.

THEOREM 1.6. *Under the above conditions on D and f we have:*

(i) $\eta_{D,f} = 0$ *if i is not divisible by N.*

(ii) *Assume that $i = N$. Any $x \in \mathbf{k}_i^\times = \mathbf{k}_N^\times$ can be thought of as an element of $\mathrm{GL}(N,\mathbf{k})$ defined up to conjugacy; abusing the notation we denote its image in $\mathrm{PGL}(N,\mathbf{k})$ by the same letter. Then for any $\pi \in \mathrm{Irr}_{\mathrm{cusp}}(\mathrm{PGL}(N,\mathbf{k}))$ we have*

$$\eta_{D,f}(\pi) = \chi_\pi(x). \tag{2}$$

Here χ_π is the character of π.

(iii) *Let now $i = aN$ with $a \geq 1$. Then*

$$\eta_{D,f}(\pi) = \chi_{\mathrm{Lift}_a(\pi)}(x) \tag{3}$$

Note that (3) provides an explicit description of the character of $\mathrm{Lift}_a(\pi)$ on the set of elliptic elements in $\mathrm{PGL}(N, \mathbf{k}_a)$.

1.5. Sketch of the proof of Theorem 1.1. The proof of Theorem 1.1 will be divided into the following steps (we sketch the proof here, since we think that it is of independent interest).

STEP 1. Since the algebra A naturally acts on V, in order to define a map $\mathbf{act} : A \to V$ it is enough to choose an element $\delta \in V$ and set $\mathbf{act}(a) = a \underset{0}{\star} \delta$ where

$\underset{0}{\star}$ stands for the action of A at 0. We then need to show that **act** is an isomorphism – i.e. that V is a free module over A of rank 1 with generator δ. We shall choose δ to be the characteristic function of the trivial G-bundle on $\mathbb{P}^1$ with the tautological trivialization at 0 and ∞.

On the other hand V is endowed with a natural involution coming from the map $z \to z^{-1}$ on $\mathbb{P}^1$ which fixes δ. If we know that **act** is an isomorphism, then we can transport this involution to a map $\iota : A \to A$. Since δ is preserved by the above involution of V we obviously have

$$a \underset{0}{\star} \delta = \iota(a) \underset{\infty}{\star} \delta$$

which immediately implies that $\iota(ab) = \iota(b)\iota(a)$, thus ι is an anti-involution of A. The verification of properties 1) and 2) after Corollary 1.2 is straightforward and we leave it to the reader.

STEP 2. We now need to show that ι is an isomorphism. Let Λ denote the coweight lattice of G and let Λ^+ be the set of dominant coweights. Both sets are endowed with a standard partial order ($\lambda > \mu$ if $\lambda - \mu$ is a sum of positive roots), and both A and V acquire a natural filtration numbered by elements of Λ^+ and the map **act** is compatible with the filtrations. We shall denote the corresponding filtration components by $A_{\leq\lambda}$ and $V_{\leq\lambda}$.

These filtrations are defined as follows. First, the double quotient $K_1\backslash G(\mathbb{K})/K_1$ maps naturally to $G(\mathbb{O})\backslash G(\mathbb{K})/G(\mathbb{O})$. It is known that the latter quotient is naturally identified with Λ^+; we shall denote by $(G(\mathbb{O})\backslash G(\mathbb{K})/G(\mathbb{O}))_{\leq\lambda}$ the union of all double cosets corresponding to $\mu \leq \lambda$. Similarly, we define $(K_1\backslash G(\mathbb{K})/K_1)_{\leq\lambda}$ to be the preimage of $(G(\mathbb{O})\backslash G(\mathbb{K})/G(\mathbb{O}))_{\leq\lambda}$ in $(K_1\backslash G(\mathbb{K})/K_1)_{\leq\lambda}$. We then define $A_{\leq\lambda}$ to be the subspace of A consisting of functions supported on $(G(\mathbb{O})\backslash G(\mathbb{K})/G(\mathbb{O}))_{\leq\lambda}$. This is an algebra filtration on A.

Similarly, the set of isomorphism classes of G-bundles on $\mathbb{P}^1$ is naturally identified with Λ^+; hence the stack $\mathrm{Bun}_G(\mathbb{P}^1)_{0,\infty}$ is stratified by elements of Λ^+ and we define $V_{\leq\lambda}$ similarly to $A_{\leq\lambda}$.

STEP 3. Since the map **act** is compatible with the above filtrations, in order to prove that **act** is an isomorphism, it is enough to show that it induces an isomorphism on the accociatd graded spaces. So, we set $A_\lambda = A_{\leq\lambda}/A_{<\lambda}$ and $V_\lambda = V_{\leq\lambda}/V_{<\lambda}$. We also denote by $\mathbf{act}_\lambda$ the corresponding map $A_\lambda \to V_\lambda$.

To each λ we can naturally associate a parabolic subgroup P_λ of G with a Levi subgroup M_λ. We also denote by P_λ^- the opposite parabolic (so $P_\lambda \cap P_\lambda^- = M_\lambda$). Then we prove the following

THEOREM 1.7.

(i) *We have natural isomorphisms*

$$V_\lambda \simeq F(((G/U_\lambda \underset{M_\lambda}{\times} G/U_\lambda))(\mathbf{k})) \tag{4}$$

and

$$A_\lambda \simeq F(((G/U_\lambda \underset{M_\lambda}{\times} G/U_\lambda^-))(\mathbf{k})). \tag{5}$$

Here $\underset{M_\lambda}{\times}$ *stands for the quotient of the product by the diagonal action on* M_λ.

(ii) *Under the above isomorphisms the map* $\mathbf{act}_\lambda$ *is the standard intertwining operator corresponding to the pair* (P_λ, P_λ^-) *acting on the 2nd factor (cf. Section 4 for precise definitions).*

STEP 4. According to Theorem 1.7 it is enough to show that given two associate parabolic subgroups P and Q of G [1] with unipotent radicals U_P, U_Q the standard intertwining operator $I_{P,Q} : F(G/U_P(\mathbf{k})) \to F(G/U_Q(\mathbf{k}))$ is an isomorphism. This is proven in [**4**] (cf. also Theorem 2.4 in [**7**]).

2. Bundles on $\mathbb{P}^1$ with trivializations and proof of (4)

2.1. The stack of G-bundles on $\mathbb{P}^1$. We denote by $\mathrm{Bun}_G(\mathbb{P}^1)$ the moduli stack of G-bundles on $\mathbb{P}^1$. It is well-known that $\mathrm{Bun}_G(\mathbb{P}^1)(\mathbf{k})$ naturally identifies with Λ^+. This identification can be described as follows. Let $\lambda \in \Lambda$; this is a homomorphism from $\mathbb{G}_m$ to the (chosen) maximal torus T in G; thus λ defines a homomorphism $\mathbb{G}_m \to G$. Applying it to the line bundle $\mathcal{O}(1)$ on $\mathbb{P}^1$ we get a G-bundle $\mathcal{O}(\lambda)$ whose isomorphism class only depends on the orbit of λ under the Weyl group W, and the corresponding map $\Lambda^+ \to \mathrm{Bun}_G(\mathbb{P}^1)(\mathbf{k})$ is a bijection. Note that for any $z \in \mathbb{P}^1(k)$ we can write canonically $\mathcal{O}(1) = \mathcal{O}(z)$. Thus the bundle $\mathcal{O}(\lambda)$ acquires a canonical trivialization away from z. In what follows we shall choose such a z so that $z \in \mathbf{k}^\times$. Some of our construction will slightly depend on this choice (in principle, one can choose $z = 1$ but there is nothing special about this choice).

Geometrically this defines a stratification of $\mathrm{Bun}_G(\mathbb{P}^1)$ – we denote by $\mathrm{Bun}_G(\mathbb{P}^1)^\lambda$ the corresponding locally closed substack of $\mathrm{Bun}_G(\mathbb{P}^1)$. Denote by $\overline{\mathrm{Bun}_G(\mathbb{P}^1)^\lambda}$ its closure. Then it is easy to see that $\mathrm{Bun}_G(\mathbb{P}^1)^\mu \subset \overline{\mathrm{Bun}_G(\mathbb{P}^1)^\lambda}$ if and only if $\lambda \leq \mu$. We set

$$\mathrm{Bun}_G(\mathbb{P}^1)^{\leq\lambda} = \bigcup_{\mu\leq\lambda} \mathrm{Bun}_G(\mathbb{P}^1)^\mu.$$

This is an open substack of $\mathrm{Bun}_G(\mathbb{P}^1)$.

2.2. Parabolic subgroups and Harder-Narasimhan filtrations. Recall that we have chosen a Cartan subgroup T of G and a Borel subgroup B containing it. Every $\lambda \in \Lambda$ defines a parabolic subgroup P_λ in G (its unipotent radical is generated by root vectors corresponding to roots α such that $\langle\lambda, \alpha\rangle > 0$). We have also the opposite parabolic P_{λ^-} which have a common Levi subgroup M_λ. We denote by $\pi_\lambda : P_\lambda \to M_\lambda$, $i_\lambda : M_\lambda \to P_\lambda$ the natural maps (similarly for π_λ^- and i_λ^-).

By construction every $\mathcal{O}(\lambda)$ is endowed with a T-structure. However, this structure is not canonical in the sense that it is not invariant under the group of automorphisms of $\mathcal{O}(\lambda)$ – in particular, it is not well-defined on a G-bundle abstractly isomorphic to $\mathcal{O}(\lambda)$. However, it is easy to see that the induced P_λ-structure is invariant under the automorphisms of $\mathcal{O}(\lambda)$. For $G = \mathrm{GL}(N)$ this is exactly the Harder-Narasimhan filtration, so for general G we shall also sometimes use this term for the P_λ structure on G-bundle isomorphic to $\mathcal{O}(\lambda)$.

[1] I.e. parabolic subgroups which have a common Levi subgroup

2.3. The main stack. Our main geometric object in this Section is the stack $\mathrm{Bun}_G(\mathbb{P}^1)_{0,\infty}$ classifying G-bundles on $\mathbb{P}^1$ with trivializations at 0 and ∞. We denote by $\mathrm{Bun}_G(\mathbb{P}^1)_{0,\infty}^{\leq\lambda}$ and $\mathrm{Bun}_G(\mathbb{P}^1)_{0,\infty}^{\lambda}$ the corresponding (open and locally closed) substacks of $\mathrm{Bun}_G(\mathbb{P}^1)_{0,\infty}$ which are preimages respectively of $\mathrm{Bun}_G(\mathbb{P}^1)^{\leq\lambda}$ and of $\mathrm{Bun}_G(\mathbb{P}^1)^{\lambda}$. As before we set V to be the space of F-valued functions on $\mathrm{Bun}_G(\mathbb{P}^1)_{0,\infty}(\mathbf{k})$ with finite support. We also set $V_{\leq\lambda}$ (resp. V_λ) to be the space of F-valued functions on $\mathrm{Bun}_G(\mathbb{P}^1)_{0,\infty}^{\leq\lambda}(\mathbf{k})$ (resp. on $\mathrm{Bun}_G(\mathbb{P}^1)_{0,\infty}^{\lambda}(\mathbf{k})$). Then $V_{\leq\lambda}$ define a filtration on V with V_λ's being the corresponding associated graded quotients.

THEOREM 2.1. *There exists a map* $\eta_\lambda : \mathrm{Bun}_G(\mathbb{P}^1)_{0,\infty}^{\lambda} \to G/U_\lambda \underset{M_\lambda}{\times} G/U_\lambda$ *such that:*

(i) η_λ *is surjective on* $\mathbf{k}$*-points*

(ii) *The fibers of* η_λ *are classifying stacks of some unipotent group.*

In particular (i) and (ii) imply that η_λ *is an isomorphism on the level of* $\mathbf{k}$*-points.*

COROLLARY 2.2. $V_\lambda \simeq F(G/U_\lambda \underset{M_\lambda}{\times} G/U_\lambda)(\mathbf{k})$ *(this is the statement of (4)).*

PROOF. Let us begin by observing that the variety $G/U_\lambda \underset{M_\lambda}{\times} G/U_\lambda$ is the moduli space of the following data: two P_λ-structures on the trivial G-torsor together with an isomorphism between the corresponding M_λ-torsors.

Now any bundle on $\mathbb{P}^1$ with isomorphism type λ is endowed with a canonical P_λ-structure (for $G = GL(N)$ this is the Harder-Narasimhan filtration). By looking at the image of this structure at 0 and ∞ we get a map from $\mathrm{Bun}_G(\mathbb{P}^1)_{0,\infty}^{\lambda}$ to $G/P_\lambda \times G/P_\lambda$. In addition, let us recall that we have chosen some $z \in \mathbb{P}^1(\mathbf{k})\backslash\{0,\infty\}$. Then we claim that the corresponding M_λ-torsors are canonically isomorphic. Indeed, it is enough to check it for $G = \mathrm{GL}(N)$. In this case, it follows from the fact the every graded piece of the associated graded of any rank N vector bundle with respect to its Harder-Narasimhan filtration is isomorphic to $\mathcal{L}^k$ for some $\mathcal{L} \in \mathrm{Pic}(\mathbb{P}^1)$, $k \in \mathbb{Z}_{>0}$. Now we have unique up to constant isomorphism $\mathcal{L} \simeq \mathcal{O}(d\cdot z)$ where $d = \deg(\mathcal{L})$, which identifies the fibers of $\mathcal{L}$ at 0 and ∞. Thus we get a map $\eta_\lambda : \mathrm{Bun}_G(\mathbb{P}^1)_{0,\infty}^{\lambda} \to G/U_\lambda \underset{M_\lambda}{\times} G/U_\lambda$.

We now need to prove (1) and (2). To prove (1) let us observe that $\mathrm{Bun}_G(\mathbb{P}^1)_{0,\infty}^{\lambda}$ is acted on transitively by the group $G \times G$, and the same is true for $G/U_\lambda \underset{M_\lambda}{\times} G/U_\lambda(\mathbf{k})$; moreover, this action is transitive on the latter.

To check (2) let us describe the stack $\mathrm{Bun}_G(\mathbb{P}^1)_{0,\infty}^{\lambda}$ explicitly. Let $\mathcal{F}$ be a G-bundle of type λ with trivializations at 0 and ∞. Let $\mathrm{Aut}(\mathcal{F})$ be the group of automorphisms of $\mathcal{F}$ (just as a bundle – i.e. the automorphisms are not supposed to preserve the trivializations). Looking at the action of automorphisms at 0 and ∞ we get a map $\mathrm{Aut}(\mathcal{F}) \to G \times G$. The stack $\mathrm{Bun}_G(\mathbb{P}^1)_{0,\infty}^{\lambda}$ is equivalent to $G \times G/\mathrm{Aut}(\mathcal{F})$.[2] Let us determine the image of $\mathrm{Aut}(\mathcal{F})$ in $G \times G$. For this it is enough to take any $\mathcal{F}$ as above, and we take $\mathcal{F} = \mathcal{O}(\lambda)$ with trivializations at 0 and ∞ coming from the trivialization of $\mathcal{O}(\lambda)$ on $\mathbb{P}^1\backslash\{z\}$. Then the image of $\mathrm{Aut}(\mathcal{F})$ in $G \times G$ is easily seen to be the subgroup of $P_\lambda \times P_\lambda$ consisting of pairs (p, p') whose images in M_λ coincide. Thus the quotient of $G \times G$ by the image of

[2]The map $\mathrm{Aut}(\mathcal{F}) \to G \times G$ is in general not injective, thus the quotient is indeed a stack.

$\mathrm{Aut}(\mathcal{F})$ is precisely $G/U_\lambda \underset{M_\lambda}{\times} G/U_\lambda$. It remains to note that the kernel of the map $\mathrm{Aut}(\mathcal{F}) \to G \times G$ is unipotent. □

3. Hecke stack for the 1st congruence subgroup

3.1. The Hecke stack in the spherical case. Let $\mathcal{H}_{\mathbb{O}}$ be the stack classifying triples $(\mathcal{F}_1, \mathcal{F}_2, \kappa)$

where

(1) $\mathcal{F}_i$ is a G-bundle on the formal disc $\mathcal{D} = \mathrm{Spec}(\mathbb{O})$.
(2) κ is an isomorphism $\mathcal{F}_1|_{\mathcal{D}^*} \to \mathcal{F}_2|_{\mathcal{D}^*}$ where $\mathcal{D}^* = \mathrm{Spec}(\mathbb{K})$.

This stack can be identified with the double quotient $G(\mathbb{O})\backslash G(\mathbb{K})/G(\mathbb{O})$. As before we have the natural bijection $\mathcal{H}_{\mathbb{O}}(\mathbf{k}) = \Lambda^+$. Here every $\lambda \in \Lambda^+$ corresponds to the element $t^\lambda = \lambda(t) \in T(\mathbb{K}) \subset G(\mathbb{K})$. In fact, the stack $\mathcal{H}_{\mathbb{O}}$ endowed with a natural stratification

$$\mathcal{H}_{\mathbb{O}} = \bigsqcup_{\lambda \in \Lambda^+} \mathcal{H}^\lambda,$$

where $\mathcal{H}^\lambda_{\mathbb{O}}$ is the (reduced) locally closed substack of $\mathcal{H}_{\mathbb{O}}$ whose only $\mathbf{k}$-point is given by the above construction. The closure $\mathcal{H}^{\leq\lambda}_{\mathbb{O}}$ of $\mathcal{H}^\lambda_{\mathbb{O}}$ is equal to the union of all $\mathcal{H}^\mu_{\mathbb{O}}$ with $\mu \leq \lambda$.

3.2. The Hecke stack for the 1st congruence subgroup. Let $\mathcal{H}$ denote the stack of 5-tuples $(\mathcal{F}_1, \mathcal{F}_2, \phi_1, \phi_2, \kappa)$ where

(1) $\mathcal{F}_i$ is a G-bundle on the formal disc $\mathcal{D} = \mathrm{Spec}(\mathbb{O})$ where $\mathbb{O} = \mathbf{k}[[t]]$.
(2) ϕ_i is the trivialization of the fiber of $\mathcal{F}_i$ at 0.
(3) κ is an isomorphism $\mathcal{F}_1|_{\mathcal{D}^*} \to \mathcal{F}_2|_{\mathcal{D}^*}$ where $\mathcal{D}^* = \mathrm{Spec}(\mathbb{K})$ where $\mathbb{K} = \mathbf{k}((t))$.

Let $\mathcal{H}^\lambda$ denote the preimage of $\mathcal{H}^\lambda_{\mathbb{O}}$ under the natural map $\mathcal{H} \to \mathcal{H}_{\mathbb{O}}$; this is a locally closed substack of $\mathcal{H}$. Similarly, we denote by $\mathcal{H}^{\leq\lambda}$ the preimage of $\mathcal{H}^{\leq\lambda}_{\mathbb{O}}$.

3.3. The algebra A and its filtration. Let A denote the space of finitely supported F-valued functions on $\mathcal{H}(\mathbf{k})$. This is an algebra with respect to convolution. We let $A_{\leq\lambda}$ be the space of functions on $\mathcal{H}^{\leq\lambda}$. It is easy to see that $A_\lambda \cdot A_\mu \subset A_{\leq\lambda+\mu}$; thus $A_{\leq\lambda}$ is a filtration on A by dominant coweights. The associated graded algebra is the direct sum of all the A_λ where A_λ is the space of F-valued functions on $\mathcal{H}^\lambda$.

We would like to describe the space A_λ. For this we are going to need the following geometric description of the stack $\mathcal{H}^\lambda$.

THEOREM 3.1. *There is a map $\mathcal{H}^\lambda \to G/U_\lambda \underset{M_\lambda}{\times} G/U^-_\lambda$ which is surjective on $\mathbf{k}$-points and whose fibers are classifying spaces of some pro-unipotent groups (in particular, this map is bijective of $\mathbf{k}$-points).*

To construct this map we need a digression on Jantzen filtrations.

3.4. Jantzen filtrations. Let $\mathcal{E}, \mathcal{E}'$ be two finitely generated torsion-free $\mathbb{O}$-modules endowed with an isomorphism $\kappa : \mathcal{E} \underset{\mathbb{O}}{\otimes} \mathbb{K} \to \mathcal{E}' \underset{\mathbb{O}}{\otimes} \mathbb{K}$. Let E, E' be the fibers of $\mathcal{E}, \mathcal{E}'$ at 0.

Set E_i be the set of all $v \in E$ such that for some (equivalently for any) $s \in \mathcal{E}$ with $s(0) = v$ we have $t^{-i}\kappa(s) \in \mathcal{E}'$. Similarly we define E_i' by replacing κ by κ^{-1}. Then it is clear that

(1) E_i is an increasing filtration on E, E_i' is an increasing filtration on E' (we are going to call them Jantzen filtrations).
(2) κ establishes a canonical isomorphism between $\mathrm{gr}_i(E)$ and $\mathrm{gr}_{-i}(E')$.

Let $N = \dim(E) = \dim(E')$. Then we can view $\mathcal{E}$ and $\mathcal{E}'$ as rank N vector bundles on $\mathcal{D}$, and κ defines a point of the double quotient stack

$$\mathrm{GL}(N, \mathbb{O})\backslash \mathrm{GL}(N, \mathbb{K}) / \mathrm{GL}(N, \mathbb{O}).$$

On the other hand, the above Jantzen fitrations define a reduction of V and V' to some parabolic subgroups P, P' of $\mathrm{GL}(N)$. Moreover, (2) above implies that the type of P' is opposite to that of P.

Recall that $\mathbf{k}$-points of $\mathrm{GL}(N, \mathbb{O})\backslash \mathrm{GL}(N, \mathbb{K}) / \mathrm{GL}(N, \mathbb{O})$ are in one-to-one correspondence with $\lambda \in \Lambda^+$ (where Λ^+ denotes the set of dominant coweights of $\mathrm{GL}(N)$).

LEMMA 3.2. *Assume that $(\mathcal{E}, \mathcal{E}', \kappa)$ has type λ. Then the Jantzen filtration on E defines a reduction of E to P_λ. Similarly, the Jantzen filtration on E' defines a reduction of E' to P_λ^-.*

As before, to prove this lemma, it is enough to look at the case $\mathcal{E} = \mathcal{E}' = \mathbb{O}^N$ and κ being the diagonal matrix with entries $(t^{n_1}, \cdots, t^{n_N})$ where $\lambda = (n_1, \cdots, n_N)$, in which case the above lemma is obvious.

Since Jantzen filtrations are completely canonical, they are functorial and are compatible with tensor products. As a corollary, we get the following

LEMMA 3.3. *Let G be as before, and let $(\mathcal{F}_1, \mathcal{F}_2, \kappa)$ be a $\mathbf{k}$-point of the stack $G(\mathbb{O})\backslash G(\mathbb{K})/G(\mathbb{O})$ of type λ. Let F_i be the fiber of $\mathcal{F}_i$ at 0. Then F_1 is endowed with a P_λ-structure, F_2 is endowed with a P_λ^--structure and the corresponding M_λ-bundles are canonically isomorphic.*

3.5. End of the proof of Theorem 3.1. It is clear that $G/U_\lambda \underset{M_\lambda}{\times} G/U_\lambda^-$ is the moduli space of the following data:

(1) A reduction of the trivial G-torsor to P_λ.
(2) A reduction of the trivial G-torsor to P_λ^-.
(3) An isomorphism between the corresponding M_λ-torsors.

Thus using Lemma 3.3 we immediately get a map $\mathcal{H}^\lambda \to G/U_\lambda \underset{M_\lambda}{\times} G/U_\lambda^-$. We now need to prove that it is surjective on $\mathbf{k}$-points and that the fibers are classifying stacks of some pro-unipotent group. The former follows again from the fact that $G(\mathbf{k}) \times G(\mathbf{k})$ acts transitively on $G/U_\lambda \underset{M_\lambda}{\times} G/U_\lambda^-(\mathbf{k})$. To prove the latter as before we note that $\mathcal{H}^\lambda$ is equivalent to $G \times G/A_\lambda$ where A_λ is the group of automorphisms of the triple $(\mathcal{F}_1, \mathcal{F}_2, \kappa)$ where $\mathcal{F}_i$ are trivial bundles and $\kappa = t^\lambda$ (it maps to $G \times G$ by looking at its actions on the fibers of $\mathcal{F}_i$ at 0). In other words, A_λ is the subgroup of $G(\mathbb{O})$ consisting of elements $g \in G(\mathbb{O})$ such that $\kappa g \kappa^{-1} \in G(\mathbb{O})$ and the map to $G \times G$ is given by $(g \mod t, \kappa g \kappa^{-1} \mod t)$. It is easy to see that the kernel of the map $A_\lambda \to G \times G$ is pro-unipotent and the image of A_λ in $G \times G$ is equal to

$$\{(p, p') \in P_\lambda \times P_\lambda^- \mid \pi_\lambda(p) = \pi_\lambda^-(p')\}.$$

which finishes the proof.

4. Hecke correspondences and proof of Theorem 1.7

The purpose of this Section is to prove Theorem 1.7. As was explained in the Introduction this implies Theorem 1.1.

4.1. Intertwining operators. We begin by recalling the definition and properties of the standard intertwining operators for induced representations of the group $G(\mathbf{k})$.

Let P and P^- be any two opposite parabolics in G – i.e. two parabolics whose intersection M is a Levi subgroup in both. Let γ (resp. γ_-) be the natural map $G \to G/U$ (resp. $G \to G/U^-$). Then we can view G as a correspondence between G/U_- and G/U and we define $\Phi_{P_-,P} : F(G/U^-(\mathbf{k})) \to F(G/U(\mathbf{k}))$ be the corresponding intertwining operator (a.k.a. "Radon transform"). This operators commutes with the natural action of $G(\mathbf{k}) \times M(\mathbf{k})$ on both sides. The following result is proven in [**4**] (cf. also [**7**]):

THEOREM 4.1. *The operator $\Phi_{P^-,P}$ is an isomorphism.*

4.2. The map act and the stack $\mathcal{C}$. The algebra $A \otimes A$ acts naturally on the space V by convolution at 0 and ∞; as in subsection 1.5 we denote these actions by $a \underset{0}{\star} v$ and $a \underset{\infty}{\star} v$.

Let $\delta \in V$ be the characteristic function of the $\mathbf{k}$-point of $\mathrm{Bun}_G(\mathbb{P}^1)_{0,\infty}$ corresponding to the trivial bundle with tautological trivializations at 0 and ∞. We define the map $\mathbf{act} : A \to V$ by setting

$$\mathbf{act}(a) = a \underset{0}{\star} \delta.$$

It is easy to see that $\mathbf{act}(a)$ is also equal to $\iota(a) \underset{\infty}{\star} (\delta)$ where ι is defined in subsection 1.2. Indeed

4.3. The space $\mathcal{C}^\lambda$. Let $\mathcal{C}^\lambda$ denote the moduli space of the following data:

(1) A G-bundle $\mathcal{F}$ on $\mathbb{P}^1$ of isomorphism class λ.
(2) Trivialization of the fiber of $\mathcal{F}$ at 0.
(3) A trivialization of $\mathcal{F}|_{\mathbb{P}^1\setminus\{0\}}$ which has type λ near 0.

We have obvious maps

$$\begin{array}{ccc} \mathcal{C}^\lambda & \longrightarrow & \mathrm{Bun}_G(\mathbb{P}^1)^\lambda_{0,\infty} \\ \big\downarrow & & \\ \mathcal{H}^\lambda & & \end{array}$$

Here the horizontal arrow is obtained by looking at (1), (2) and (3) above and using only the image of (3) at ∞, and the vertical arrow is obtained by restricting all of the data to the formal disc near 0 (and by observing that the trivial bundle is endowed with a canonical trivialization of its fiber at 0). Thus in view of the previous sections we get natural maps

$$\begin{array}{ccc} \mathcal{C}^\lambda & \xrightarrow{\beta} & G/U_\lambda \underset{M_\lambda}{\times} G/U_\lambda \\ {\scriptstyle\alpha}\big\downarrow & & \\ G/U_\lambda \underset{M_\lambda}{\times} G/U_\lambda^-. & & \end{array}$$

Note that

$$G/U_\lambda \underset{M_\lambda}{\times} G/U_\lambda = G\times G/(U_\lambda\times U_\lambda)\cdot\Delta M_\lambda$$

and

$$(G/U_\lambda\times G/U_\lambda^-)/M_\lambda = G\times G/(U_\lambda\times U_\lambda^-)\cdot\Delta M_\lambda.$$

We now want to describe $\mathcal{C}^\lambda$ explicity. First, consider the embedding $P_\lambda\hookrightarrow G\times G$ which sends p to $(p, i_\lambda\circ\pi_\lambda(p))$. Let P'_λ denote the image of P_λ under this embedding. Note that P'_λ is contained in both $(U_\lambda\times U_\lambda)\cdot\Delta M_\lambda$ and in $(U_\lambda\times U_\lambda^-)\cdot\Delta M_\lambda$. Thus we get natural maps

$$\begin{array}{ccc} (G\times G)/P'_\lambda & \xrightarrow{\ \delta\ } & G/U_\lambda \underset{M_\lambda}{\times} G/U_\lambda \\ \gamma\Big\downarrow & & \\ G/U_\lambda \underset{M_\lambda}{\times} G/U_\lambda^-. & & \end{array}$$

THEOREM 4.2. *There is a $G\times G$-equivariant isomorphism $\mathcal{C}^\lambda\simeq(G\times G)/P'_\lambda$ under which α goes to γ and β goes to δ.*

PROOF. Since $(G\times G)/P'_\lambda$ is acted on transitively by $G\times G$ in order to construct the above isomorphism it is enough to check the following things:

(a) The two maps $\mathcal{C}^\lambda\to G/P_\lambda$ given by composing either α or β with the natural (first) projection to G/P_λ coincide

(b) The fibers of these maps are G-torsors with respect to the action of the 2nd copy of G (corresponding to changing the trivialization of $\mathcal{F}$ on $\mathbb{P}^1\backslash\{0\}$ by a constant map $\mathbb{P}^1\backslash\{0\}\to G$). Moreover, the fiber over 1 is canonically the trivial torsor (i.e. it is equal to G)

(c) Any $p\in P_\lambda$ acts on the fiber over $1\in G/P_\lambda$ (which is naturally isomorphic to G by the last sentence of (b)) by the right multiplication by $i_\lambda\pi_\lambda(p^{-1})$.

First, we claim that $\mathcal{C}^\lambda$ is acted on transitively by $G\times G$. Indeed, for this it is enough to check that a trivialization of $\mathcal{O}(\lambda)$ on $\mathbb{P}^1\backslash\{0\}$ of type λ is unique up to constant change of trivialization. This is a special case of the modular description of transversal slices to $G(\mathbb{O})$-orbits in the affine Grassmannian (cf. the proof of Theorem 2.8 in [**2**]).

Now (a) follows from the fact that the fiber of the Harder-Narasimhan flag of $\mathcal{F}$ at 0 is equal to the Jantzen flag at 0 w.r.to the trivialization on $\mathbb{P}^1\backslash\{0\}$. In view of the above transitivity in order to check this it is enough to assume that $\mathcal{F}=\mathcal{O}(\lambda)$ with the standard trivialization; in this case it is a straightforward calculation.

Let us check (b). Again, it follows from the proof of Theorem 2.8 in [**2**] that the space $\overline{\mathcal{C}}^\lambda$ consisting just of data (1) and (3) from Subsection 4.3 is isomorphic to G/P_λ, where the action of G is given by changing the trivialization by a constant map. More precisely, the claim is that G acts transitively on such data and given any $(\mathcal{F},\kappa)\in\overline{\mathcal{C}}^\lambda$ the group of all $g\in\mathrm{Aut}(\mathcal{F})$ which commute with κ maps injectively to the group $\mathrm{Aut}(\mathcal{F}_0)$ (automorphisms of the fiber of $\mathcal{F}$ at 0) and its image consists of those automorphisms of $\mathcal{F}_0$ which preserve the restriction of the Harder-Narasimhan flag of $\mathcal{F}$ to 0. This implies that if we add to $(\mathcal{F},\kappa)$ a trivialization of $\mathcal{F}_0$ such that the restriction of the Harder-Narasimhan flag of $\mathcal{F}$ to 0 gives a fixed P_λ-structure on the corresponding trivial G-torsor, then G still acts transitively but with trivial stabilizer. This is precisely the first assertion of (b). For the 2nd assertion we just need to say that we have the bundle $\mathcal{O}(\lambda)$ with a natural trivilization on $\mathbb{P}^1\backslash\{0\}$ and on $\mathbb{P}^1\backslash\{z\}$ which gives rise to a point of $\mathcal{C}^\lambda$ which sits over $1 \in G/P_\lambda$, which trivializes the corresponding torsor. Part (c) is then straightforward. □

COROLLARY 4.3. *Let* $\mathbf{k} = \mathbb{F}_q$. *Then* $\mathcal{C}^\lambda$ *(viewed as a correspondence) defines an isomorphism between* $\mathbb{C}[\mathcal{H}^\lambda(\mathbf{k})]$ *and* $\mathbb{C}[\mathrm{Bun}_G(\mathbb{P}^1)^\lambda_{0,\infty}(\mathbf{k})]$.

5. Proof of Theorems 1.3 and 1.6

5.1. Proof of Theorem 1.3. First of all, it is clear that assertion (i) of Theorem 1.3 follows from assertion (ii). Also it is enough to prove (ii) in the case $F = \overline{\mathbb{Q}}_\ell$ (here ℓ is prime to the characteristic of $\mathbf{k}$). In this case (ii) follows from the (weak form of the) global Langlands correspondence for functional fields proved by L. Lafforgue. Namely, the following statement is contained in [**8**].

Let $\pi = \otimes' \pi_\nu$ be an irreducible automorphic representation G for the global field $E = \mathbf{k}(\mathbb{P}^1)$ (here ν runs over places of E) such that

(a) π_ν is unramfied by $\nu \neq 0, \infty$ (i.e. $\pi_\nu^{G(\mathbb{O}_{E_\nu})} \neq 0$ for $\nu \neq 0, \infty$);
(b) $\pi_\nu^{K_{1,\nu}} \neq 0$ for $\nu = 0, \infty$ where $K_{1,\nu}$ denotes the 1st congruence subgroup of $G(E_\nu)$.

Let $\pi_1(\mathbb{P}^1)_{0,\infty}$ denote the fundamental group of $\mathbb{P}^1\backslash\{0,\infty\}$ tamely ramified at 0 and ∞. Then there exists a homomorphism $\rho : \pi_1(\mathbb{P}^1)_{0,\infty} \to G^\vee(\overline{\mathbb{Q}}_\ell)$ such that the Satake parameter of π_ν for $\nu \neq 0, \infty$ (which a priori is a semi-simple conjugacy class in $G^\vee(\overline{\mathbb{Q}}_\ell)$) is equal to $\rho(\mathrm{Fr}_\nu)$ (where Fr_ν is the Frobenius element at the place ν which is well defined as an element of $\pi_1(\mathbb{P}^1)_{0,\infty}$ up to conjugacy).

The group $\Gamma = \pi_1(\mathbb{P}^1)_{0,\infty}$ is isomorphic to the semidirect product of $\widehat{\mathbb{Z}}$ (with generator Fr, corresponding to the Frobenius element at the point $1 \in \mathbf{k}^\times$) and the group $\Gamma_0 = \varprojlim \mathbf{k}_i^\times$ (this is the corresponding geometric fundamental group), such that $\mathrm{Fr}\,\gamma\,\mathrm{Fr}^{-1} = \gamma^q$ for $\gamma \in \Gamma_0$. Thus a homomorphism $\Gamma \to G^\vee(\overline{\mathbb{Q}}_\ell)$ is given by a pair (s, u) as in Subsection 1.1. The proof then follows from the following two observations:

(1) If $x \in \mathbf{k}_i^\times$ and x does not belong to $\mathbf{k}_j$ for any $j < i$ then $\rho(\mathrm{Fr}_x)$ is conjugate to $s^i u(y)$ where y as in Theorem 1.3.
(2) Every irreducible representation of A does appear as π_0 for some π as above (this follows immediately from Theorem 1.1).

5.2. Proof of Lemma 1.5. Assume that G is semi-simple. Then it is clear that $P_\lambda \neq G$ for any $\lambda \neq 0$. Hence Lemma 1.5 follows from (4).

5.3. Proof of Theorem 1.6(i). For the remainder of this Section we set $G = \mathrm{PGL}(N)$ and we use freely the notation from Subsection 1.4. The set of connected components of the stack $\mathrm{Bun}_G(\mathbb{P}^1)$ (and thus of $\mathrm{Bun}_G(\mathbb{P}^1)_{0,\infty}$) is naturally identified $\mathbb{Z}/N\mathbb{Z}$. Thus the space V acquires a natural $\mathbb{Z}/N\mathbb{Z}$ grading $V = \oplus_{d\in\mathbb{Z}/N\mathbb{Z}} V_d$. We claim that $(V_d)_{\mathrm{cusp}} = 0$ unless $d = 0$ – this again follows immediately from (4). On the other hand, the operator $h_{D,f}$ maps V_d to $V_{d+\bar{i}}$ where $\bar{i}$ is the image of i in $\mathbb{Z}/N\mathbb{Z}$, which shows that $h_{D,f}$ acts by 0 on V_{cusp} unless $\bar{i} = 0$.

5.4. Proof of Theorem 1.6(ii). Here we give a direct geometric proof of Theorem 1.6(ii). In the next subsection we shall give a proof of Theorem 1.6(iii) which contains Theorem 1.6(ii) as a special case – but that proof will be indirect (it will use Theorem 1.3 (and thus it will rely on our knowledge of the global Langlands correspondence) and the formula for the character of Deligne-Lusztig representation from [**5**]).

First of all, let us recall the precise definition of the operator $h_{D,f}$ in our case. Let D for the moment be any subset of $\overline{\mathbf{k}}^{\times}$ invariant under the Galois group of $\overline{\mathbf{k}}$ over $\mathbf{k}$. Let us first work with $G = \mathrm{GL}(N)$ rather than $\mathrm{PGL}(N)$. Then we can consider triples $(\widetilde{\mathcal{F}_1}, \widetilde{\mathcal{F}_2}, \widetilde{\kappa})$ where $\widetilde{\mathcal{F}_i}$ is a locally free sheaf on $\mathbb{P}^1$ of rank N trivialized at 0 and ∞ and $\widetilde{\kappa}: \widetilde{\mathcal{F}_1} \to \widetilde{\mathcal{F}_2}$ is an injective map such that $\widetilde{\mathcal{F}_2}/\widetilde{\mathcal{F}_1} = \mathcal{O}_D$ (the structure sheaf of D). For $\mathrm{PGL}(N)$ we consider the quotient of this stack by $\mathrm{Pic}(\mathbb{P}^1)$ acting diagonally. We denote the resulting stack by $\mathcal{H}_{D,N}$ and we denote its typical point by $(\mathcal{F}_1, \mathcal{F}_2, \kappa)$ (here each $\mathcal{F}_i$ is a $\mathrm{PGL}(N)$-bundle on $\mathbb{P}^1$ trivialized at 0 and ∞ and κ is an isomorphism between $\mathcal{F}_1$ and $\mathcal{F}_2$ away from D (the isomorphism is not required to respect the trivializations at 0 and ∞)). It can be viewed as a correspondence from $\mathrm{Bun}_G(\mathbb{P}^1)_{0,\infty}$ to $\mathrm{Bun}_G(\mathbb{P}^1)_{0,\infty}$. We also denote by $\mathcal{H}^0_{D,N}$ the locus where both $\mathcal{F}_i$ are trivial. This is a scheme (rather than a stack). It can be viewed as a correspondence from G to G.

Then $h_{D,f}$ is the operator induced by the above correspondence multiplied by $(-q^{-1/2})^{(N-1)\deg(D)}$. We now specialize to the case when D is the Frobenius orbit of some $x \in \mathbf{k}_i^{\times}$ (which doesn't belong to any $\mathbf{k}_j$ with $j < i$). In this case $\deg(D) = i$. In view of the previous subsection we only care about the case when i is divisible by N. In this case the number $i(N-1)$ is even and hence

$$(-q^{-1/2})^{(N-1)\deg(D)} = q^{-\frac{iN}{2}}.$$

To prove Theorem 1.6(ii) we need the following

LEMMA 5.1. *Let D be as above. Assume in addition that $i = N$. Then x defines a conjugacy class in $G = \mathrm{PGL}(N)$; we shall denote this class by $\Omega_x \subset G$. Then as a correspondence from G to G the scheme $\mathcal{H}_{D,N}$ is isomorphic to the subset of $G \times G$ consisting of elements (g_1, g_2) such that $g_1 g_2^{-1} \in \Omega_x$.*

PROOF. First, let us formulate the analogous statement over $\overline{\mathbf{k}}$ and also replace $\mathrm{PGL}(N)$ by $\mathrm{GL}(N)$. Then we geet the following statement. Let $(a_1, \cdots, a_N) \in \overline{\mathbf{k}}^{\times}$ be such that $a_j \neq a_j$ for $i \neq j$. Let $\Omega_{a_1,\cdots,a_N} \subset \mathrm{GL}(N, \mathbf{k})$ be the orbits of the diagonal matrix $(-a_1, \cdots, -a_N)$ in the group $\mathrm{GL}(N, \mathbf{k})$. Let Z denote the space of injective maps $\phi: \mathcal{O}^N \to \mathcal{O}(1)^N$ such that

$$\mathcal{O}(1)^N/\phi(\mathcal{O}^N) = \bigoplus_{i=0}^{N} \mathcal{O}_{a_i}.$$

We can think about ϕ as an $N \times N$ matrix with entries in $\mathbf{k}[t]$ which is invertible away from $a_1, \cdots, a_N$ and such that $\phi(t)/t$ is regular at ∞ and its value at ∞ is invertible. Then Z is isomorphic to the subset of $\mathrm{GL}(N) \times \mathrm{GL}(N)$ consisting of pairs (g_1, g_2) such that $g_1 g_2^{-1} \in \Omega_{a_1,\cdots,a_N}$, where the map to $G \times G$ is given by $(\phi(0), \frac{\phi(t)}{t}(\infty))$. The proof is as follows. The space of subsheaves $\mathcal{E} \subset \mathcal{O}(1)^N$ such that $\mathcal{O}(1)^N/\mathcal{E} = \bigoplus_{i=0}^{N} \mathcal{O}_{a_i}$ is naturally isomorphic to $(\mathbb{P}^{N-1})^N$. The open subset corresponding to the condition that $\mathcal{E}$ is trivial is equal to the complement to all diagonals. Thus the group $\mathrm{GL}(N)$ acts transitively on it (the action is by automorphisms of $\mathcal{O}(1)^N$). Thus it follows that if in addition we choose a trivialization of $\mathcal{E}$ (which is what we need to get precisely the scheme Z) we get a scheme isomorphic to $\mathrm{GL}(N)\times\Omega$ where Ω is some adjoint orbit in $\mathrm{GL}(N)$. It remains to remark that the diagonal matrix $\phi(t)=(t-a_1,\cdots,t-a_N)$ lies in Z and in this case $\phi(0)=(-a_1,\cdots,-a_N)$ and $\frac{\phi(t)}{t}(\infty)$ is the identity matrix. □

Let finish the proof of Theorem 1.6(ii). If π is an irreducible representation of $G(\mathbf{k})$ then the element $\sum\limits_{g\in\Omega_x} g\in F[G(\mathbf{k})]$ acts in π by by a scalar $c(\pi,x)$. Lemma 5.1 implies that if π is cuspidal then $\eta_{D,f}(\pi)=q^{-\frac{N(N-1)}{2}}c(\pi,x)$. On the other hand, it is clear (by computing the trace of $\sum_{g\in\Omega_x} g$ in π) that

$$|\Omega_x(\mathbf{k})|\chi_\pi(x)=\dim(\pi)\cdot c(\pi,x).$$

But it is easy to see that

$$|\Omega_x(\mathbf{k})|=|\,\mathrm{GL}(N,\mathbf{k})|/|\mathbf{k}_N^\times|=\prod_{j=1}^{N-1}(q^N-q^j)$$

On the other hand, it follows from [**5**] that if π is cuspidal and irreducible then

$$\dim(\pi)=\prod_{j=1}^{N-1}(q^j-1)=q^{-\frac{N(N-1)}{2}}\prod_{j=1}^{N-1}(q^N-q^j)=q^{-\frac{N(N-1)}{2}}|\Omega_x(\mathbf{k})|.$$

Hence

$$\eta_{D,f}(\pi)=q^{-\frac{N(N-1)}{2}}c(\pi,x)=\eta_{D,f}(\pi)=q^{-\frac{N(N-1)}{2}}\chi_\pi(x)\frac{|\Omega_x(\mathbf{k})|}{\dim(\pi)}=\chi_\pi(x)$$

which finishes the proof.

5.5. Proof of Theorem 1.6(iii). Assume first that $x\in\mathbf{k}_N^\times$ and let $[\pi]=(-1)^{N-1}R(T,\theta)$ where $T(\mathbf{k})=\mathbf{k}_N^\times/\mathbf{k}^\times$ and $\theta:T(\mathbf{k})\to F^\times$ is a character such that $\theta^{q^j}\neq\theta$ for all $0<j<N$ (the latter condition is equivalent to cuspidality of π)[3]. Let us view π as a representation of $G(\mathbb{O})$ via the evaluation map $G(\mathbb{O})\to G(\mathbf{k})$. Then $\mathrm{ind}_{G(\mathbb{O})}^{G(\mathbb{K})}(\pi)$ (here ind stands for compact induction) is irreducible and thus $(\mathrm{ind}_{G(\mathbb{O})}^{G(\mathbb{K})}(\pi))^{K_1}$ is an irreducible (obviously non-zero) representation of A. The corresponding parameters (s,u) are described as follows.

LEMMA 5.2. *Under the above conditions* $u:\Gamma_0=\lim\limits_{\leftarrow}\mathbf{k}_i^\times\to G^\vee(F)=\mathrm{SL}(N,F)$ *factorizes through* $\mathbf{k}_N^\times$ *and it is equal to the diagonal matrix*

$$(-1)^{N-1}(\theta,\theta^q,\theta^{q^2},\cdots,\theta^{q^{N-1}}).$$

Also s *is given by the cyclic permutation* $(1\to 2\to\cdots\to N\to 1)$.

The proof easily follows from the suitable generalization of [**3**] (specifically, the arguments from the proof Theorem 2.2.8. in *loc. cit.*) to the ramified setting. This is relatively straightforward, but we postpone it for another publication.

[3]The sign $(-1)^{N-1}$ is needed in order to turn $R(T,\theta)$ into actual representation (as opposed to a virtual representation)

So, $s^N = 1$ and the RHS of (1) is equal to

$$(-1)^{N-1} \sum_{i=0}^{N-1} \theta(x)^{q^i}.$$

This is precisely the value of the character of $(-1)^{N-1} R(T, \theta)$ at x.

We now proceed to the case of general x; i.e. we assume that $x \in \mathbf{k}_i^\times$ with $i = aN$ and that x does not belong to any $\mathbf{k}_j$ with $j < i$. Then again $s^i = 1$ and since $\mathbf{k}_N \subset \mathbf{k}_i$ the RHS of (1) becomes just equal to

$$(-1)^{N-1} \sum_{i=0}^{N-1} \theta(\mathrm{Norm}_{i,N}(x))^{q^i},$$

which is again the character of the above representation π.

Acknowledgments

We thank R. Bezrukavnikov and M. Finkelberg for very useful discussions.

References

[1] A. Braverman, G. Dobrovolska, and M. Finkelberg, *Gaiotto-Witten superpotential and Whittaker D-modules on monopoles*, Adv. Math. **300** (2016), 451–472, DOI 10.1016/j.aim.2016.03.024. MR3534838

[2] A. Braverman and M. Finkelberg, *Semi-infinite Schubert varieties and quantum K-theory of flag manifolds*, J. Amer. Math. Soc. **27** (2014), no. 4, 1147–1168, DOI 10.1090/S0894-0347-2014-00797-9. MR3230820

[3] A. Braverman and D. Gaitsgory, *Geometric Eisenstein series*, Invent. Math. **150** (2002), no. 2, 287–384, DOI 10.1007/s00222-002-0237-8. MR1933587

[4] R. Dipper and J. Du, *Harish-Chandra vertices*, J. Reine Angew. Math. **437** (1993), 101–130, DOI 10.1515/crll.1993.437.101. MR1212254

[5] P. Deligne and G. Lusztig, *Representations of reductive groups over finite fields*, Ann. of Math. (2) **103** (1976), no. 1, 103–161, DOI 10.2307/1971021. MR393266

[6] V. Lafforgue, *Chtoucas pour les groupes réductifs et paramétrisation de Langlands globale* (French), J. Amer. Math. Soc. **31** (2018), no. 3, 719–891, DOI 10.1090/jams/897. MR3787407

[7] R. B. Howlett and G. I. Lehrer, *On Harish-Chandra induction and restriction for modules of Levi subgroups*, J. Algebra **165** (1994), no. 1, 172–183, DOI 10.1006/jabr.1994.1104. MR1272585

[8] V. Lafforgue, *Chtoucas pour les groupes réductifs et paramétrisation de Langlands globale* (French), J. Amer. Math. Soc. **31** (2018), no. 3, 719–891, DOI 10.1090/jams/897. MR3787407

Department of Mathematics, University of Toronto,Toronto ON, Canada M5S 2E4 and Perimeter Institute for Theoretical Physics, Waterloo, Ontario, Canada, N2L 2Y5
Email address: `braval@math.toronto.edu`

Hebrew University of Jerusalem, Institute of Mathematics, 9190401 Jerusalem, Israel
Email address: `kazhdan@math.huji.ac.il`

Contemporary Mathematics
Volume **823**, 2025
https://doi.org/10.1090/conm/823/16496

Discrete Poisson hardcore 1D model and applications

Ivan Cherednik

Dedicated to the memory of Igor Krichever.

ABSTRACT. We suggest and study new hardcore Poisson-type distributions originated in sequences of non-overlapping subsegments in a given one of the lengths from some finite list, which are governed by a variant of the Matérn II process in 1D. The boundary effects are important: the last subsegment is allowed to go beyond the endpoint of the initial one. This is somewhat analogous to the interlacing sequences due to Kerov and others, but the corresponding Young diagrams are finite and the basic point process is different in our setup. When the lengths of subsegments are comparable with that of the initial segment, the classical limit becomes a new "truncated" Poisson-type distribution. The normal-type classical limit is of interest too. A variant is a random walk when the number of steps satisfies the classical Poisson distribution or our truncated one. Then the Bessel I-functions naturally occur, which provides an interesting probabilistic interpretation of their basic properties. Among applications, our truncated Poisson distribution can be used to model reinfections in epidemics.

1. Introduction

The main general result of this paper is the definition and study of new Poisson-type distributions for sequences of non-overlapping subsegments of lengths $\{L_1, \cdots, L_s\}$ in a segment of length N. They depend only on their multiplicities $\{r_i\}$ (not on their order) subject to the inequality $\sum_{i=1}^{s} r_i L_i < N + \max\{L_i\}$. This is somewhat similar to the interlacing sequences $x_1 \le y_1 \le x_2 \le \ldots \le y_{n-1} \le x_n$ due to Kerov and others, a starting point for far-reaching applications, including Jack-Whittaker-Macdonald functions and the corresponding stochastic point processes. Basically, $y_i - x_i \ge L$ for random $L \in \{L_i\}$ in our approach, but the distribution depends only on N and $\{L_i, r_i\}$ due to our choice of the point process.

The creation of these subsegments is governed by a discrete variant of the time-ordered Matérn II process in 1D. If the corresponding uniform random variable takes the value inside the previously created subsegment, then a new subsegment will not be created at this point. It is natural to allow the last subsegment to go beyond the initial segment, which significantly influences the considerations.

2020 *Mathematics Subject Classification.* Primary 05A15, 05A18, 33C10, 60C05, 60G55, 60E05, 62P10, 91F99.

Key words and phrases. Hardcore Poisson point processes, Matérn processes, reinfections, Young diagrams, stochastic processes, Bessel functions.

Partially supported by NSF grant DMS–1901796.

Provided $\lim_{N\to\infty} L_i/N = \nu_i > 0$, the quasi-classical limit of this model required a procedure which seems new even for one L. The continuous distribution we obtain in (7) is a certain truncated version of the classical *Poisson distribution*; we think it is new. It is not any straightforward truncation via the usage of relative probabilities; such truncations were considered in the literature.

A variant is the probability distribution for random walks with steps ± 1 at the points such that their number satisfies the classical Poisson distribution or our one. In the simplest case, we arrive at $\{e^{-\alpha} I_r(\alpha)\}$, where $-\infty < r < \infty$, for the Bessel I-functions, which provides a probabilistic interpretation of basic properties of I-functions. The restriction of this approach to Catalan (non-negative) paths is considered too.

The contents of the sections. Section 2,5 are devoted to the discrete theory with one L: The passage to the normal distribution is right before Section 3, which is devoted to the Poisson-type limit. The Poisson-Catalan distribution and related Bessel functions are discussed in Section 4. Two-processes and processes with any number of L_i are considered in Sections 6,7,8, concluded in Section 9.

Reinfections. One of the applications of our truncated Poisson distribution is to model *reinfections* in epidemics, aimed at the Covid-19 pandemic. This is actual for any epidemics, but this one lasted nonstop for a long period of time and multiple reinfections were quite frequent. Reinfections during one (seasonal) cycle are rare for flu and cold. Also, Covid-19 is the first major epidemic, when we have exceptionally good data, which resulted in brand new Bessel-type models in for the total number of infections. Bessel functions occur in the paper, but because of different reasons.

Combinatorially, subsegments of one or several different lengths become protective immunity intervals; they are placed in a bigger segment, which is the whole epidemic cycle. Practically, no greater than 2-3 different major strains can be strongly present simultaneously. The case of 1 strain is when $s = 1$.

The Poisson distribution can be expected if the immunity and the duration of the disease are disregarded, but this is unsatisfactory: immunity matters. There were not many papers on modeling reinfections; see e.g. [**ADDP**], which was SIR-based. The SIR-type models appeared not really applicable to Covid-19. Bessel-type models in [**Ch1, Ch2**] provided an almost exact match for many waves of Covid-19. These papers are not related to this work; we mention them to put the application to reinfections into perspective.

Our main assumptions for reinfections and similar processes are as follows. The main is that people are exposed to the infection *uniformly* during the cycle of the epidemic (N days) with probability β per day. To be more exact, the curves of the total number of detected infections are of Bessel-type according to [**Ch1, Ch2**]. However, the uniformity can be assumed statistically for the purpose of reinfections: Covid-19 lasted almost non-stop for several years with many consecutive waves and in many countries.

The 2nd hypothesis is that the impact of the vaccinations and undetected and asymptomatic cases can be addressed via diminishing the *susceptible population* of a country and the corresponding changing β. The 3rd hypothesis is a minor one: we disregard the duration of the disease. It is simply added to the immunity interval.

We generally make $\beta = \frac{\alpha}{N}$ for some parameter α. Practically, α can be determined if the total number of non-infected people (during the whole cycle) and those with exactly 1 infection are known as we will see. In the absence of immunity, α can be estimated as the total number of infected people divided by the size of the population in the area. Here and in our model (with immunity), $e^{-\alpha}$ is the number of *noninfected people* (during the whole cycle of epidemic) divided by the population of the country. Other probabilities depend on $\nu = \lim_{N\to\infty} L/N$.

Poisson hardcore point processes. The random variables that determine the creation of subsegments are governed by the time-ordered *Matérn process II* in its *lattice* variant. Generally, there are very many processes where the distances between neighboring objects must be greater than some constant. This is a one-dimensional process. The points and subsegments are created consecutively; if a point belongs to the subsegment created before, it is disregarded.

Beyond reinfections, examples of *Matérn processes* are *forestry, ecology, cellular networks, vehicular networks, trading equities in stock markets*, etc. The last 2 are one-dimensional and time-ordered. Also, see [**Gi**] on *Tonks gas.* In statistical physics, "small systems" are of this type, those far from thermodynamic equilibrium. Almost any networks have *refractoriness*: excited agents cannot be immediately re-excited. Poisson-type processes are those with relatively small number of possible states.

There is a vast literature on Matérn processes I,II,III, mostly in 2D. See e.g. [**KD**] on vehicular networks and references there. Modeling *vehicular networks*, clear 1D processes, is somewhat similar to modeling reinfections. The intervals between cars are counterparts of the immunity intervals.

In such and similar examples, the intervals between objects are frequently assumed small vs. the domains where they are considered. Accordingly, the *edge effects* are mostly ignored. This is different for epidemics and the applications mentioned above: the immunity intervals can be comparable with the duration of the epidemic cycle.

We begin with the following discrete (combinatorial) setting: configurations of non-overlapping subsegments of the length $L+1$ with possible gaps between them (can be empty) in a given segment of the length N. Both, L and N, are necessary in the combinatorial setting; one can expect that only $\lim_{N\to\infty} L/N$ occurs in the limit (see below). In the lattice version, such and similar processes are *integrable* in the following sense: exact formulas can be obtained in terms of binomial coefficients, which can be extended to subsegments of any lengths. We make the formulas as explicit as possible; see Corollary 5.3.

Since the subsegments are created randomly, the last one must be allowed to go beyond N. This assumption naturally results in an additional term, a certain sum of $L+1$ binomial coefficients. A challenge is to calculate the limit of this sum when the sizes of the subsegments are comparable with that of the initial segment. This requires some algebraic transformations. Namely, we need a formula where the number of terms depends only on the number of subsegments (not on L).

For one length L, the continuous limit is when $\lim_{N\to\infty} L/N = \nu > 0$. Accordingly, the resulting process has Ceiling$[1/\nu]$ states: possible numbers of subsegments. The corresponding probabilities and our method are already interesting in this case.

Several lengths. A natural extension is from one L to $\{L_i, 1 \le i \le s\}$. We do this in 2 stages: first for $s = 2$ and then for any s. Interestingly, the formulas remain quite compact, and the limiting procedure for one L can be used almost without modifications for several L_i. See Theorems 7.1 and 7.2.

A system of subsegments naturally results in a Young diagram of order $< N + \max\{L_i\}$ with $\sum_i r_i$ rows. Namely, it is a collection of intervals from 0 to the *endpoints* of the corresponding subsegments. Equivalently, these diagrams are those with consecutive gaps $\ge L$ for $L \in \{L_i\}$.

For instance, the systems of subsegments of one length L can be interpreted as *minimal gap partitions* with the consecutive gaps at least L considered in [**BY**]. We do not see any connection with the assumptions and formulas there and in [**Ke, BO, 01**].

The key difference is our choice of the corresponding point process, which is the time-ordered Matérn II process in 1D. The L_i-subsegments are generally created with their own probabilities. We obtain an interesting multiparametric distribution, which depends only on N and $\{r_i, L_i\}$ subject to $\sum_i r_i L_i < N + \max\{L_i\}$. So the consecutive gaps in our Young diagrams can be assumed *exactly* $L \in \{L_i\}$ in this setting and only the multiplicities r_i of L_i in our diagrams matter (not the order of the corresponding subsegments). The distributions we obtain are quite explicit, as well as their classical Poisson-type limits. One can generally calculate the corresponding correlation functions; we provide examples in this paper.

Poisson-type random walks. There can be other applications of the model we consider. For instance, one can interpret L_i as some *jumps* of the energy function or similar functions with some given probabilities and assume that the energy constantly increases by 1 (a fixed unit) for some time-interval unless for these jumps. I.e. we consider "significant jumps", those by L_i, upon some background linear increase of energy.

It is natural in this approach to allow the jumps to be by $\pm L_i$, not only by $+L_i$ as for reinfections and similar processes, i.e. consider the corresponding *random walks* on top of a certain linear trend.

In the quasi-classical limit $\lim_{N\to\infty} L_i/N = \nu_i$, it leads to interesting multi-variable Bessel-type functions. For one L and when $\nu \to 0$, i.e. when the segments become points, one arrives at the random walk where the steps ± 1 are with probabilities $p, q = 1 - p$, and the number of such steps is subject to the Poisson α-distribution. The output (the sum of such ± 1) for $p = \frac{1}{2} = q$ is r with probability $e^{-\alpha} I_r(\alpha)$ in terms of the Bessel functions I_r. It is reasonable here to allow only *positive* partial sums of the steps during this process, i.e. to impose the Catalan condition for the paths; the energy of the system cannot drop below zero (or so). See Theorem 4.1.

Bessel functions and Bessel processes occur in many situations. For instance, see Section 2.5 in [**So**], though the corresponding *determinantal process* for Young diagrams and *poissonized Plancherel measure* are quite different from our ones. See also [**Kat**].

Main findings. Using the interpretation via reinfections for the sake of concreteness, we find the probabilities π_r for r infections during the total period of N days in terms of the immunity interval L and the probability $\beta = \alpha/N$ to be exposed to the infection during 1 day. Then we calculate $\pi'_r = \lim_{N\to\infty} \pi_r$ as $\lim_{N\to\infty} L/N = \nu > 0$.

As far as we know, our distributions $\{\pi_r\}$ and $\{\pi'_r\}$ are new, as well as their application to reinfections of epidemics and similar processes. The number of sates is bounded in our approach: $r \le$Ceiling$[N/L]$. The edge effects are important; we allow one of the subsegments to go beyond N, the right endpoint. Stock markets are such too: there can be open positions after the end of the considered period. Mathematically, the corresponding sum of all probabilities will not be 1 without the edge effects.

This approach can be smoothly extended to any number of lengths: $\{L_i, 1 \le i \le s\}$ with the corresponding probabilities. The formulas remain reasonably compact; the limiting procedure is basically the same as for one L. See Theorems 7.1 and 7.2.

Our combinatorial distributions have a very reasonable *normal approximation.* In the case of one L, it can be expressed in terms of the classical integral $\int e^{-(z-z^{-1})}dz$.

An interesting variant here is as follows. Let the "energy function" jump by $\pm L_i$ with probabilities $\beta p_i, q_i = \beta(1-p_1)$ and increase by 1 otherwise with probability $(1-\beta)$. This is motivated by physics, networks and share-prices. When $N \to \infty$ and $\nu_i \to 0$, i.e. the segments L_i become points, we arrive at the random 1D walk with jumps up and down with their number satisfying some Poisson-type distribution or our truncated one. The formula is in terms of Bessel I-functions. Catalan-type (non-negative) paths are of obvious interest here. Only $s = 1$ (one L) is considered in Theorem 4.1.

Integrability. Generally, models are integrable when there are explicit formulas, at least for basic processes, and rich symmetries are present. The works of *Igor Krichever* greatly contributed to our understanding of integrability, including deep applications well beyond differential and difference equations. Low-dimensional geometry-topology, classical algebraic geometry, theory of special functions are among such applications.

One-dimensional combinatorial statistical models can be frequently "solved" directly in terms of the binomial coefficients, recurrence relations and generating functions. Even if they are solvable in this sense, the integrability of the corresponding continuous modes is not granted. At least in statistical physics, only the limits $N \to \infty$ matter. The classical example is the way to obtain the Poisson and normal distributions from the binomial distribution. We show that even basic combinatorial problems with minor twists (subsegments with boundary conditions) can lead to non-trivial continuous theories.

For relatively simple hardcore models considered in the paper, the corresponding Poisson limits appear interesting and new "truncated" Poisson-type distributions. The limit to the normal distribution was managed too. We mention that similar considerations result in "integrable" families of orthogonal polynomials. This direction is left beyond the present work, but we systematically provide and discuss "multiparticle" generalizations.

2. Hardcore Poisson-type processes

If the immunity factor is omitted, the distribution of reinfections is as follows. Assume that an epidemic lasts N days and $\beta = \alpha/N$ is the probability to be infected during one day. Then the probability to be infected r times during N days and its continuous limit are given by the classical Poisson distribution and its combinatorial counterpart. Namely:

$$p_r = \binom{N}{r}\beta^r(1-\beta)^{N-r} \text{ for } r = 0,1,2,\ldots,\ p'_r = \lim_{N\to\infty} p_r = \frac{\alpha^r}{r!e^{-\alpha}}, \tag{1}$$

where e is the Euler's number.

Adding immunity. It will be convenient here and below to continue using the reinfection interpretation and related terminology. Assume that an individual infected at day x cannot be infected again for days $x+1, x+2, \ldots, x+L$, i.e. $L < N$ is the duration of the immunity interval.

Let π_r be the probability of r infections during N days for $r = 0,1,\ldots$. If $1 \le x_1 < x_2 < x_3 < \ldots < x_r \le N$ are the infection days, then $x_1 < x_2 - L < x_3 - 2L \ldots x_r - (r-1)L$ and there are 2 cases:

(a) $x_r + L \le N$, and, otherwise, (b) $x_r + L > N$.

Here x_i are days of the actual infections: when the disease begins. The potential infections are when an individual was exposed to the infection, with probability β per 1 day. Due to the immunity, only some exposures result in actual infections. Potential infections will continue to occur during the periods $x_i+1,\ldots,x_i+L$ for $1 \le i < r$ and during the end period $x_r+1,\ldots,\min\{x_r+L, N\}$. They will be non-actual. This means that these periods can be removed from the consideration when counting the probabilities. Switching to $x'_i = x_i - (i-1)L$ for $1 \le i \le r$, the probability of the event "$x'_1 < x'_2 < \ldots < x'_r$" is $\beta^r(1-\beta)^{N-rL-r}$ for (a) and $\beta^r(1-\beta)^{(x_r-(r-1)L-r)}$ for (b).

Let $P(r,N,\beta) \stackrel{\text{def}}{=} \binom{N-Lr}{r}\beta^r(1-\beta)^{N-Lr-r}$; it is 0 if $r < 0$ or when $N < Lr+r$. We obtain the following straightforward formula:

$$\pi_r = P(r,N,\beta) + \beta\sum_{i=1}^{L} P(r-1, N-i, \beta),\ r = 0,1,\ldots. \tag{2}$$

To give an example: for $L = 1$, $\pi_r = \beta^r(1-\beta)^{N-2r}\left(\binom{N-r+1}{r} - \beta\binom{N-r}{r-1}\right)$.

One has: $\sum_{r=0}^{\infty}\pi_r = 1$, where the inequality $r \le \frac{N+L}{L+1}$ are sufficient in this sum. This is a combinatorial identity, which immediately follows from the probabilistic definition of π_r. Obviously, $\pi_0 = p_0 = (1-\beta)^N$.

Let us rewrite the formula for π_r without the L-summation. We will omit a simple proof.

THEOREM 2.1. *Let $F_r(X) = \frac{d^{r-1}}{dX^{r-1}}\left(X^{N-Lr}\,\frac{1-X^L}{1-X}\right)$ (the $(r-1)$th derivative), where $r > 0, L > 0$ and $N - Lr > 0$. Then*

$$\pi_r = \binom{N-Lr}{r}\beta^r(1-\beta)^{N-Lr-r} + \frac{\beta^r}{(r-1)!}F_r(X \mapsto 1-\beta). \quad \square \tag{3}$$

The number of terms in this formula depends only on r (not on L), which will be the key when the limit $\lim_{N\to\infty} L/N \to \nu > 0$ is performed. Before this, let us discuss the limit to the normal distribution.

Normal approximation. The connection to the normal distribution is not the main line for us. This is a must for any models and it appeared with some interesting steps in our case. This is when $\lim_{N\to\infty} L/N = 0$ for fixed $0 < \beta < 1$. Then $P(r, N, \beta)$ in (2) coincides with $P(r-1, N-i, \beta)$ there for any i there provided that $r - \beta(N - rL) \sim 0$. Thus, π_r can be approximated as $N \gg 0$ by the following $\widetilde{\pi}_r$:

$$\widetilde{\pi}_r = (1+\beta L)\frac{e^{-(r-\beta u)^2/(2u\beta(1-\beta))}}{\sqrt{2\pi u\beta(1-\beta)}}, \text{ where } u \stackrel{\text{def}}{=\!=} N - Lr, \tag{4}$$

$$u = \frac{N}{1+\beta L} + O(\sqrt{N}),\ \ r = \beta u + O(\sqrt{u}) = \frac{\beta N}{1+\beta L} + O(\sqrt{N}).$$

Here u is the number of points in $[1, N]$ not covered by the subsegments plus i such that $0 \le i \le L-1$ if the last subsegment goes beyond the endpoint by i points (the edge effect). The conditions for u and r are standard ones needed to obtain the normal distribution.

Let us check that the formulas for $\widetilde{\pi}_r$ really give a *distribution* when $0 \le r < \infty$ and $u = N - Lr$ as $N \to \infty$. We will check that

$$\begin{aligned} I_N &= (1+\beta L)\int_0^{N/L} \frac{e^{-(r-\beta u)^2/(2u\beta(1-\beta))}}{\sqrt{2\pi u\beta(1-\beta)}}dr \\ &= \frac{1+\beta L}{L}\int_0^{N} \frac{e^{-(r-\beta u)^2/(2u\beta(1-\beta))}}{\sqrt{2\pi u\beta(1-\beta)}}du \to 1, \end{aligned} \tag{5}$$

where $dr = -\frac{1}{L}du$. Let $u = v^2$, $du = 2vdv$, $a = \frac{1+\beta L}{L}, b = \frac{N}{L}$, $c = \beta(1-\beta)$. Consider a, b, c as arbitrary positive parameters. Then

$$I_N = 2a\int_0^{\sqrt{N}} \frac{e^{-(av-bv^{-1})^2/(2c)}}{\sqrt{2\pi c}}dv. \tag{6}$$

This is a standard integral, but let us provide its calculation. It does not actually depend on b as $N \gg 0$, so we can make the upper limit ∞. Let $z = \frac{b}{a}v^{-1}$, $v = \frac{b}{a}z^{-1}$, $dv = -\frac{b}{a}z^{-2}$, and $w = az - b/z$. Then I_∞ (when $\sqrt{N} \mapsto \infty$) equals $2b\int_0^\infty \frac{e^{-(bz^{-1}-az)^2/(2c)}}{\sqrt{2\pi c}}\frac{dz}{z^2}$. Combining the latter formula with the one for I_N above as $N \to \infty$ (changing v by z) we obtain the required:

$$I_\infty = \int_0^\infty \frac{e^{-(bz^{-1}-az)^2/(2c)}}{\sqrt{2\pi c}}(a + \frac{b}{z^2})dz = \int_{-\infty}^\infty \frac{e^{-w^2/(2c)}}{\sqrt{2\pi c}}dw = 1,$$

where $dw/dz = a + b/z^2 > 0$ for $0 < z < \infty$; its range is $-\infty < w < \infty$.

3. Poisson-type limit

When L is fixed and $\beta = \alpha/N$, then the limit $\lim_{N\to\infty}\pi_r = p'_r$ becomes the classical Poisson distribution from (1). The main case for us is when $\lim_{n\to\infty} L/N = \nu > 0$.

Let us begin with the first 4 cases of (3):

$$\pi_0 = (1-\beta)^N,\ \pi_1 = \beta(N-L)\,(1-\beta)^{N-L-1} + (1-(1-\beta)^L)\,(1-\beta)^{N-L},$$
$$\pi_2 = \beta^2\binom{N-2L}{2}\,(1-\beta)^{N-2L-2} + \beta(N-2L)\,\big(1-(1-\beta)^L\big)\,(1-\beta)^{N-2L-1}$$
$$+\Big(1-(1-\beta)^L - \beta L(1-\beta)^{L-1}\Big)(1-\beta)^{N-2L},$$
$$\pi_3 = \frac{\beta^3}{6}\Bigg((N-3L)(N-3L-1)(N-3L-2)x^{N-3L-3} +$$
$$3\Bigg(-2Lx^{L-1}\,\big(\frac{x^{N-3L}}{\beta^2} + \frac{(N-3L)x^{N-3L-1}}{\beta}\big) - \frac{L(L-1)x^{N-2L-2}}{\beta} +$$
$$(1-x^L)\,\Big(\frac{2x^{N-3L}}{\beta^3} + \frac{2(N-3L)x^{N-3L-1}}{\beta^2} + \frac{(N-3L)(N-3L-1)x^{N-3L-2}}{\beta}\Big)\Bigg)\Bigg).$$

We set $x \overset{\text{def}}{=} 1-\beta$ in the formula for π_3. Here $N \ge L$, $N \ge 2L$ and $N \ge 3L$ correspondingly; generally, $r \le \frac{N}{L}$. Recall that $\pi_r > 0$ for $N \ge L(r-1)+r$, i.e. for $r \le \frac{N+L}{L+1}$. For instance, $\pi_1 = 1-(1-\beta)^N$ for *any* L such that $L > N$.

Recall that π_0 does not depend on L: the duration of the immunity interval does not affect those non-infected.

Let $\beta \overset{\text{def}}{=} \alpha/N$ and $\lim_{N\to\infty} L/N = \nu \ge 0$. Then $\pi'_r = \lim_{N\to\infty}\pi_r$ for $r = 0,1,2$ are as follows:

$$\pi'_0 = e^{-\alpha},\ \ \pi'_1 = e^{(\nu-1)\alpha}\big(1+\alpha(1-\nu)\big) - e^{-\alpha},$$
$$\pi'_2 = e^{(2\nu-1)\alpha}\left(1+\alpha(1-2\nu) + \frac{\alpha(1-2\nu)^2}{2}\right) - e^{(\nu-1)\alpha}\left(1+\alpha(1-\nu)\right).$$

We assume here that $r\nu \le 1$. Generally, under this assumption:

$$\text{(7)}\quad \pi'_r = e^{(r\nu-1)\alpha}\left(1+\alpha(1-r\nu) + \frac{\alpha^2(1-r\nu)^2}{2} + \ldots + \frac{\alpha^r(1-r\nu)^r}{r!}\right)$$
$$- e^{((r-1)\nu-1)\alpha}\left(1+\alpha(1-(r-1)\nu) + \ldots + \frac{\alpha^{r-1}(1-(r-1)\nu)^{r-1}}{(r-1)!}\right).$$

The last value of r here is $r_\flat =$Floor$[\frac{1}{\nu}]$, where Floor$[x]$ is the integer part of x. The initial inequality $r \le \frac{N+L}{L+1}$ gives that the last nonzero π'_r is for $r_\sharp =$Ceiling$[N/L]$, which is $r_\flat+1$ if $\frac{1}{\nu}$ is not an integer, and $\pi_\flat$ if $1/\nu \in \mathbb{Z}_+$. Indeed, $\frac{N}{L}+1 > \frac{N+L}{L+1} > \frac{N}{L}$. When $r_\sharp = r_\flat+1$, one has:

$$\pi'_{r_\flat+1} = 1 - e^{(r_\flat\nu-1)\alpha}\left(1+\alpha(1-r_\flat\nu) + \frac{\alpha^2(1-r_\flat\nu)^2}{2} + \ldots + \frac{\alpha^{r_\flat}(1-r_\flat\nu)^{r_\flat}}{(r_\flat)!}\right).$$

The positivity of π'_r for $0 \le r \le r_\flat+1$ can be readily seen from these formulas, though it of course follows from the origin of π'_r. Here we have calculated $\pi'_{r_\flat+1}$ directly from the definition. Alternatively, it can be obtained from the identity

$\sum_{r=0}^{r_\sharp} \pi'_r = 1$. This sum is obviously 1 (the telescoping summation), which holds *a priori* because $\sum_{r=0}^{r_\sharp} \pi_r = 1$, which is due to the definition of π_r. We arrive at the following theorem.

THEOREM 3.1. *Assume that $N \to \infty$, $\beta = \alpha/N$, $\lim_{N\to\infty} L/N = \nu$ for some $\alpha > 0$ and $0 \le \nu \le 1$. Then formula (7) holds for any $0 \le r \le r_\flat$, as well as the boundary formula for $\pi'_{r_\flat+1}$ if $\frac{1}{\nu}$ is not an integer.*

PROOF. We can simplify F_r from (3) considered in the limit as follows. Let $\Phi_r \stackrel{\text{def}}{=\!=} \mathcal{B}^{r-1}(\frac{X^{N-Lr}-X^{N-L(r-1)}}{1-X})$ for the following *formal* differentiation $\mathcal{B}$ of the ring generated by X^{N-sL} and $1/(1-X)$ *treated as independent symbols*:

$$\mathcal{B}^p(X^{N-sL}) = (1-s\nu)^p X^{N-sL},\ \mathcal{B}^p\Big(\frac{1}{1-X}\Big) = \frac{\beta^p\, p!}{(1-X)^{p+1}},\ p = 0,1,2,\ldots.$$

This differentiation is $\beta\, d/dX$ with the following simplifications due to taking the limit. First, we replace $\binom{N-sL}{r}$ by $\frac{N^r}{r!}$ for $s \le r$ due to $r \ll N - sL$. Second, $X^{N-sL\,\pm p}$ is replaced by X^{N-sL} for $0 \le p \le r$ because these powers will be finally evaluated at $X = 1-\beta = 1-\frac{\alpha}{N} \to 1$.

We obtain that $\pi'_r = \frac{\alpha^r(1-r\nu)^r}{r!}\, e^{(r\nu-1)\alpha} + \frac{1}{(r-1)!}\,\Phi'_r$, where Φ'_r is the limit of $\beta\Phi_r$ after the evaluation $1-X \mapsto \beta$, $X^{N-Ls} \mapsto e^{\alpha(s\nu-1)}$.

One has: $\Phi_r = \sum_{s=0}^{r-1}\binom{r-1}{s}\,\mathcal{B}^s\Big(X^{N-Lr} - X^{N-L(r-1)}\Big)\,\mathcal{B}^{r-1-s}\Big(\frac{1}{1-X}\Big) =$

$$\sum_{s=0}^{r-1} \beta^{r-s-1}\,\frac{(r-1)!}{s!}\left(\frac{\alpha^s(1-r\nu)^s X^{N-Lr} - \alpha^s(1-(r-1)\nu)^s X^{N-L(r-1)}}{(1-X)^{r-s}}\right).$$

Finally, $\frac{\beta^{r-s-1}}{(1-X)^{r-s}} \mapsto \beta^{-1}$, $\lim_{N\to\infty}(1-\beta)^{N-Ls} = e^{\alpha(s\nu)-1}$, and

$$\Phi'_r = \sum_{s=0}^{r-1}\frac{(r-1)!}{s!}\Big(\alpha^s(1-r\nu)^s e^{\alpha(r\nu-1)} - \alpha^s(1-(r-1)\nu)^s e^{\alpha((r-1)\nu-1)}\Big).$$

This gives formula (7). The calculation is similar for $\pi_{r_\sharp}$. □

The ν-dependence of π'_r is interesting. For instance, since π'_0 does not depend on ν, π'_1 increases if ν increases. Indeed, the chances of reinfections (counted by π'_r for $r \ge 2$) diminish. Similarly, $\pi_{r_\sharp}$ decreases if present (if $1/\nu$ is not an integer). Generally, we have the following straightforward corollary.

COROLLARY 3.2. *Let $\delta_r = r\alpha\, e^{(r\nu-1)\alpha}\,\frac{\alpha^r(1-r\nu)^r}{r!}$ for $0 \le r \le r_\flat$, and $\delta_{-1} = \delta_{r_\sharp} = 0$. Then $d\pi'_r/d\nu = \delta_r - \delta_{r-1}$ for $0 \le r \le r_\sharp$. In particular, $d\pi'_r/d\nu \ge 0$ for $1 \le r \le \frac{1}{\nu}$ if and only if $\nu\alpha\, e^{\nu\alpha} \ge (r-1)(1+z)^{r-1}z$ for $z = \frac{\nu}{1-r\nu}$. Otherwise, this derivative is negative.* □

Practically, triple reinfections ($r = 4$) are hardly possible for one cycle of any epidemic. Though there can be other random processes of this kind where big r make sense. The distribution in (7) is a discretization of the Poisson distribution, which occurs when $\nu \to 0$. Accordingly, (2) is its discretization with 2 parameters, L and N. One more parameter can be added to (2) by switching to the q-binomial coefficients there, which we will not discuss.

4. Poisson-Catalan distribution

Let us assume that some physics system is subject to N random events of 3 sorts: (i) adding $1/N$ to its *energy* E with probability $(1-\frac{\alpha}{N})$, (ii) adding L/N to E with probability $\frac{\alpha p}{N}$ and (iii) subtracting L/N from E with probability $\frac{\alpha q}{N}$ for $p, q = 1-p \geq 0$.

We begin with $E = 0$ and consider below *Catalan paths*: all intermediate balances during $r+2m$ steps with $r+m$ events for $+L/N$ and m events with $-L/N$ must be non-negative. This is a pure combinatorial condition, which is quite reasonable when $L \gg 1$ to ensure that E always remains non-negative.

Equivalently, we switch in π_r and π'_r from (7) to $r_m = r+2m$, $N_m = N + 2mL$, $\nu_m = \frac{L}{N_m} = \frac{\nu}{1+2m\nu}$, $\beta_m = (1+2m\nu)\,\beta$ and $\alpha_m = (1+2m\nu)\,\alpha$. The resulting π'_{r+2m} will be denoted $\pi'_{r,m}$; $\mathcal{P}'_{r,m}$ will be the *Catalan-Poisson probability* for the corresponding Catalan paths.

THEOREM 4.1.

(i) *Let* $\lim_{N\to\infty} L/N = \nu > 0$. *Then one has for* $m \geq 0$ *and* $0 \leq r \leq$ *Ceiling* $[1/\nu]$ *:*

$$\mathcal{P}'_{r,m} = \pi'_{r,m}\, p^{r+m} q^m \binom{r+2m}{m} \frac{r+1}{r+m+1}. \tag{8}$$

(ii) *Let* $\mathcal{P}^0_{r,m} \stackrel{\mathbf{def}}{=\!=} \lim_{\nu\to 0} \mathcal{P}'_{r,m}$. *Then* $\pi'_{r,m}$ *becomes the Poisson distribution* $\frac{\alpha^{r+2m}e^{-\alpha}}{(r+2m)!}$ *and we obtain that* $\mathcal{P}^0_{r,m} = \frac{\alpha^{r+2m}p^{r+m}q^m e^{-\alpha}}{m!(r+m)!}\frac{r+1}{r+m+1}$. *Furthermore, let* $p = \frac{1}{2} = q$ *and* $\sigma^0_r \stackrel{\mathbf{def}}{=\!=} \sum_{m=0}^{\infty} \mathcal{P}^0_{r,m}$. *Then* $\sigma^0_r = \frac{r+1}{\alpha/2} e^{-\alpha} I_{r+1}(\alpha)$ *for the* hyperbolic Bessel function *defined as* $I_u(z) = \sum_{m=0}^{\infty} \frac{(z/2)^{2m+u}}{m!\Gamma(m+u+1)}$. *Note that* $I_0(z) = \sum_{m=0}^{\infty} \frac{(z/2)^{2m}}{(m!)^2}$ *and* $I_1(z) = \frac{dI_0(z)}{dz}$.

(iii) *Let* $p = 1/2 = q$, $\nu \to 0$ *as in (ii) and* σ^0_* *be the probability* $\sum_{r=0}^{\infty} \sigma^0_r$ *of obtaining a Catalan path with* $m \geq 0$ *(for any* $r \geq 0$*). Then* $\sigma^0_* = e^{-\alpha}\big(I_0(\alpha)+I_1(\alpha)\big)$. *Also,* $T = 1$ *for the average value* $T = \sum_{r=0}^{\infty}(r+1)\,\sigma^0_r$ *of* $(r+1)$ *with respect to the distribution* $\{\sigma^0_r\}$.

(iv) *Setting* $\nu \to 0$, *let us consider* all *paths, disregarding now the Catalan condition and allowing* r *to be negative. Let* $p_r(m)$ *be the corresponding probability with* m *steps* -1 *and the output* r. *Then:*

$$p_r(m) = p^{r+m} q^m \binom{r+2m}{m} \frac{e^{-\alpha}}{(r+2m)!} \quad \textit{for } r \geq -m, \textit{ and}$$

$$P_r = \sum_{m=0}^{\infty} p_r(m) \;\textit{equals}\; e^{-\alpha}\Big(\frac{p}{q}\Big)^{r/2} I_r\big(2\alpha(pq)^{1/2}\big). \tag{9}$$

In particular, $\sum_{r=-\infty}^{\infty} P_r = 1$.

PROOF.

(i,ii) The number of sequences $\{a_i = \pm 1, 1 \leq i \leq n\}$ such that $\sum_{i=1}^{k} a_i \geq 0$ for any $1 \leq k \leq n$ and $\sum_{i=1}^{n} a_i = r \geq 0$ is as follows: $C_n^{0\to r} = \binom{n}{(n-r)/2}\frac{r+1}{(n+r)/2+1}$, where $n-r$ must be even. This is a standard formula in the theory of Catalan numbers. For $n = r+2m$, we obtain $C^{0\to r}_{r+2m} = \binom{r+2m}{m}\frac{r+1}{r+m+1}$. Then we multiply the latter expression by

$p^{r+m}q^m$ and by $\pi'_{r,m}$. This proves (i); the justification of (ii) is straightforward. We will switch now to (iv).

(iv) To allow any sequences of ± 1, we must omit the terms $\frac{r+1}{m+r+1}$. Then $\sum_{r=\infty}^{\infty} P_r = 1$, which is the classical identity

$$e^{\frac{\alpha}{2}(t+\frac{1}{t})} = \sum_{r=-\infty}^{\infty} I_r(\alpha)t^r, \text{ where } I_r(\alpha) = I_{-r}(\alpha). \tag{10}$$

Indeed, $p_r(m) + p_{-r}(m)$ for $r > 0$ equals

$$(p^{r+m}q^m + q^{r+m}p^m)\binom{r+2m}{m}\frac{e^{-\alpha}}{(r+2m)!} = (pq)^{\frac{r+2m}{2}}\Big(\big(\frac{p}{q}\big)^{\frac{r}{2}} + \big(\frac{q}{p}\big)^{\frac{r}{2}}\Big)\frac{e^{-\alpha}}{m!(r+m)!}.$$

The sum of these terms and the one for $r = 0$ is

$$P_r = e^{-\alpha} I_0(2\alpha(pq)^{1/2}) + e^{-\alpha}\sum_{r=1}^{\infty}\Big(\big(\frac{p}{q}\big)^{r/2} + \big(\frac{q}{p}\big)^{r/2}\Big) I_r(2\alpha(pq)^{1/2})$$

$$= \exp\left(-\alpha + \alpha(pq)^{1/2}\Big(\big(\frac{p}{q}\big)^{1/2} + \big(\frac{q}{p}\big)^{1/2}\Big)\right) = 1 \text{ due to } p + q = 1.$$

Since we know that the sum must be 1, *this proves formula* (10). Indeed, we can make $t = (p/q)^{1/2}$ arbitrary here.

(iii) Differentiating (10): $e^{\frac{\alpha}{2}(t+\frac{1}{t})-\alpha} = \sum_{r=0}^{\infty} \mathcal{P}_r^0 \frac{t^{r+1}-t^{-r-1}}{t-t^{-1}}$. By making here $t \to 1$, we arrive at the formula $T = 1$. Indeed, $\frac{t^{r+1}-t^{-r-1}}{t-t^{-1}} = t^r + t^{r-2} + \cdots + t^{-r}$. To obtain σ_*^0, observe that the latter sum contains 1 for even r and t for odd r; then use (10) for $r = 0, 1$.

Actually, we do not need (10) for *(iii)*. The total number of *all* Catalan paths with n steps is $\binom{n}{n/2}$ for even n and $\binom{n}{(n-1)/2}$ for odd n. It is $C_n^{0\to *} = \sum_{r=0}^{n} C_n^{0\to r}$, where $n - r$ is even, in the notation above. Thus, the total probability to obtain a Catalan path is

$$\sum_{r=0}^{\infty} \sigma_r^0 = \sum_{n=0}^{\infty} e^{-\alpha}\frac{\alpha^n}{n!}\binom{n}{\text{Floor}[n/2]} = e^{-\alpha}\big(I_0(\alpha) + I_1(\alpha)\big).$$

Then, $2^n - \binom{n}{\text{Floor}[n/2]} = \sum_{r=0}^{\infty} r\, C_n^{0\to r}$ for even $n - r$, since the ideal roulette is a 0-sum game, and $2^n = \sum_{r=0}^{\infty}(r+1)\, C_n^{0\to r}$ using the formula above for $C_n^{0\to *}$. This results in $T = 1$. □

PROPOSITION 4.2. *Let $p = 1/2 = q$ and $\nu = 0$ as in Part (iii) of Theorem 4.1. Now the starting level will be $k \geq 0$ and the corresponding probability of Catalan paths will be $\sigma_*^0\langle k\rangle = e^{-\alpha}\sum_{n=0}^{\infty}\frac{\alpha^n}{n!}C_n^{k\to *}$. Then*

$$C_n^{k\to *} = \sum_{n=0}^{\infty}\left(\binom{n}{\text{Floor}[\frac{n-k}{2}]} + \binom{n}{\text{Floor}[\frac{n-k}{2}]+1} + \cdots + \binom{n}{\text{Floor}[\frac{n+k}{2}]}\right).$$

Accordingly, $\sigma_*^0\langle k\rangle = e^{-\alpha}\big(I_{-k}(\alpha) + I_{-k+1} + \cdots + I_{k+1}(\alpha)\big)$. □

5. Generating functions

We will provide the generating function for $\pi_r(N, L)$. We need now the notation $\pi_r(N, L)$ instead of π_r, though L will be fixed here and below. Let $G(t, u) \stackrel{\text{def}}{=} \sum_{N=0}^{\infty} t^N \pi_r(N, L)t^N u^r$.

THEOREM 5.1. $G(t,u)=\frac{1}{1-t}+\frac{\beta\,(u-1)\,t}{(1-t)(1-(1-\beta)t-\beta ut^{L+1})}$.

PROOF. First of all, let us calculate $G^\circ_N=\sum_{r=0}^{\infty}P(r,N,\beta)u^r=\sum_{r=0}^{\infty}u^r\beta^r(1-\beta)^{N-Lr-r}\binom{N-Lr}{r}$. This is the classical problem about tiling the segment with N boxes by $(L+1)$-minos, sequences of $L+1$ consecutive boxes, and with 1-minos. The case of dimers and monomers in a 2D square is important in statistical physics, though there are no exact 2D formulas in the presence of monomers. Here we count the tilings with the weights as above. For $u=1,\beta=1/2,L=1$: $G^\circ_N=f_{N+1}/2^N$ for the Fibonacci numbers f_N. Generally:

$$G^\circ_N=(1-\beta)G^\circ_{N-1}+\beta uG^\circ_{N-L-1},\ G^\circ_0=1,\ G^\circ_i=(1-\beta)^i\ \text{ for }\ 1\le i\le L.$$

For instance, $G^\circ_{L+1}=(1-\beta)^{L+1}+\beta u$. Using the standard facts in the theory of generating functions or a straightforward consideration:

$$G^\circ(t,u)=\sum_{N=0}^{\infty}G^\circ_N t^N=\frac{1}{1-(1-\beta)t-\beta ut^{L+1}}.$$

Due to formula (2), $G_N=\sum_{r=0}^{\infty}\pi_r(N,L)u^r$ satisfies the same recurrence as for G°_N, but with different initial conditions. Namely, $G(t,u)=(1+\beta ut+\beta ut^2+\ldots+\beta ut^L)\,G^\circ(t,u)$. Finally,

$$\begin{aligned}G(t,u)&=\frac{1-t+\beta ut(1-t^L)}{(1-t)(1-(1-\beta)t-\beta ut^{L+1})}\\ &=\frac{1-(1-\beta)t-\beta ut^{L+1}+\beta ut-\beta t}{(1-t)(1-(1-\beta)t-\beta ut^{L+1})}=\frac{1}{1-t}\Big(1+\frac{\beta\,(u-1)\,t}{(1-(1-\beta)t-\beta ut^{L+1})}\Big).\ \square\end{aligned}$$

For $u=0$: $\sum_{N=0}^{\infty}\pi_0(N,L)=\frac{1}{1-(1-\beta)t}$, which we know without any calculations. For $u=1$: $\sum_{r=0}^{\infty}\pi_r(N,L)t^N=\frac{1}{1-t}$, which gives a combinatorial proof of the identities $\sum_{r=0}^{\infty}\pi_r(N,L)=1$ for any N,L.

Explicit formulas. The theorem readily gives that

$$\pi_r(N,L)=\frac{d^r}{du^r}\Big(\frac{1}{1-t}\Big(1+\frac{\beta\,(u-1)\,t}{1-(1-\beta)t-\beta ut^{L+1}}\Big)\Big)(u\mapsto 0). \tag{11}$$

Performing the differentiation, we obtain the following "telescopic-type" presentation of π_r.

COROLLARY 5.2. *Let* $\Pi_r(L)=\frac{\beta^r t^{(L+1)r-L}}{(1-t)(1-(1-\beta)t)^r}$ *for* $r\ge 1$ *and* $\Pi_0(L)=\frac{1}{1-t}$. *Then* $\sum_{N=0}^{\infty}\pi_r(N,L)t^N=\Pi_r(L)-\Pi_{r+1}(L)$ *for* $r\ge 1$. *This immediately gives that* $\sum_{r=0}^{\infty}\pi_r(N,L)=1$ *for any* N,L. $\square$

One can argue directly or use this corollary to make the formulas for π_r quite explicit: directly expressed in terms of the binomial coefficients. The sums that occur can be readily calculated using the standard combinatorial identities. We arrive at the following formulas.

COROLLARY 5.3. *Provided that* $N-L(r-1)-r\ge 0$,

$$\pi_r(N,L)=\beta^r\sum_{s=0}^{m}\binom{s+r-1}{r-1}(1-\beta)^{\max\{m-L,s\}},\ m=N-L(r-1)-r,$$

where there are $L+1$ *powers of* $(1-\beta)$ *with integral positive coefficients for* $m\ge L$. *These powers are* $(1-\beta)^{m-L+k}$ *for* $k=0,1,\ldots,L$ *and the coefficients depend only* *on k, r unless $k = 0$. For $m < L$, these powers are $(1-\beta)^k$, where $0 \le k \le m$ and all coefficients depend only on k, r.* □

6. Two-processes

From the viewpoint of epidemics, it is quite possible that several strains (point processes) can be present simultaneously. They can be generally with different immunity intervals L and β. In physics, propagation of segments of different lengths with different probabilities is quite possible too.

Let β_1 and β_2 be the probabilities of being infected by strain 1 and strain 2 during 1 day, assuming that the simultaneous infections by 1 and 2 are negligible. We continue using the interpretation via reinfections to simplify the terminology.

We set $\beta_0 = \beta_1 + \beta_2$. The corresponding immunity intervals after these infections will be L_1, L_2. Let $\pi_{r_1,r_2}(N, L_1, L_2)$ be the probability to have r_1 cases for strain 1 and r_2 for 2. Accordingly, we need to calculate the generating function $G = \sum_{N=0}^{\infty} G_N t^N$, where $G_N = \sum_{r_1,r_2=0}^{\infty} u_1^{r_1} u_2^{r_2}\, \pi_{r_1,r_2}(N, L_1, L_2)$.

Similar to the above consideration, the basic combinatorial problem is now to count the number of coverings of an N-segment by non-overlapping $(L_1 + 1)$-subsegments, $(L_2 + 1)$-subsegments, and 1-subsegments (monomers). One of the subsegment (the last) can go through N, the endpoint of the N-segment. Then G_N satisfies the recurrence relation $G_N = (1-\beta_0)G_{N-1} + \beta_1 u_1 G_{N-1-L_1} + \beta_2 u_2 G_{N-1-L_2}$, and:

$$
\begin{aligned}
G =& \frac{1 + u_1\beta_1\, t\,(1 + t + \ldots + t^{L_1-1}) + u_2\beta_2\, t\,(1 + t + \ldots + t^{L_2-1})}{1 - (1-\beta_0)t - u_1\beta_1 t^{L_1+1} - u_2\beta_2 t^{L_2+1}} \\
=& \frac{1 - t + u_1\beta_1\, t\,(1 - t^{L_1}) + u_2\beta_2\, t\,(1 - t^{L_2})}{(1-t)(1 - (1-\beta_0)t - u_1\beta_1 t^{L_1+1} - u_2\beta_2 t^{L_2+1})} \\
=& \frac{1}{1-t}\left(1 + \frac{\beta_1\,(u_1 - 1)t + \beta_2\,(u_2-1)t}{1 - (1-\beta_0)t - u_1\beta_1 t^{L_1+1} - u_2\beta_2 t^{L_2+1}}\right). \qquad (12)
\end{aligned}
$$

Particular cases. Using that the dependence of u_1, u_2 is linear in the numerator and denominator, it is not difficult to perform the necessary u-differentiations and calculate the generating functions with fixed r_1, r_2. For instance, let $u_1 = u = u_2$, $\pi_r(N, L_1, L_2)$ be the probability that $r_1 + r_2 = r$, and $\mathcal{P}_r(t) = \sum_{N=0}^{\infty} \pi_r(N, L_1, L_2) t^N$. Then

$$
\mathcal{P}_1(t) = \frac{t(\beta_1+\beta_2)\left(1 - t(1-\beta_1-\beta_2) - \beta_1 t^{L_1+1} - \beta_2 t^{L_2+1}\right)}{(1-t)(1-t(1-\beta_1-\beta_2))^2}.
$$

This is for 1 infection by any strain (from two). When $L_1 = L = L_2$ and $\beta = \beta_0 = \beta_1 + \beta_2$, we arrive at the case of one type of infection. For $r > 1$, this is generally not true: even if $L_1 = L_2$, the order of the strains in their sequences matters. An explicit combinatorial formula for $N > L_2 \ge L_1$ is as follows:

$$
\begin{aligned}
&\beta_0^{-1}\pi_1(N, L_1 \le L_2) = x^{N-L_2-1}\Big(x^0(N - L_2)(1-\beta_1) + x^1\big(1 - (N - L_2 + 1)\beta_1\big) \\
&+ x^2\big(1 - (N - L_2)\beta_1\big) + \ldots + x^{L_2-L_1-1}\big(1 - (N - L_1 - 1)\beta_1\big) \\
&+ x^{L_2-L_1} + x^{L_2-L_1+1} + \ldots + x^{L_2}\Big), \quad \text{where } x \stackrel{\text{def}}{=\!=} 1 - \beta_1 - \beta_2.
\end{aligned}
$$

It becomes somewhat simpler combinatorially in terms of x, β_2:

$$\beta_0^{-1}\pi_1(N, L_1 \le L_2) \;=\; x^{N-L_2-1}\Big(x^0(N-L_2)\beta_2 + x^1(N-L_2+1)\beta_2$$
$$+x^2(N-L_2+2)\beta_2 + \ldots + x^{L_2-L_1-1}(N-L_1-1)\beta_2 + x^{L_2-L_1}(N-L_1)$$
$$+x^{L_2-L_1+1} + \ldots + x^{L_2}\Big), \text{ where } x \stackrel{\mathbf{def}}{=\!=} 1-\beta_1-\beta_2,\ N > L_2 \ge L_1.$$

There are $L_2{+}1$ powers of x here; the terms $x^1\big(1-(N-L_2+1)\beta_1\big)$, $x^1(N-L_2+1)\beta_2$ are present only if $L_2 > L_1 + 1$. For $N \le L_2$, the number of terms is N: they are exactly the *top* N terms in the formulas above. The sums of the binomial coefficients in this formula can be readily calculated, which is useful for obtaining the limits as $N \to \infty$, when $N\beta_i \to \alpha_i$, $L_i/N \to \nu_i$ for $i = 1, 2$. We note that by setting $\beta_1 = 0, \beta = \beta_2, L = L_2$ in the 1st formula, we obtain the 2nd where $\beta_2 = 0, \beta = \beta_1, L = L_1$. This is our formula for $\pi_1(N, L)$.

Similarly, one calculates $\pi_{\langle 1\rangle} = \pi_{1,0}$ and $\pi_{\langle 2\rangle} = \pi_{0,1}$, which are the coefficients of u_1 and u_2 of G; $\pi_{\langle i\rangle}$ depends only on L_i. For $N > L_i$:

$$\pi_{\langle i\rangle}(N, L_i) \;=\; x^{N-L_i-1}\beta_i\Big(x^0(N-L_i) + x^1 + \ldots + x^{L_i}\Big),\ i = 1, 2.$$

where $x \stackrel{\mathbf{def}}{=\!=} 1-\beta_1-\beta_2$. Accordingly, the top N terms must be taken if $N \le L_i$. Obviously, $\pi_1(N, L_1 \le L_2) = \pi_{\langle 1\rangle}(N, L_1) + \pi_{\langle 2\rangle}(N, L_2)$. In the limit $\beta_i N \to \alpha_i$ and $L_i/N \to \nu_i$, we obtain for $\alpha_0 = \alpha_1 + \alpha_2$:

$$\pi'_{\langle i\rangle} = \lim_{N\to\infty} \pi_{\langle i\rangle} = e^{\alpha_0(\nu_i-1)}\alpha_i\Big(1 - \nu_i + \frac{1-e^{-\alpha_0\nu_i}}{\alpha_0}\Big),\ i = 1, 2.$$

Similar to Corollary 3.2 for $r = 1$, we get:

$$\frac{d\pi'_{\langle i\rangle}}{d\,\nu_i} = \alpha_i\alpha_0 e^{\alpha_0(\nu_i-1)}(1-\nu_i).$$

We see that $\pi'_{\langle i\rangle}$ increases in terms of the corresponding ν_i for $0 \le \nu_i < 1$ and fixed $\alpha_i > 0$. This could be expected: the greater ν_i the smaller the total number of the corresponding reinfections. This is because π'_0 does not depend on the immunity intervals.

7. Generalizations

Let us now provide the generalizations of Theorems 2.1, 3.1 to the case of any L_i for $i = 1, 2, \ldots, s$ and the corresponding β_i. We set $\beta_0 = \beta_1 + \ldots \beta_s$, $\beta_i = \alpha_i/N$ and $\alpha_0 = \alpha_1 + \ldots \alpha_s$. Let $\pi_{r_1,\ldots,r_s}$ (we may add $[N]$ here) be the probability as above, where r_i is the number of segments of the length $L_i + 1$ in $[1, N]$ corresponding to β_i. One has: $\pi_{0,\ldots,0} = (1-\beta_0)^N$. We will assume below that $r \stackrel{\mathbf{def}}{=\!=} r_1 + \ldots + r_s$ is positive. Note that in the formulas below, we set $\beta_i^{r_i} = 1 = \alpha_i^{r_i}$ for $r_i = 0$ even when $\beta_i = 0 = \alpha_i$.

The generating function in (12) can be readily extended to any number of simultaneous processes (strains).

For $G \stackrel{\mathbf{def}}{=\!=} \sum_{N=0}^{\infty}\sum_{r_1,r_2,\ldots,r_s=0}^{\infty} t^N\, u_1^{r_1}u_2^{r_2}\cdots u_s^{r_s}\,\pi^{[N]}_{r_1,r_2,\ldots,r_s}$,

$$G = \frac{1}{1-t}\Big(1 + \frac{\sum_i \beta_i\,(u_i-1)\,t}{1-(1-\beta_0)\,t - \sum_i u_i\beta_i\,t^{L_i+1}}\Big).$$

THEOREM 7.1. *Setting* $\sigma = N - \sum_{i=1}^{s} r_i L_i$*, for* $\sigma \geq r$*:*

$$\pi_{r_1,\ldots,r_s} = \frac{(\beta_1^{r_1}\cdots\beta_s^{r_s})(1-\beta_0)^{\sigma-r}\,\sigma!}{(r_1!\cdots r_s!)\,(\sigma-r)!} \tag{13}$$

$$+\frac{\beta_1^{r_1}\cdots\beta_s^{r_s}}{r_1!\cdots r_s!}\frac{d^{r-1}}{dX^{r-1}}\left(\sum_{k=1}^{s} r_k X^{\sigma}\frac{1-X^{L_k}}{1-X}\right)(X\mapsto 1-\beta_0).$$

Provided that $\beta_i > 0$ *for any* i*, the nonzero* $\pi_{r_1,\ldots,r_s}$ *are exactly when* $\sigma - r \geq L_k$ *for at least one* k*. Accordingly, for* $\sigma < r$*:*

$$\pi_{r_1,\ldots,r_s} = \frac{\beta_1^{r_1}\cdots\beta_s^{r_s}}{r_1!\cdots r_s!}\frac{d^{r-1}}{dX^{r-1}}\left(\sum_{k=1}^{s} r_k X^{r-1}\frac{1-X^{\sigma+L_k-r+1}}{1-X}\right)(X\mapsto 1-\beta_0),$$

where the summation is only over k *such that* $L_k \geq r-\sigma$*.* □

THEOREM 7.2. *We assume that* $L_k/N \to \nu_k$ *for* $1 \leq k \leq s$ *as* $N \to \infty$*. Let* $\pi'_{r_1,\ldots,r_s} = \lim_{N\to\infty}\pi_{r_1,\ldots,r_s}$*,* $\mu = 1-(\nu_1 r_1 + \ldots + \nu_s r_s)$ *and* $\mu_k = \mu + \nu_k$*. Then* $\pi'_{0,\ldots,0} = e^{-\alpha_0}$*. For* $r > 0$ *and* $\mu \geq 0$*:*

$$\pi'_{r_1,\ldots,r_s} = \frac{\mu^r\alpha_1^{r_1}\cdots\alpha_s^{r_s}}{r_1!\cdots r_s!}e^{-\alpha_0\mu} \tag{14}$$

$$+\frac{\alpha_1^{r_1}\cdots\alpha_s^{r_s}\,(r-1)!}{\alpha_0^{r}\,r_1!\cdots r_s!}\sum_{k=1}^{s}\sum_{i=0}^{r-1}\frac{r_k}{i!}\left(e^{-\alpha_0\mu}(\alpha_0\mu)^i - e^{-\alpha_0\mu_k}(\alpha_0\mu_k)^i\right)$$

$$=\frac{\alpha_1^{r_1}\cdots\alpha_s^{r_s}\,(r-1)!}{\alpha_0^{r}\,r_1!\cdots r_s!}\sum_{k=1}^{s}\left(\sum_{i=0}^{r}\frac{r_k}{i!}e^{-\alpha_0\mu}(\alpha_0\mu)^i - \sum_{i=0}^{r-1}\frac{r_k}{i!}e^{-\alpha_0\mu_k}(\alpha_0\mu_k)^i\right).$$

Additionally, the nonzero values occur if $-\nu_k < \mu \leq 0$ *for at least one* k*:*

$$\pi'_{r_1,\ldots,r_s} = \frac{\alpha_1^{r_1}\cdots\alpha_s^{r_s}\,(r-1)!}{\alpha_0^{r}\,r_1!\cdots r_s!}\sum_{k=1}^{s}\left(r_k - \sum_{i=0}^{r-1}\frac{r_k}{i!}e^{-\alpha_0\mu_k}(\alpha_0\mu_k)^i\right), \tag{15}$$

where the summation is over k *such that* $\mu_k = \mu+\nu_k > 0$*. If* $\mu = 0$ *here, then* (15) *coincides with* (14)*; recall that* $0^0 = 1$*.*

PROOF. For the first term in (13), we have:

$$\frac{(\beta_1^{r_1}\cdots\beta_s^{r_s})(1-\beta_0)^{\sigma-r}\,\sigma!}{(r_1!\cdots r_s!)\,(\sigma-r)!} \to \frac{\mu^r\alpha_1^{r_1}\cdots\alpha_s^{r_s}}{r_1!\cdots r_s!}e^{-\alpha_0\mu}.$$

Following the proof of Theorem 3.1, the remaining summation in (13) becomes the limit of

$$\sum_{k=1}^{s}\sum_{i=0}^{r-1}\frac{r_k\beta_1^{r_1}\cdots\beta_s^{r_s}}{r_1!\cdots r_s!}\binom{r-1}{i}\left(\sigma^i-(\sigma+L_k)^i\right)\frac{(r-1-i)!}{\beta_0^{r-i}}.$$

Then we perform the substitutions $\beta_i = \alpha_i/N$ and $L_i = \nu_i N$, $\beta_0^i\sigma^i = (\alpha_0\mu)^i$, $\beta_0^i(\sigma+L_k)^i = (\alpha_0\mu_k)^i$ and tend N to ∞.

When $\mu < 0$, the above sum becomes:

$$\sum_{k=1}^{s}\sum_{i=0}^{r-1}\frac{r_k\beta_1^{r_1}\cdots\beta_s^{r_s}}{r_1!\cdots r_s!}\binom{r-1}{i}\left((r-1)\cdots(r-i)-(\sigma+L_k)^i\right)\frac{(r-1-i)!}{\beta_0^{r-i}},$$

where $(r-1)\cdots(r-i)\beta_0^i$ vanishes in the limit unless $i=0$. The summation is over k such that $\sigma+L_k \geq r$, which gives $\mu_k > 0$. □

Assuming that $0<\nu_1\le\nu_2\le\ldots\le\nu_s$, we can define $\kappa\langle\mu\rangle=\min\{k\mid \mu+\nu_k>0\}$ for $-\nu_s<\mu\le 0$; the corresponding $\{r_k\}$ are called *extreme*. The condition $\mu_k=\mu+\nu_k>0$ in (15) becomes $k\ge\kappa\langle\mu\rangle$ and this formula for $-\nu_s<\mu\le 0$ reads:

$$\pi'_{r_1,\ldots,r_s}=\frac{\alpha_1^{r_1}\cdots\alpha_s^{r_s}\,(r-1)!}{\alpha_0^r\, r_1!\cdots r_s!}\sum_{k=\kappa\langle\mu\rangle}^{s}\Bigl(r_k-\sum_{i=0}^{r-1}\frac{r_k}{i!}e^{-\alpha_0\mu_k}(\alpha_0\mu_k)^i\Bigr).\tag{16}$$

The extreme configurations have little to do with epidemics, but can have potential applications in physics and networks. For instance, let us estimate $\pi'_{r_1,\ldots,r_s}$ at the *resonances*: when $\mu=0$. Then $\kappa=1$ and the Taylor formulas gives that for some $0<\epsilon_k<\nu_k$:

$$\frac{\alpha_1^{r_1}\cdots\alpha_s^{r_s}}{r_1!\cdots r_s!}e^{-\alpha_0\nu_s}<\pi'_{r_1,\ldots,r_s}=\frac{\alpha_1^{r_1}\cdots\alpha_s^{r_s}}{r_1!\cdots r_s!}\sum_{k=1}^{s}\frac{r_k}{r}e^{-\alpha_0\epsilon_k}<\frac{\alpha_1^{r_1}\cdots\alpha_s^{r_s}}{r_1!\cdots r_s!}.$$

The dependence on μ is of interest. It takes values in the set $\{\mu\}=\{1-\sum_{k=1}^{s}r_k\nu_k>-\nu_s\mid r_k\ge 0\}$; we can define $\pi'(\mu)$ as the sum of $\pi'_{r_1,\ldots,r_s}$ over the configurations with $1-\sum_{k=1}^{s}r_k\nu_k=\mu$. The sequence $\{r_k\}$ is uniquely determined by μ if ν_k are in a general position. For rational $\nu_k=\ell_k/N$: $\{\mu\}=\{N-\sum_{k=1}^{s}r_k\ell_k>-\ell_s\mid r_k\ge 0\}/N$. The extreme μ are then for $\{r_k\}$ such that $N+\ell_s>\sum_{k=1}^{s}r_k\ell_k\ge N$, which is linked to the Frobenius coin problem.

8. Special cases

Let us discuss *extreme configurations* further. There is a natural way to obtain a distribution of probabilities *only on the extreme configurations.* We set $\rho_i=\alpha_i/\alpha_0$ and consider $\widetilde{\pi}'_{r_1,\ldots,r_s}\stackrel{\mathbf{def}}{=\!=}\lim_{\alpha_0\to\infty}\pi'_{r_1,\ldots,r_s}$. For $r>0$:

$$\widetilde{\pi}'_{r_1,\ldots,r_s}=\frac{\rho_1^{r_1}\cdots\rho_s^{r_s}r!}{r_1!\cdots r_s!}\sum_{k=\kappa\langle\mu\rangle}^{s}\frac{r_k}{r},\quad \mu=1-\sum_{k=1}^{s}r_k\nu_k,\tag{17}$$

for extreme $\{r_k\}$, and 0 otherwise. For finite α_0, all $\mu>0$ occur; generally, the "defects" (gaps) play a very significant role.

The combinatorial counterpart of (17) is straightforward. We assume that β_0 is significantly larger than $1/N$ and that the powers $(1-\beta_0)^{\nu_k N}$ can be disregarded. This means that we cover $[1,N]$ by L_k-segments *without gaps* between them (the last segment can go beyond N). The lengths of the segments are here $\ell_k\ge 1$, not (L_k+1) as in Theorem 7.1. Then $\beta_0=1$ and we set $\rho_k=\beta_k$. Let $0<\ell_1\le\ell_2\le\ldots\le\ell_s$; as above, $\sigma=N-\sum_{k=1}^{s}r_k\ell_k$, $\sigma_k=\sigma+\ell_k$. Then, provided that $-\ell_s<\sigma\le 0$,

$$\widetilde{\pi}_{r_1,\ldots,r_s}=\frac{\rho_1^{r_1}\cdots\rho_s^{r_s}r!}{r_1!\cdots r_s!}\sum_{k=æ\langle\sigma\rangle}^{s}\frac{r_k}{r},\quad æ\langle\sigma\rangle=\min\{k\mid\sigma_k>0\},$$

which coincides with (17) in the limit $\ell_i/N\to\nu_i$. We obtain that $\sum_{\{r_k\}}\widetilde{\pi}_{r_1,\ldots,r_s}=1$ and deduce from it that $\sum_{\{r_k\}}\widetilde{\pi}'_{r_1,\ldots,r_s}=1$.

Correlation functions. As above, we will cover $[1,N]$ by $r=r_1+\ldots+r_s$ segments of lengths $\ell_1,\ldots,\ell_s$ (with probabilities $\rho_1,\ldots,\rho_s$). The corresponding m-point function is defined for $m\ge 1$, $1\le k_i\le s$, and $1\le N_1\le N_2\le\ldots\le N_m$. It is the probability $\mathcal{P}^{k_1,\ldots,k_m}_{N_1,\ldots,N_m}(r_1,\cdots,r_s)$ of such coverings subject to the following:

$$N_1\in X[k_1]\ \text{and}\ \ N_i+\ell_{k_1}+\ldots\ell_{k_{i-1}}\in X[k_i]\ \text{for any}\ 1<i\le m,$$

where $X[k]$ is the corresponding consecutive segment of the configuration; k denotes its length: $|X[k]| = \ell_k$. For instance, the 1-point function is when $X[k_1]$ covers N_1 such that $1 \leq N_1 \leq$.

Let $r_k^{(i)} \geq 0$ for $i = 1, 2, \ldots, m+1$ be the number of segments of types $k = 1, 2, \ldots, s$ strictly between $X[k_{i-1}]$ and $X[k_i]$, before $X[k_1]$ for $i = 1$, and (strictly) after X_{k_m}. Thus, $r_k - \sum_{i=1}^{m+1} r_k^{(i)}$ is the number of $k_i = k$ in the sequence $\{k_i, 1 \leq i \leq m\}$; this relation will be imposed below. We set $\varsigma^{(i)} \stackrel{\text{def}}{=\!=} \sum_{j=1}^{i}(r_1^{(j)}\ell_1 + \ldots + r_s^{(j)}\ell_s)$ for $1 \leq i \leq m+1$. For instance, $\varsigma^{(m+1)} = \sum_{i=1}^{m} r_i\ell_i - (\ell_{k_1} + \ldots + \ell_{k_m})$. We obtain the following formula in terms of the multinomial coefficients $\binom{a+b+c+\ldots}{a,\, b,\, c,\, \ldots}$:

$$\mathcal{P}_{N_1,\ldots,N_m}^{k_1,\ldots,k_m}(r_1, \ldots, r_s) = \sum \rho_1^{r_1} \cdots \rho_s^{r_s} \binom{r_1^{(1)} + \ldots + r_s^{(1)}}{r_1^{(1)},\ \ldots,\ r_s^{(1)}} \cdots \tag{18}$$

$$\cdots \binom{r_1^{(m)} + \ldots + r_s^{(m)}}{r_1^{(m)},\ \ldots,\ r_s^{(m)}} \binom{r_1^{(m+1)} + \ldots + r_k^{(m+1)} + \ldots + r_s^{(m+1)} - 1}{r_1^{(m+1)},\ \ldots,\ r_k^{(m+1)} - 1,\ \ldots,\ r_s^{(m+1)}},$$

where the summation is over $1 \leq k \leq s$ and $\{r_k^{(i)} \geq 0\}$ such that

$$N_1 + \ell_{k_1} > \varsigma^{(1)} \geq N_1, N_2 + \ell_{k_2} > \varsigma^{(2)} \geq N_2,\ \ldots,\ N_m + \ell_{k_m} > \varsigma^{(m)} \geq N_m,$$

and $N + \ell_k > \varsigma^{(m+1)} + \ell_{k_1} + \ldots + \ell_{k_m} \geq N$, equivalently, $N \in X[k]$.

The last term in the product occurs only if $r_1^{(m+1)} + \ldots + r_s^{(m+1)} \geq 1$, i.e. when the last segment is not fixed in the m-point function.

The m-point function serving Theorem 7.1 is when $\rho_k = \beta_k$, $\ell_k = L_k + 1$, and we add $\ell_0 = 1$, $\rho_0 = 1 - \beta_0$. It is $\mathcal{P}_{N_1,\ldots,N_m}^{k_1,\ldots,k_m}(r_0, r_1, \ldots, r_s)$, where now $k_i \geq 0$. We can omit r_0 here, the number of positions not covered by the segments, and fix only the corresponding numbers of segments: $r_1, \ldots, r_s$. We will not discuss the limit as $N \to \infty$, providing only the following simple example.

Let $s = 1$, so we have only 1 length L. Accordingly $\beta = \alpha/N$, and we use below the formulas for π'_r from Theorem 3.1 in terms of β, N, L: $\alpha \mapsto \beta N, \nu \mapsto L/N$. We will write $\pi'_r(N; L, \beta)$.

Then for $m = 1$, the limit $N \to \infty$ of $\mathcal{P}^0 N_1(r)$, which is when the position $N_1 \leq N$ is not covered by a segment, is

$$\mathcal{P}'^{\,0}_{N_1}(r) = \sum_{i_1, i_2} e^{(i_1 L - N_1)\beta} \pi'_{i_2}(N - N_1; L, \beta) \ \text{ for}$$

$$0 \leq i_i < \frac{N_1 + 1}{L+1},\ 0 \leq i_2 \leq \frac{N_1 + 1}{L+1},\ i_1 + i_2 = r.$$

Here we do not need the limiting procedure; this is a direct calculation with probabilities. Also, $\mathcal{P}'^{\,1}_{N_1}(r)$, the 1-point function when N_1 is covered by an $(L+1)$-segment, is $\pi'_r(N; L, \beta) - \mathcal{P}'^{\,0}_{N_1}(r)$.

9. Some perspectives

Considering *ensembles* of segments of various lengths (L_k+1) inside $[1, N]$ links our paper to *stochastic processes*, namely to *Whittaker-type processes* based on the distances between *neighboring* particles. See [**BC**].

Namely, L_k with the corresponding multiplicities r_k are some substitutes for these distances with one reservation: recall that we allow "defects", the gaps between our segments. This analogy becomes even more direct if the gaps are considered covered by segments of length $L_0 = 1$ created with probability β_0 $(= 1 - \alpha/N$ for $s = 1)$.

The usage of *Matérn II statistics* is the simplest possible. Recall that for consecutive time moments the segment $[i, i + L_k]$ can be created at any $i = 1, 2, \ldots, N$ with probability β_k unless i belongs to the previously created segment. The latter is called *thinning.* We allow only finitely many possible L_k in the combinatorial setting: $1 \le k \le s$. Generally, their number can be infinite as $N \to \infty$.

The creation of segments is a random process, which may depend on their type/size and can generally depend on the whole configuration. Different kinds of statistical assumptions can be used here. Some combinatorial formulas can be always expected in dimension one. However, the classical limits can be very sophisticated, which is not the case for the models we considered in this paper. Even adding a simple boundary condition at the endpoint N required some efforts as $N \to \infty$.

There is another natural link of this kind: to the so-called *interlacing sequences* $x_1 < y_1 < x_2 < y_2 < \ldots < x_{n-1} < y_n < x_{n+1}$. They can be considered as sequences of n non-overlapping segments $[x_i, y_i]$ of lengths $y_i - x_i$ inside the segment $[x_1, x_{n+1}]$. The corresponding *transitional probabilities* are associated with the t-residues of the function $F(t) = \frac{(t-y_1)(t-y_2)\ldots(t-y_n)}{(t-x_1)(t-x_2)\ldots(t-x_{n+1})}$ at its poles: $x_1, \ldots, x_{n+1}$. See [**Ke, BO, 01**].

In our approach, these segments are basically $[x_i, y_i = x_i + L_{k_i}]$, where the lengths L_{k_i} are from some (finite) list and their order do not influence the resulting distribution. More generally, $y_i - x_i \ge L_{k_i}$, but the additional points in $[x_i, y_i]$ do not influence our distributions. The creation of points and segments is governed by Matérn II process. We calculated the generating function $G(t)$; its coefficient of $t^N u_1^{r_1} u_2^{r_2} \cdots$ is the probabilistic measure of the sequence of segments that are *gaps* of the corresponding *Young diagram.* The sum of the lengths of these segments is $\le N$; r_k gives the number of rows of length $L_k + 1$.

For instance, one can take here $L_i = i$ for $i = 1, 2, \ldots, n$, i.e. allow all possible lengths. Then our approach becomes really close to the theory of the transitional probabilities. The denominator of $G(t)$ is basically that in $F(t)$, considered with undetermined coefficients; they are our $u_i \beta_i$. The numerator of $G(t)$, a polynomial of degree n, incorporates the edge effects. It is not "generic" in this approach; this is different.

The points x_i, y_i do not occur in our formulas; the formulas the corresponding distribution is in terms of N, $\{r_k\}$ the probabilities $\{\beta_k\}$, and the lengths $\{L_k\}$. The classical theory results in the distribution of probabilities for Young diagrams related to the celebrated hook-formula and *Jack polynomials.* The probabilities of our Young diagrams from (13) and (14) are of different nature: for different point processes. Recall that we can assume that the consecutive gaps are $L \in \{L_i\}$.

A variant of our approach is for "Poisson-Catalan processes," where $\pm L_i$ are considered as the jumps up and down of the "energy function". Theorem 4.1 is stated only for one L (for $s = 1$). The probabilities $\mathcal{P}'_{r,m}$ there can be extended to multi-dimensional (type A_s) Catalan paths, where the steps are at the points subject to the Poisson-type distribution from Theorem 7.2. Combinatorially, one needs here formulas for the number of standard Young tableau for a given skew

Young diagram, for instance the *Naruse hook-length formula*. See e.g. [**MPP**]. Certain multi-dimension Bessel-type functions occur here, which are interesting to study.

Acknowledgments

The author thanks very much Alexei Borodin for various important discussions and shaping this paper. He also thanks Evgeny Feigin for his help. Many thanks to Anton Dzhamay and the referee.

References

[ABX] Z. Al-Aly, B. Bowe, Y. Xie, *Outcomes of SARS-CoV-2 Reinfection*, Preprint, doi.org/10.21203/rs.3.rs-1749502/v1 (2022).

[ADDP] F. Arruda, S. Das, C. Dias, D. Pastore, *Modelling and optimal control of multi strain epidemics, with application to COVID-19*, PLoS ONE 16: 9 (2021), e0257512; doi.org/10.1371/journal.pone.0257512.

[BY] L. V. Bogachev and Y. V. Yakubovich, *Limit shape of minimal difference partitions and fractional statistics*, Comm. Math. Phys. **373** (2020), no. 3, 1085–1131, DOI 10.1007/s00220-019-03513-5. MR4061406

[BC] A. Borodin and I. Corwin, *Macdonald processes*, Probab. Theory Related Fields **158** (2014), no. 1-2, 225–400, DOI 10.1007/s00440-013-0482-3. MR3152785

[BO] A. Borodin and G. Olshanski, *Point processes and the infinite symmetric group*, Math. Res. Lett. **5** (1998), no. 6, 799–816, DOI 10.4310/MRL.1998.v5.n6.a9. MR1671191

[Ch1] I. Cherednik, *Momentum managing epidemic spread and Bessel functions*, Chaos Solitons Fractals **139** (2020), 110234, 19, DOI 10.1016/j.chaos.2020.110234. MR4159617

[Ch2] I. Cherednik, *Modeling the waves of Covid-19*, Acta Biotheoretica 70, 8 (2022); doi.org/10.1007/s10441-021-09428-w.

[Gi] P. Giaquinta, *Entropy and ordering of hard rods in one dimension*, Entropy 2008, 10, 248–260; doi.org/10.3390/e10030248.

[Kat] M. Katori, *Bessel processes, Schramm-Loewner evolution, and the Dyson model*, SpringerBriefs in Mathematical Physics, vol. 11, Springer, [Singapore], 2015. MR3445783

[Ke] S. V. Kerov, *Anisotropic Young diagrams and symmetric Jack functions* (Russian, with Russian summary), Funktsional. Anal. i Prilozhen. **34** (2000), no. 1, 51–64, 96, DOI 10.1007/BF02467066; English transl., Funct. Anal. Appl. **34** (2000), no. 1, 41–51. MR1756734

[KD] K. Koufos, C. Dettmann, *Moments of interference in vehicular networks with hardcore headway distance*, IEEE Transactions on Wireless Communications, 17 :12 (2018), 8330–8341; doi.org/10.1109/TWC.2018.2876241.

[MPP] A. H. Morales, I. Pak, and G. Panova, *Hook formulas for skew shapes I. q-analogues and bijections*, J. Combin. Theory Ser. A **154** (2018), 350–405, DOI 10.1016/j.jcta.2017.09.002. MR3718070

[01] G. Akemann, J. Baik, and P. Di Francesco (eds.), *The Oxford handbook of random matrix theory*, Oxford University Press, Oxford, 2015. Paperback edition of the 2011 original [MR2920518]. MR3410165

[So] A. Soshnikov, *Determinantal random point fields* (Russian, with Russian summary), Uspekhi Mat. Nauk **55** (2000), no. 5(335), 107–160, DOI 10.1070/rm2000v055n05ABEH000321; English transl., Russian Math. Surveys **55** (2000), no. 5, 923–975. MR1799012

(I. Cherednik) Department of Mathematics, UNC Chapel Hill, North Carolina 27599
Email address: `chered@email.unc.edu`

Contemporary Mathematics
Volume **823**, 2025
https://doi.org/10.1090/conm/823/16497

Computational approach to the Schottky problem

Eddy B. de Leon, Jörg Frauendiener, and Christian Klein

This paper is dedicated to the memory of Igor M. Krichever.

Abstract. We present a computational approach to the classical Schottky problem based on Fay's trisecant identity for genus $g \geq 4$. For a given Riemann matrix $\mathbb{B}$ (symmetric matrices with a positive definite imaginary part), the Fay identity establishes linear dependence of secants in the Kummer variety if and only if the Riemann matrix corresponds to a Jacobian variety as shown by Krichever. The theta functions in terms of which these secants are expressed depend on the Abel maps of four arbitrary points on a Riemann surface. However, there is no concept of an Abel map for general Riemann matrices. To establish linear dependence of the secants, four components of the vectors entering the theta functions can be chosen freely. The remaining components are determined by a Newton iteration to minimize the residual of the Fay identity. If this residual vanishes within the finite numerical precision for a generic choice of input data, then the Riemann matrix is with this numerical precision the period matrix of a Riemann surface. The algorithm is compared in genus 4 for some examples to the Schottky-Igusa modular form, known to give the Jacobi locus in this case. It is shown that the same residuals are achieved by the Schottky-Igusa form and the approach based on the Fay identity in this case. In genera 5, 6 and 7, we discuss known examples of Riemann matrices and perturbations thereof for which the Fay identity is not satisfied.

1. Introduction

The classical Schottky problem is concerned with identifying Riemann matrices as period matrices of some compact Riemann surface of genus $g \geq 4$ among all symmetric matrices with positive definite imaginary part forming the Siegel half-space $\mathbb{H}^g$. In this paper, we present a computational approach to this problem based on Fay's trisecant identity [**15**], which is an identity between theta functions depending on the Abel maps of four arbitrary distinct points on a Riemann surface. As shown by Krichever [**22**], this identity only holds on Jacobians and thus gives a sharp criterion to identify the *Jacobi locus* in $\mathbb{H}^g$, namely, the subset whose elements

2020 *Mathematics Subject Classification.* 14H42,14K25,14Q05.
Key words and phrases. Schottky problem, secant identities, computational aspects.
This work was partially supported by the EIPHI Graduate School (contract ANR-17-EURE-0002)the Bourgogne Franche-Comté Region, the European fund FEDER, and by the European Union Horizon 2020 research and innovation program under the Marie Sklodowska-Curie RISE 2017 grant agreement no. 778010 IPaDEGAN. JF thanks for the support by the CNRS for an extended stay in Dijon and the hospitality of the IMB. We thank S. Grushevsky, D. Korotkin and B. Sturmfels for helpful discussions and hints.

are period matrices of some suitable Riemann surface of genus g. The idea is to decide for a given Riemann matrix $\mathbb{B} \in \mathbb{H}^g$, and having no information about the Abel map, whether it is possible to choose the arguments of the theta functions, which are g-dimensional vectors, such that the trisecant identity is satisfied. This is equivalent to the following optimization problem: can the vectors, corresponding to Abel maps if $\mathbb{B}$ is in the Jacobi locus, be selected in a way such that the residual of the Fay identity vanishes with numerical precision δ[1]? This would indicate that $\mathbb{B}$ is in the Jacobi locus with precision δ and that the vectors for which Fay's identity is satisfied define Abel maps.

The Schottky problem essentially goes back to Riemann; see [**13, 19, 28**] for additional references and a more detailed history of the problem. The Siegel half-space $\mathbb{H}^g$ of Riemann matrices has (complex) dimension $g(g+1)/2$. But the Jacobi locus of Riemann matrices, being period matrices of Riemann surfaces of genus g, has the (complex) dimension $3g-3$. Thus the Schottky problem to identify such matrices is non-trivial for $g \geq 4$. Schottky solved the problem in genus 4 via what is now known as the Schottky-Igusa modular form [**30**] (a rigorous proof was given by Igusa in [**21**]), a modular invariant built from *theta constants*, theta functions with argument equal to zero. The Schottky-Igusa form vanishes exactly on the Jacobi locus. A generalization of this form to a higher genus is not known, but see for instance [**3**] for attempts in this direction. Farkas, Grushevsky and Salvati Manni gave a solution to the weak Schottky problem in terms of modular invariants for arbitrary genus in [**13**]. This means that their modular invariant vanishes on the Jacobi locus, but as was shown for genus 5 by Donagi [**11**], vanishing of the invariant for some matrix $\mathbb{B}$ does not guarantee this matrix is related to a Jacobian.

Algebro-geometric approaches to integrable systems, see for instance [**4, 12**], not only allowed to construct quasi-periodic solutions to integrable partial differential equations (PDEs), but also proved useful in the context of the Schottky problem. Krichever's solution of the Kadomtsev-Petviashvili (KP) equation in terms of multi-dimensional theta functions on arbitrary compact Riemann surfaces raised the question of the computability of the solutions. An effective parametrization of the solutions will not work for $g \geq 4$ exactly because of the Schottky problem. This led Novikov to conjecture that Krichever's formula gives a solution to the KP equation if and only if the matrix $\mathbb{B}$ is in the Jacobi locus. Shiota proved this conjecture in [**31**]. Since it was stressed in [**27**] that many integrable PDEs can be solved in terms of theta functions as a consequence of Fay's identity (1.2), Welters [**35**] conjectured that this identity holds as the Novikov conjecture just on Jacobians. This was proven by Krichever [**22**] and is the basis of the algorithm to be presented in this paper.

To state Fay's identity, we consider four arbitrary points on a Riemann surface $\mathcal{R}$ of genus g: $P_1, P_2, P_3, P_4 \in \mathcal{R}$. The *Abel map*

$$\begin{aligned} \alpha : \mathcal{R} &\to Jac(\mathcal{R}), \\ P &\mapsto \left(\int_{P_0}^{P} \omega_1, \dots, \int_{P_0}^{P} \omega_g \right) \quad \bmod \Lambda, \end{aligned}$$

[1]In principle, the algorithm works for any chosen numerical accuracy. We work here with double precision, i.e., a maximal accuracy of the order $\delta \sim 10^{-16}$; in practice, due to rounding errors, a maximal precision of $\delta \sim 10^{-12}$ can be reached in this setting.

is a bijective map from the surface into the Jacobian $Jac(\mathcal{R}) = \mathbb{C}^g/\Lambda$, where Λ is the full-rank lattice formed by the periods of the holomorphic one-forms on $\mathcal{R}$. Moreover, we consider a normalized basis $\{\omega_1, \ldots, \omega_g\}$ of differentials with respect to a symplectic basis of the first homology group $\{a_1, b_1, \ldots, a_g, b_g\}$, i.e., $\int_{a_j} \omega_k = \delta_{jk}$. With this choice, the matrix of b-periods, which has components $\mathbb{B}_{jk} = \int_{b_j} \omega_k$, is symmetric and with positive definite imaginary part. Thus, the lattice Λ takes the form

$$\Lambda = \mathbb{Z}^g + \mathbb{B}\mathbb{Z}^g = \{\lambda \in \mathbb{C}^g \quad | \quad \lambda = \mathrm{m} + \mathbb{B}\,\mathrm{n}; \quad \mathrm{m}, \mathrm{n} \in \mathbb{Z}^g\}.$$

The Abel map is extended to any divisor $D = \sum_{k=1}^N n_k P_k$ on $\mathcal{R}$ (a divisor is a formal symbol, $n_k \in \mathbb{Z}$, $k = 1, \ldots, N$) via $\alpha(\sum_{k=1}^N n_k P_k) = \sum_{k=1}^N n_k \alpha(P_k)$. Following [**14**] we introduce the cross ratio function (Θ^* is a theta function with a nonsingular odd characteristic, see (2.6))

$$\lambda_{1234} = \frac{\Theta^*(\alpha(P_1) - \alpha(P_2))\Theta^*(\alpha(P_3) - \alpha(P_4))}{\Theta^*(\alpha(P_1) - \alpha(P_4))\Theta^*(\alpha(P_3) - \alpha(P_2))}, \tag{1.1}$$

which is a function on $\mathcal{R}^4$ that vanishes for $P_1 = P_2$ and $P_3 = P_4$ and has poles for $P_1 = P_4$ and $P_2 = P_3$. Then the following identity holds:

$$\begin{aligned} 0 =&\lambda_{3124}\, \Theta(\mathrm{u} + \int_{P_2}^{P_3})\, \Theta(\mathrm{u} + \int_{P_1}^{P_4}) + \lambda_{3214}\, \Theta(\mathrm{u} + \int_{P_1}^{P_3})\, \Theta(\mathrm{u} + \int_{P_2}^{P_4}) \\ &- \Theta(\mathrm{u})\, \Theta(\mathrm{u} + \int_{P_2}^{P_3} + \int_{P_1}^{P_4}), \end{aligned} \tag{1.2}$$

for all $\mathrm{u} \in \mathbb{C}^g$, where we use the notation $\int_P^Q = \alpha(Q - P)$ as well as $\int_{P_1+P_2}^{Q_1+Q_2} = \int_{P_1}^{Q_1} + \int_{P_2}^{Q_2} = \alpha(Q_1 + Q_2 - P_1 - P_2)$.

With the binary addition theorem (2.9) for theta functions, we get for (1.2)

$$\lambda_{3124}\Theta\begin{bmatrix}\epsilon/2\\0\end{bmatrix}\left(\int_{P_2+P_4}^{P_1+P_3}, 2\mathbb{B}\right) + \lambda_{3214}\Theta\begin{bmatrix}\epsilon/2\\0\end{bmatrix}\left(\int_{P_1+P_4}^{P_2+P_3}, 2\mathbb{B}\right) - \Theta\begin{bmatrix}\epsilon/2\\0\end{bmatrix}\left(\int_{P_1+P_2}^{P_3+P_4}, 2\mathbb{B}\right) = 0, \tag{1.3}$$

for all $\epsilon \in \mathbb{Z}^g/2\mathbb{Z}^g$, i.e., the g dimensional vectors ϵ with components $\epsilon_i = 0, 1$, $i = 1, \ldots, g$. Thus the system (1.3) consists of 2^g equations. We put

$$X := \frac{1}{2}\int_{P_1+P_2}^{P_3+P_4}, \quad Y := \frac{1}{2}\int_{P_2+P_3}^{P_1+P_4}, \quad Z := \frac{1}{2}\int_{P_2+P_4}^{P_1+P_3}, \tag{1.4}$$

which means that the 2^g equations (1.3) take the form

$$F_\epsilon(X, Y, Z) := c_1\Theta\left[\begin{smallmatrix}\epsilon/2\\0\end{smallmatrix}\right](2Z, 2\mathbb{B}) + c_2\Theta\left[\begin{smallmatrix}\epsilon/2\\0\end{smallmatrix}\right](2Y, 2\mathbb{B}) - \Theta\left[\begin{smallmatrix}\epsilon/2\\0\end{smallmatrix}\right](2X, 2\mathbb{B}), \tag{1.5}$$

where

$$c_1 = \frac{\Theta^*(Y - X)\Theta^*(Y + X)}{\Theta^*(Y + Z)\Theta^*(Y - Z)}, \quad c_2 = \frac{\Theta^*(X + Z)\Theta^*(X - Z)}{\Theta^*(Y + Z)\Theta^*(Y - Z)}. \tag{1.6}$$

In this paper we use equation (1.3) in the following way: we choose X, Y, Z to be in the fundamental domain given by

$$U_{\mathbb{B}} := \{\mathrm{z} \in \mathbb{C}^g \quad | \quad \mathrm{z} = \mathrm{p} + \mathbb{B}\,\mathrm{q}, \quad p_i, q_i \in [-1/2, 1/2], \quad i = 1, \ldots, g\}. \tag{1.7}$$

In other words, an arbitrary point of the Jacobian can be given in terms of the (p_i, q_i), which are called *characteristics*. Thus, the function F with components (1.5) has the form $F : U_{\mathbb{B}}^3 \subset \mathbb{C}^{3g} \to \mathbb{C}^{2^g}$, where $U_{\mathbb{B}}^3$ is a compact subset, and it is known from Fay's identity, see [**27**], that if $\mathbb{C}^g/\Lambda$ is a Jacobian the set of zeros of F is 4-dimensional, and it is parametrized according to (1.4). However, we are interested in general $\mathbb{C}^g/\Lambda$ where the notion of an Abel map is unknown; therefore, the zeros of F cannot be determined through (1.4). Thus, if we fix for X, Y, Z the first components and for instance the second component of X, the function F can be seen as a function of $\mathbf{x}$, the vector built from the remaining components of X, Y, Z of dimension $3g-4$. According to Krichever's theorem, this function vanishes for generic values of $X_{1,2}$, Y_1 and Z_1 for a certain choice of the vector $\mathbf{x}$ if and only if $\mathbb{B}$ is in the Jacobi locus. In the context of the Schottky problem, the task is to find the zero of $F(X,Y,Z)$ in the space of dimension $3g-4$ by varying $\mathbf{x}$. If the residual is smaller than the numerical accuracy δ, then the matrix is within numerical precision in the Jacobi locus. Since F is a locally holomorphic function of X, Y, Z, it is possible to find the zero with a standard Newton iteration as will be shown for examples in genus $g \leq 7$.

The paper is organized as follows: In section 2, we summarize basic definitions regarding theta functions, the Siegel fundamental domain and the Schottky-Igusa form. In section 3 we present the numerical algorithms. In section 4, we consider known examples in genus 4 and compare our algorithm to the Schottky-Igusa form. We show that both the Schottky-Igusa form and the approach based on the Fay identity yield the same residuals for the studied examples. This means they identify the Jacobi locus with the same numerical accuracy. In section 5, we consider examples in genera 5, 6 and 7. We add some concluding remarks in section 6. A pseudo code for the algorithm is given in the appendix.

2. Preliminaries

In this section, we summarize some theoretical facts needed in the following to numerically address the Schottky problem.

2.1. Jacobians and principally polarized Abelian varieties. Let $\mathbb{H}^g$ be the space of $g\times g$ complex symmetric matrices with positive definite imaginary part called the Siegel half-space. The Riemann theta function is given by the series

$$\Theta(\mathrm{z},\mathbb{B}) = \sum_{\mathrm{N}\in\mathbb{Z}^g} \exp(\pi \mathrm{i}\langle \mathrm{N},\mathbb{B}\mathrm{N}\rangle + 2\pi\mathrm{i}\langle \mathrm{N},\mathrm{z}\rangle) \tag{2.1}$$

with $\mathrm{z}\in\mathbb{C}^g$ and where $\langle\cdot,\cdot\rangle$ denotes the Euclidean scalar product $\langle \mathrm{N},\mathrm{z}\rangle = \sum_{i=1}^g \mathrm{N}_i\mathrm{z}_i$. Since $\mathbb{B}\in\mathbb{H}^g$, the series (2.1) converges uniformly for all $\mathrm{z}\in\mathbb{C}^g$; thus it is an entire function.

The Siegel space parametrizes the set of principally polarized Abelian varieties (PPAV) via the assignment $\mathbb{B}\mapsto(\mathcal{A}_\mathbb{B},\Theta_\mathbb{B})$, where the complex torus has the form $\mathcal{A}_\mathbb{B} = \mathbb{C}^g/(\mathbb{Z}^g+\mathbb{B}\mathbb{Z}^g)$ and $\Theta_\mathbb{B}$ is the divisor of the theta function (2.1) (the set of zeros of the theta function), which is a well-defined subvariety of $\mathcal{A}_\mathbb{B}$ because of its quasi-periodicity properties (2.8). In the following, we refer to the Riemann matrix $\mathbb{B}$ and the PPAV $(\mathcal{A}_\mathbb{B},\Theta_\mathbb{B})$ interchangeably.

In the special case where $\mathbb{B}$ is the period matrix of some Riemann surface $\mathcal{R}$ (which is only unique up to a modular transformation), its associated PPAV is known as the Jacobian $Jac(\mathcal{R})$ of $\mathcal{R}$.

2.2. Symplectic transformations. Two Riemann matrices $\mathbb{B}, \tilde{\mathbb{B}} \in \mathbb{H}^g$ can define isomorphic PPAVs if they define equivalent complex tori and if they are principally polarized. $\mathcal{A}_{\tilde{\mathbb{B}}}$ and $\mathcal{A}_{\mathbb{B}}$ are equivalent as complex tori if there exists an invertible homomorphism $\mathbb{C}^g/\tilde{\Lambda} \to \mathbb{C}^g/\Lambda$. This is equivalent to the existence of a linear transformation $\mathcal{M} : \mathbb{C}^g \to \mathbb{C}^g$ with $\mathcal{M}(\tilde{\Lambda}) = \Lambda$ and a matrix

$$R = \begin{pmatrix} A & B \\ C & D \end{pmatrix} \in M_{2g}(\mathbb{Z}) \tag{2.2}$$

such that

$$\mathcal{M}(\tilde{\mathbb{B}}, \mathbb{I}_g) = (\mathbb{B}, \mathbb{I}_g) R^\top, \tag{2.3}$$

holds. Additionally, in order for the principal polarization to be preserved, the condition

$$\begin{pmatrix} A & B \\ C & D \end{pmatrix}^\top \begin{pmatrix} 0_g & \mathbb{I}_g \\ -\mathbb{I}_g & 0_g \end{pmatrix} \begin{pmatrix} A & B \\ C & D \end{pmatrix} = \begin{pmatrix} 0_g & \mathbb{I}_g \\ -\mathbb{I}_g & 0_g \end{pmatrix} \tag{2.4}$$

must be satisfied [**5**], which means that $R \in \mathrm{Sp}(2g, \mathbb{Z})$. Thus, two matrices in $\mathbb{H}^g$ conjugated under the action of the symplectic (also called modular) group $\mathrm{Sp}(2g, \mathbb{Z})$ define equivalent PPAVs, where the action of $\mathrm{Sp}(2g, \mathbb{Z})$ on $\mathbb{H}^g$ is given by (2.3), which is equivalent to

$$\tilde{\mathbb{B}} = R \cdot \mathbb{B} = (A\mathbb{B} + B)(C\mathbb{B} + D)^{-1}, \tag{2.5}$$

which we call a modular transformation. Thus, given $\mathbb{B} \in \mathbb{H}^g$ we can choose the most convenient symplectically equivalent $\tilde{\mathbb{B}}$ in order to perform the computations as efficiently as possible. For this we will consider Siegel's fundamental domain.

Siegel [**32**] gave the following fundamental domain for the modular group:

DEFINITION 2.1. Siegel's fundamental domain is the subset of $\mathbb{H}^g$ such that $\mathbb{B} = X + \mathrm{i}Y \in \mathbb{H}^g$ satisfies:

(1) $|X_{nm}| \leq 1/2$, $n, m = 1, \dots, g$,
(2) Y is in the fundamental region of Minkowski reductions [**25, 26**],
(3) $|\det(C\mathbb{B} + D)| \geq 1$ for all C, D (2.4).

2.3. Theta functions. The multi-dimensional theta functions with characteristics are defined as the series,

$$\Theta\left[\begin{smallmatrix} \mathrm{p} \\ \mathrm{q} \end{smallmatrix}\right](\mathrm{z}, \mathbb{B}) = \sum_{\mathrm{N} \in \mathbb{Z}^g} \exp\left\{\pi \mathrm{i} \langle \mathrm{N} + \mathrm{p}, \mathbb{B}(\mathrm{N} + \mathrm{p})\rangle + 2\pi \mathrm{i} \langle \mathrm{N} + \mathrm{p}, \mathrm{z} + \mathrm{q}\rangle\right\}, \tag{2.6}$$

with the *characteristics* p, $\mathrm{q} \in \mathbb{R}^g$. Analogously to (2.1), the theta function with characteristics is an entire function on $\mathbb{C}^g$. If $\mathbb{B}$ is the period matrix of some Riemann surface $\mathcal{R}$, a characteristic is called *singular* if the corresponding theta function vanishes identically on $\mathcal{R}$.

Of special interest are half-integer characteristics with $2\mathrm{p}, 2\mathrm{q} \in \mathbb{Z}^g$. Such a characteristic is called *even* if $4\langle \mathrm{p}, \mathrm{q}\rangle = 0 \bmod 2$ and *odd* otherwise. It can be easily shown that theta functions with odd (even) characteristics are odd (even) functions of the argument z. The theta function with characteristic is related to the Riemann theta function Θ, the theta function with zero characteristic $\Theta := \Theta\left[\begin{smallmatrix} 0 \\ 0 \end{smallmatrix}\right]$, via

$$\Theta\left[\begin{smallmatrix} \mathrm{p} \\ \mathrm{q} \end{smallmatrix}\right](\mathrm{z}, \mathbb{B}) = \Theta(\mathrm{z} + \mathbb{B}\mathrm{p} + \mathrm{q}) \exp\left\{\pi \mathrm{i} \langle \mathrm{p}, \mathbb{B}\mathrm{p}\rangle + 2\pi \mathrm{i} \langle \mathrm{p}, \mathrm{z} + \mathrm{q}\rangle\right\}. \tag{2.7}$$

Note that from now on we will suppress the second argument of the theta functions if it is just $\mathbb{B}$ for ease of representation. From its definition, a theta function has the periodicity properties

$$\Theta\left[\begin{smallmatrix}\mathrm{p}\\ \mathrm{q}\end{smallmatrix}\right](\mathrm{z}+\mathrm{e}_j) = \mathrm{e}^{2\pi \mathrm{i} \mathrm{p}_j}\Theta\left[\begin{smallmatrix}\mathrm{p}\\ \mathrm{q}\end{smallmatrix}\right](\mathrm{z}), \quad \Theta\left[\begin{smallmatrix}\mathrm{p}\\ \mathrm{q}\end{smallmatrix}\right](\mathrm{z}+\mathbb{B}\mathrm{e}_j) = \mathrm{e}^{-2\pi \mathrm{i}(z_j+q_j)-\mathrm{i}\pi B_{jj}}\Theta\left[\begin{smallmatrix}\mathrm{p}\\ \mathrm{q}\end{smallmatrix}\right](\mathrm{z}), \tag{2.8}$$

where e_j is a vector in $\mathbb{Z}^g$ consisting of zeros except for a 1 in the j-th position. Theta functions satisfy the *binary addition theorem*, see for instance [**34**]

$$\begin{aligned} &\Theta\left[\begin{smallmatrix}\alpha\\ \gamma\end{smallmatrix}\right](z_1+z_2,\mathbb{B})\Theta\left[\begin{smallmatrix}\beta\\ \delta\end{smallmatrix}\right](z_1-z_2,\mathbb{B}) = \\ &\sum_{\epsilon\in\mathbb{Z}^g/2\mathbb{Z}^g} \Theta\left[\begin{smallmatrix}(\alpha+\beta)/2+\epsilon/2\\ \gamma+\delta\end{smallmatrix}\right](2z_1,2\mathbb{B})\Theta\left[\begin{smallmatrix}(\alpha-\beta)/2+\epsilon/2\\ \gamma-\delta\end{smallmatrix}\right](2z_2,2\mathbb{B}). \end{aligned} \tag{2.9}$$

The action of the modular group on theta functions is known, see for instance [**4, 15, 27**]. One has

$$\Theta\left[\begin{smallmatrix}\tilde{\mathrm{p}}\\ \tilde{\mathrm{q}}\end{smallmatrix}\right](\mathcal{M}^{-1}\mathrm{z},\tilde{\mathbb{B}}) = k\sqrt{\det(\mathcal{M})}\exp\left(\frac{1}{2}\sum_{i\le j} z_i z_j \frac{\partial}{\partial \mathbb{B}_{ij}}\ln\det\mathcal{M}\right)\Theta\left[\begin{smallmatrix}\mathrm{p}\\ \mathrm{q}\end{smallmatrix}\right](\mathrm{z}), \tag{2.10}$$

where $\tilde{\mathbb{B}}$ is given by (2.5) and k is a constant with respect to z, and where

$$\mathcal{M} = C\mathbb{B}+D, \quad \begin{pmatrix}\tilde{\mathrm{p}}\\ \tilde{\mathrm{q}}\end{pmatrix} = \begin{pmatrix} D & -C\\ -B & A\end{pmatrix}\begin{pmatrix}\mathrm{p}\\ \mathrm{q}\end{pmatrix} + \frac{1}{2}\begin{pmatrix}\mathrm{diag}(CD^\top)\\ \mathrm{diag}(AB^\top)\end{pmatrix}, \tag{2.11}$$

with diag denoting the diagonal of the corresponding matrices.

2.4. Schottky-Igusa modular form. The Schottky-Igusa modular form [**21, 30**] is a polynomial of degree 16 in the theta constants $\Theta\left[\begin{smallmatrix}\mathrm{p}\\ \mathrm{q}\end{smallmatrix}\right](0)$. We follow here the presentation in [**8**]: choose the characteristics

$$\mathrm{p}^{(1)} = \mathrm{q}^{(1)} = \frac{1}{2}\begin{pmatrix}1\\0\\1\\0\end{pmatrix}, \quad \mathrm{p}^{(2)} = \frac{1}{2}\begin{pmatrix}0\\0\\0\\1\end{pmatrix}, \quad \mathrm{q}^{(2)} = \frac{1}{2}\begin{pmatrix}1\\0\\0\\0\end{pmatrix},$$

and

$$\mathrm{p}^{(3)} = \frac{1}{2}\begin{pmatrix}0\\0\\1\\1\end{pmatrix}, \quad \mathrm{q}^{(3)} = \frac{1}{2}\begin{pmatrix}1\\0\\1\\1\end{pmatrix}.$$

Consider a rank 3 subgroup N of $\mathbb{Z}^8/(2\mathbb{Z}^8)$ generated by

$$\mathrm{n}_1 = \frac{1}{2}\begin{pmatrix}0&1\\0&1\\0&1\\1&0\end{pmatrix}, \quad \mathrm{n}_2 = \frac{1}{2}\begin{pmatrix}0&0\\0&0\\1&0\\1&1\end{pmatrix}, \quad \mathrm{n}_3 = \frac{1}{2}\begin{pmatrix}0&1\\0&0\\1&1\\0&1\end{pmatrix}.$$

We consider the product of theta constants

$$\pi_i = \prod_{(\mathrm{pq})\in(\mathrm{p}^{(i)}\mathrm{q}^{(i)})+N} \Theta\left[\begin{smallmatrix}\mathrm{p}\\ \mathrm{q}\end{smallmatrix}\right](0) \tag{2.12}$$

and define the Schottky-Igusa modular form as

$$\Sigma(\mathbb{B}) = \pi_1^2+\pi_2^2+\pi_3^2-\pi_1\pi_2-\pi_1\pi_3-\pi_2\pi_3. \tag{2.13}$$

It was shown in [**21**, **30**] that this modular form vanishes exactly on the Jacobi locus. Our numerical approach is equivalent to (2.13) in the sense that the only input we require is the matrix $\mathbb{B} \in \mathbb{H}^g$.

2.5. Kummer variety. For the geometrical interpretation of Fay's identity, we need to introduce the Kummer variety, which is the embedding of the complex torus into projective space via level-two theta functions (theta functions of double period).

DEFINITION 2.2. Let $(\mathcal{A}_\mathbb{B}, \Theta_\mathbb{B})$ be an indecomposable PPAV. Thus, its Kummer variety is the image of the map

$$\begin{aligned} \mathrm{Kum}: \mathcal{A}_\mathbb{B}/\sigma &\to \mathbb{P}^{2^g-1}, \\ Z &\mapsto \left[\Theta\left[\begin{smallmatrix}\epsilon/2\\0\end{smallmatrix}\right](2Z, 2\mathbb{B})\right]_{\epsilon\in\mathbb{Z}^g/2\mathbb{Z}^g}, \end{aligned}$$

where $\sigma(Z) = -Z$. This map is an embedding by Lefschetz' theorem [**5**].

Jacobian varieties are known to be indecomposable, thus we will also require this condition on general $\mathbb{B} \in \mathbb{H}^g$, i.e., $\mathbb{B}$ must not be symplectically equivalent to a matrix of the form $\mathrm{diag}[\mathbb{B}', \mathbb{B}'']$ with $\mathbb{B}' \in \mathbb{H}^{g'}$, $\mathbb{B}'' \in \mathbb{H}^{g''}$ and $1 \le g'' \le g'$. Recall that a PPAV $(\mathcal{A}_\mathbb{B}, \Theta_\mathbb{B})$ is indecomposable if there do not exist lower dimensional PPAVs $(\mathcal{A}_{\mathbb{B}'}, \Theta_{\mathbb{B}'})$, $(\mathcal{A}_{\mathbb{B}''}, \Theta_{\mathbb{B}''})$ such that $\mathcal{A}_\mathbb{B} \cong \mathcal{A}_{\mathbb{B}'} \times \mathcal{A}_{\mathbb{B}''}$ and $\Theta_\mathbb{B} \cong \mathcal{A}_{\mathbb{B}'} \times \Theta_{\mathbb{B}''} + \Theta_{\mathbb{B}'} \times \mathcal{A}_{\mathbb{B}''}$ (see [**21**]).

DEFINITION 2.3. Let $[V_1]$, $[V_2]$, $[V_3]$ be points in the projective space $\mathbb{P}^n$. They are said to be trisecant or collinear points in $\mathbb{P}^n$ if their representatives $V_1, V_2, V_3 \in \mathbb{C}^{n+1}$ are linearly dependent, i.e., there exist non-zero $c_1, c_2 \in \mathbb{C}$ such that $V_3 = c_1 V_1 + c_2 V_2$.

In particular, we say the Kummer variety of $(\mathcal{A}_\mathbb{B}, \Theta_\mathbb{B})$ admits trisecant points if there exist $\mathrm{Kum}(X)$, $\mathrm{Kum}(Y)$, $\mathrm{Kum}(Z) \in \mathbb{P}^{2^g-1}$ satisfying Definition 2.3.

3. Numerical approaches

In this section, we outline the numerical approaches to be used in the context of the Schottky problem.

3.1. Computation of theta functions. The standard way to compute the series (2.6) is to approximate it by a sum, $|N_i| \le \mathcal{N}_\delta$, $i = 1, \ldots, g$, where the constant $\mathcal{N}_\delta$ is chosen such that all omitted terms in (2.6) are smaller in modulus than the aimed-at accuracy δ. In contrast to [**10**] and [**2**], we do not give a specific bound for each $|N_i|$, $i = 1, \ldots, g$, i.e., we sum over a hyper-cube of dimension g instead of an ellipsoid. The reason for this is that it does not add much to the computational cost, but that it simplifies a parallelization of the computation of the theta function in which we are interested. Taking into account that we can choose z in the fundamental domain of the Jacobian because of (2.8), we get for the Riemann theta function the estimate

$$\mathcal{N}_\delta > \sqrt{-\frac{\ln\delta}{\pi y_{\min}}} + \frac{1}{2} \tag{3.1}$$

Here $y_{\min}$ is the length of the shortest vector in the lattice defined by the imaginary part of the Riemann matrix $\mathbb{B} = X + \mathrm{i}Y$: let $Y = T^\top T$, i.e., T be the Cholesky decomposition of Y, then T defines a lattice, i.e., a discrete additive subgroup of $\mathbb{R}^g$, of the form

$$\mathcal{L}(t_1, \dots, t_g) = \{TN \mid N \in \mathbb{Z}^g\}, \tag{3.2}$$

where $T = [t_1, t_2, \dots, t_g] \in \mathbb{R}^{g\times g}$ has rank g.

The greater the norm of the shortest lattice vector, the more rapid will be the convergence of the theta series. Note that, in general, the convergence of a theta series contrary to popular belief can be very slow, see for instance the discussion in [**10**]. Siegel showed that the length of $y_{\min} \geq \sqrt{3}/2$ in the Siegel fundamental domain. Unfortunately, no algorithm is known to construct a symplectic transformation for a general Riemann matrix to this fundamental domain. But Siegel [**32**] gave an algorithm to achieve this approximately. A problem in this context is the Minkowski ordering. Just as the identification of the shortest lattice vector, this is a problem for which only algorithms are known whose time grows exponentially with the dimension g. Therefore, the implementation of the Siegel algorithm in [**10**] uses an approximation to these problems known as the LLL algorithm [**23**]. Whereas this algorithm is fast, it is not very efficient. Therefore, in [**17**], Siegel's algorithm was implemented via an exact identification of the shortest lattice vector which leads to a more efficient computation of the theta functions. Thus, for all Riemann matrices to be considered in this paper we always first apply the method outlined in [**17**] in order to obtain faster convergent theta series. In practice, this means that an accuracy of the order $\delta \sim 10^{-12}$ can be reached with $\mathcal{N}_\delta = 5$.

Note that derivatives of theta functions are computed in an analogous way as the theta function itself since

$$\frac{\partial\Theta\left[\begin{smallmatrix}\mathrm{p}\\ \mathrm{q}\end{smallmatrix}\right](\mathrm{z},\mathbb{B})}{\partial \mathrm{z}_i} = 2\pi\mathrm{i}\sum_{\mathrm{N}\in\mathbb{Z}^g}(N_i+p_i)\exp\left\{\pi\mathrm{i}\left\langle \mathrm{N}+\mathrm{p}, \mathbb{B}\left(\mathrm{N}+\mathrm{p}\right)\right\rangle + 2\pi\mathrm{i}\left\langle \mathrm{N}+\mathrm{p}, \mathrm{z}+\mathrm{q}\right\rangle\right\}. \tag{3.3}$$

Let $\mathcal{T}(z)$ be the array

$$\mathcal{T}(z) = \exp\left\{\pi\mathrm{i}\left\langle \mathrm{N}+\mathrm{p}, \mathbb{B}\left(\mathrm{N}+\mathrm{p}\right)\right\rangle + 2\pi\mathrm{i}\left\langle \mathrm{N}+\mathrm{p}, \mathrm{z}+\mathrm{q}\right\rangle\right\}, \quad \mathrm{N} \in [-\mathcal{N}_\delta, \mathcal{N}_\delta]^g$$

The construction of this array is the most expensive part of the routine that computes the theta function, but it allows the immediate computation of the gradient. Let us define the $(2\mathcal{N}_\delta + 1)^g \times g$ matrix M_{N} by

$$M_{\mathrm{N}} = [\mathrm{N}^{(1)} + \mathrm{p}_1, \cdots, \mathrm{N}^{(g)} + \mathrm{p}_g]$$

as the matrix whose row vectors contain all the N considered in the sum approximating the theta function plus the characteristic p. We use $\mathrm{N}^{(i)}$ to denote the column vector of length $(2\mathcal{N}_\delta + 1)^g$ containing the i-th components of all N in the same order as in the definition of $\mathcal{T}(z)$. Thus, the gradient (given as a row vector) is easily obtained through the matrix multiplication

$$\nabla\Theta\left[\begin{smallmatrix}\mathrm{p}\\ \mathrm{q}\end{smallmatrix}\right](\mathrm{z},\mathbb{B}) \approx 2\pi\mathrm{i}\mathcal{T}(z)^\top * M_{\mathrm{N}}. \tag{3.4}$$

3.2. Newton iteration. Newton's iteration can be used to find zeros of locally holomorphic functions of the form $f : \mathbb{C}^M \to \mathbb{C}^N$ if the initial iterate $\mathbf{x}^{(0)} \in \mathbb{C}^M$ is sufficiently close to a zero. With the choice $\mathbf{x}^{(0)}$, we find new iterates via

$$\mathbf{x}^{(n+1)} = \mathbf{x}^{(n)} - \mathrm{Jac}(F(\mathbf{x}^{(n)}))^{-1}F(\mathbf{x}^{(n)}), \quad n = 0, 1, \dots; \tag{3.5}$$

where $\mathrm{Jac}(F(\mathbf{x}^{(n)}))$ denotes the Jacobian matrix of F. However, the zero set must be discrete, otherwise $\mathrm{Jac}(F(\mathbf{x}^{(n)}))$ becomes singular.
The function we are interested in is

$$\begin{aligned} &F : U_{\mathbb{B}}^3 \to \mathbb{C}^{2^g}, \\ &(X, Y, Z) \mapsto c_1(X, Y, Z)\vec{\Theta}(Z) + c_2(X, Y, Z)\vec{\Theta}(Y) - \vec{\Theta}(X), \end{aligned} \tag{3.6}$$

where $\vec{\Theta}(Z) := \left(\Theta\left[\begin{smallmatrix}\epsilon/2\\0\end{smallmatrix}\right](2Z, 2\mathbb{B})\right)_{\epsilon\in\mathbb{Z}^g/2\mathbb{Z}^g}$ is the representative of the Kummer point $\mathrm{Kum}(Z)$ in $\mathbb{C}^{2^g}$ and $U_{\mathbb{B}}$ is the fundamental domain (1.7) of $\mathcal{A}_{\mathbb{B}}$. Additionally, we need to impose the condition that the zeros must be non-trivial.

DEFINITION 3.1 (Trivial zeros). We say that (X, Y, Z) is a trivial zero if $X \neq \pm Z$, $Y \neq \pm Z$ or $X \neq \pm Y$, since $F(X, Y, Z) = 0$ for such cases regardless of the nature of $\mathbb{B} \in \mathbb{H}^g$.

With the aforementioned constraint, $F(X, Y, Z) = 0$ if and only if $\mathrm{Kum}(X)$, $\mathrm{Kum}(Y)$, $\mathrm{Kum}(Z)$ are trisecant points, none of which coincide. Therefore, Welters-Krichever's result [**22, 35**] can be stated in terms of F.

THEOREM 3.2 (Trisecant theorem). *Let $(\mathcal{A}_{\mathbb{B}}, \Theta_{\mathbb{B}})$ be an indecomposable principally polarized Abelian variety. Then, it is the Jacobian of some Riemann surface $\mathcal{R}$ of genus g if and only if the function F has non-trivial zeros.*

This implies that the zero set of the constrained F is either empty or it is the four-dimensional set given by the parametrization (1.4). However, even if $\mathbb{B}$ turns out to be in the Jacobi locus, we still have to determine X, Y, Z without an Abel map, since this is generally unknown.

Since Theorem 3.2 only requires the existence of one zero to conclude that $\mathbb{B}$ is in the Jacobi locus $\mathcal{J}_g$, we can add four constraints to F in such a way that the zero set of the constrained function is discrete, which will allow us to use Newton's iteration (3.5). A simple way to do this is by considering its intersection with the zero set of four functions of the form $f(V) = V_j - c$, where $V = (X, Y, Z) \in U_{\mathbb{B}}^3$ and $c \in \mathbb{C}$. This is equivalent to fixing four components of the vector V in the iterative process. Having chosen starting vectors satisfying the non-triviality condition, we fix their first components so that their iterates $X^{(n)}$, $Y^{(n)}$ and $Z^{(n)}$ remain different along the whole iteration and, in addition, fix any other component, e.g., X_2. Thus, we obtain a function of the form $F|_W : W \subset \mathbb{C}^{3g-4} \to \mathbb{C}^{2^g}$, where W is the compact subset

$$W = \{(X, Y, Z) \in U_{\mathbb{B}}^3 \subset \mathbb{C}^{3g} \quad | \quad X_1 = X_1^{(0)}, X_2 = X_2^{(0)}, Y_1 = Y_1^{(0)}, Z_1 = Z_1^{(0)}\}. \tag{3.7}$$

For simplicity, we drop the subscript in $F|_W$. Let us denote by $\mathbf{x}$ the remaining $3g-4$ components of $V \in U_{\mathbb{B}}^3$, i.e., $\mathbf{x} \in W$. The task is to find a possible zero of the function $F(\mathbf{x})$, i.e., to decide whether it can have values smaller than the specified accuracy δ. If this is the case, then the Riemann matrix is regarded as lying on the Jacobi locus within the given numerical precision.

Thus, after choosing some initial vector $V^{(0)} \in U_{\mathbb{B}}^3$, fixing four components to obtain W, setting the initial iterate $\mathbf{x}^{(0)}$, we numerically identify a zero of F by applying the standard Newton iteration (3.5). Notice that $\mathrm{Jac}(F(\mathbf{x}^{(n)}))$ is composed of $3g - 4$ linearly independent components of the full Jacobian matrix of (3.6). Since F is locally holomorphic in the vectors X, Y, and Z, the Jacobian can be directly computed via the derivatives $\frac{\partial F}{\partial \mathbf{x}_i}$, $i = 1, \dots, 3g-4$. In practice we compute the full Jacobian with respect to all components of X, Y, and Z; keep the necessary components and then compute the *Newton step* $\mathrm{Jac}(F(\mathbf{x}^{(n)}))^{-1}F(\mathbf{x}^{(n)})$ with the Matlab command '\'. This means that the overdetermined linear system $\mathrm{Jac}(F(\mathbf{x}^{(n)}))\mathbf{y} = F(\mathbf{x}^{(n)})$ is solved for $\mathbf{y}$ in a least squares sense. In our context, this has the advantage that all equations in (1.5) are satisfied as required to a certain accuracy if the iteration converges.

REMARK 3.3. If only $3g-4$ equations of $F = 0$ are used for the Newton iteration, the latter in general converges to a value of $\mathbf{x}$ for which the remaining components of F do not vanish. Thus, it is important that all 2^g equations are used in the iteration.

We recall that a Newton iteration has quadratic convergence which means that if $\tilde{\mathbf{x}}$ is the wanted zero of F, then $\|\mathbf{x}^{(n+1)} - \tilde{\mathbf{x}}\| \propto \|\mathbf{x}^{(n)} - \tilde{\mathbf{x}}\|^2$, provided $\mathbf{x}^{(n)}$ is in the so-called *basin of attraction*, which is the subset of W for which the iteration converges. Loosely speaking the number of correct digits in the iteration doubles in each step of the iteration. However, note that the convergence of the Newton iteration is local and, therefore, depends on the starting point $\mathbf{x}^{(0)}$. If this is not adequately chosen, the iteration may fail to converge.

The iteration is stopped either when $\|\mathbf{x}^{(n+1)} - \mathbf{x}^{(n)}\| < \delta$ or the residual of F is smaller than δ — in which case we speak of convergence — or else after 100 iterations, in which case the iteration is deemed to not have converged. The latter typically indicates an inadequate choice of the initial iterate, and the iteration can simply be restarted with another initial vector. When the iteration stops, we not only check whether the residual of F in (1.5) is below the aimed-at accuracy δ which would indicate that the considered Riemann matrix is in the Jacobi locus. We also compute the *singular value decomposition*[2] of the matrix formed by the three vectors $\vec{\Theta}(X)$, $\vec{\Theta}(Y)$, $\vec{\Theta}(Z)$. If the modulus of the smallest singular value of this matrix, denoted by Δ in the following, is smaller than δ, the vectors are linearly dependent and the Riemann matrix is in the Jacobi locus within numerical accuracy.

REMARK 3.4. It is important that we used in (1.6) ratios of theta functions instead of a form of the Fay identity free of denominators. The reason for this is that odd theta functions have co-dimension one zero sets, and the Newton iteration would converge to all factors in front of $\vec{\Theta}(X)$, $\vec{\Theta}(Y)$, $\vec{\Theta}(Z)$ being zero. The same would happen if we obtain the constants from three of the equations (1.5) in the usual definition of linear dependence of vectors. This is the reason why we also check the linear dependence of the vectors via an SVD. The many zeros of the theta functions in the denominators of (1.6) can also lead to a slow convergence of the Newton iteration for the first iterates.

[2]The singular-value decomposition (SVD) of an $m \times n$-matrix M with complex entries is given by $M = U\Sigma V^\dagger$; here U is an $m \times m$ unitary matrix, $V^\dagger$ denotes the conjugate transpose of V, an $n \times n$ unitary matrix, and the $m \times n$ matrix Σ is diagonal (as defined for a rectangular matrix); the non-negative numbers on the diagonal of Σ are called the singular values of M.

Since the convergence of the Newton iteration depends on the choice of the initial iterate $\mathbf{x}^{(0)}$, we must consider a strategy to assure convergence. We have the following choices:

(a) Take initial vectors of the form with

$$V^{(0)} = \frac{\ell}{2}\left(\mathrm{p}^{(1)} + \mathbb{B}\mathrm{q}^{(1)}, \mathrm{p}^{(2)} + \mathbb{B}\mathrm{q}^{(2)}, \mathrm{p}^{(3)} + \mathbb{B}\mathrm{q}^{(3)}\right), \tag{3.8}$$

where $0 < \ell < 1$ and $\mathrm{p}^{(i)}, \mathrm{q}^{(i)} \in \mathbb{Z}^g/2\mathbb{Z}^g$ with $\mathrm{q}^{(i)} \neq 0$. We require $\ell \neq 0, 1$ since the Jacobian matrix of the Kummer map is singular at half-period vectors. Thus, this choice will prevent the iterates from being too close to such singular points.

(b) Take completely random vectors $X^{(0)}$, $Y^{(0)}$, $Z^{(0)}$ in the fundamental domain satisfying the non-triviality conditions.

(c) Take two random initial vectors and the third one arbitrarily close to one of them, e.g., $V^{(0)} = (X^{(0)}, \ell X^{(0)}, Z^{(0)})$ with $\ell \approx 1$ or $V^{(0)} = (X^{(0)}, X^{(0)} + \varepsilon, Z^{(0)})$ with $\|\varepsilon\| \approx 0$. We cannot simply set $Y = Z$, $X = Y$ or $X = Z$ because of the non-triviality condition. However, from Fay's identity, we know that there exists $Y \neq X$ in the neighborhood of X making F vanish.

For reproducibility purposes, we choose initial vectors according to (3.8) with specific values of ℓ. Thus, for a given Riemann matrix $\mathbb{B}_0 \in \mathbb{H}^g$, we perform the following steps:

(i) Given $\mathbb{B}_0 \in \mathbb{H}^g$, look for a modular transformation R such that $\mathbb{B} = R \cdot \mathbb{B}_0$ is approximately in the Siegel fundamental domain (the important point is to ensure that the shortest lattice vector has length greater than $\sqrt{3}/2$). This is done with the algorithm [**17**].

(ii) Set a value for ℓ in (3.8), $\Delta\ell$, $n_{\max}$ and the precision δ. For the examples presented here, we used $\ell = 0.10$, $\Delta\ell = 0.10$, $n_{\max} = 100$ and $\delta = 10^{-10}$.

(iii) Choose starting vectors $X^{(0)}, Y^{(0)}, Z^{(0)} \in U_{\mathbb{B}}$ in the form (3.8) with $\mathrm{p}^{(1)} = \mathrm{q}^{(1)} = e_{g-2}$, $\mathrm{p}^{(2)} = \mathrm{q}^{(2)} = e_{g-1}$, $\mathrm{p}^{(3)} = \mathrm{q}^{(3)} = e_g$ and fix the components X_1, X_2, Y_1, Z_1.

(iv) Set up the function $F : W \subset \mathbb{C}^{3g-4} \to \mathbb{C}^{2^g}$.

(v) Perform Newton's iteration (3.5) until $\|\mathbf{x}^{(n+1)} - \mathbf{x}^{(n)}\| < \delta$ or the maximum number of iterations $n_{\max}$ is reached or $\|F(\mathbf{x}^{(n)})\| < \delta$. Keep the vectors $X^{(n)}$, $Y^{(n)}$, $Z^{(n)}$ in the fundamental domain at every step.

(vi) If $\Delta^{(N)} := \min(\mathrm{svd}(\vec{\Theta}(X^N), \vec{\Theta}(Y^N), \vec{\Theta}(Z^N))) < \delta$, where N is the iteration at which the Newton iteration stopped, we conclude that $\mathbb{B}$ is in the Jacobi locus with precision δ and stop the computations.

(vii) If the iteration did not converge, replace ℓ by $\ell + \Delta\ell$ and go back to step (iii) and perform a new iteration. Stop if $\ell > \ell_{\max}$ and conclude with a precision δ that $\mathbb{B}$ is not in the Jacobi locus. We used $\ell_{\max} = 0.5$ for our examples.

We show these steps more explicitly in the pseudo code Algorithm 1.

4. Examples in genus 4

In this section we study known examples in genus 4 since the Schottky-Igusa form gives the identification of the Jacobi locus in genus 4. This provides interesting tests for our approach. We start by studying the stability of the code by considering the Riemann matrix of Bring's curve, as well as images of the Abel map (which are computed numerically). We use the algorithm described above, but with perturbations of the Abel images as initial iterates. We expect quadratic convergence from the beginning since the initial iterates are in the vicinity of zeros.

4.1. Bring's curve. Bring's curve is the curve with the highest number of automorphisms in genus 4, see [**6**] for the computation of its Riemann matrix. It can be defined by the algebraic curve

$$\{(x,y) \in \mathbb{C}^2 \quad | \quad xy^5 + x + x^2y^2 - x^4y - 2y^3 = 0\}. \tag{4.1}$$

For a given algebraic curve, the Riemann matrix can be computed via the symbolic approach by Deconinck and van Hoeij [**9**] implemented in Maple or in Sage [**33**] or the purely numerical approach [**16**], see also the respective chapters in [**7**]. We use here the approach of [**16**] and find after applying Siegel's algorithm in the form [**17**] the Riemann matrix

```
RieMat =

-0.5000 + 0.8685i -0.0000 + 0.0649i -0.5000 - 0.2678i 0.5000 - 0.2678i
0.0000 + 0.0649i -0.5000 + 0.8685i 0.5000 + 0.2678i -0.5000 + 0.2678i
-0.5000 - 0.2678i 0.5000 + 0.2678i -0.0000 + 1.0714i 0.5000 - 0.2678i
0.5000 - 0.2678i -0.5000 + 0.2678i 0.5000 - 0.2678i -0.5000 + 0.8685i.
```

Note that the accuracy of the computed matrix is estimated to be better than 10^{-14}, but for the ease of readability, we give only four digits here. For this matrix, we get for the Schottky-Igusa form $|\Sigma| = 1.04 \times 10^{-15}$, i.e., a value of the order of the rounding error as expected. This is to ensure that we are indeed testing with a matrix $\mathbb{B}$ in the Jacobi locus.

To test the computation of the vectors X, Y, Z satisfying the Fay identity, we consider the points covering the point $z = 2.5$ in the complex plane on the sheets 1 to 4 of Bring's curve. The code [**16**] gives

```
AbelMap =

-0.7052 + 0.3692i 0.0545 + 0.0278i 0.0293 + 0.0775i -0.0607 + 0.1180i
0.1286 - 0.2662i 0.2747 + 0.2456i -0.4068 + 0.4113i 0.0318 + 0.2067i
-0.4351 + 0.2906i -0.2108 + 0.0422i 0.0250 + 0.2906i 0.0823 - 0.2451i
0.4519 - 0.6915i 0.0718 + 0.0915i -0.0126 - 0.0140i 0.0487 + 0.0493i.
```

We use these vectors (the columns of the matrix) and set $X^{(0)}$, $Y^{(0)}$, $Z^{(0)}$ as the linear combinations given by (1.4) for this particular example, rather than using the choice (3.8) as we will do in general. The residual of the function F in (1.5) for this Abel map is of the order of 10^{-10}, and the minimal singular value of the matrix with the vectors $\vec{\Theta}(X)$, $\vec{\Theta}(Y)$ and $\vec{\Theta}(Z)$ is $\Delta \sim 10^{-11}$. This indicates that the Abel map is computed to an accuracy of the order of 10^{-11}. After one Newton iteration, the difference between the new and old vector is of the order 10^{-11}, and the residual of $F \sim 10^{-14}$ and $\Delta \sim 10^{-14}$. This shows that the iteration is stable (the numerical error in the Abel map is 'corrected' by the Newton iteration), and that a similar residual of F is reached in this example as for the Schottky-Igusa form.

The stability of the iteration is also shown by perturbing the initial vector. We multiply the above $\mathbf{x}$ by a factor 1.1, i.e., we keep the entries $X_1^{(0)}$, $X_2^{(0)}$, $Y_1^{(0)}$, $Z_1^{(0)}$ and multiply the remaining ones by 1.1. After 7 iterations, we get the same residual, minimal singular value Δ and final vector $\mathbf{x}$ as before. The same behavior is observed for $0.9\mathbf{x}$ as the initial iterate.

It is known that Matlab timings are not very precise since they strongly depend on how many precompiled commands are used in the coding, but they provide an indication of actual computing times for a given task. On a standard laptop, the above examples take a few seconds.

To finish the tests with Bring's curve, we choose the same $X_1^{(0)}, X_2^{(0)}, Y_1^{(0)}, Z_1^{(0)}$ as before, but the remaining components are chosen randomly with the condition that $X^{(0)}$, $Y^{(0)}$, $Z^{(0)}$ must be in the fundamental domain. The algorithm finally produces a residual for F smaller than 10^{-12}, but finds a vector $\mathbf{x}$ different from the one produced by the Abel map. This is because the zero set of $F|_W$, with W given by (3.7), does not necessarily have a unique element.

4.2. Family of genus 4 Riemann matrices. We now turn to the family of Riemann matrices studied in [**8**], which correspond to the genus 4 algebraic curves

$$\{(x,y) \in \mathbb{C}^2 \quad | \quad y^6 = x(x+1)(x-t)\}, \tag{4.2}$$

parametrized by $t \in \mathbb{P}^1 \backslash \{0, 1, \infty\}$. Their period matrices are given by [**20**], i.e., $\mathbb{B}_\tau = A^{-1}B$, where

$$(A|B) = \begin{pmatrix} \tau & \tau & 0 & -\tau-1 & 1 & 1 & 0 & -1 \\ \zeta^2-1 & 1 & -\zeta^2+1 & 1 & 1 & -\zeta^2 & \zeta^2 & -\zeta^2+1 \\ \zeta^3-\zeta & -\zeta^3 & -2\zeta^3+2\zeta^2+\zeta-1 & \zeta^2-\zeta & 1 & \zeta^2-1 & \zeta^3-\zeta^2-2\zeta+2 & \zeta^2 \\ -\zeta^3+\zeta & \zeta^3 & 2\zeta^3+2\zeta^2-\zeta-1 & \zeta^2+\zeta & 1 & \zeta^2-1 & -\zeta^3-\zeta^2+2\zeta+2 & \zeta^2 \end{pmatrix}, \tag{4.3}$$

with $\zeta = e^{2\pi \mathrm{i}/12}$ and $t = \mu(\tau)$ for some $\mu : \mathbb{H} \to \mathbb{P}^1$.

Let us first look at the convergence rate of the iteration for one specific Riemann matrix, e.g., $\mathbb{B}_\tau$ with $\tau = 1 + \mathrm{i}$. In Figure 1 we observe that as soon as the vector $\mathbf{x}^{(n)}$ falls into a basin of attraction, it converges to a zero very rapidly. Recall that N is the step at which the iteration stops. Thus, $\|F(\mathbf{x}^{(N)})\|$ is the best residual achieved by the iteration corresponding to the starting vectors $X^{(0)}, Y^{(0)}, Z^{(0)}$. For this test, we chose the starting vectors according to Algorithm 1. In contrast, notice that with a small perturbation of the form

$$\mathbb{B}_{\tau,s} = \mathbb{B}_\tau + s \cdot \mathrm{diag}[2,3,5,7], \quad \text{with } s = 0.01; \tag{4.4}$$

the smallest value of $\|F(\mathbf{x}^{(N)})\|$ is above 10^{-4} (although this lower bound will depend on how small s is). Thus, for these particular examples there are at least six orders of magnitude of difference in the smallest attained residuals.

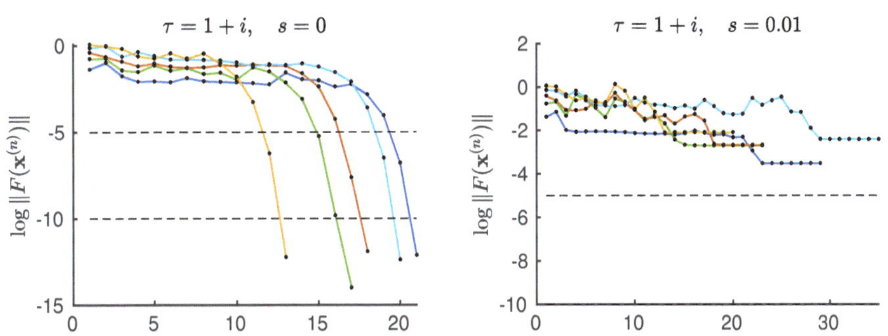

FIGURE 1. Residual of the function F, corresponding to the matrix $\mathbb{B}_{\tau,s}$, in dependence of the n-th iteration.

REMARK 4.1. Using Algorithm 1, but with randomly chosen initial vectors, the iterations converge with a frequency of approximately 90% when $\mathbb{B}$ is in the Jacobi locus. We attain the residuals shown in Figure 2 for 1000 different tests for the same matrix $\mathbb{B}_\tau$, i.e., 1000 iterations with different randomly chosen initial vectors $X^{(0)}, Y^{(0)}, Z^{(0)}$ attained $\Delta^{(N)} < 10^{-10}$ 896 times. In contrast, none of the tests attained a value below 10^{-5} when the considered matrix is $\mathbb{B}_{\tau,s}$ with $s = 0.1$.

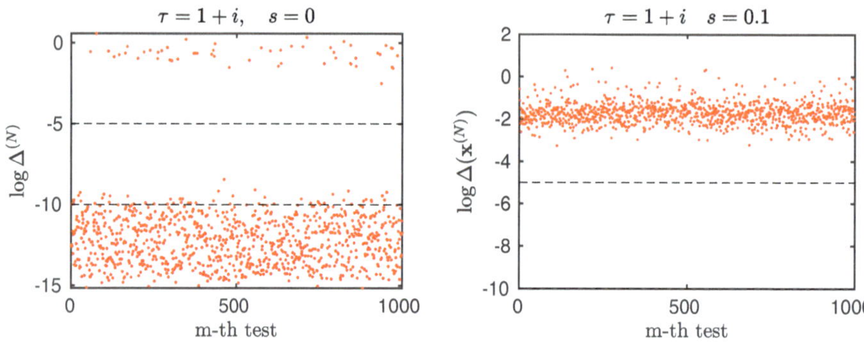

FIGURE 2. Attained values of $\Delta^{(N)}$ for randomly chosen initial vectors. The left-hand side corresponds to $\mathbb{B}_{\tau,s}$ with $s = 0$ and the right-hand side to $s = 0.1$.

Let us consider the family of Riemann matrices of the form (4.4) parametrized by $s \in [-1/2, 1/2]$ and $\tau = x + \mathrm{i}$, with $x \in [0, 1]$. We expect the Schottky-Igusa form of every $\mathbb{B}_{\tau,s}$ and the residual $\|F(\mathbf{x}^{(N)})\|$ (or its associated minimum singular value) obtained with Algorithm 1 to be within the same order of magnitude. This is indeed what we observe in Figure 3. With this algorithm, we conclude with precision $\delta = 10^{-10}$ that the Riemann matrices $\mathbb{B}_{\tau,s}$ with $s \neq 0$ are not in the Jacobi locus, agreeing with the Schottky-Igusa form.

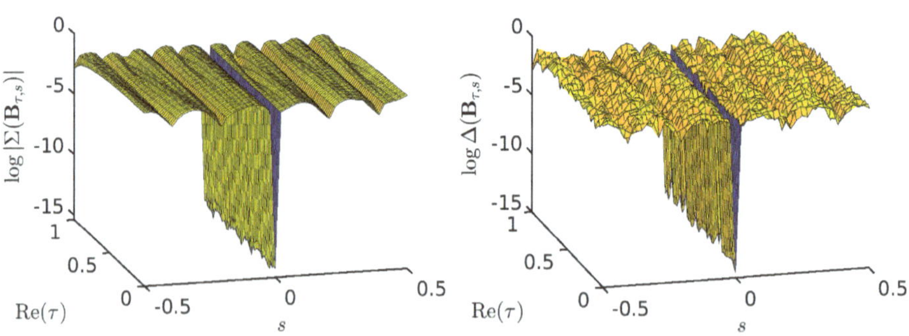

FIGURE 3. On the left the Schottky-Igusa form and on the right the minimum singular value $\Delta^{(N)}$, both in dependence of the Riemann matrices $\mathbb{B}_{\tau,s}$.

Note that our approach produces numerically essentially the same residuals as the Schottky-Igusa form and is thus of the same practical relevance in deciding whether a matrix is in the Jacobi locus or not with given numerical accuracy. In both cases the input is only the Riemann matrix $\mathbb{B}$, but in the Fay approach an initial iterate for the vectors corresponding to Abel maps for $\mathbb{B}$ in the Jacobi locus.

The matrices $\mathbb{B}_\tau$ considered above are in exact form, thus the smallest residual is expected to be zero to machine precision, i.e., approximately 10^{-14} with Matlab. However, the smallest $\|F(\mathbf{x}^{(N)})\|$ might be considered as an indicative of the precision of its input matrix $\mathbb{B}$, if this is not in exact form. For example, let us consider the matrices $\mathbb{B}_{\tau,s} = \mathbb{B}_\tau + s(M + \mathrm{i}M)$ where M is the symmetric matrix with coefficients $M_{jk} = (j+k)/5$, $\tau = 1 + \mathrm{i}$ and $s \in [10^{-15}, 10^{-1}]$ are small perturbations. Thus, $\mathbb{B}_{\tau,s}$ can be seen as the Riemann matrices of the curve (4.2) with an accuracy of the order s. We remove the stopping criterion $\|F(\mathbf{x}^{(N)})\| < \delta$ for this particular example in order to visualize the smallest residual we can achieve with Matlab's precision.

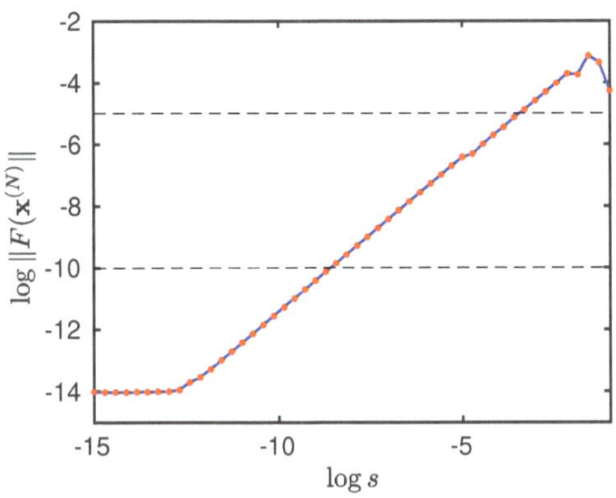

FIGURE 4. Smallest residuals of period matrices with accuracy s.

It can be observed in Figure 4 that we attain a residual to machine precision if the matrices are within a precision of approximately 10^{-13} and, in particular, the residual is below $\delta = 10^{-10}$ if the Riemann matrix is within a precision of the order δ. This is important, since some of the Riemann matrices in higher dimensions are only known numerically.

5. Examples in higher genus

We perform similar tests with matrices in higher genus such as, for example, the period matrices of hyperelliptic curves of arbitrary genus given by [**29**]. In this paper we only show examples corresponding to the curve

$$\{(x,y) \in \mathbb{C}^2 \quad | \quad y^2 = x(x^{2g+1} - 1)\}, \tag{5.1}$$

but similar results are obtained with the period matrices of other hyperelliptic curves. The period matrix of (5.1) is

$$\mathbb{B}_{jk} = 1 + \frac{1}{\tau_1} \sum_{l=1}^{j} \tau_l \tau_{k-j+l} \quad \text{for } 1 \leq j \leq k \leq g, \tag{5.2}$$

where $\tau_1 = (-1)^g \zeta_{2g+1}^{g^2}$ with $\zeta_m^n = \exp(2\pi i m/n)$ and

$$\tau_{j+1} = \frac{\tau_1}{1+\zeta_{2g+1}^j}\left(1 - \sum_{l=2}^{j} \zeta_{2g+1}^{g-j+l-1} \tau_l \tau_{j-l+2}\right), \quad j = 1, \dots, g-1.$$

We use these matrices for tests up to $g = 7$, but we also consider period matrices of non-hyperelliptic curves. For $g = 6$ the Fermat curve

$$\{[X : Y : Z] \in \mathbb{P}^2 \quad | \quad X^m + Y^m = Z^m\}, \tag{5.3}$$

with $m = 5$, provides an example. With the numerical approach [**16**], we get for the Fermat curve (5.3) after applying the algorithm [**17**] the Riemann matrix

```
RieMat =

Columns 1 through 4

-0.3735 + 0.9276i -0.3574 + 0.4580i -0.4578 + 0.3092i -0.2891 + 0.3705i
-0.3574 + 0.4580i 0.1365 + 1.0006i -0.0161 + 0.4697i 0.1104 - 0.1415i
-0.4578 + 0.3092i -0.0161 + 0.4697i 0.3474 + 1.0079i -0.2630 - 0.3894i
-0.2891 + 0.3705i 0.1104 - 0.1415i -0.2630 - 0.3894i -0.3152 + 1.1305i
0.0905 + 0.4390i -0.4616 + 0.4201i 0.3635 + 0.5382i -0.3735 - 0.2479i
-0.4417 - 0.1605i -0.3313 - 0.3020i 0.2891 - 0.3705i -0.1725 + 0.0496i

Columns 5 through 6

0.0905 + 0.4390i -0.4417 - 0.1605i
-0.4616 + 0.4201i -0.3313 - 0.3020i
0.3635 + 0.5382i 0.2891 - 0.3705i
-0.3735 - 0.2479i -0.1725 + 0.0496i
-0.4839 + 1.0692i -0.3796 - 0.0685i
-0.3796 - 0.0685i -0.4095 + 0.8023i.
```

For this curve, we could again compute an Abel map, but we are interested also in perturbations of this Riemann matrix not in the Jacobi locus.
Analogously to (4.4), we add diagonal perturbations of the form $\mathbb{B}_s = \mathbb{B} + s \cdot \mathrm{diag}[2, \dots, g+1]$ with $s \in [-1/2, 1/2]$. As in the previous example, we observe several orders of magnitude of difference between the case $s = 0$ and the cases $s \neq 0$, as shown in Figure 5. Although a similar behavior to Figure 4 is to be expected for small values of s.

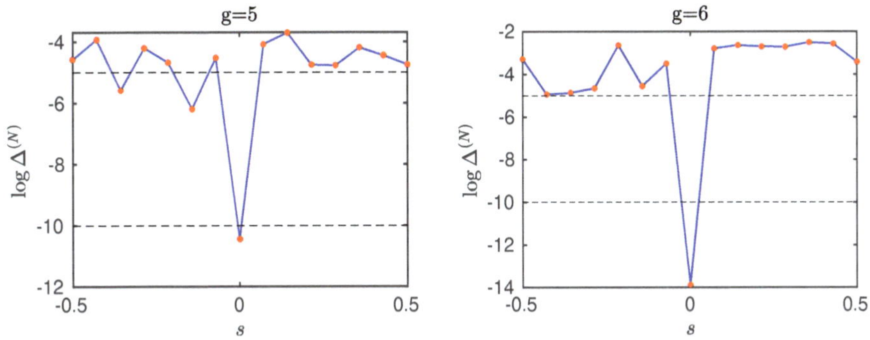

FIGURE 5. Minimum $\Delta^{(N)}$ obtained for the matrices $\mathbb{B}_s \in \mathbb{H}^g$ for $g = 5$ and $g = 6$.

The Fricke-Macbeath curve [**18, 24**] is a curve of genus 7 with the maximal number $84(g-1) = 504$ of automorphisms. It can be defined via the algebraic curve

$$(5.4) \quad \{(x,y) \in \mathbb{C}^2 \quad | \quad 1 + 7yx + 21y^2x^2 + 35x^3y^3 + 28x^4y^4 + 2x^7 + 2y^7 = 0\}.$$

After applying the algorithm from [**17**], the code [**16**] leads to the Riemann matrix

```
RieMat =

Columns 1 through 4

0.3967 + 1.0211i 0.0615 - 0.1322i 0.0000 - 0.0000i 0.4609 + 0.2609i
0.0615 - 0.1322i 0.3967 + 1.0211i -0.3553 + 0.5828i 0.3386 - 0.1933i
0.0000 - 0.0000i -0.3553 + 0.5828i 0.2894 + 1.1656i 0.0905 + 0.2450i
0.4609 + 0.2609i 0.3386 - 0.1933i 0.0905 + 0.2450i 0.3967 + 1.0211i
-0.3553 + 0.5828i -0.4776 + 0.1287i -0.4776 + 0.1287i -0.4776 + 0.1287i
-0.1838 - 0.3219i -0.2743 - 0.5669i 0.3871 - 0.3736i 0.0167 - 0.3895i
0.3386 - 0.1933i 0.3386 - 0.1933i -0.1223 - 0.4541i 0.0615 - 0.1322i

Columns 5 through 7

-0.3553 + 0.5828i -0.1838 - 0.3219i 0.3386 - 0.1933i
-0.4776 + 0.1287i -0.2743 - 0.5669i 0.3386 - 0.1933i
-0.4776 + 0.1287i 0.3871 - 0.3736i -0.1223 - 0.4541i
-0.4776 + 0.1287i 0.0167 - 0.3895i 0.0615 - 0.1322i
0.2894 + 1.1656i -0.1671 - 0.7115i 0.0905 + 0.2450i
-0.1671 - 0.7115i 0.4414 + 1.2784i -0.3386 + 0.1933i
0.0905 + 0.2450i -0.3386 + 0.1933i 0.3967 + 1.0211i.
```

The accuracy of this matrix is estimated to be of the order of 10^{-10}.

Fast convergence when $\mathbf{x}^{(n)}$ falls into a basin of attraction is still observed in higher genus. For example, Figure 6 shows the convergence of the residual of F corresponding to the period matrix of Fermat curve for $g = 6$ and the period matrix of the Fricke curve for $g = 7$. The dimension of the domain of F increases linearly with g, thus it might take more steps for the iteration to find a basin of attraction, but when it does, fast convergence is assured.

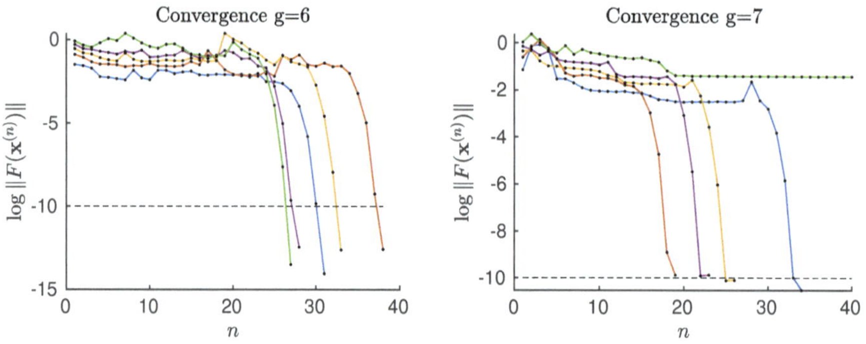

FIGURE 6. Residual of the function F in dependence of the n-th iteration.

It is worth mentioning that the convergence is slower if the matrix $\mathbb{B}$ corresponds to the hyperelliptic example, as we can see in Figure 7 for the period matrices of the curve (5.1) with $g = 6$ and $g = 7$. This might be due to the extra symmetries in their matrix elements, which makes the convergence slower. However, these are just special cases amongst all the possible matrices in $\mathbb{H}^g$.

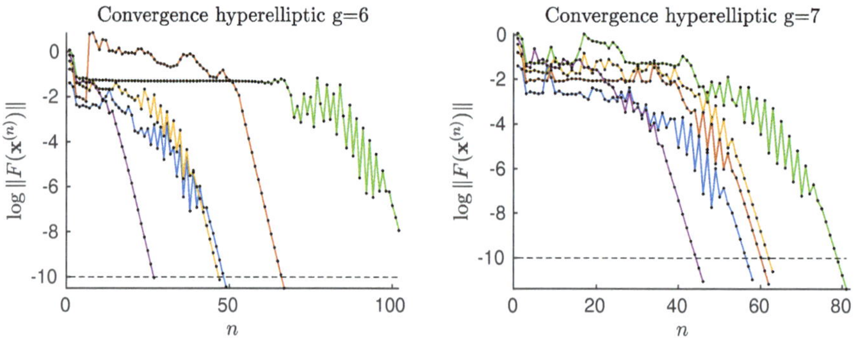

FIGURE 7. Residual of the function F in dependence of the n-th iteration.

The computation per step becomes more expensive as g increases since the number of summands in the approximation of every theta function is $(2\mathcal{N}_\delta + 1)^g$. Besides that, we need to compute 2^g even and 6 odd theta functions in each step. This increases the computational time of the whole algorithm. As a reference, the total time of the tests for the period matrices of non-hyperelliptic curves with $g = 4, 5, 6$ presented in Figures 1 and 6 consisting of 5 iterative processes are

g	time
4	6 s
6	28 min
7	19 h

.

However, since it suffices finding only one zero, the computations can also be terminated as soon as we find a vector with $\|F(\mathbf{x}^{(n)})\| < \delta$, thus reducing these times. On the other hand, the total computational time of the five iterative processes for the period matrices of hyperelliptic curves are

g	time
4	14 s
5	98 s
6	56 min
7	30 h

6. Outlook

In this paper, we have shown how the Fay identity can be used to identify the Jacobi locus for small genera in a computational context. Concretely, the Fay identity in the form (1.5) is considered for a given Riemann matrix $\mathbb{B}$ and its fulfillment is probed by using a Newton iteration. Four components of the vectors X, Y and Z are fixed in order to avoid convergence to a trivial solution where two or more of the vectors coincide. This leads to an iteration to find the zeros of a function F depending on the vector $\mathbf{x}$ with $3g-4$ complex components. The starting point for the iteration is the choice (3.8).

The approach uses a finite precision δ, and the answer to whether a matrix is in the Jacobi locus is thus given with this accuracy. In this paper, we have only studied the case of so-called double precision leading in practice to a $\delta \sim 10^{-12}$. However, the code is set up in a way that inclusion of multi-precision packages is straightforward allowing essentially arbitrary precision.

We have considered in this paper examples up to $g = 7$. This limit is imposed by the size of memory on the used standard computers for the computation of the theta functions. Since the algorithm [**17**] allows the computation of theta functions and their derivatives with a truncation parameter $\mathcal{N}_\delta = 5$ in double precision, these computations are very efficient. In order to take advantage of Matlab's vectorization algorithms, the exponentials of the theta function (2.6) are stored in one array with 11^g components. If one wants to go to higher values of g, the memory limitations can be avoided by not storing these values and applying loops to compute them. The Jacobian matrix in (3.5) is of complex dimension $2^g \times (3g-4)$, but its computation is also fully parallelizable. It will be the subject of further research which genera can be reasonably studied with this approach.

We note that the algorithm produces as a by-product three vectors being linear combinations of the Abel maps of four points if the Riemann matrix is identified to be in the Jacobi locus. The resulting vectors are just auxiliary to test whether the Fay identity is satisfied. It is an interesting question whether they can be used to identify the defining algebraic curves in a similar way as in [**1**].

Appendix A. Algorithm

Algorithm 1 Algorithm to find trisecant points

```
procedure TRISECANT(𝔹)          ▷ Find one trisecant in the Kummer variety of X_𝔹
    𝔹 ← siegeltrans(𝔹)                              ▷ Find Siegel's transform of 𝔹
    δ ← 10^-10                                                 ▷ Set precision δ
    ℓ ← 0.1
    Δℓ ← 0.1
    max_iter ← 100                              ▷ Set maximum number of iterations
    Δ_N ← [ ]
    while ℓ ≤ 0.5 do
        X ← ℓ/2 · (e_{g-2} + 𝔹 * e_{g-2})
        Y ← ℓ/2 · (e_{g-1} + 𝔹 * e_{g-1})
        Z ← ℓ/2 · (e_g + 𝔹 * e_g)
        x ← [X(3 : g); Y(2 : g); Z(2 : g)]                        ▷ Initial vector
        V ← [X(1 : 2); Y(1); Z(1)]               ▷ Fixed components X_1, X_2, Y_1, Z_1
        [F, JF, x, res, Δ] ← fay(x, V, 𝔹)
        ε ← −JF\F
        iter ← 1
        while ||ε|| > δ do                                      ▷ Newton's iteration
            x ← x + ε
            [F, JF, x, res, Δ] ← fay(x, W, 𝔹)
            ε ← −JF\F
            if res < δ or iter > max_iter then
                End Newton's iteration
            end if
            iter ← iter + 1
        end while
        Δ_N ← [Δ_N, Δ]                            ▷ Store the smallest Δ for each ℓ
        Δ_min ← min(Δ_N)
        if Δ_min < δ then
            End algorithm
        end if
        ℓ ← ℓ + Δℓ
    end while
    return res, Δ_min
end procedure
```

The following is the routine computing the function $F : W \subset \mathbb{C}^{3g-4} \to \mathbb{C}^{2^g}$, where the fixed components X_1, X_2, Y_1, Z_1 are entered as parameters.

Algorithm 2 Algorithm to compute the function F

procedure FAY($\mathbf{x}, V, \mathbb{B}$)
  $X \leftarrow [V(1:2); \mathbf{x}(1:g-2)]$
  $Y \leftarrow [V(3); \mathbf{x}(g-1:2g-3)]$
  $Z \leftarrow [V(4); \mathbf{x}(2g-2:2g-4)]$
  $X \leftarrow [[X]]$ ▷ Keep X, Y, Z in the fundamental domain
  $Y \leftarrow [[Y]]$
  $Z \leftarrow [[Z]]$
  $\mathbf{x} \leftarrow [X(3:g); Y(2:g); Z(2:g)]$
  $[\lambda_1, \nabla_Y \lambda_1, \nabla_Z \lambda_1] \leftarrow \mathtt{lambdafun}(Y, Z, \mathbb{B})$
  $[\lambda_2, \nabla_Z \lambda_2, \nabla_X \lambda_2] \leftarrow \mathtt{lambdafun}(Z, X, \mathbb{B})$
  $[\lambda_3, \nabla_X \lambda_3, \nabla_Y \lambda_3] \leftarrow \mathtt{lambdafun}(X, Y, \mathbb{B})$
  $[KX, JKX] \leftarrow \mathtt{kummer}(X, \mathbb{B})$
  $[KY, JKY] \leftarrow \mathtt{kummer}(Y, \mathbb{B})$
  $[KZ, JKZ] \leftarrow \mathtt{kummer}(Z, \mathbb{B})$
  $F \leftarrow KX + \frac{\lambda_2}{\lambda_1} \cdot KY + \frac{\lambda_3}{\lambda_1} \cdot KZ$
  $J_X F \leftarrow JKX + \frac{1}{\lambda_1} \cdot KY * \nabla_X \lambda_2 + \frac{1}{\lambda_1} \cdot KZ * \nabla_X \lambda_3$
  $J_Y F \leftarrow \frac{\lambda_2}{\lambda_1} \cdot JKY + \frac{1}{\lambda_1} \cdot KZ * \nabla_Y \lambda_3 - \frac{1}{(\lambda_1)^2} \cdot [\lambda_2 \cdot KY + \lambda_3 \cdot KZ] * \nabla_Y \lambda_1$
  $J_Z F \leftarrow \frac{1}{\lambda_1} \cdot KY * \nabla_Z \lambda_2 + \frac{\lambda_3}{\lambda_1} \cdot JKZ - \frac{1}{(\lambda_1)^2} \cdot [\lambda_2 \cdot KY + \lambda_3 \cdot KZ] * \nabla_Z \lambda_1$
  $JF = [J_X F, J_Y F, J_Z F]$
  $JF(:, [1, 2, g+1, 2g+1]) \leftarrow [\ \ ]$ ▷ Deleting columns corresponding to X_1, X_2, Y_1, Z_1
  res $\leftarrow \|F\|$ ▷ Residual of F
  $\Delta \leftarrow \min(\mathrm{svd}(KX, KY, KZ))$ ▷ Linear dependence of $KX\ KY$, KZ
  return $F, JF, \mathbf{x}$, res, Δ
end procedure

Here $\mathtt{kummer}(X, \mathbb{B})$ is the routine computing $\vec{\Theta}(X)$ (which is expressed as a column vector) and its Jacobian matrix $J\vec{\Theta}(X)$. The operation '$*$' is the matrix multiplication and '$\cdot$' is the multiplication by a scalar.

Algorithm 3 Algorithm to compute the Kummer map

procedure KUMMER($X, \mathbb{B}$)
  for $I = 1 : 2^g$ **do**
    Set characteristic ϵ_I ▷ $\epsilon_I \in \{0,1\}^g$ in column vector form
    $factor \leftarrow \exp\left(\frac{1}{2}\pi\mathrm{i}\langle \epsilon_I, \mathbb{B} * \epsilon_I \rangle + 2\pi\mathrm{i}\langle \epsilon_I, X \rangle\right)$
    $[t, \nabla t] \leftarrow \mathtt{thetagrad}(2X + \mathbb{B} * \epsilon_I, 2\mathbb{B}, 0)$
    $K(I) \leftarrow factor \cdot t$
    $JK(I, :) \leftarrow 2 \cdot factor \cdot (\nabla t + \pi\mathrm{i} \cdot t \cdot \epsilon_I^\top)$
  end for
  return K, JK
end procedure

The routine $\texttt{thetagrad}(X, \mathbb{B}, C)$ computes the multidimensional theta function $\Theta[C](X, \mathbb{B})$ defined in (2.6) with characteristic $C = \left[\begin{smallmatrix} \mathrm{p} \\ \mathrm{q} \end{smallmatrix}\right]$, as well as its gradient $\nabla\Theta[C](X, \mathbb{B})$ (which is expressed as a row vector) through (3.4). We used the relation (2.7) to compute the even theta functions more efficiently in terms of the zero-characteristic theta function, since in practice we enter the array containing $\langle \mathrm{N}, \mathbb{B}\mathrm{N}\rangle$ as an input, rather than computing $\langle(\mathrm{N} + \epsilon_I/2), \mathbb{B}(\mathrm{N} + \epsilon_I/2)\rangle$ for every $\epsilon_I \in Z^g/2\mathbb{Z}^g$, which would take up more memory space. Finally, the routine to compute the function $\lambda(a, b) := \Theta^*(a+b, \mathbb{B})\Theta^*(a-b, \mathbb{B})$ with and odd characteristic $2\delta_o \in \mathbb{Z}^{2g}/2\mathbb{Z}^{2g}$, as well as its gradients $\nabla_a\lambda(a, b)$, $\nabla_b\lambda(a, b)$ is the following:

Algorithm 4 Algorithm to compute the lambda coefficients

```
procedure LAMBDAFUN(a, b, B)
    δ_o ← [e_1/2; e_1/2]                    ▷ Default odd characteristic
    [t_p, ∇t_p] ← thetagrad(a + b, B, δ_o)
    [t_m, ∇t_m] ← thetagrad(a − b, B, δ_o)
    λ ← t_p · t_m
    ∇_a λ ← t_m · ∇t_p + t_p · ∇t_m
    ∇_b λ ← t_m · ∇t_p − t_p · ∇t_m
    return λ, ∇_a λ, ∇_b λ
end procedure
```

As before, since we use the same odd characteristic throughout the whole algorithm, we enter $\langle(\mathrm{N} + e_1/2), \mathbb{B}(\mathrm{N} + e_1/2)\rangle$ as an input in the computation of the odd theta function $\Theta^*(\mathrm{z}, \mathbb{B})$ and its gradient.

References

[1] D. Agostini, T. Ö. Çelik, and D. Eken, *Numerical reconstruction of curves from their Jacobians*, Arithmetic, geometry, cryptography, and coding theory 2021, Contemp. Math., vol. 779, Amer. Math. Soc., Providence, RI, 2022, pp. 1–12, DOI 10.1090/conm/779/15667. MR4445767

[2] D. Agostini and L. Chua, *Computing theta functions with Julia*, J. Softw. Algebra Geom. **11** (2021), no. 1, 41–51, DOI 10.2140/jsag.2021.11.41. MR4285763

[3] E. Arbarello and C. De Concini, *On a set of equations characterizing Riemann matrices*, Ann. of Math. (2) **120** (1984), no. 1, 119–140, DOI 10.2307/2007073. MR750718

[4] E.D. Belokolos, A.I. Bobenko, V.Z. Enolskii, A.R. Its, V.B. Matveev. Algebro-geometric approach to nonlinear integrable equations. Springer, Berlin (1994)

[5] C. Birkenhake and H. Lange, *Complex abelian varieties*, 2nd ed., Grundlehren der mathematischen Wissenschaften [Fundamental Principles of Mathematical Sciences], vol. 302, Springer-Verlag, Berlin, 2004, DOI 10.1007/978-3-662-06307-1. MR2062673

[6] H. W. Braden and T. P. Northover, *Bring's curve: its period matrix and the vector of Riemann constants*, SIGMA Symmetry Integrability Geom. Methods Appl. **8** (2012), Paper 065, 20, DOI 10.3842/SIGMA.2012.065. MR2988029

[7] A. I. Bobenko and C. Klein (eds.), *Computational approach to Riemann surfaces*, Lecture Notes in Mathematics, vol. 2013, Springer, Heidelberg, 2011, DOI 10.1007/978-3-642-17413-1. MR2920500

[8] L. Chua, M. Kummer, and B. Sturmfels, *Schottky algorithms: classical meets tropical*, Math. Comp. **88** (2019), no. 319, 2541–2558, DOI 10.1090/mcom/3406. MR3957905

[9] B. Deconinck and M. van Hoeij, *Computing Riemann matrices of algebraic curves*, Phys. D **152/153** (2001), 28–46, DOI 10.1016/S0167-2789(01)00156-7. Advances in nonlinear mathematics and science. MR1837895

[10] B. Deconinck, M. Heil, A. Bobenko, M. van Hoeij, and M. Schmies, *Computing Riemann theta functions*, Math. Comp. **73** (2004), no. 247, 1417–1442, DOI 10.1090/S0025-5718-03-01609-0. MR2047094

[11] R. Donagi, *Big Schottky*, Invent. Math. **89** (1987), no. 3, 569–599, DOI 10.1007/BF01388986. MR903385

[12] B. A. Dubrovin, *Theta-functions and nonlinear equations* (Russian), Uspekhi Mat. Nauk **36** (1981), no. 2(218), 11–80. MR616797

[13] H. M. Farkas, S. Grushevsky, and R. Salvati Manni, *An explicit solution to the weak Schottky problem*, Algebr. Geom. **8** (2021), no. 3, 358–373, DOI 10.14231/ag-2021-009. MR4206440

[14] H. M. Farkas and I. Kra, *Riemann surfaces*, Graduate Texts in Mathematics, vol. 71, Springer-Verlag, New York-Berlin, 1980. MR583745

[15] J. D. Fay, *Theta functions on Riemann surfaces*, Lecture Notes in Mathematics, Vol. 352, Springer-Verlag, Berlin-New York, 1973. MR335789

[16] J. Frauendiener and C. Klein, *Computational approach to compact Riemann surfaces*, Nonlinearity **30** (2017), no. 1, 138–172, DOI 10.1088/1361-6544/30/1/138. MR3604607

[17] J. Frauendiener, C. Jaber, and C. Klein, *Efficient computation of multidimensional theta functions*, J. Geom. Phys. **141** (2019), 147–158, DOI 10.1016/j.geomphys.2019.03.011. MR3937404

[18] R. Fricke, *Ueber eine einfache Gruppe von 504 Operationen* (German), Math. Ann. **52** (1899), no. 2-3, 321–339, DOI 10.1007/BF01476163. MR1511059

[19] S. Grushevsky, *The Schottky problem*, Current developments in algebraic geometry, Math. Sci. Res. Inst. Publ., vol. 59, Cambridge Univ. Press, Cambridge, 2012, pp. 129–164. MR2931868

[20] S. Grushevsky and M. Möller, *Explicit formulas for infinitely many Shimura curves in genus 4*, Asian J. Math. **22** (2018), no. 2, 381–390, DOI 10.4310/ajm.2018.v22.n2.a12. MR3824574

[21] J.-i. Igusa, *On the irreducibility of Schottky's divisor*, J. Fac. Sci. Univ. Tokyo Sect. IA Math. **28** (1981), no. 3, 531–545 (1982). MR656035

[22] I. Krichever, *Characterizing Jacobians via trisecants of the Kummer variety*, Ann. of Math. (2) **172** (2010), no. 1, 485–516, DOI 10.4007/annals.2010.172.485. MR2680424

[23] A. K. Lenstra, H. W. Lenstra Jr., and L. Lovász, *Factoring polynomials with rational coefficients*, Math. Ann. **261** (1982), no. 4, 515–534, DOI 10.1007/BF01457454. MR682664

[24] A. M. Macbeath, *On a curve of genus 7*, Proc. London Math. Soc. (3) **15** (1965), 527–542, DOI 10.1112/plms/s3-15.1.527. MR177342

[25] H. Minkowski, *Über die positiven quadratischen Formen und über kettenbruchähnliche Algorithmen*, J. Reine Angew. Math. **107** (1891), 278–297, DOI 10.1515/crelle-1891-189110721. MR1581684

[26] H. Minkowski. Gesammelte Abhandlungen 1, pp. 145-148, 153-156, 217-218 (Leipzig-Berlin: Teubner 1911).

[27] D. Mumford, *Tata lectures on theta. I*, Progress in Mathematics, vol. 28, Birkhäuser Boston, Inc., Boston, MA, 1983, DOI 10.1007/978-1-4899-2843-6. MR688651

[28] H. E. Rauch and H. M. Farkas, *Theta functions with applications to Riemann surfaces*, Williams & Wilkins Co., Baltimore, MD, 1974. MR352108

[29] B. Schindler, *Period matrices of hyperelliptic curves*, Manuscripta Math. **78** (1993), no. 4, 369–380, DOI 10.1007/BF02599319. MR1208647

[30] F. Schottky, *Zur Theorie der Abelschen Functionen von vier Variabeln* (German), J. Reine Angew. Math. **102** (1888), 304–352, DOI 10.1515/crll.1888.102.304. MR1580147

[31] T. Shiota, *Characterization of Jacobian varieties in terms of soliton equations*, Invent. Math. **83** (1986), no. 2, 333–382, DOI 10.1007/BF01388967. MR818357

[32] C. L. Siegel, *Topics in complex function theory. Vol. III: Abelian functions and modular functions of several variables*, Wiley Classics Library, John Wiley & Sons, Inc., New York, 1989. A Wiley-Interscience Publication. MR1013364

[33] C. Swierczewski and B. Deconinck, *Computing Riemann theta functions in Sage with applications*, Math. Comput. Simulation **127** (2016), 263–272, DOI 10.1016/j.matcom.2013.04.018. MR3501304

[34] I. A. Taĭmanov, *Secants of abelian varieties, theta functions and soliton equations* (Russian), Uspekhi Mat. Nauk **52** (1997), no. 1(313), 149–224, DOI 10.1070/RM1997v052n01ABEH001743; English transl., Russian Math. Surveys **52** (1997), no. 1, 147–218. MR1453569

[35] G. E. Welters, *A criterion for Jacobi varieties*, Ann. of Math. (2) **120** (1984), no. 3, 497–504, DOI 10.2307/1971084. MR769160

INSTITUT DE MATHÉMATIQUES DE BOURGOGNE, UNIVERSITÉ DE BOURGOGNE, 9 AVENUE ALAIN SAVARY, 21078 DIJON CEDEX, FRANCE

Current address: Technische Universität München, Boltzmannstr. 3, 85748 Garching b. München

Email address: eddybrandon11@hotmail.com

DEPARTMENT OF MATHEMATICS AND STATISTICS, UNIVERSITY OF OTAGO, P.O. BOX 56, DUNEDIN 9010, NEW ZEALAND

Email address: joerg.frauendiener@otago.ac.nz

INSTITUT DE MATHÉMATIQUES DE BOURGOGNE, UNIVERSITÉ DE BOURGOGNE, 9 AVENUE ALAIN SAVARY, 21078 DIJON CEDEX, FRANCE AND INSTITUT UNIVERSITAIRE DE FRANCE

Email address: Christian.Klein@u-bourgogne.fr

Contemporary Mathematics
Volume **823**, 2025
https://doi.org/10.1090/conm/823/16498

On strong zero-dispersion asymptotics for Benjamin-Ono soliton ensembles

Elliot Blackstone, Louise Gassot, and Peter D. Miller

To the memory of Igor Krichever, for the enduring legacy of his scientific influence.

ABSTRACT. A soliton ensemble is a particular kind of approximation of the solution of an initial-value problem for an integrable equation by a reflectionless potential that is well adapted to singular asymptotics like the small-dispersion limit. We study soliton ensembles for the Benjamin-Ono equation by using reasonable hypotheses to develop local approximations that capture highly oscillatory features of the solution and hence provide more information than weak convergence results that are easier to obtain. These local approximations are deduced independently from empirically-observed but unproven distributions of eigenvalues of two related matrices, one Hermitian and another non-Hermitian. We perform careful numerical experiments to study the asymptotic behavior of the eigenvalues of these matrices in the small-dispersion limit, and formulate conjectures reflecting our observations. Then we apply the conjectures to construct the local approximations of slowly varying profiles and rapidly oscillating profiles as well. We show that the latter profiles are consistent with the predictions of Whitham modulation theory as originally developed for the Benjamin-Ono equation by Dobrokhotov and Krichever.

1. Introduction

The Benjamin-Ono (BO) equation [**Ben67, Ono75**] is given by

$$u_t + 2uu_x + \epsilon\mathcal{H}[u_{xx}] = 0,\ x \in \mathbb{R},\ t > 0, \tag{1.1}$$

2020 *Mathematics Subject Classification.* Primary 35R09, 35C20; Secondary 35Q53, 35C08, 35C07.

Key words and phrases. Benjamin-Ono equation, small-dispersion limit.

The first author was supported in part by NSF Grant DMS-1812625.

The second author's work was conducted within the France 2030 framework program, the Centre Henri Lebesgue ANR-11-LABX-0020-01.

The third author was partially supported by the National Science Foundation under grants DMS-1812625 and DMS-2204896, and some of this work was done with the support of a Leverhulme Trust Visiting Professorship at Bristol University, UK..

The authors wish to thank Alejandro Uribe, Alix Deleporte, and San Vũ Ngọc for useful discussions.

where $\epsilon \geq 0$ is a parameter and $\mathcal{H}$ is the classical Hilbert transform[1]

$$\mathcal{H}[f](x) = \frac{1}{\pi}\text{P.V.}\int_{-\infty}^{\infty} \frac{f(y)}{y-x}\,\mathrm{d}y. \tag{1.2}$$

The *zero-dispersion limit* refers to the analysis of the solution $u = u(x,t)$ of the Cauchy problem for (1.1) where $u(x,0) = u_0(x)$ is an initial condition independent of ϵ. For convenience, we will assume that u_0 is an admissible initial condition in the sense of [**MX11**, Definition 3.1].

The BO equation (1.1) is an asymptotic model, derived in a small-amplitude and long-wave limit, for internal water waves propagating in one direction. It applies to gravity-driven motions of the pycnocline separating a lower-density upper fluid layer from a higher-density lower fluid layer in the situation that the lower layer is assumed to be infinitely deep. The solution $u(x,t)$ is a measure of the vertical displacement of the interface at position x and time t. The parameter $\epsilon > 0$ is a measure of the relative strength of dispersion compared to nonlinear effects. When $\epsilon \ll 1$, the local solution of the Cauchy problem for the inviscid Burgers equation

$$u_t^{\mathrm{B}} + 2u^{\mathrm{B}}u_x^{\mathrm{B}} = 0 \tag{1.3}$$

with the same initial data $u^{\mathrm{B}}(x,0) = u_0(x)$ is expected to be a good approximation of $u(x,t)$ as long as $u^{\mathrm{B}}(x,t)$ remains smooth. The solution of (1.3) satisfying $u^{\mathrm{B}}(x,0) = u_0(x)$ is given implicitly by

$$u^{\mathrm{B}}(x,t) = u_0(x - 2tu^{\mathrm{B}}(x,t)). \tag{1.4}$$

However, for typical initial data u_0, a gradient catastrophe occurs in $u^{\mathrm{B}}(x,t)$ at a finite time $t = t_b$, beyond which the dispersion term in (1.1) can no longer be neglected. See Figure 1, right panel, for solutions of Burgers' equation before and after $t = t_b$. Numerical experiments show that the effect of small dispersion is to generate a dispersive shock wave, that is, a train of waves with wavelength proportional to ϵ but with amplitude that is not small. See Figure 2 below. Such waves are described at a formal level by Whitham modulation theory [**Whi99**]. That theory starts from a family of exact solutions of (generally multiphase) waves parameterized by arbitrary constant amplitude, wavelength, and wave-shape parameters, and based on multiphase averaging of local conservation laws or a variational principle, posits a system of modulation equations which are partial differential equations governing slowly-varying fields replacing the constant parameters.

For the BO equation, nonlocality makes it challenging both to properly define the multiphase wave solutions and to determine their modulation equations. Both of these problems were solved by Igor Krichever in a joint work with Dobrokhotov [**DK91**] (see also the review by Krichever [**Kri91**]) that has had a large and lasting influence on the subject. One of the remarkable observations in that work is that there exist Riemann invariant variables for the modulation equations in which they take the form of a collection of $2P+1$ independent copies of the inviscid Burgers equation (1.3), for some $P = 0, 1, 2, \ldots$ (the number of phases in the modulating wave), see Figure 1, left panel. By contrast, in the Riemann-invariant form of the modulation equations for the Korteweg-de Vries (KdV) equation, one replaces the

[1]This definition follows the "physicist's" sign convention, as used for instance by Matsuno in his textbook on the BO equation [**Mat84**, Eqn. (3.2)]. However, more mathematical works (e.g., [**Gé24**]) define the Hilbert transform with the opposite sign.

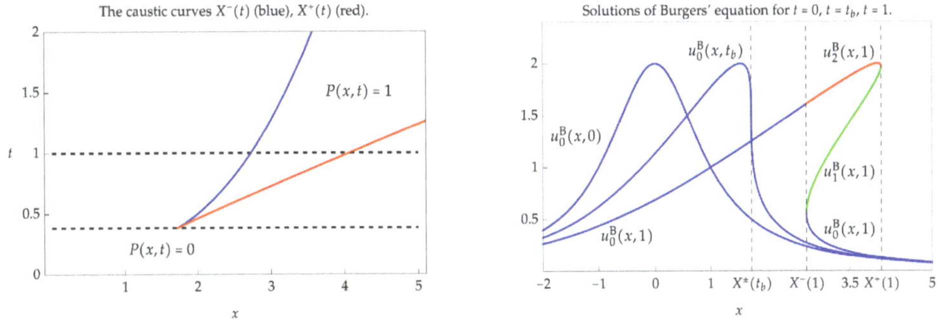

FIGURE 1. The single and triple-valued region for Burgers' equation corresponding to $u_0(x) = 2(1+x^2)^{-1}$ are separated by the caustic curves $X^-(t)$, $X^+(t)$ for $t \geq t_b$, see left panel. The curves $X^{\pm}(t)$ are the double roots of (1.4). In the right panel, the solutions of Burgers' equation are plotted for $t = 0, t_b, 1$. The blue, green, red curves are $u_0^{\mathrm{B}}(x,t)$, $u_1^{\mathrm{B}}(x,t)$, $u_2^{\mathrm{B}}(x,t)$, respectively. COLOR ONLINE.

characteristic speed of a Riemann invariant by a complicated expression involving all of the invariants and represented as a ratio of hyperelliptic integrals, see [**FFM80**].

One approach to describing the asymptotic behavior of $u(x,t)$ is to adopt a reasonable topology of convergence in which there is a limiting function, denoted $\overline{u}(x,t)$, as $\epsilon \to 0$. Following the seminal work of Lax and Levermore [**LL83**] on a corresponding small-dispersion limit for the KdV equation, there has been some progress in proving convergence of $u(x,t)$ (or a suitable surrogate, see below) to a limit $\overline{u}(x,t)$ in the weak $L^2(\mathbb{R})$ topology with respect to x, uniformly on compact intervals of t. This has been done both in the setting of $x \in \mathbb{R}$ [**MX11**] and on the torus (periodic boundary conditions) [**Gas23b**]. The results of [**MX11**] have recently been strengthened by Gérard [**Gé24**] using a different approach. In both the periodic and non-periodic cases, the formula for the weak limit $\overline{u}(x,t)$ is remarkably simple. Indeed, let $u^{\mathrm{B}} = u_n^{\mathrm{B}}(x,t)$, $n = 0, \ldots, 2P$, $u_m^{\mathrm{B}}(x,t) \leq u_n^{\mathrm{B}}(x,t)$ for $m < n$, denote the generically distinct solutions of the implicit equation (1.4), which correspond to the "sheets" above a given point $(x,t) \in \mathbb{R}^2$ of the multi-valued solution of the inviscid Burgers equation (1.3) with initial data $u^{\mathrm{B}}(x,0) = u_0(x)$ (the number P depends on (x,t)), see Figure 1. Then the weak limit of $u(x,t)$ is given by the alternating sum

$$\overline{u}(x,t) = \sum_{n=0}^{2P} (-1)^n u_n^{\mathrm{B}}(x,t). \tag{1.5}$$

This result is far simpler than the corresponding result for KdV obtained in [**LL83**]. The weak convergence of $u(x,t)$ can be extended [**MX12**] to an infinite number of conserved "local" densities (the differential algebra of fields has to be augmented with Hilbert transforms), but even with this additional control, the topology of convergence is insufficient to capture the waveform and phase of the dispersive shock wave that forms in $u(x,t)$ for $t > t_b$. Indeed, the wild oscillations are simply averaged out upon integration in x against a test function in $L^2(\mathbb{R})$. The formula

(1.5) suggests that the oscillations occupy the part of the (x,t)-plane where $P = P(x,t) \geq 1$, and numerical experiments support this assertion as well. Moreover, the dispersive shock wave is expected to be described by the one-phase solution (assuming $P = 1$). See [**KS21**, Section 2.6] for an overview of corresponding results in the context of the KdV equation.

The surrogate for $u(x,t)$ that was mentioned above is an approximation that we will call in this paper a *soliton ensemble*. It is a family denoted $\tilde{u}(x,t)$ of exact solutions of the BO equation (1.1) associated to the specified function $x \mapsto u_0(x)$ via a systematic small-ϵ approximation of the scattering data for the BO Lax operator with potential u_0. In this approximation, one firstly replaces the true discrete eigenvalues $\lambda < 0$ with approximate ones obtained from a type of quantization rule (see (1.15) below), and makes a similar approximation of the auxiliary phase constant γ associated with each eigenvalue (see (1.16) below). Secondly, one neglects the reflection coefficient defined for $\lambda > 0$. The exact solution of BO corresponding to this modified scattering data for each $\epsilon > 0$ is precisely the function $\tilde{u}(x,t)$ (see (1.18) below). It is a "nonlinear superposition" of a large number, proportional to ϵ^{-1}, of solitons, which combine coherently to yield an approximation of the given data u_0 in the strong L^2 sense when $t = 0$, see [**MX11**, Corollary 1.2]. The idea of using such a soliton ensemble to analyze the small-dispersion limit originated in the Lax-Levermore theory of the KdV equation [**LL83**]. In both the BO and KdV cases, the soliton ensemble $\tilde{u}(x,t)$ is expressed explicitly in terms of finite determinants of size proportional to ϵ^{-1}. While soliton ensembles have proven to be useful to analyze the small-dispersion limit of the BO equation in the sense of weak convergence, in this paper we will study them from the point of view of strong convergence, aiming to capture the oscillatory profile of the dispersive shock wave.

Soliton ensembles may appear to be similar to soliton gases such as studied in [**GGJM21**], as both are constructed from a large number of solitons. Soliton gases are limiting solutions in which the number of solitons is actually infinite while the limit process in their construction does not tie the number of solitons to a small parameter in the equation. Moreover, soliton gases are not specifically constructed to match any particular initial condition u_0 at $t = 0$. Hence, one can perhaps think of a soliton ensemble as a type of soliton gas consisting of an arbitrarily large but finite number of solitons that is adapted to the small-dispersion limit and designed to approximate the solution of a specific Cauchy problem.

1.1. Definition of soliton ensembles. Define the Cauchy transforms

$$\mathcal{C}_{\pm}[f](x) := \lim_{\delta \to 0^+} \frac{1}{2\pi \mathrm{i}} \int_{-\infty}^{\infty} \frac{f(y)}{y - (x \pm \mathrm{i}\delta)}\, \mathrm{d}y, \tag{1.6}$$

noting that $\pm\mathcal{C}_{\pm}$ are self-adjoint orthogonal projections from $L^2(\mathbb{R})$ onto the Hardy space

$$H^{\pm}(\mathbb{R}) = \left\{ f \text{ holomorphic on } \mathbb{C}^{\pm} : \|f\|_{H^{\pm}}^2 = \sup_{y \in (0,\infty)} \int_{\mathbb{R}} |f(x \pm \mathrm{i}y)|^2\, \mathrm{d}x < \infty \right\},$$

where $\mathbb{C}^{\pm}$ denotes the upper($+$)/lower($-$) half plane. The inverse-scattering transform solution of the Cauchy problem for the BO equation (1.1) is based on the self-adjoint Lax operator

$$\mathcal{L} := -\mathrm{i}\epsilon \frac{\partial}{\partial x} - \mathcal{C}_+ \circ u \circ \mathcal{C}_+ \tag{1.7}$$

acting on a domain dense in $H^+(\mathbb{R})$, wherein u denotes the operator of multiplication by a bounded function $u : \mathbb{R} \to \mathbb{R}$. It has been proved [**Wu16**] that if $u \in L^1(\mathbb{R}) \cap L^\infty(\mathbb{R})$ and $x \mapsto xu(x)$ is in $L^2(\mathbb{R})$, then $\mathcal{L}$ has only finitely many eigenvalues, all negative real numbers $\lambda_1 < \lambda_2 < \cdots < \lambda_N < 0$ with unit geometric multiplicity. To each eigenvalue $\lambda_j < 0$, there corresponds a real number called a phase constant, denoted γ_j. Under suitable conditions on u [**Wu17**] there is also a complex-valued function $\lambda \mapsto \beta(\lambda)$ defined for $\lambda > 0$ called the reflection coefficient. The collection of N pairs (λ_j, γ_j) and the function β are said to constitute the scattering data associated with the initial data $u = u_0$ for (1.1) in the inverse-scattering transform first proposed by Fokas and Ablowitz [**FA83**]. The inversion of the transform has not yet been fully justified, but if $\beta(\lambda) \equiv 0$, the procedure reduces to finite-dimensional linear algebra, resulting in the formula

$$(1.8) \qquad u(x,t) = 2\epsilon\frac{\partial}{\partial x}\mathrm{Im}\left[\log\left(\det\left(\mathbb{I} + \frac{\mathrm{i}}{\epsilon}\mathbf{A}(x,t)\right)\right)\right],$$

valid for any choice of branch of the complex logarithm, where $\mathbb{I}$ denotes the $N \times N$ identity matrix, and $\mathbf{A}(x,t)$ is the $N \times N$ Hermitian matrix with elements

$$(1.9) \qquad A_{jk}(x,t) = \begin{cases} \dfrac{2\mathrm{i}\epsilon\sqrt{\lambda_j\lambda_k}}{\lambda_j - \lambda_k}, & j \neq k, \\ -2\lambda_j(x + 2\lambda_j t + \gamma_j), & j = k. \end{cases}$$

This formula represents a multi-soliton solution of (1.1), and it was first derived by Matsuno [**Mat79**] using Hirota's bilinear method. The inversion formula has then been proved on the N-soliton manifolds in [**Sun21**, Eq. (1.19)].

In two papers [**Mat82, Mat82**], Matsuno used formal arguments to investigate the asymptotic behavior of the scattering data described above for fixed u_0 in the limit $\epsilon \to 0$. The main results of [**Mat82, Mat82**] were asymptotic formulæ for $|\beta(\lambda)|^2$ for $\lambda > 0$ and for the distribution of eigenvalues $\lambda_j < 0$, of which there are in general a large number $N \sim \epsilon^{-1}$. These results were rigorously proven in [**MW16**] under the additional assumption that u_0 is a rational function for which $u_0(x) = -\lambda$ has generically either two or zero real solutions x given $\lambda \in \mathbb{R}$, and for such u_0 an asymptotic formula for the phase constant γ_j associated with a given eigenvalue λ_j was also rigorously established. We now restate these asymptotic results in the special case that u_0 is a smooth positive function with a single critical point (the maximizer) and with sufficient decay as $x \to \pm\infty$. For such u_0, we define positive constants $L, M > 0$ by

$$(1.10) \qquad L := \max_{x\in\mathbb{R}} u_0(x), \quad M := \frac{1}{2\pi}\int_{\mathbb{R}} u_0(x)\,\mathrm{d}x,$$

and for $-L < \lambda < 0$ define the turning points $x = x_\pm(\lambda)$, $x_-(\lambda) < x_+(\lambda)$ as the two roots of $u_0(x) = -\lambda$. Then the following are true:

- the reflection coefficient $\beta(\lambda)$ defined for $\lambda > 0$ vanishes in the limit $\epsilon \to 0$;
- there are $N = M/\epsilon + \mathcal{O}(1)$ eigenvalues in the interval $-L < \lambda < 0$, and the number of eigenvalues with $-L < a < \lambda < b < 0$ is $N(a,b)$ satisfying

$$(1.11) \qquad N(a,b) = \frac{1}{\epsilon}\int_a^b F(\lambda)\,\mathrm{d}\lambda + \mathcal{O}(1),$$

where the density of eigenvalues near λ is $\epsilon^{-1}F(\lambda)$ with

$$F(\lambda) := \frac{1}{2\pi}(x_+(\lambda) - x_-(\lambda)), \quad -L < \lambda < 0; \tag{1.12}$$

- if λ_j is an eigenvalue that converges to a number $\lambda \in (-L, 0)$ as $\epsilon \to 0$, then also $\gamma_j \to \gamma(\lambda)$, where

$$\gamma(\lambda) := -\frac{1}{2}(x_+(\lambda) + x_-(\lambda)), \quad -L < \lambda < 0. \tag{1.13}$$

In the periodic case, analogues of formulas (1.11), (1.12) and (1.13) should also hold [**Mol20**], and have been proved for bell-shaped initial data in [**Gas23a**].

Based on these results, we will now define the soliton ensemble for the BO equation (1.1) associated with a positive initial condition u_0 of the type described above. Firstly, we define the exact number of approximate eigenvalues by setting

$$N(\epsilon) := \left\lfloor \frac{M}{\epsilon} \right\rfloor. \tag{1.14}$$

Then, we define $N(\epsilon)$ approximate eigenvalues $\tilde{\lambda}_j \in (-L, 0)$, $j = 1, \ldots, N(\epsilon)$ by quantizing the density formula (1.12):

$$\int_{-L}^{\tilde{\lambda}_j} F(\lambda)\,\mathrm{d}\lambda = \epsilon\left(j - \frac{1}{2}\right), \quad j = 1, \ldots, N(\epsilon). \tag{1.15}$$

Finally, we define corresponding phase constants by setting

$$\tilde{\gamma}_j := \gamma(\tilde{\lambda}_j), \quad j = 1, \ldots, N(\epsilon), \tag{1.16}$$

where $\gamma(\cdot)$ is defined by (1.13). Then we neglect the reflection coefficient and define an exact multi-soliton solution of the BO equation (1.1) using the approximate discrete data. In detail, first define the elements of the $N(\epsilon) \times N(\epsilon)$ Hermitian matrix $\tilde{\mathbf{A}}(x, t)$ as (compare with (1.9))

$$\tilde{A}_{jk}(x, t) = \begin{cases} \dfrac{2\mathrm{i}\epsilon\sqrt{\tilde{\lambda}_j\tilde{\lambda}_k}}{\tilde{\lambda}_j - \tilde{\lambda}_k}, & j \neq k, \\ -2\tilde{\lambda}_j(x + 2\tilde{\lambda}_j t + \tilde{\gamma}_j), & j = k. \end{cases} \tag{1.17}$$

Then set (compare with (1.8))

$$\tilde{u}(x, t) := 2\epsilon\frac{\partial}{\partial x}\mathrm{Im}\left[\log\left(\det\left(\mathbb{I} + \frac{\mathrm{i}}{\epsilon}\tilde{\mathbf{A}}(x, t)\right)\right)\right]. \tag{1.18}$$

The family of functions $\tilde{u}(x, t)$ parameterized by $\epsilon > 0$ constitute the soliton ensemble for the BO equation (1.1) for the Cauchy data u_0. See Figure 2.

Note that (1.15) implies that

$$\int_{-L}^{\tilde{\lambda}_1} F(\lambda)\,\mathrm{d}\lambda = \frac{\epsilon}{2}. \tag{1.19}$$

So in the sense measured by differences of the left-hand side of (1.15), the first approximate eigenvalue is half the distance to $-L$ as the remaining approximate eigenvalues are to their nearest neighbors. For an argument appearing below in

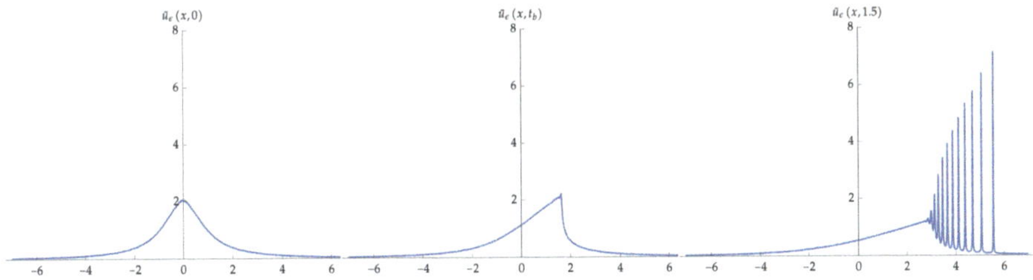

FIGURE 2. Time evolution of $\tilde{u}(x,t)$ for $u_0(x) = 2(1+x^2)^{-1}$ and $\epsilon = 0.02$.

Section 3, we will want to ensure a similar condition for the last approximate eigenvalue, namely that

$$\int_{\tilde{\lambda}_{N(\epsilon)}}^{0} F(\lambda)\,\mathrm{d}\lambda = \frac{\epsilon}{2}. \tag{1.20}$$

It is easy to see that this condition holds precisely when $\epsilon = \epsilon_N := M/N$ for an integer $N \in \mathbb{Z}_{>0}$, in which case (1.11) returns $N(\epsilon_N) = N$. For convenience, we will therefore assume below that ϵ tends to zero within this specific discrete sequence.

Also, since the rest of this paper is concerned only with the ϵ-dependent function $\tilde{u}(x,t)$, henceforth we will drop all tildes for simplicity.

1.2. Alternate formulæ for the soliton ensemble $u(x,t)$. Let $\alpha_k(x,t)$, $k = 1, \dots, N(\epsilon)$, denote the real eigenvalues of $\mathbf{A}(x,t)$. The proof of weak convergence of $u(x,t)$ to $\overline{u}(x,t)$ given in [**MX11**] is based on expressing $u(x,t)$ in the form

$$u(x,t) = \frac{\partial I}{\partial x}(x,t), \tag{1.21}$$

where

$$I(x,t) := 2\epsilon \mathrm{Im}\left[\log\left(\det\left(\mathbb{I} + \frac{\mathrm{i}}{\epsilon}\mathbf{A}(x,t)\right)\right)\right] = 2\epsilon \sum_{k=1}^{N(\epsilon)} \arctan(\epsilon^{-1}\alpha_k(x,t)). \tag{1.22}$$

Here to define $I(x,t)$ properly, we resolve the ambiguity of the logarithm by taking the principal branch of the arctangent with values in $(-\frac{1}{2}\pi, \frac{1}{2}\pi)$ in the sum. The quantity $I(x,t)$ then resembles a Riemann sum for an integral of a discontinuous integrand $\arctan(\epsilon^{-1}\alpha) \to \frac{1}{2}\pi\mathrm{sgn}(\alpha)$ against the distribution of eigenvalues of $\mathbf{A}(x,t)$, and in [**MX11**] this (x,t)-dependent distribution is calculated and used to prove locally uniform convergence of $I(x,t)$ to an antiderivative of $\overline{u}(x,t)$. The weak nature of the convergence of $u(x,t)$ to $\overline{u}(x,t)$ can then be attributed to the x-derivative in (1.21)–(1.22).

Our aim in this article is to build upon the method of [**MX11**] by observing that if it is desired to improve the nature of the convergence of $u(x,t)$ as $\epsilon \to 0$, one should differentiate first in (1.21)–(1.22) and only then analyze $u(x,t)$. To this end, explicitly differentiating $I(x,t)$ with respect to x in (1.21)–(1.22) (assuming differentiability of the eigenvalues $\alpha_k(x,t)$), we obtain

$$u(x,t) = \sum_{k=1}^{N(\epsilon)} \frac{2\epsilon^2 \alpha_{k,x}(x,t)}{\alpha_k(x,t)^2 + \epsilon^2}, \quad \alpha_{k,x}(x,t) := \frac{\partial \alpha_k}{\partial x}(x,t). \tag{1.23}$$

Rather than diagonalizing $\mathbf{A}(x,t)=\mathbf{A}(x,t)^\dagger$ first and then differentiating, one could try to take advantage of the fact that the matrix elements of $\mathbf{A}(x,t)$ depend on x in a very simple way. Indeed, notice that

$$\mathbb{I}+\mathrm{i}\epsilon^{-1}\mathbf{A}(x,t)=\mathrm{i}\epsilon^{-1}\mathbf{D}\left[x\mathbb{I}-\mathbf{B}(t)-\mathrm{i}\epsilon\mathbf{D}^{-2}\right]\mathbf{D}, \tag{1.24}$$

where $\mathbf{D}$ is a Hermitian $N(\epsilon)\times N(\epsilon)$ diagonal matrix given by

$$\mathbf{D}:=\mathrm{diag}\left(\sqrt{-2\lambda_1},\dots,\sqrt{-2\lambda_{N(\epsilon)}}\right) \tag{1.25}$$

and $\mathbf{B}(t)$ is a Hermitian $N(\epsilon)\times N(\epsilon)$ matrix with elements

$$B_{jk}(t)=\begin{cases}\dfrac{-\mathrm{i}\epsilon}{\lambda_j-\lambda_k}, & j\neq k,\\ -2\lambda_j t-\gamma_j, & j=k.\end{cases} \tag{1.26}$$

Thus, for any branch of the complex logarithm,

$$\frac{\partial}{\partial x}\log\left(\det\left(\mathbb{I}+\frac{\mathrm{i}}{\epsilon}\mathbf{A}(x,t)\right)\right)=\frac{\partial}{\partial x}\log\left(\det\left(x\mathbb{I}-\mathbf{B}(t)-\mathrm{i}\epsilon\mathbf{D}^{-2}\right)\right). \tag{1.27}$$

If we let $\sigma_k(t)=\mu_k(t)+\mathrm{i}\nu_k(t)$, where $\mu_k(t),\nu_k(t)\in\mathbb{R}$, denote the complex eigenvalues of the non-Hermitian but x-independent matrix

$$\mathbf{C}(t):=\mathbf{B}(t)+\mathrm{i}\epsilon\mathbf{D}^{-2}, \tag{1.28}$$

then (1.18) can be written as

$$\begin{aligned}u(x,t)&=2\epsilon\mathrm{Im}\left[\frac{\partial}{\partial x}\log(\det(x\mathbb{I}-\mathbf{B}(t)-\mathrm{i}\epsilon\mathbf{D}^{-2})\right]\\&=2\epsilon\mathrm{Im}\left[\frac{\partial}{\partial x}\sum_{k=1}^{N(\epsilon)}\log(x-\sigma_k(t))\right]\\&=\mathrm{Im}\left[\sum_{k=1}^{N(\epsilon)}\frac{2\epsilon}{x-\sigma_k(t)}\right]\\&=\sum_{k=1}^{N(\epsilon)}\frac{2\epsilon\nu_k(t)}{(x-\mu_k(t))^2+\nu_k(t)^2}.\end{aligned} \tag{1.29}$$

By substituting the formula (1.29) into the BO equation (1.1) and using the fact that

$$\mathcal{H}[u(\cdot,t)](x)=\sum_{k=1}^{N(\epsilon)}\frac{-2\epsilon(x-\mu_k(t))}{(x-\mu_k(t))^2+\nu_k(t)^2},$$

it can be verified that the $\sigma_k(t)$'s satisfy the system of differential equations

$$\sigma_k'(t)=\sum_{j=1}^{N(\epsilon)}\frac{2\mathrm{i}\epsilon}{\sigma_k(t)-\sigma_j(t)^*}-\sum_{\substack{j=1\\j\neq k}}^{N(\epsilon)}\frac{2\mathrm{i}\epsilon}{\sigma_k(t)-\sigma_j(t)}. \tag{1.30}$$

This was first discovered in [**CLP79**, Eq. (7)], wherein (1.30) was shown to be equivalent to the famous Calogero-Moser N-body system. Let $\mathbf{w}_k(t)=(w_{k,1},\dots,w_{k,N(\epsilon)})^\top$ be the normalized eigenvector of $\mathbf{C}(t)$ corresponding to the eigenvalue $\sigma_k(t)$. Clearly

$$\sigma_k(t)=\mathbf{w}_k(t)^\dagger\mathbf{C}(t)\mathbf{w}_k(t)=\mathbf{w}_k(t)^\dagger\left(\mathbf{B}(t)+\mathrm{i}\epsilon\mathbf{D}^{-2}\right)\mathbf{w}_k(t), \tag{1.31}$$

where † denotes the conjugate transpose. By comparing real and imaginary parts of (1.31), we have

$$\mu_k(t) = -\sum_{j=1}^{N(\epsilon)} (2t\lambda_j + \gamma(\lambda_j))|w_{k,j}(t)|^2 - \mathrm{i}\epsilon \sum_{j=1}^{N(\epsilon)} \sum_{\substack{s=1\\ s\neq j}}^{N(\epsilon)} \frac{w_{k,j}(t)^* w_{k,s}(t)}{\lambda_s - \lambda_j}, \tag{1.32}$$

$$\nu_k(t) = -\frac{\epsilon}{2} \sum_{j=1}^{N(\epsilon)} \frac{|w_{k,j}(t)|^2}{\lambda_j}. \tag{1.33}$$

Since $-L < \lambda_1 < \cdots < \lambda_j < \cdots < \lambda_{N(\epsilon)} < 0$, using $|w_{k,1}(t)|^2 + \cdots + |w_{k,N(\epsilon)}(t)|^2 = 1$ we obtain

$$0 < \frac{1}{2} L \cdot \epsilon < \nu_k(t) < \frac{1}{2} \cdot \frac{\epsilon}{|\lambda_{N(\epsilon)}|}, \quad k = 1, \ldots, N(\epsilon). \tag{1.34}$$

Likewise, using also $|a||b| \leq \frac{1}{2}(a^2 + b^2)$,

$$|\mu_k(t)| \leq \left(2tL + \sup_{-L<\lambda<0} |\gamma(\lambda)|\right) + \frac{\epsilon N(\epsilon)}{\min_{s\neq j} |\lambda_s - \lambda_j|}, \quad k = 1, \ldots, N(\epsilon). \tag{1.35}$$

Below in (3.85) it is shown that if $u_0(x)$ decays algebraically as $x \to \pm\infty$, i.e., $u_0(x) \sim Cx^{-2p}$ for some $p \geq 1$, then assuming that $\epsilon = \epsilon_N = M/N$ so that (1.20) holds one finds that $|\lambda_{N(\epsilon)}| \gtrsim \epsilon^{2p/(2p-1)}$ and that $\min_{s\neq j} |\lambda_s - \lambda_j| \gtrsim \epsilon^{2p/(2p-1)}$, in which case the imaginary parts of the eigenvalues have uniform bounds of the form

$$\frac{1}{2} L \cdot \epsilon < \nu_k(t) \lesssim \epsilon^{-1/(2p-1)}, \quad k = 1, \ldots, N(\epsilon), \tag{1.36}$$

which is equivalent to the rescaled form

$$\frac{1}{2} L \cdot \frac{1}{\ln(\epsilon^{-1})} < \frac{\nu_k(t)}{\epsilon \ln(\epsilon^{-1})} \lesssim \frac{\epsilon^{-2p/(2p-1)}}{\ln(\epsilon^{-1})}, \quad k = 1, \ldots, N(\epsilon), \tag{1.37}$$

and similarly assuming also that $\gamma(\cdot)$ is a bounded function (it vanishes identically if $u_0(\cdot)$ is even),

$$|\mu_k(t)| \lesssim \epsilon^{-2p/(2p-1)}, \quad k = 1, \ldots, N(\epsilon). \tag{1.38}$$

1.3. Outline of the paper. The purpose of this work is to report the results of several numerical experiments undertaken to study the small-ϵ asymptotic behavior of the real eigenvalues $\{\alpha_k(x,t)\}$ and the complex eigenvalues $\{\sigma_k(t)\}$ of the matrices $\mathbf{A}(x,t)$ and $\mathbf{C}(t)$ respectively. We recall that $u(x,0) = u_0(x)$ is an admissible initial condition (see [**MX11**, Definition 3.1]). For simplicity, we assume further that u_0 has exactly one inflection point to the right of the maximizer. Consequently, we always have that for generic (x,t) with $t > 0$, $P(x,t) = 0$ or $P(x,t) = 1$. Moreover, for fixed $t > t_b$ the multi-valued region (in x) for Burgers' equation is an interval $(X^-(t), X^+(t))$, see Figure 1. For simplicity, all of our numerical experiments will be for the soliton ensemble associated with the initial condition

$$u_0(x) = \frac{2}{1+x^2}. \tag{1.39}$$

For this initial condition, the breaking time for the inviscid Burgers (dispersionless, i.e., $\epsilon = 0$) approximation of the BO equation (1.1) is exactly $t = t_b = 2\sqrt{3}/9 \approx 0.3849$. When $t > t_b$, the dispersive terms are expected to be important and form a highly-oscillatory dispersive shock wave (see Figure 2, right panel) in the t-dependent interval of x on which the method of characteristics predicts a triple-valued solution of the dispersionless approximation.

As will be seen, the numerical experiments suggest that when ϵ is small, the complex eigenvalues $\{\sigma_k(t)\}$ and the real eigenvalues $\{\alpha_k(x,t)\}$ that are so small as to contribute substantially to the sum in (1.23) are evidently distributed in a regular fashion. After formulating reasonable conjectures based on the numerical observations, we prove that they imply many of the properties of the dispersive shock wave. These properties go beyond the weak convergence result that $u(x,t) \rightharpoonup \overline{u}(x,t)$ with limit $\overline{u}(x,t)$ given by (1.5). On the other hand, in general the soliton ensemble only approximates u_0 in the strong $L^2(\mathbb{R})$ sense when $t=0$, and translation in x is continuous on $L^2(\mathbb{R})$. Therefore it is possible that $u(\cdot,0)$ is nearly a translation of $u_0(\cdot)$ by a fraction of the wavelength of the dispersive shock wave that appears for $t>t_b$, leading to $\mathcal{O}(1)$ errors for the solution of the Cauchy problem when measured in $L^\infty(\mathbb{R})$. In this sense, our main results below, Theorems 2.11 and 3.8, apply to the soliton ensemble but would require phase corrections to apply also to the solution of the Cauchy problem for (1.1) with initial data u_0.

Our study of the complex eigenvalues of $\mathbf{C}(t)$ and how their asymptotic properties imply the most important features of $u(x,t)$ via (1.29) is presented in Section 2. Then in Section 3 we give a parallel analysis logically independent of Section 2 for the eigenvalues of $\mathbf{A}(x,t)$ and the formula (1.23). A completely different approach to strong asymptotics for the BO equation (1.1) that is based instead on a remarkable formula of Gérard [**Gé23**] is the subject of a forthcoming work [**BGGM24**].

To end this introduction, we emphasize that this paper falls into the category of experimental mathematics. Our results are not entirely definitive as they are partly based on numerical experiments. However in our study of this topic we have uncovered numerous interesting connections with traditional areas of mathematical analysis that we want to report, in hopes of stimulating further research. Indeed, different parts of this paper are related to the theory of singular asymptotics of dispersive nonlinear waves, to the spectra of large structured matrices, and to geometric quantization and semiclassical analysis.

2. Asymptotic properties of the complex eigenvalues $\sigma_k(t)$ and their implications

This section of the paper concerns the nonhermitian matrix $\mathbf{C}(t)$. We present numerical experiments leading to conjectures about the distribution of its eigenvalues, and then we show how those conjectures lead to strong asymptotics of $u(x,t)$ consistent with Whitham modulation theory.

2.1. Numerical experiments. Given an initial condition $u_0(\cdot)$, a time $t\geq 0$, and a value of ϵ, it is straightforward to construct the nonhermitian matrix $\mathbf{C}(t)$ and to numerically extract its eigenvalues $\{\sigma_k(t)\}$ with high accuracy. The first observation is that, roughly speaking, most of the complex eigenvalues are close to the real axis. However depending on the initial condition, there can be some "outliers" that do not follow this rule, as shown in Figure 3.

What evidently distinguishes the outlier eigenvalues from the rest is that they have small real parts and imaginary parts that are possibly large instead of small.

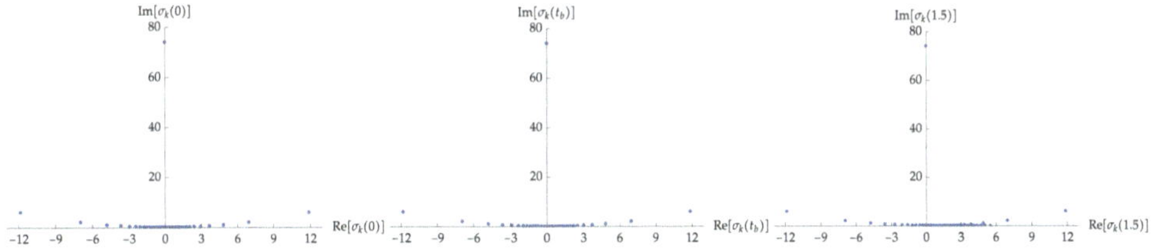

FIGURE 3. Outlier eigenvalues for the initial condition $u_0(x) = 2(1+x^2)^{-1}$ with $\epsilon = 0.02$.

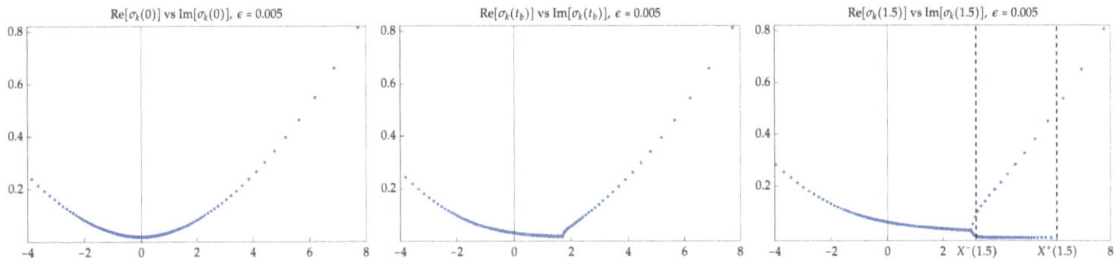

FIGURE 4. Time evolution of $\sigma_k(t)$ for $u_0(x) = 2(1+x^2)^{-1}$ and $\epsilon = 0.005$.

We will make the following definition:

DEFINITION 2.1 (Outliers). For fixed $\beta > 0$ and $B > 0$ independent of t but possibly depending on the initial condition $u_0(\cdot)$, an eigenvalue $\sigma = \mu + \mathrm{i}\nu$ will be called an outlier if $\nu \geq \beta$ and $|\mu| \leq B$.

For suitable initial conditions $u_0(\cdot)$, this definition is saying that the outliers include in particular any eigenvalues with bounded real part that saturate the upper bound given in (1.36). Let $S_o \subset \{\sigma_k(t)\}$ denote the subset of outliers, and let the remaining eigenvalues constitute the "bulk" S_b, so that $S_b \sqcup S_0 = \{\sigma_k(t)\}$. The numerics suggest that there are relatively few outliers compared to the total number $N(\epsilon) \sim \epsilon^{-1}$ of eigenvalues. Hence we formulate the following conjecture:

CONJECTURE 1 (Outliers). There is an exponent $0 \leq p < 1$ and a constant $K > 0$ independent of t such that $|S_o| \leq K\epsilon^{-p}$ holds for all $\epsilon > 0$ sufficiently small.

In Proposition 2.4 below we will use this to estimate the contribution of the outliers to the sum (1.29). The bound $|\mu| \leq B$ on the real part is part of Definition 2.1 because there are evidently some eigenvalues with both real and imaginary parts that are large when ϵ is small (see the upward trend with increasing real part of the points near the real line in the plots shown in Figures 3 and 4), and as they apparently lie along certain curves and there can be many of them, it is more natural to include them in S_b.

Next, we examine the bulk S_b of the complex eigenvalues, which requires a different scale for the imaginary part coordinate. As shown in Figure 4, these bulk eigenvalues appear to be distributed along curves in the complex upper half-plane, and these curves have interesting dynamical behavior as $t \geq 0$ varies. From these and similar plots it seems clear that when $t \leq t_b$, the bulk eigenvalues arrange themselves along a single curve in the upper half-plane; however as soon as $t > t_b$, a second curve bifurcates from the former curve into the part of the upper half-plane below. We therefore further partition S_b into an "upper branch" subset $S_U \subseteq S_b$ and a complementary "lower branch" subset $S_L := S_b \setminus S_U$ (when $t \leq t_b$, $S_L = \emptyset$). For $t > t_b$ fixed, both S_U, S_L are non-empty and judging from numerics both have cardinality $N_U(\epsilon)$, $N_L(\epsilon)$ proportional to $N(\epsilon)$. Moreover, numerics suggest that the real parts of the eigenvalues belonging to S_L lie within the interval $(X^-(t), X^+(t))$. For example, when $t = 1.5$, $X^-(t) \simeq 3.20$ and $X^+(t) \simeq 6.04$, see Figure 4, right panel.

From the plots, it appears that the real parts of the eigenvalues on either branch are locally equally spaced. We formulate the following conjecture to express this observation in detail.

CONJECTURE 2 (Real parts). For all $t \geq 0$ and $\epsilon > 0$, the eigenvalues comprising S_U have distinct real parts; similarly for S_L when $t > t_b$. Assume that the points of S_U and S_L are indexed by increasing real part, i.e., $\mu_{U,1}(t) < \mu_{U,2}(t) < \cdots < \mu_{U,N_U(\epsilon)}(t)$ for all $t \geq 0$ and $\mu_{L,1}(t) < \mu_{L,2}(t) < \cdots < \mu_{L,N_L(\epsilon)}(t)$ for all $t > t_b$. Then, the sets $\{\mu_{U,k}(t)\}$, $\{\mu_{L,k}(t)\}$ are approximate samplings of two respective ϵ-independent functions $\mu_U : (0,1) \times [0,\infty) \to \mathbb{R}$, $\mu_L : (0,1) \times (t_b,\infty) \to (X^-(t), X^+(t))$. More precisely, with $'$ denoting differentiation with respect to the first argument,

$$\mu_{U,k}(t) = \mu_U(y_k,t) + \mathcal{O}(\epsilon^2 \mu_U'(y_k,t)), \quad y_k := \frac{k-\frac{1}{2}}{N_U(\epsilon)} \tag{2.1}$$

and, for $t > t_b$,

$$\mu_{L,k}(t) = \mu_L(y_k,t) + \mathcal{O}(\epsilon^2 \mu_L'(y_k,t)), \quad y_k := \frac{k-\frac{1}{2}}{N_L(\epsilon)} \tag{2.2}$$

where the error terms are uniform for $k = 1,\ldots,N_U(\epsilon)$ and $k = 1,\ldots,N_L(\epsilon)$ respectively. The functions μ_U, μ_L have the following properties.

Invertibility: $\mu_L(\cdot,t)$, $\mu_U(\cdot,t)$ are strictly increasing surjective functions for any t on their respective domains, so there exists an inverse function $\mu_{L,U}^{-1}(\cdot,t)$, in the sense that $\mu_{L,U}^{-1}(\mu_{L,U}(y,t),t) = y$.

Smoothness and bounds for μ_U: For $0 \leq t < t_b$, $y \mapsto \mu_U(y,t)$ is of class $C^2((0,1))$. For $t \geq t_b$, $y \mapsto \mu_U(y,t)$ of class $C^2((0,1) \setminus \{y^-(t)\})$ where $y^-(t) := \mu_U^{-1}(X^-(t),t)$, and is Lipschitz continuous on $(0,1)$. In both cases there are exponents $q_\pm > 0$ and constants $0 < c < C$ such that

$$-Cy^{-q_-} < \mu_U(y,t) < -cy^{-q_-} \quad \text{and} \quad \mu_U'(y,t) = \mathcal{O}(y^{-(q_-+1)}),$$

as $y \downarrow 0$, and

$$c(1-y)^{-q_+} < \mu_U(y,t) < C(1-y)^{-q_+} \quad \text{and} \quad \mu_U'(y,t) = \mathcal{O}((1-y)^{-(q_++1)}),$$

as $y \uparrow 1$.

Smoothness and bounds for μ_L: For $t > t_b$, $y \mapsto \mu_L(y,t)$ extends by continuity to $y \in [0,1]$ and is of class $C^2((0,1))$.

All estimates involving μ_U are uniform for bounded $t \geq 0$, in which case $\epsilon N_U(\epsilon)$ has a finite nonzero limit as $\epsilon \to 0$. Likewise all estimates involving μ_L are uniform for bounded $t > t_b$ with $t - t_b$ bounded below by a positive quantity, in which case also $\epsilon N_L(\epsilon)$ has a finite nonzero limiting value.

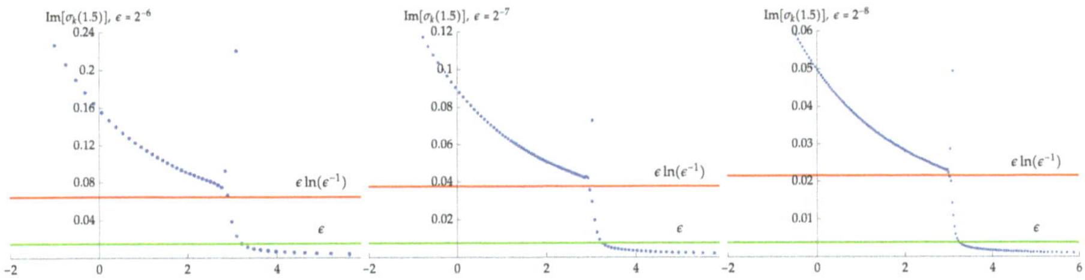

FIGURE 5. Upper and lower branch scaling for $\epsilon = 2^{-6}, 2^{-7}, 2^{-8}$ and $t = 1.5$. The horizontal axis is $\mathrm{Re}[\sigma_k]$ in each plot.

Note that $\mu_{U,L}(y_{k+1}, t) - \mu_{U,L}(y_k, t) \approx \mu'_{U,L}(y_k, t)/N_{U,L}(\epsilon)$. Since $N_{U,L}(\epsilon)$ are inversely proportional to ϵ, equations (2.1) and (2.2) assert that the sampling error is ϵ times a uniform multiple of the local spacing.

In particular, the power-law behavior of $\mu_U(y, t)$ asserted in this conjecture is strongly supported by numerical experiments. For the initial condition (1.39), we estimated the exponents $q_\pm$ from slopes of best-fit lines in plots of $\ln(|\mu_{U,1}(t)|)$ and $\ln(\mu_{U,N_U(\epsilon)}(t))$ versus $\ln(\epsilon)$ over the range from 2^{-6} through 2^{-12} and obtained

t	q_-	q_+
0	1.00021	1.00021
t_b	1.00023	1.00019
1.5	1.00028	1.00013

Therefore, for this initial condition, the exponents $q_\pm$ appear to be approximately equal to 1 regardless of whether $t < t_b$, $t = t_b$, or $t > t_b$. We also found that the linear fit of the data is extremely accurate over the full range of scales.

In order to formulate similar conjectures regarding the imaginary parts of the eigenvalues on the upper and lower branches, it is necessary first to notice that unlike the real parts, the imaginary parts of eigenvalues on both branches are asymptotically small as $\epsilon \to 0$, with different asymptotic scales. See Figure 5. These plots strongly suggest that the imaginary parts of the eigenvalues in S_L scale proportionally with ϵ, while those on the upper branch are somewhat larger. In fact, the plots in Figure 5 provide good evidence that the scaling of imaginary parts of eigenvalues in S_U is proportional to $\epsilon \ln(\epsilon^{-1})$. Observing log-type growth/decay via numerics is notoriously difficult. However, we are fortunate in that the conditional results we will prove in Section 2.2 below are rather insensitive to the precise scale of the imaginary parts of eigenvalues on the upper branch. It will be enough that they scale as $\delta(\epsilon)$ lying in the asymptotic range $\epsilon \ll \delta(\epsilon) \ll 1$.

CONJECTURE 3 (Upper branch imaginary parts). There is a scale $\delta(\epsilon)$ in the asymptotic range $\epsilon \ll \delta(\epsilon) \ll 1$ such that the set $\{\nu_{U,k}(t)\}$ is the approximate sampling of a scaled ϵ-independent function $\nu_U : (0,1) \times [0, \infty) \to (0, \infty)$, i.e., as $\epsilon \to 0$,

$$\nu_{U,k}(t) = \delta(\epsilon)\nu_U(y_k, t)(1 + o(1)), \quad y_k := \frac{k - \frac{1}{2}}{N_U(\epsilon)} \tag{2.3}$$

holds uniformly over all indices $k = 1, \ldots, N_U(\epsilon)$. The function ν_U has the following properties.

Lower bound: there exists a constant $c_U > 0$ such that $c_U < \nu_U(y, t)$.

Edge behavior: there are exponents $0 < r_\pm < 2q_\pm + 1$ (see Conjecture 2) and constants $0 < c < C$ such that

$$cy^{-r_-}(1-y)^{-r_+} < \nu_U(y,t) < Cy^{-r_-}(1-y)^{-r_+}, \quad y \to 0, 1.$$

Smoothness: For $0 \leq t < t_b$, $y \mapsto \nu_U(y,t)$ is of class $C^1((0,1))$. For $t \geq t_b$, $y \mapsto \nu_U(y,t)$ is absolutely continuous on $(0,1)$ and is of class $C^1((0,1) \setminus \{y^-(t)\})$, where $y^-(t)$ is as defined in Conjecture 2.

All estimates are uniform for bounded $t \geq 0$.

Again, the power-law behavior of $\nu_U(y,t)$ as $y \to 0, 1$ is supported by numerics. For the same initial condition (1.39), by determining the scaling of the quantity $\nu_{U,k}(t)/(\epsilon \ln(\epsilon^{-1}))$ for $k = 1$ and $k = N_U(\epsilon)$ with ϵ, we obtained the following best-fit values of $r_\pm$, again with very accurate approximation over the whole range of scales:

t	r_-	r_+
0	1.83460	1.83460
t_b	1.83456	1.83464
1.5	1.83445	1.83475

Comparing with the experimental values of $q_\pm$, we see that indeed the inequalities $0 < r_\pm < 2q_\pm + 1$ are evidently satisfied.

Finally, we offer a conjecture summarizing our numerical observations concerning the imaginary parts of eigenvalues on the lower branch if $t > t_b$.

CONJECTURE 4 (Lower branch imaginary parts). The set $\{\nu_{L,k}(t)\}$ is the approximate sampling of a scaled ϵ-independent function $\nu_L : (0,1) \times (t_b, \infty) \to (0, \infty)$, i.e., as $\epsilon \to 0$,

$$\nu_{L,k}(t) = \epsilon \nu_L(y_k, t)(1 + o(1)), \quad y_k := \frac{k - \frac{1}{2}}{N_L(\epsilon)} \tag{2.4}$$

holds uniformly over all indices $k = 1, \ldots, N_L(\epsilon)$. The function ν_L has the following properties, in which $\delta(\epsilon)$ is the scale from Conjecture 3, here asserted to have the additional property that

$$\int_0^\epsilon \frac{\delta(y)}{y} \, dy < \infty \tag{2.5}$$

holds for $\epsilon > 0$ sufficiently small.

Lower bound: there exists a constant $c_L > 0$ such that $c_L < \nu_L(y,t)$.
Left edge behavior: There exist constants $c, C > 0$ such that

$$c\delta(y)y^{-1} < \nu_L(y,t) < C\delta(y)y^{-1}, \quad y \downarrow 0.$$

Smoothness: For $t > t_b$, $y \mapsto \nu_L(y,t)$ is of class $C^1((0,1))$.

All estimates are uniform for bounded $t > t_b$ with $t - t_b$ bounded below by a positive quantity.

REMARK 2.2. The statement that $\nu_L(y,t) \sim \delta(y)y^{-1}$ as $y \downarrow 0$ is intended to capture the phenomenon that the lower branch bifurcates from the upper branch at $y = 0$ (according to the parametrization of the lower branch; it is instead the point $y^-(t)$ in that of the upper branch) and the imaginary parts of the eigenvalues on the upper branch are asymptotically large compared to those of the lower branch eigenvalues. The assertion of integrability of $\delta(y)y^{-1}$ at $y = 0$ is reasonable given the numerical evidence that $\delta(\epsilon) \sim \epsilon \ln(\epsilon^{-1})$; see Figure 5.

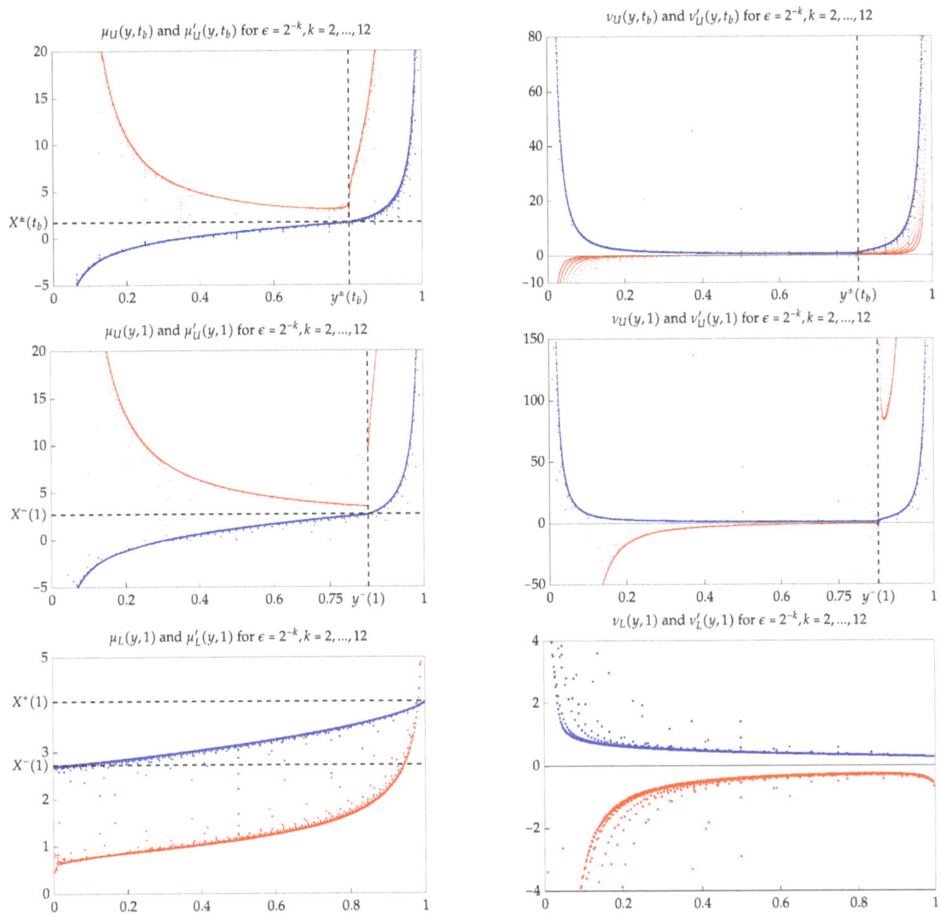

FIGURE 6. The real and rescaled imaginary parts of the eigenvalues (blue) in the sets $\{\sigma_{U,k}(t)\} = S_U$ and $\{\sigma_{L,k}(t)\} = S_L$, obtained from the matrix $\mathbf{C}(t)$ for the initial condition (1.39), plotted versus the parameter $y \in (0,1)$ for $t = t_b$ (upper two panels) and $t = 1 > t_b$ (lower four panels). Also shown are the numerical derivatives with respect to y (red). COLOR ONLINE.

The approximate sampling properties asserted in Conjectures 2, 3 and 4 are easily illustrated. After computing the eigenvalues of $\mathbf{C}(t)$ for a given small value of ϵ, one omits the outliers and partitions the bulk into $S_b = S_U \sqcup S_L$ based on the size of the imaginary parts, and these sets of complex numbers are then ordered by increasing real parts. Setting $N_U(\epsilon) := |S_U|$, each point $\sigma_{U,k}(t) = \mu_{U,k}(t) + \mathrm{i}\nu_{U,k}(t)$ is then assigned a value of the parameter $y = y_k := (k - \frac{1}{2})/N_U(\epsilon)$. Likewise, if $t > t_b$, setting $N_L(\epsilon) := |S_L|$, each point $\sigma_{L,k}(t) = \mu_{L,k}(t) + \mathrm{i}\nu_{L,k}(t)$ is assigned a value of the parameter $y = y_k := (k - \frac{1}{2})/N_L(\epsilon)$. Assuming the hypothetical scale $\delta(\epsilon) = \epsilon \ln(\epsilon^{-1})$, the points $\{(y_k, \mu_{U,k}(t))\}$ and $\{(y_k, \nu_{U,k}/\delta(\epsilon))\}$, and if $t > t_b$, $\{(y_k, \mu_{L,k}(t))\}$ and $\{(y_k, \nu_{L,k}(t)/\epsilon)\}$ can be plotted on the same axes for a range of scales for ϵ; these are shown in blue in the panels of Figure 6. One can clearly observe the blue points condensing onto limiting fixed curves as $\epsilon \to 0$. These limiting curves should be the graphs of the sampling functions $y \mapsto \mu_U(y,t)$, $y \mapsto \nu_U(y,t)$, and for $t > t_b$, $y \mapsto \mu_L(y,t)$ and $y \mapsto \nu_L(y,t)$. It is also straightforward to compute from the data difference quotient approximations (numerical derivatives) as follows:

$$\mu_U'\left(\frac{k-\frac{1}{2}}{N_U(\epsilon)},t\right) \simeq \frac{\mu_{U,k+1}(t)-\mu_{U,k}(t)}{N_U(\epsilon)^{-1}}, \quad \nu_U'\left(\frac{k-\frac{1}{2}}{N_U(\epsilon)},t\right) \simeq \frac{\nu_{U,k+1}(t)-\nu_{U,k}(t)}{N_U(\epsilon)^{-1}\epsilon\ln(\epsilon^{-1})},$$
$$\mu_L'\left(\frac{k-\frac{1}{2}}{N_L(\epsilon)},t\right) \simeq \frac{\mu_{L,k+1}(t)-\mu_{L,k}(t)}{N_L(\epsilon)^{-1}}, \quad \nu_L'\left(\frac{k-\frac{1}{2}}{N_L(\epsilon)},t\right) \simeq \frac{\nu_{L,k+1}(t)-\nu_{L,k}(t)}{N_L(\epsilon)^{-1}\epsilon}.$$

These are plotted against y as red points in the same figure. The red points also condense onto limiting curves, although it is clear that the limiting derivative curves for the data obtained from the upper-branch eigenvalues are discontinuous at $y = y^-(t)$ whenever $t > t_b$.

We have not yet been able to prove these conjectures. We are aware of formal techniques applicable to nonselfadjoint eigenvalue problems for differential equations with analytic coefficients, and such methods have been applied to deduce curves in the complex plane that attract eigenvalues in a semiclassical limit similar to $N(\epsilon) \to \infty$ (see, e.g., [**Mil01**]). Unfortunately, the nonhermitian matrix $\mathbf{C}(t)$ does not have fixed bandwidth, so making an analogy with differential or difference equations is challenging. Even though $\mathbf{C}(t)$ does have an approximate Toeplitz structure, we are not aware of methods in the theory of Toeplitz quantization for nonhermitian matrices that would be sufficiently powerful to prove the above conjectures. We will however apply elements of Toeplitz quantization to the Hermitian matrix $\mathbf{A}(x,t)$ in Section 3 below.

2.2. Conditional small-ϵ asymptotics of the sum (1.29)**.** Now we will use the conjectures inspired by numerical experiments to study the soliton ensemble for the BO equation with initial data $u_0(x)$. We can split the sum (1.29) into three parts corresponding to outliers in S_o, upper branch eigenvalues in S_U, and lower branch eigenvalues in S_L, i.e.,

$$u(x,t) = u_U(x,t) + u_L(x,t) + u_o(x,t), \tag{2.6}$$

where

$$\begin{aligned} u_U(x,t) &:= \sum_{\sigma(t)\in S_U} \frac{2\epsilon\nu(t)}{(x-\mu(t))^2+\nu(t)^2} = \sum_{k=1}^{N_U(\epsilon)} \frac{2\epsilon\nu_{U,k}(t)}{(x-\mu_{U,k}(t))^2+\nu_{U,k}(t)^2} \\ u_L(x,t) &:= \sum_{\sigma(t)\in S_L} \frac{2\epsilon\nu(t)}{(x-\mu(t))^2+\nu(t)^2} = \sum_{k=1}^{N_L(\epsilon)} \frac{2\epsilon\nu_{L,k}(t)}{(x-\mu_{L,k}(t))^2+\nu_{L,k}(t)^2} \end{aligned} \tag{2.7}$$

and $u_o(x,t) := u(x,t) - u_U(x,t) - u_L(x,t)$ is the sum over the remaining outlier eigenvalues. It will turn out that $u_o(x,t)$ is of lower order compared to the other components and can be ignored (see Proposition 2.4 below). The other two terms evidently make quite different contributions to the overall sum, as can be seen in Figure 7, which compares $u_U(x,t)$ and $u_L(x,t)$ with $u(x,t)$ for a value of t that exceeds the breaking time t_b. These plots suggest that the rapid oscillations (which occur only after the breaking time) come from the term $u_L(x,t)$ while the smooth background comes instead from $u_U(x,t)$.

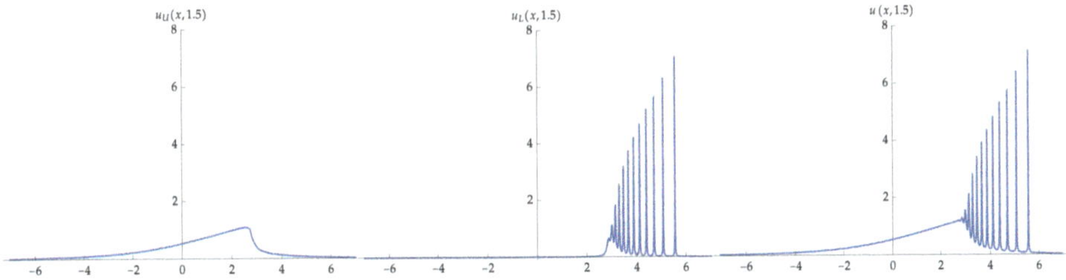

FIGURE 7. The two essential components $u_U(x,t)$ and $u_L(x,t)$ (see (2.7)) of $u(x,t)$ compared with $u(x,t)$ itself for $u_0(x) = 2(1+x^2)^{-1}$, $t = 1.5$, and $\epsilon = 0.02$.

First, we formulate a lemma showing that replacing the eigenvalues with their respective sampling functions in either of the two summands (2.7) produces a relatively small error term that is uniform with respect to the sum index.

LEMMA 2.3 (Sampling error). *Assume that Conjecture 2 holds and fix constants $K > 0$ and $\eta > 0$. Then in the limit $\epsilon \to 0$, the summand in $u_U(x,t)$ can be written in the form*

$$\frac{2\epsilon\nu_{U,k}(t)}{(x-\mu_{U,k}(t))^2+\nu_{U,k}(t)^2} = \frac{2\epsilon\delta(\epsilon)\nu_U(y_k,t)}{(x-\mu_U(y_k,t))^2+\delta(\epsilon)^2\nu_U(y_k,t)^2}(1+o(1)), \tag{2.8}$$

where $y_k := \frac{k-\frac{1}{2}}{N_U(\epsilon)}$, provided also Conjecture 3 holds, where the error term is uniform for $0 \le t \le K$, $|x| \le K$, and $k = 1, \ldots, N_U(\epsilon)$. Likewise, as $\epsilon \to 0$, the summand in $u_L(x,t)$ can be written in the form

$$\frac{2\epsilon\nu_{L,k}(t)}{(x-\mu_{L,k}(t))^2+\nu_{L,k}(t)^2} = \frac{2\epsilon^2\nu_L(y_k,t)}{(x-\mu_L(y_k,t))^2+\epsilon^2\nu_L(y_k,t)^2}(1+o(1)), \tag{2.9}$$

where $y_k := \frac{k-\frac{1}{2}}{N_L(\epsilon)}$, provided also Conjecture 4 holds, where the error term is uniform for $t_b + \eta \le t \le K$, $X^-(t) + \eta \le x \le X^+(t) - \eta$ and $k = 1, \ldots, N_L(\epsilon)$.

PROOF. For convenience we suppress the dependence on t, which should be taken in different intervals depending on which summand is considered. According to Conjectures 3 and 4, we have $\nu_{U,k} = \delta(\epsilon)\nu_U(y_k)(1+o(1))$ with $y_k = (k - \frac{1}{2})/N_U(\epsilon)$ uniformly for $k = 1, \ldots, N_U(\epsilon)$, and $\nu_{L,k} = \epsilon\nu_L(y_k)(1+o(1))$ with $y_k = (k - \frac{1}{2})/N_L(\epsilon)$ uniformly for $k = 1, \ldots, N_L(\epsilon)$. Therefore, it remains to get a corresponding uniform $o(1)$ relative error estimate for the denominator of each summand.

In other words, we want to obtain uniform $o(1)$ estimates for

$$Q_{U,k} := \left|\frac{[(x-\mu_{U,k})^2+\nu_{U,k}^2] - [(x-\mu_U(y_k))^2+\delta(\epsilon)^2\nu_U(y_k)^2]}{(x-\mu_U(y_k))^2+\delta(\epsilon)^2\nu_U(y_k)^2}\right|, \tag{2.10}$$

where $y_k := \frac{k-\frac{1}{2}}{N_U(\epsilon)}$, and

$$Q_{L,k} := \left|\frac{[(x-\mu_{L,k})^2+\nu_{L,k}^2] - [(x-\mu_L(y_k))^2+\epsilon^2\nu_L(y_k)^2]}{(x-\mu_L(y_k))^2+\epsilon^2\nu_L(y_k)^2}\right|, \tag{2.11}$$

where $y_k := \frac{k-\frac{1}{2}}{N_L(\epsilon)}$. Let us write $\mu_{U,k} = \mu_U(y_k) + \Delta\mu_{U,k}$, $\mu_{L,k} = \mu_L(y_k) + \Delta\mu_{L,k}$, $\nu_{U,k} = \delta(\epsilon)\nu_U(y_k)(1+\Delta\nu_{U,k})$, and $\nu_{L,k} = \epsilon\nu_L(y_k)(1+\Delta\nu_{L,k})$. We can assume that $|\Delta\nu_{U,k}| < 1$ and $|\Delta\nu_{L,k}| < 1$, so

(2.12)
$$Q_{U,k} \leq \frac{2|x-\mu_U(y_k)||\Delta\mu_{U,k}| + |\Delta\mu_{U,k}|^2 + 3\delta(\epsilon)^2\nu_U(y_k)^2|\Delta\nu_{U,k}|}{(x-\mu_U(y_k))^2 + \delta(\epsilon)^2\nu_U(y_k)^2} \leq R_{U,k} + 3|\Delta\nu_{U,k}|$$

where

$$R_{U,k} := \frac{2|x-\mu_U(y_k)||\Delta\mu_{U,k}| + |\Delta\mu_{U,k}|^2}{(x-\mu_U(y_k))^2 + \delta(\epsilon)^2\nu_U(y_k)^2}, \quad y_k := \frac{k-\frac{1}{2}}{N_U(\epsilon)} \tag{2.13}$$

and

(2.14)
$$Q_{L,k} \leq \frac{2|x-\mu_L(y_k)||\Delta\mu_{L,k}| + |\Delta\mu_{L,k}|^2 + 3\epsilon^2\nu_L(y_k)^2|\Delta\nu_{L,k}|}{(x-\mu_L(y_k))^2 + \epsilon^2\nu_L(y_k)^2} \leq R_{L,k} + 3|\Delta\nu_{L,k}|$$

where

$$R_{L,k} := \frac{2|x-\mu_L(y_k)||\Delta\mu_{L,k}| + |\Delta\mu_{L,k}|^2}{(x-\mu_L(y_k))^2 + \epsilon^2\nu_L(y_k)^2}, \quad y_k := \frac{k-\frac{1}{2}}{N_L(\epsilon)}. \tag{2.15}$$

Since $\Delta\nu_{U,k} = o(1)$ and $\Delta\nu_{L,k} = o(1)$ both hold uniformly for $k = 1, \dots, N_U(\epsilon)$ and $k = 1, \dots, N_L(\epsilon)$ respectively, it is enough to estimate $R_{U,k}$ and $R_{L,k}$.

First we consider $R_{U,k}$. If k is such that $|x-\mu_U(y_k)| \geq 1$, then by neglecting $\delta(\epsilon)^2\nu_U(y_k)^2$ in the denominator,

$$R_{U,k} \leq 2\frac{|\Delta\mu_{U,k}|}{|x-\mu_U(y_k)|} + \left(\frac{|\Delta\mu_{U,k}|}{|x-\mu_U(y_k)|}\right)^2. \tag{2.16}$$

But then we have the estimate by Conjecture 2

$$\frac{|\Delta\mu_{U,k}|}{|x-\mu_U(y_k)|} = \mathcal{O}\left(\frac{\epsilon^2\mu_U'(y_k)}{|x-\mu_U(y_k)|}\right) \tag{2.17}$$

which is uniform for k in the range $1 \leq k \leq N_U(\epsilon)$. This is $\mathcal{O}(\epsilon^2)$ as long as y_k is bounded away from the endpoints $0, 1$. However, according to Conjecture 2, $\mu_U'(y)/\mu_U(y) = \mathcal{O}(y^{-1})$ as $y \downarrow 0$ while $\mu_U'(y)/\mu_U(y) = \mathcal{O}((1-y)^{-1})$ as $y \uparrow 1$. Using these estimates and the fact that y_k and $1-y_k$ are both greater than or equal to $\frac{1}{2}N_U(\epsilon)^{-1}$ shows that

$$|x-\mu_U(y_k)| \geq 1 \implies \frac{|\Delta\mu_{U,k}|}{|x-\mu_U(y_k)|} = \mathcal{O}(\epsilon^2 N_U(\epsilon)) = \mathcal{O}(\epsilon) \implies R_{U,k} = \mathcal{O}(\epsilon)$$

holds uniformly over the indicated subset of indices k. On the other hand, if $|x-\mu_U(y_k)| < 1$, then by neglecting $(x-\mu_U(y_k))^2$ in the denominator

$$R_{U,k} < \frac{2|\Delta\mu_{U,k}| + |\Delta\mu_{U,k}|^2}{\delta(\epsilon)^2\nu_U(y_k)^2} \leq \frac{2|\Delta\mu_{U,k}| + |\Delta\mu_{U,k}|^2}{\delta(\epsilon)^2 c_U^2}, \tag{2.18}$$

where $\nu_U(y) \geq c_U$ holds for $0 < y < 1$ by Conjecture 3. Since $\mu'(y)$ is uniformly bounded if $\mu(y)$ itself is, $|x-\mu_U(y_k)| < 1$ controls $\mu'(y_k)$ uniformly for bounded x, and hence $\Delta\mu_{U,k} = \mathcal{O}(\epsilon^2\mu'(y_k)) = \mathcal{O}(\epsilon^2)$. Therefore

$$|x-\mu_U(y_k)| < 1 \implies R_{U,k} = \mathcal{O}\left(\frac{\epsilon^2}{\delta(\epsilon)^2}\right) = o(1) \tag{2.19}$$

holds uniformly for bounded x and in the indicated range of indices k because $\epsilon \ll \delta(\epsilon) \ll 1$ as asserted in Conjecture 3. This shows that $R_{U,k} = o(1)$ holds uniformly for $k = 1, \ldots, N_U(\epsilon)$ if x is uniformly bounded. Hence also $Q_{U,k} = o(1)$ in the same sense.

Next, we consider $R_{L,k}$. If k is such that $|x - \mu_L(y_k)| \geq \epsilon$, then neglecting $\epsilon^2\nu_L(y_k)^2$ from the denominator as before, $|x - \mu_L(y_k)| \geq \epsilon$ implies that

$$R_{L,k} \leq 2\frac{|\Delta\mu_{L,k}|}{|x-\mu_L(y_k)|} + \left(\frac{|\Delta\mu_{L,k}|}{|x-\mu_L(y_k)|}\right)^2 \leq 2\frac{|\Delta\mu_{U,k}|}{\epsilon} + \left(\frac{|\Delta\mu_{U,k}|}{\epsilon}\right)^2. \tag{2.20}$$

Since $\mu_L'(y)$ is uniformly bounded on $0 < y < 1$, we use the uniform estimate $\Delta\mu_{L,k} = \mathcal{O}(\epsilon^2\mu_L'(y_k)) = \mathcal{O}(\epsilon^2)$ from Conjecture 2 to obtain $R_{L,k} = \mathcal{O}(\epsilon)$ as a uniform bound for indices satisfying the indicated condition. Conversely, if $|x - \mu_L(y_k)| < \epsilon$, we omit $(x - \mu_L(y_k))^2$ from the denominator instead and obtain

$$|x-\mu_L(y_k)| < \epsilon \implies R_{L,k} < \frac{2\epsilon|\Delta\mu_{L,k}| + |\Delta\mu_{L,k}|^2}{\epsilon^2\nu_L(y_k)^2} \leq \frac{2\epsilon|\Delta\mu_{L,k}| + |\Delta\mu_{L,k}|^2}{\epsilon^2 c_L^2}, \tag{2.21}$$

using $\nu_L(y) \geq c_L$ for $0 < y < 1$ as follows from Conjecture 4. Again using $\Delta\mu_{L,k} = \mathcal{O}(\epsilon^2)$ and combining with the result for $|x - \mu_L(y_k)| \geq \epsilon$ shows that the bound $R_{L,k} = \mathcal{O}(\epsilon)$ holds uniformly for all indices $k = 1, \ldots, N_L(\epsilon)$. Hence also $Q_{L,k} = o(1)$ holds in the same sense. □

We now investigate the small-ϵ asymptotics of each sum, starting with the outlier sum $u_o(x,t)$.

PROPOSITION 2.4 (Outlier sum). Let $\beta > 0$ be as in Definition 2.1, and assume that Conjecture 1 holds with some constant $K > 0$ and exponent $0 \leq p < 1$. Then, $1 - p > 0$, and

$$u_o(x,t) = \sum_{\sigma_k(t)\in S_o} \frac{2\epsilon\nu_k(t)}{(x-\mu_k(t))^2 + \nu_k(t)^2} = \mathcal{O}(\epsilon^{1-p}) \tag{2.22}$$

for sufficiently small $\epsilon > 0$ where the error term is uniform with respect to x, t.

PROOF. The sum (2.22) is positive because all its terms are positive, so

$$\sum_{\sigma_k(t)\in S_o} \frac{2\epsilon\nu_k(t)}{(x-\mu_k(t))^2 + \nu_k(t)^2} \leq \sum_{\sigma_k(t)\in S_o} \frac{2\epsilon}{\nu_k(t)} \leq \tilde{K}\epsilon^{1-p}, \tag{2.23}$$

where $\tilde{K} = 2K/\beta > 0$ is independent of x, t, ϵ, as desired. □

Next we consider the upper sum $u_U(x,t)$. The main idea here is that the sum resembles a Riemann sum for an integral, which in turn has an integrand involving a highly-peaked function that approximates a Dirac delta. The first observation we make is that the conjectured asymptotic behavior of $\mu_U(y,t)$ and $\nu_U(y,t)$ near the endpoints $y = 0, 1$ is enough to neglect the contributions of the extreme eigenvalues on the upper branch.

LEMMA 2.5 (Real parts close to x). *Assume that Conjectures 2 and 3 both hold. Then, uniformly for bounded $t \geq 0$ and bounded $x \in \mathbb{R}$, and arbitrary fixed $\eta > 0$,*

$$u_U(x,t) = \sum_{\substack{\sigma_k(t)\in S_U \\ |\mu_k(t)-x|<\eta}} \frac{2\epsilon\nu_k(t)}{(x-\mu_k(t))^2 + \nu_k(t)^2} + \mathcal{O}(\delta(\epsilon)), \quad \epsilon \to 0. \tag{2.24}$$

PROOF. We estimate the complementary part of $u_U(x,t)$ by first invoking Lemma 2.3 to obtain (denoting $y_k := (k-\frac{1}{2})/N_U(\epsilon)$)

$$\begin{aligned}\sum_{\substack{\sigma_k(t)\in S_U\\|\mu_k(t)-x|\geq\eta}}&\frac{2\epsilon\nu_k(t)}{(x-\mu_k(t))^2+\nu_k(t)^2}\\&=(1+o(1))\sum_{\substack{\sigma_k(t)\in S_U\\|\mu_k(t)-x|\geq\eta}}\frac{2\epsilon\delta(\epsilon)\nu_U(y_k,t)}{(x-\mu_U(y_k,t))^2+\delta(\epsilon)^2\nu_U(y_k,t)^2}\end{aligned}\tag{2.25}$$

where we used the fact that the terms are all positive. But neglecting $\delta(\epsilon)^2\nu_U(y_k,t)^2$ from the denominator and using Conjectures 2 and 3 we get

$$\begin{aligned}\sum_{\substack{\sigma_k(t)\in S_U\\\mu_k(t)\leq x-\eta}}\frac{2\epsilon\delta(\epsilon)\nu_U(y_k,t)}{(x-\mu_U(y_k,t))^2+\delta(\epsilon)^2\nu_U(y_k,t)^2}&\leq\sum_{\substack{\sigma_k(t)\in S_U\\\mu_k(t)\leq x-\eta}}\frac{2\epsilon\delta(\epsilon)\nu_U(y_k,t)}{(x-\mu_U(y_k,t))^2}\\&\lesssim\sum_{\substack{\sigma_k(t)\in S_U\\\mu_k(t)\leq x-\eta}}\frac{2\epsilon\delta(\epsilon)y_k^{-r_-}}{y_k^{-2q_-}}.\end{aligned}\tag{2.26}$$

Letting $N_-(\epsilon)=\mathcal{O}(\epsilon^{-1})$ denote the greatest index k for which $\mu_k(t)\leq x-\eta$, this latter sum is just

$$\begin{aligned}\sum_{k=1}^{N_-(\epsilon)}\frac{2\epsilon\delta(\epsilon)y_k^{-r_-}}{y_k^{-2q_-}}&=2\epsilon\delta(\epsilon)N_U(\epsilon)^{-(2q_--r_-)}\sum_{k=1}^{N_-(\epsilon)}(k-\tfrac{1}{2})^{-(2q_--r_-)}\\&\lesssim 2\epsilon\delta(\epsilon)N_U(\epsilon)^{-(2q_--r_-)}N_-(\epsilon)^{1-(2q_--r_-)},\end{aligned}\tag{2.27}$$

because $2q_--r_->-1$, which is $\mathcal{O}(\delta(\epsilon))$ as desired. The estimate of the part of the upper bound in (2.25) with $\mu_k(t)\geq x+\eta$ is similar, using $2q_+-r_+>-1$. □

Next, we have a Lemma that will help us compute the remaining terms in the upper sum $u_U(x,t)$.

LEMMA 2.6. *Suppose that $x\in\mathbb{R}$ and $\eta>0$. Let $f:[x-\eta,x+\eta]\to\mathbb{R}$ be a continuous function with positive lower bound $f(z)\geq c_1>0$ that is differentiable with uniformly bounded derivative: $|f'(z)|\leq c_2$. The function*

$$D_\delta(z;x):=\frac{1}{\pi}\cdot\frac{\delta f(z)}{(z-x)^2+\delta^2f(z)^2}\tag{2.28}$$

with parameter $\delta>0$ is an approximate delta function in the sense that for any bounded Lipschitz continuous function $g:[x-\eta,x+\eta]\to\mathbb{R}$,

$$\int_{x-\eta}^{x+\eta}D_\delta(z;x)g(z)\,\mathrm{d}z=g(x)+\mathcal{O}(\delta\ln(\delta^{-1})),\quad\delta\to 0.\tag{2.29}$$

PROOF. Setting $s(z) := (z-x)/(\delta f(z))$, we have

$$\begin{aligned}\int_{x-\eta}^{x+\eta} D_\delta(z;x)\,\mathrm{d}z &= \int_{x-\eta}^{x+\eta} \frac{1}{s(z)^2+1}\left(s'(z)+\frac{(z-x)f'(z)}{\delta f(z)^2}\right)\frac{\mathrm{d}z}{\pi} \\ &= \int_{-\eta/(\delta f(x-\eta))}^{\eta/(\delta f(x+\eta))} \frac{1}{s^2+1}\frac{\mathrm{d}s}{\pi} + \int_{x-\eta}^{x+\eta}\frac{\delta(z-x)f'(z)}{(z-x)^2+\delta^2 f(z)^2}\frac{\mathrm{d}z}{\pi} \\ &= 1+\mathcal{O}(\delta)+\int_{x-\eta}^{x+\eta}\frac{\delta(z-x)f'(z)}{(z-x)^2+\delta^2 f(z)^2}\frac{\mathrm{d}z}{\pi}, \quad \delta\to 0,\end{aligned} \tag{2.30}$$

where on the last line we used the positive lower bound for $f(z)$. But using also the upper bound for $|f'(z)|$,

$$\begin{aligned}\left|\int_{x-\eta}^{x+\eta}\frac{\delta(z-x)f'(z)}{(z-x)^2+\delta^2 f(z)^2}\frac{\mathrm{d}z}{\pi}\right| &\le \frac{c_2\delta}{\pi}\int_{x-\eta}^{x+\eta}\frac{|z-x|\,\mathrm{d}z}{(z-x)^2+\delta^2 c_1^2} \\ &= \frac{c_2\delta}{\pi}\int_{-(\delta c_1)^{-1}}^{(\delta c_1)^{-1}}\frac{|w|\,\mathrm{d}w}{w^2+1} \\ &= \mathcal{O}(\delta\ln(\delta^{-1})), \quad \delta\to 0.\end{aligned} \tag{2.31}$$

Therefore,

$$\int_{x-\eta}^{x+\eta} D_\delta(z;x)g(z)\,\mathrm{d}z = g(x)+\mathcal{O}(\delta\ln(\delta^{-1}))+\int_{x-\eta}^{x+\eta} D_\delta(z;x)(g(z)-g(x))\,\mathrm{d}z.$$

Let $c_g>0$ denote the Lipschitz constant of g, so that $|g(z)-g(x)|\le c_g|z-x|$. Then

$$\begin{aligned}\left|\int_{x-\eta}^{x+\eta} D_\delta(z;x)(g(z)-g(x))\,\mathrm{d}z\right| &\le c_g\int_{x-\eta}^{x+\eta} D_\delta(z;x)|z-x|\,\mathrm{d}z \\ &\le \frac{Cc_g\delta}{\pi}\int_{x-\eta}^{x+\eta}\frac{|z-x|\,\mathrm{d}z}{(z-x)^2+\delta^2c_1^2}.\end{aligned} \tag{2.32}$$

where C denotes the finite positive maximum value of $f(z)$ on $[x-\eta, x+\eta]$. This upper bound is again $\mathcal{O}(\delta\ln(\delta^{-1}))$, so the proof is finished. □

Now we may compute the upper branch sum.

PROPOSITION 2.7 (Upper branch sum). *Assume that Conjectures 2 and 3 hold. Then,*

$$u_U(x,t) := \sum_{\sigma_k(t)\in S_U}\frac{2\epsilon\nu_k(t)}{(x-\mu_k(t))^2+\nu_k(t)^2} = \psi_U(x,t)+o(1), \quad \epsilon\to 0, \tag{2.33}$$

where

$$\psi_U(x,t) := \frac{2\pi\epsilon N_U(\epsilon)}{\mu_U'(\mu_U^{-1}(x,t),t)}, \quad \mu_U'(y,t) := \partial_y\mu_U(y,t), \tag{2.34}$$

and the $o(1)$ error term is uniform on bounded subsets of $(x,t)\in\mathbb{R}\times[0,\infty)$ for which (x,t) is bounded away from the curve $(X^-(\tau),\tau)$ parametrized by $\tau\ge t_b$.

REMARK 2.8. The product $\epsilon N_U(\epsilon)$ converges as $\epsilon\to 0$ to a finite nonzero value depending on t only. In fact, for $0\le t\le t_b$, $\epsilon N_U(\epsilon)\to M$ defined in (1.10) as $\epsilon\to 0$.

PROOF. Thanks to Lemma 2.5 and the fact that $\delta(\epsilon) = o(1)$ as $\epsilon \to 0$ it suffices to study the sum over eigenvalues $\sigma_k(t) \in S_U$ for which $|\mu_k(t) - x| < \eta$ for any suitable $\eta > 0$. A suitable value for η will be specified at the end of the proof. Letting $k = k_-, \ldots, k_+$ denote the index range for which $|\mu_k(t) - x| < \eta$ holds, applying Lemma 2.3 and using the positivity of the summand, the contributing terms can be written as
(2.35)

$$\sum_{\substack{\sigma_k(t) \in S_U \\ |\mu_k(t) - x| < \eta}} \frac{2\epsilon \nu_k(t)}{(x - \mu_k(t))^2 + \nu_k(t)^2} = (1 + o(1)) \sum_{k=k_-}^{k_+} \frac{2\epsilon\delta(\epsilon)\nu(y_k)}{(x - \mu(y_k))^2 + \delta(\epsilon)^2 \nu(y_k)^2},$$

where $y_k := (k - \frac{1}{2})/N_U(\epsilon)$, and for simplicity we are abbreviating $\mu(y) := \mu_U(y,t)$ and $\nu(y) := \nu_U(y,t)$. Since we have a sum of a sampling of a continuous function over a large number of grid points because $k_+ - k_- \gtrsim \epsilon^{-1}$, we can apply the Euler-Maclaurin formula in the form

$$\sum_{k=k_-}^{k_+} h(k) = \int_{k_-}^{k_+} h(k)\,\mathrm{d}k + \frac{h(k_+) + h(k_-)}{2} + \int_{k_-}^{k_+} h'(k)(k - \lfloor k \rfloor - \tfrac{1}{2})\,\mathrm{d}k \tag{2.36}$$

with $h(k)$ a function defined on the real interval $[k_-, k_+]$ by

$$h(k) := \frac{2\epsilon\delta(\epsilon)\nu(y_k)}{(x - \mu(y_k))^2 + \delta(\epsilon)^2\nu(y_k)^2}, \quad y_k := \frac{k - \frac{1}{2}}{N_U(\epsilon)}, \quad k \in [k_+, k_-]. \tag{2.37}$$

Now, when $k = k_\pm$, $\mu(y_k) \approx x \pm \eta$, so certainly $|x - \mu(y_{k_\pm})| \geq \frac{1}{2}\eta$. Therefore $|h(k_\pm)| \leq 8\epsilon\delta(\epsilon)\nu(y_{k_\pm})\eta^{-1}$. Since $\nu(y)$ is a continuous function of y on any closed subinterval of $(0,1)$, and is independent of ϵ, we easily obtain

$$\frac{h(k_+) + h(k_-)}{2} = \mathcal{O}(\epsilon\delta(\epsilon)), \quad \epsilon \to 0. \tag{2.38}$$

The last integral on the right-hand side of (2.36) can be estimated using that $|k - \lfloor k \rfloor - \frac{1}{2}| \leq \frac{1}{2}$ holds for all $k \in \mathbb{R}$, and hence with a change of the integration variable we get

$$\left| \int_{k_-}^{k_+} h'(k)(k - \lfloor k \rfloor - \tfrac{1}{2})\,\mathrm{d}k \right| \leq \frac{1}{2} \int_{k_-}^{k_+} |h'(k)|\,\mathrm{d}k = \frac{1}{2} \int_{y_{k_-}}^{y_{k_+}} \left| \frac{\mathrm{d}}{\mathrm{d}y} h(N_U(\epsilon)y + \tfrac{1}{2}) \right| \mathrm{d}y.$$

Thus, using (2.37) we have

$$\left| \int_{k_-}^{k_+} h'(k)(k - \lfloor k \rfloor - \tfrac{1}{2})\,\mathrm{d}k \right| \leq \epsilon\delta(\epsilon) \int_{y_{k_-}}^{y_{k_+}} \left| \frac{\mathrm{d}}{\mathrm{d}y} \frac{\nu(y)}{(x - \mu(y))^2 + \delta(\epsilon)^2\nu(y)^2} \right| \mathrm{d}y.$$

Doing the differentiation, we get a larger upper bound as a sum of three terms:

$$\int_{y_{k_-}}^{y_{k_+}} \left| \frac{\mathrm{d}}{\mathrm{d}y} \frac{\nu(y)}{(x - \mu(y))^2 + \delta(\epsilon)^2\nu(y)^2} \right| \mathrm{d}y \leq I_1 + I_2 + I_3, \tag{2.39}$$

where

$$
\begin{aligned}
I_1 &:= \int_{y_{k_-}}^{y_{k_+}} \frac{|\nu'(y)|\,\mathrm{d}y}{(x-\mu(y))^2+\delta(\epsilon)^2\nu(y)^2},\\
I_2 &:= \int_{y_{k_-}}^{y_{k_+}} \frac{2\delta(\epsilon)^2\nu(y)^2|\nu'(y)|\,\mathrm{d}y}{[(x-\mu(y))^2+\delta(\epsilon)^2\nu(y)^2]^2},\\
I_3 &:= \int_{y_{k_-}}^{y_{k_+}} \frac{2|\mu(y)-x|\nu(y)\mu'(y)\,\mathrm{d}y}{[(\mu(y)-x)^2+\delta(\epsilon)^2\nu(y)^2]^2},
\end{aligned}
\tag{2.40}
$$

where to write I_3 we used the fact that $\nu(y)>0$ and $\mu'(y)>0$. Now, according to Conjecture 3, we have a lower bound of the form $\nu(y)\geq c_U>0$, and as noted above $\nu(y)$ attains a finite maximum on $[y_{k_-},y_{k_+}]$ that we denote by C. Likewise, Conjecture 3 asserts the absolute continuity of the derivative $\nu'(y)$ on $(0,1)$, and hence

$$
I_1 \leq \frac{1}{c_U^2\delta(\epsilon)^2}\int_{y_{k_-}}^{y_{k_+}} |\nu'(y)|\,\mathrm{d}y \quad\text{and}\quad I_2 \leq \frac{2C^2}{c_U^4\delta(\epsilon)^2}\int_{y_{k_-}}^{y_{k_+}} |\nu'(y)|\,\mathrm{d}y.
\tag{2.41}
$$

Since $y_{k_\pm}\approx\mu^{-1}(x\pm\eta)$, and the latter values do not depend on ϵ, it is clear that $I_1+I_2=\mathcal{O}(\delta(\epsilon)^{-2})$ as $\epsilon\to 0$. To estimate I_3, we first use the upper and lower bounds on $\nu(y)$ to obtain

$$
I_3 \leq C\int_{y_{k_-}}^{y_{k_+}} \frac{2|\mu(y)-x|\mu'(y)\,\mathrm{d}y}{[(\mu(y)-x)^2+\delta(\epsilon)^2c_U^2]^2}.
\tag{2.42}
$$

We next split the integral at the point $y=\mu^{-1}(x)$; then $\mu(y)>x$ holds for $\mu^{-1}(x)<y<y_{k_+}$ while $\mu(y)<x$ holds for $y_{k_-}<y<\mu^{-1}(x)$. Therefore,

$$
\begin{aligned}
\int_{y_{k_-}}^{y_{k_+}} \frac{2|\mu(y)-x|\mu'(y)\,\mathrm{d}y}{[(\mu(y)-x)^2+\delta(\epsilon)^2c_U^2]^2} &= \left(\int_{\mu^{-1}(x)}^{y_{k_+}} - \int_{y_{k_-}}^{\mu^{-1}(x)}\right)\frac{2(\mu(y)-x)\mu'(y)\,\mathrm{d}y}{[(\mu(y)-x)^2+\delta(\epsilon)^2c_U^2]^2}\\
&= \left(\int_{x}^{\mu(y_{k_+})} - \int_{\mu(y_{k_-})}^{x}\right)\frac{2(\mu-x)\,\mathrm{d}\mu}{[(\mu-x)^2+\delta(\epsilon)^2c_U^2]^2}.
\end{aligned}
$$

Carrying out the explicit integration and using the result in (2.42) gives

$$
I_3 \leq \frac{2C}{\delta(\epsilon)^2c_U^2} - \frac{C}{(\mu(y_{k_+})-x)^2+\delta(\epsilon)^2c_U^2} - \frac{C}{(\mu(y_{k_-})-x)^2+\delta(\epsilon)^2c_U^2} \leq \frac{2C}{\delta(\epsilon)^2c_U^2}
$$

so also $I_3=\mathcal{O}(\delta(\epsilon)^{-2})$ as $\epsilon\to 0$. Combining the estimates and using $\epsilon\ll\delta(\epsilon)\ll 1$ shows that

$$
\begin{aligned}
\sum_{k=k_-}^{k=k_+} h(k) - \int_{k_-}^{k_+} h(k)\,\mathrm{d}k &= \mathcal{O}(\epsilon\delta(\epsilon)) + \mathcal{O}(\epsilon\delta(\epsilon)^{-1})\\
&= \mathcal{O}(\epsilon\delta(\epsilon)^{-1}) = o(1), \quad \epsilon\to 0.
\end{aligned}
\tag{2.43}
$$

It only remains to consider the integral

$$
\begin{aligned}
\int_{k_-}^{k_+} h(k)\,\mathrm{d}k &= N_U(\epsilon)\int_{y_{k_-}}^{y_{k_+}} h(N_U(\epsilon)y+\tfrac{1}{2})\,\mathrm{d}y \\
&= 2\epsilon N_U(\epsilon)\int_{y_{k_-}}^{y_{k_+}} \frac{\delta(\epsilon)\nu(y)\,\mathrm{d}y}{(x-\mu(y))^2+\delta(\epsilon)^2\nu(y)^2} \\
&= \frac{2\pi\epsilon N_U(\epsilon)}{\pi}\int_{\mu(y_{k_-})}^{\mu(y_{k_+})} \frac{\delta(\epsilon)\nu(\mu^{-1}(z))}{(x-z)^2+\delta(\epsilon)^2\nu(\mu^{-1}(z))^2}\frac{\mathrm{d}z}{\mu'(\mu^{-1}(z))},
\end{aligned}
\tag{2.44}
$$

where on the last line we changed variables by the increasing map $z=\mu(y)$. Since $\mu(y_{k_\pm})=x\pm\eta+\mathcal{O}(\epsilon)$ as $\epsilon\to 0$ and the integrand is $\mathcal{O}(\delta(\epsilon))$ near the limits of integration,

$$
\int_{k_-}^{k_+} h(k)\,\mathrm{d}k = \frac{2\pi\epsilon N_U(\epsilon)}{\pi}\int_{x-\eta}^{x+\eta} \frac{\delta(\epsilon)\nu(\mu^{-1}(z))}{(x-z)^2+\delta(\epsilon)^2\nu(\mu^{-1}(z))^2}\frac{\mathrm{d}z}{\mu'(\mu^{-1}(z))}+\mathcal{O}(\epsilon\delta(\epsilon))
$$

because $\epsilon N_U(\epsilon)=\mathcal{O}(1)$. Finally, we appeal to Lemma 2.6 with $f(z):=\nu(\mu^{-1}(z))$ and $g(z):=\mu'(\mu^{-1}(z))^{-1}$, which will satisfy the required hypotheses provided we now choose $\eta>0$ appropriately. If $0\le t<t_b$, we will take $\eta=1$, but if $t\ge t_b$, since $x\neq X^-(t)$, we will take $\eta<\frac{1}{2}|x-X^-(t)|$. This choice guarantees that, according to Conjectures 2 and 3, $f(z)$ is continuous with a positive lower bound and is continuously differentiable (hence having bounded derivative) for $z\in[x-\eta,x+\eta]$; likewise $g(z)$ is continuously differentiable (hence Lipschitz) on the same interval. We conclude that

$$
\begin{aligned}
\int_{k_-}^{k_+} h(k)\,\mathrm{d}k &= \frac{2\pi\epsilon N_U(\epsilon)}{\mu'(\mu^{-1}(x))}+\mathcal{O}(\delta(\epsilon)\ln(\delta(\epsilon)^{-1})) \\
&= \frac{2\pi\epsilon N_U(\epsilon)}{\mu'(\mu^{-1}(x))}+o(1),\quad \epsilon\to 0,
\end{aligned}
\tag{2.45}
$$

because $\epsilon\ll\delta(\epsilon)\ll 1$. Combining with (2.35), (2.37), and (2.43) completes the proof. □

Now we can turn our attention to the sum over the lower branch of eigenvalues, assuming that (x,t) is inside the triple-valued region for Burgers' equation.

PROPOSITION 2.9 (Lower branch sum). Fix $t>t_b$, and let x_0 be fixed in the interior of the triple-valued region for Burgers' equation, i.e., $X^-(t)<x_0<X^+(t)$. Assume that Conjectures 2 and 4 hold. Then,

$$
u_L(x,t)=\frac{\psi_L(x_0,t)\sinh(\phi_L(x_0,t))}{\cosh(\phi_L(x_0,t))-\cos(\psi_L(x_0,t)\epsilon^{-1}(x-x_0)+2\pi p(x_0))}+o(1)
\tag{2.46}
$$

holds uniformly in the limit $\epsilon\to 0$ for $x-x_0=\mathcal{O}(\epsilon^{3/4})$ where $p(x_0)$ is a uniformly bounded (bound independent of x_0) phase shift, see (2.61) below, and

$$
\psi_L(x,t):=\frac{2\pi\epsilon N_L(\epsilon)}{\mu_L'(\mu_L^{-1}(x,t),t)},\quad \phi_L(x,t):=\psi_L(x,t)\nu_L(\mu_L^{-1}(x,t),t).
\tag{2.47}
$$

PROOF. For brevity, we write $\nu(y):=\nu_L(y,t)$ and $\mu(y):=\mu_L(y,t)$ for the sampling functions defined in Conjectures 2 and 4 as t is fixed and we are only concerned with the lower branch. For $u_L(x,t)$ defined by (2.7), we first apply Lemma 2.3 to get

$$u_L(x,t) = (1+o(1)) \sum_{k=1}^{N_L(\epsilon)} \frac{2\epsilon^2\nu(y_k)}{(x-\mu(y_k))^2+\epsilon^2\nu(y_k)^2}, \quad y_k := \frac{k-\frac{1}{2}}{N_L(\epsilon)} \tag{2.48}$$

because the terms are all positive. We will work with the explicit sum and then deal later with the multiplicative factor $1+o(1)$. The leading-order contribution to the explicit sum will come from the terms where $\mu(y_k)$ is close to x_0 which we specify by the condition $|\mu(y_k)-x_0| \leq 2\epsilon^r$ where r is an exponent with $0 < r < \frac{1}{2}$. Indeed, under the complementary condition $|\mu(y_k)-x_0| > 2\epsilon^r$, we have $|x-\mu_k(y)| \geq ||\mu(y_k)-x_0|-|x-x_0|| = |\mu(y_k)-x_0|-|x-x_0| \geq \epsilon^r$ for $\epsilon > 0$ sufficiently small, because $x-x_0 = \mathcal{O}(\epsilon^{3/4}) = o(\epsilon^r)$ holds for $r < \frac{1}{2}$. Therefore, neglecting the term $\epsilon^2\nu(y_k)^2$ in the denominator and extending the sum over the full range of k, the sum of complementary terms is estimated as follows:

$$\sum_{\substack{k=1 \\ |\mu(y_k)-x_0|>2\epsilon^r}}^{N_L(\epsilon)} \frac{2\epsilon^2\nu(y_k)}{(x-\mu(y_k))^2+\epsilon^2\nu(y_k)^2} \leq 2\epsilon N_L(\epsilon)\cdot\epsilon^{1-2r}\sum_{k=1}^{N_L(\epsilon)}\frac{\nu(y_k)}{N_L(\epsilon)}. \tag{2.49}$$

The last sum is a Riemann sum for the integral $\int_0^1 \nu(y)\,dy$ which is finite according to Conjecture 4. Since $\epsilon N_L(\epsilon) = \mathcal{O}(1)$ as $\epsilon \to 0$ and $r < \frac{1}{2}$, the sum (2.49) of complementary terms tends to zero with ϵ.

The remaining terms in the explicit sum on the right-hand side of (2.48) have indices k in the set $\mathscr{I}^r := \{j = 1, 2, \ldots, N_L(\epsilon) : |\mu(y_j)-x_0| \leq 2\epsilon^r\}$. Next we show that for the terms in the explicit sum with indices in $\mathscr{I}^r$, we can replace $\nu(y_k)$ in the summand with $\nu(\mu^{-1}(x_0))$ because $r > 0$. Indeed, consider the difference

$$\begin{aligned} D &:= \sum_{k\in\mathscr{I}^r}\left[\frac{2\epsilon^2\nu(y_k)}{(x-\mu(y_k))^2+\epsilon^2\nu(y_k)^2} - \frac{2\epsilon^2\nu(\mu^{-1}(x_0))}{(x-\mu(y_k))^2+\epsilon^2\nu(\mu^{-1}(x_0))^2}\right] \\ &= 2\epsilon^2\sum_{k\in\mathscr{I}^r}\frac{[(x-\mu(y_k))^2-\epsilon^2\nu(y_k)\nu(\mu^{-1}(x_0))](\nu(y_k)-\nu(\mu^{-1}(x_0)))}{[(x-\mu(y_k))^2+\epsilon^2\nu(y_k)^2][(x-\mu(y_k))^2+\epsilon^2\nu(\mu^{-1}(x_0))^2]}. \end{aligned} \tag{2.50}$$

Since the function $x \mapsto \nu(\mu^{-1}(x))$ is continuously differentiable near $x = x_0$, the condition $|\mu(y_k)-x_0| \leq 2\epsilon^r$ implies that $|\nu(y_k)-\nu(\mu^{-1}(x_0))| \lesssim \epsilon^r$, and therefore

$$|D| \lesssim \epsilon^{2+r}\sum_{k\in\mathscr{I}^r}\frac{(x-\mu(y_k))^2+\epsilon^2\nu(y_k)\nu(\mu^{-1}(x_0))}{[(x-\mu(y_k))^2+\epsilon^2\nu(y_k)^2][(x-\mu(y_k))^2+\epsilon^2\nu(\mu^{-1}(x_0))^2]}. \tag{2.51}$$

Furthermore, according to Conjecture 4 we have the lower bound $\nu(y) \geq c_L > 0$, and as ν is a C^1 function near $\mu^{-1}(x_0) > 0$, we have also a local upper bound: $\nu(y) \leq K$, so

$$|D| \lesssim \epsilon^{2+r}\sum_{k\in\mathscr{I}^r}\frac{(x-\mu(y_k))^2}{[(x-\mu(y_k))^2+\epsilon^2 c_L^2]^2} + \epsilon^{4+r}\sum_{k\in\mathscr{I}^r}\frac{1}{[(x-\mu(y_k))^2+\epsilon^2 c_L^2]^2}. \tag{2.52}$$

Let $\mathscr{I}_0^r := \{k \in \mathscr{I}^r : |x - \mu(y_k)| \leq \epsilon\}$. Then the cardinality of $\mathscr{I}_0^r$ is $|\mathscr{I}_0^r| = \mathcal{O}(1)$ as $\epsilon \downarrow 0$, so

$$\begin{aligned}\sum_{k\in\mathscr{I}_0^r} \frac{(x-\mu(y_k))^2}{[(x-\mu(y_k))^2+\epsilon^2c_L^2]^2} &\leq \sum_{k\in\mathscr{I}_0^r} \frac{\epsilon^2}{\epsilon^4 c_L^4} \lesssim \epsilon^{-2},\\ \sum_{k\in\mathscr{I}_0^r} \frac{1}{[(x-\mu(y_k))^2+\epsilon^2c_L^2]^2} &\leq \sum_{k\in\mathscr{I}_0^r} \frac{1}{\epsilon^4 c_L^4} \lesssim \epsilon^{-4},\end{aligned} \tag{2.53}$$

and therefore the terms indexed by $\mathscr{I}_0^r$ contribute $\mathcal{O}(\epsilon^r)$ to the right-hand side of the estimate (2.52). Then for the remaining indices in $\mathscr{I}^r$ we can write

$$\begin{aligned}\sum_{k\in\mathscr{I}^r\setminus\mathscr{I}_0^r} \frac{(x-\mu(y_k))^2}{[(x-\mu(y_k))^2+\epsilon^2c_L^2]^2} &\leq \sum_{k\in\mathscr{I}^r\setminus\mathscr{I}_0^r} \frac{1}{(x-\mu(y_k))^2},\\ \sum_{k\in\mathscr{I}^r\setminus\mathscr{I}_0^r} \frac{1}{[(x-\mu(y_k))^2+\epsilon^2c_L^2]^2} &\leq \sum_{k\in\mathscr{I}^r\setminus\mathscr{I}_0^r} \frac{1}{(x-\mu(y_k))^4}.\end{aligned} \tag{2.54}$$

Then, for either exponent $p = 2, 4$, we can use the positive lower bound on $\mu'(y) = \mu_L'(y,t)$ implied by Conjecture 2 to get

$$\sum_{k\in\mathscr{I}^r\setminus\mathscr{I}_0^r} \frac{1}{(x-\mu(y_k))^p} \lesssim N_L(\epsilon) \sum_{k\in\mathscr{I}^r\setminus\mathscr{I}_0^r} \frac{\mu'(y_k)}{(x-\mu(y_k))^p}\cdot\frac{1}{N_L(\epsilon)}. \tag{2.55}$$

Since $1/N_L(\epsilon)$ is exactly the spacing of the points y_k, the latter sum is a Riemann sum for an integral, and hence

$$\sum_{k\in\mathscr{I}^r\setminus\mathscr{I}_0^r} \frac{1}{(x-\mu(y_k))^p} \lesssim N_L(\epsilon)\int_{|\mu-x|\geq\epsilon} \frac{\mathrm{d}\mu}{(x-\mu)^p}, \tag{2.56}$$

wherein the integration is extended to $\mu = \pm\infty$ as a finite upper bound since $p = 2, 4$. Performing the integration shows that

$$\sum_{k\in\mathscr{I}^r\setminus\mathscr{I}_0^r} \frac{1}{(x-\mu(y_k))^p} \lesssim N_L(\epsilon)\epsilon^{1-p} \lesssim \epsilon^{-p}, \quad p = 2, 4. \tag{2.57}$$

Combining with (2.54) shows that also the terms with indices in $\mathscr{I}^r\setminus\mathscr{I}_0^r$ contribute $\mathcal{O}(\epsilon^r)$ to the right-hand side of the estimate (2.52) so $D \to 0$ as $\epsilon \to 0$.

Therefore, it remains to analyze the sum

$$\Sigma := \sum_{k\in\mathscr{I}^r} \frac{2\epsilon^2\nu(\mu^{-1}(x_0))}{(x-\mu(y_k))^2+\epsilon^2\nu(\mu^{-1}(x_0))^2}, \tag{2.58}$$

which can be written in the form

$$\Sigma = \frac{2}{\nu(\mu^{-1}(x_0))}\sum_{k\in\mathscr{I}^r} L\left(\frac{x-\mu(y_k)}{\epsilon\nu(\mu^{-1}(x_0))}\right), \quad L(z) := \frac{1}{z^2+1}. \tag{2.59}$$

Let $k_0 \in \mathscr{I}^r$ be the index such that $|x_0 - \mu(y_k)|$ is minimized. The spacing of the points $\mu(y_k)$ near x_0 is approximately $N_L(\epsilon)^{-1}\mu'(y_{k_0})$. More precisely, since $\mu(y)$ is twice continuously differentiable according to Conjecture 2 with derivative $\mu'(y_{k_0}) > 0$,

$$\mu(y_{k_0}) - \mu(y_k) = \mu'(y_{k_0})\frac{k_0-k}{N_L(\epsilon)} + \mathcal{O}\left(\left(\frac{k_0-k}{N_L(\epsilon)}\right)^2\right) \tag{2.60}$$

We next express x_0 in terms of the phase shift $p(x_0)$, which is defined by the relation

$$x_0 = \mu(y_{k_0}) + \mu'(y_{k_0})\frac{p(x_0)}{N_L(\epsilon)}. \tag{2.61}$$

The phase shift $p(x_0)$ must be bounded with an upper bound on $|p(x_0)|$ close to $\frac{1}{2}$ because k_0 minimizes the distance between x_0 and $\mu(y_k)$. Combining (2.60) and (2.61), we write the argument of L in the summand of (2.59) as

$$z = \frac{x - \mu(y_k)}{\epsilon\nu(\mu^{-1}(x_0))} = \tau + \frac{k_0 - k}{\Omega(x_0;\epsilon)} + \mathcal{O}\left(\frac{(k_0-k)^2}{N_L(\epsilon)}\right), \tag{2.62}$$

where

$$\tau := \frac{x - x_0}{\epsilon\nu(\mu^{-1}(x_0))} + \frac{p(x_0)}{\Omega(x_0;\epsilon)}, \tag{2.63}$$

and $\Omega(x_0;\epsilon)$ is the quantity

$$\Omega(x_0;\epsilon) := \frac{\epsilon N_L(\epsilon)\nu(\mu^{-1}(x_0))}{\mu'(\mu^{-1}(x_0))}, \tag{2.64}$$

which has a finite nonzero limit $\Omega(x_0;0)$ as $\epsilon \to 0$. We notice that the error term in the argument of L will be small of order $\mathcal{O}(\epsilon^{1-2q})$ for indices $k \in \mathscr{I}^r$ for which $|k - k_0| \le \epsilon^{-q}$, where q is any exponent with $0 < q < \frac{1}{2}$. Using also $r < \frac{1}{2}$, these terms will be a small fraction of the total cardinality $|\mathscr{I}^r| \sim \epsilon^{r-1}$ because $q + r < 1$. Since $z \mapsto L(z)$ obviously has a uniformly bounded derivative,

$$\begin{aligned} \sum_{\substack{k\in\mathscr{I}^r \\ |k-k_0|\le\epsilon^{-q}}} L\left(\frac{x-\mu(y_k)}{\epsilon\nu(\mu^{-1}(x_0))}\right) &= \sum_{\substack{k\in\mathscr{I}^r \\ |k-k_0|\le\epsilon^{-q}}} \left[L\left(\tau + \frac{k_0-k}{\Omega(x_0;\epsilon)}\right) + \mathcal{O}(\epsilon^{1-2q})\right] \\ &= \sum_{\substack{k\in\mathscr{I}^r \\ |k-k_0|\le\epsilon^{-q}}} L\left(\tau + \frac{k_0-k}{\Omega(x_0;\epsilon)}\right) + \mathcal{O}(\epsilon^{1-3q}) \end{aligned} \tag{2.65}$$

as $\epsilon \to 0$. To ensure that the error term in (2.65) is small we will now further constrain q by assuming $q < \frac{1}{3}$. Now, for k in the complementary part of $\mathscr{I}^r$ where $|k - k_0| > \epsilon^{-q}$, we see that if we can guarantee the condition $x - x_0 = o(\epsilon^{1-q})$ as $\epsilon \to 0$, we have $\tau = o((k_0 - k)/\Omega(x_0;\epsilon))$ because also $p(x_0)/\Omega(x_0;\epsilon)$ is bounded. Therefore,

$$z = \frac{x - \mu(y_k)}{\epsilon\nu(\mu^{-1}(x_0))} = \frac{k_0 - k}{\Omega(x_0;\epsilon)}(1 + o(1)), \tag{2.66}$$

so for these terms the argument z of L satisfies $|z| \gtrsim \epsilon^{-q}$. Since $L(z) = \mathcal{O}(z^{-2})$ as $z \to \infty$,

$$\sum_{\substack{k\in\mathscr{I}^r \\ |k-k_0|>\epsilon^{-q}}} L\left(\frac{x-\mu(y_k)}{\epsilon\nu(\mu^{-1}(x_0))}\right) = \sum_{\substack{k\in\mathscr{I}^r \\ |k-k_0|>\epsilon^{-q}}} \mathcal{O}(\epsilon^{2q}) = \mathcal{O}(\epsilon^{2q+r-1}), \tag{2.67}$$

because these terms constitute the dominant fraction of those in $\mathscr{I}^r$, and $|\mathscr{I}^r| \sim \epsilon^{r-1}$. For the same reasons, we also have

$$\sum_{\substack{k\in\mathscr{I}^r \\ |k-k_0|>\epsilon^{-q}}} L\left(\tau + \frac{k_0-k}{\Omega(x_0;\epsilon)}\right) = \mathcal{O}(\epsilon^{2q+r-1}), \tag{2.68}$$

so combining (2.65), (2.67), and (2.68) we obtain

$$\sum_{k\in\mathscr{I}^r} L\left(\frac{x-\mu(y_k)}{\epsilon\nu(\mu^{-1}(x_0))}\right) = \sum_{k\in\mathscr{I}^r} L\left(\tau + \frac{k_0-k}{\Omega(x_0;\epsilon)}\right) + \mathcal{O}(\epsilon^{1-3q}) + \mathcal{O}(\epsilon^{2q+r-1})$$

as $\epsilon\to 0$. To guarantee that the term $\mathcal{O}(\epsilon^{2q+r-1})$ is negligible as $\epsilon\to 0$, we must put a lower bound on q, namely $q>\frac{1}{2}(1-r)$ which also implies $q>\frac{1}{4}$ because $r<\frac{1}{2}$. With this inequality on q we can verify that because $x-x_0=\mathcal{O}(\epsilon^{3/4})$ it also holds that $x-x_0=o(\epsilon^{1-q})$, exactly as presumed above.

Now that the argument of $L(z):=(z^2+1)^{-1}$ is linear in the index k, the corresponding infinite series summing over $k\in\mathbb{Z}$ is explicitly convergent [**GR07**, 1.445.9]:

$$\begin{aligned}\sum_{k\in\mathbb{Z}} L\left(\tau+\frac{k_0-k}{\Omega(x_0;\epsilon)}\right) &= \sum_{j\in\mathbb{Z}} L\left(\tau+\frac{j}{\Omega(x_0;\epsilon)}\right)\\ &= \frac{\pi\Omega(x_0;\epsilon)\sinh(2\pi\Omega(x_0;\epsilon))}{\cosh(2\pi\Omega(x_0;\epsilon))-\cos(2\pi\Omega(x_0;\epsilon)\tau)}.\end{aligned}\tag{2.69}$$

Hence if $0<r<\frac{1}{2}$ and $\frac{1}{4}<\frac{1}{2}(1-r)<q<\frac{1}{3}$, the bound $x-x_0=\mathcal{O}(\epsilon^{3/4})$ yields

$$\sum_{k\in\mathscr{I}^r} L\left(\frac{x-\mu(y_k)}{\epsilon\nu(\mu^{-1}(x_0))}\right) = \frac{\pi\Omega(x_0;\epsilon)\sinh(2\pi\Omega(x_0;\epsilon))}{\cosh(2\pi\Omega(x_0;\epsilon))-\cos(2\pi\Omega(x_0;\epsilon)\tau)} + o(1),\quad \epsilon\to 0,\tag{2.70}$$

and referring back to (2.58)–(2.59), this implies that Σ is given by

$$\begin{aligned}\Sigma &= \sum_{k\in\mathscr{I}^r} \frac{2\epsilon^2\nu(\mu^{-1}(x_0))}{(x-\mu(y_k))^2+\epsilon^2\nu(\mu^{-1}(x_0))^2}\\ &= \frac{2\pi\Omega(x_0;\epsilon)}{\nu(\mu^{-1}(x_0))}\frac{\sinh(2\pi\Omega(x_0;\epsilon))}{\cosh(2\pi\Omega(x_0;\epsilon))-\cos(2\pi\Omega(x_0;\epsilon)\tau)} + o(1),\quad \epsilon\to 0.\end{aligned}\tag{2.71}$$

Under the same conditions on r we have already seen that $D=o(1)$ (see (2.50)) and that the right-hand side of (2.49) is $o(1)$. Combining these results with (2.48) and noting that $2\pi\Omega(x_0;\epsilon)=\phi_L(x_0,t)$ and $2\pi\Omega(x_0;\epsilon)\tau=\psi_L(x_0,t)\epsilon^{-1}(x-x_0)+2\pi p(x_0)$ (see (2.47)) completes the proof. □

REMARK 2.10. The identity in (2.69) is a BO analogue of a corresponding identity for the KdV equation in which a periodic traveling wave solution given by the square of a Jacobi elliptic function can be expressed as a linear sum of translates of solitons proportional to the square of the hyperbolic secant function. See [**DLMF**, 22.11.14].

Combining Propositions 2.4, 2.7, and 2.9 gives us the complete result.

THEOREM 2.11 (Asymptotic expansion of the soliton ensemble). *Let u_0 be an admissible initial condition with one inflection point to the right of the maximizer, and let $u(x,t)$ denote the corresponding soliton ensemble solving* (1.1). *Suppose either that $0\leq t<t_b$, or that $t\geq t_b$ and $x\in\mathbb{R}\setminus[X^-(t),X^+(t)]$, and that Conjectures* 1, 2, *and* 3 *hold. Then*

$$u(x,t)=\psi_U(x,t)+o(1),\quad \epsilon\to 0\tag{2.72}$$

with the error term being uniform on compact subsets of the indicated domain. On the other hand, if $t > t_b$ is fixed and $x_0 \in (X^-(t), X^+(t))$, and Conjectures 1, 2, 3, and 4 hold, then as $\epsilon \to 0$,

(2.73)
$$u(x,t) = \psi_U(x,t) + \frac{\psi_L(x_0,t)\sinh(\phi_L(x_0,t))}{\cosh(\phi_L(x_0,t)) - \cos(\psi_L(x_0,t)\epsilon^{-1}(x-x_0) + 2\pi p(x_0))} + o(1)$$

with the error term being uniform for $x - x_0 = \mathcal{O}(\epsilon^{3/4})$.

2.3. Comparison with expectations of Whitham modulation theory. The explicit terms on the right-hand side of (2.73) give, for each fixed (x_0, t) a periodic function of x that is the profile f of an exact traveling wave solution $u(x,t) = f(x-ct)$ of the BO equation (1.1); see [**Ben67**, **Ono75**]. The periodic wave has wavelength proportional to ϵ, and the approximation asserted in Theorem 2.11 is valid for a range of values of $x - x_0$ that includes a large number ($\mathcal{O}(\epsilon^{-1/4})$) of wavelengths. Rapidly oscillatory time dependence of the wave enters implicitly via the phase $p(x_0)$. Resolving the time dependence of $p(x_0)$ would require further conjectures regarding that of the eigenvalues $\sigma_k(t)$ going beyond the scope of our paper. However, as the parameters (x_0, t) vary within the domain $t > t_b$ with $X^-(t) < x_0 < X^+(t)$, the parameters of the periodic wave vary as well, which means that the soliton ensemble is actually a relatively slowly modulated periodic wavetrain.

According to the formal Whitham modulation theory for the BO equation developed by Dobrokhotov and Krichever in [**DK91**], modulated periodic waves should have the form given in (2.73), but the quantities ψ_U, ψ_L, and ϕ_L should be given as functions of (x_0, t) in terms of three Riemann invariants solving a system of uncoupled Burgers equations. We identify these Riemann invariants with the branches of the multi-valued solution of Burgers' equation with initial data u_0. Matching the formula (2.73) with the Whitham theory requires the following explicit identifications:

$$\psi_U(x,t) = u_0^{\mathrm{B}}(x,t), \tag{2.74}$$

and, for (x,t) in the multi-valued Burgers region,

$$\psi_L(x,t) = u_2^{\mathrm{B}}(x,t) - u_1^{\mathrm{B}}(x,t),\ \phi_L(x,t) = \frac{1}{2}\ln\left(\frac{u_2^{\mathrm{B}}(x,t) - u_0^{\mathrm{B}}(x,t)}{u_1^{\mathrm{B}}(x,t) - u_0^{\mathrm{B}}(x,t)}\right). \tag{2.75}$$

We do not have direct proof of either (2.74) or (2.75), although we do have solid numerical evidence that both are true. Indeed, the numerical approximation of the multiscale soliton ensemble $u(x,t)$, and the slowly varying modulation parameters $\psi_{L,U}(x_0,t)$, and $\phi_L(x_0,t)$ obtained from the eigenvalues $\sigma_k(t)$ are plotted along with the expressions above in Figure 8. This shows that the distribution of complex eigenvalues $\sigma_k(t)$ of $\mathbf{C}(t)$ indeed produces an approximate formula (2.73) for the BO soliton ensemble associated with the initial data u_0 that is fully consistent with Whitham modulation theory. We formalize these expectations in a final conjecture.

CONJECTURE 5 (Connection with Whitham modulation theory). The function $\psi_U(x,t)$ defined in terms of the scaling functions of the upper-branch eigenvalues by (2.34) and the functions $\psi_L(x,t)$ and $\phi_L(x,t)$ defined in terms of those of the lower-branch eigenvalues by (2.47) are related to the generally multi-valued solution of Burgers' equation with initial data u_0 by (2.74)–(2.75).

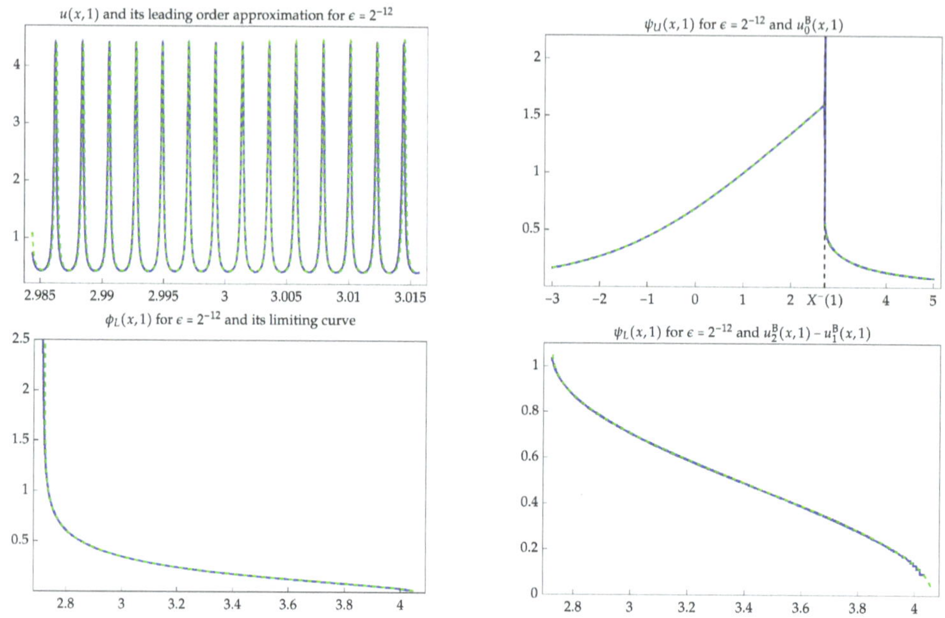

FIGURE 8. The theoretical asymptotic expressions for the soliton ensemble $u(x,t)$ ((2.73) with $x_0 = 3$) and modulation fields $\psi_{U,L}(x,t)$, and $\phi_L(x,t)$ (see (2.74)–(2.75)) shown with dashed green curves compared with the soliton ensemble for fixed nonzero $\epsilon = 2^{-12}$ (see (1.29)) and corresponding numerical approximations of the modulation fields obtained from eigenvalues of $\mathbf{C}(t)$ and the coincident discretizations of the functions $\mu_{U,L}(y)$, $\mu'_{U,L}(y)$, and $\nu_L(y)$, shown with blue curves. COLOR ONLINE.

3. Asymptotic properties of the real eigenvalues $\alpha_k(x,t)$ and their implications

In this section, we study the implications of a different formula for $u(x,t)$ in terms of the eigenvalues of the matrix $\mathbf{A}(x,t)$. The approach in this section is completely independent of that presented in Section 2.

3.1. Estimates on derivatives of the eigenvalues and the importance of small eigenvalues.

Let $\alpha_k(x,t)$, $k = 1,\dots,N$, denote the eigenvalues of $\mathbf{A}(x,t)$ and let $\mathbf{u}_k(x,t)$ denote the corresponding orthonormalized eigenvectors. Recall the formula (1.23) for $u(x,t)$ in terms of the eigenvalues of $\mathbf{A}(x,t)$ and their derivatives with respect to x. By differentiation of the relation $\mathbf{A}(x,t)\mathbf{u}_k(x,t) = \alpha_k(x,t)\mathbf{u}_k(x,t)$ and noting that the coordinates (x,t) only occur in the diagonal entries of $\mathbf{A}(x,t)$, we find that the x and t derivatives of $\alpha_k(x,t)$ can be expressed in terms of the eigenvector $\mathbf{u}_k(x,t)$ as

$$(3.1)\quad \frac{\partial \alpha_k}{\partial x}(x,t) = \sum_{j=1}^{N(\epsilon)} (-2\lambda_j)|u_{k,j}(x,t)|^2, \quad \frac{\partial \alpha_k}{\partial t}(x,t) = \sum_{j=1}^{N(\epsilon)} (-4\lambda_j^2)|u_{k,j}(x,t)|^2.$$

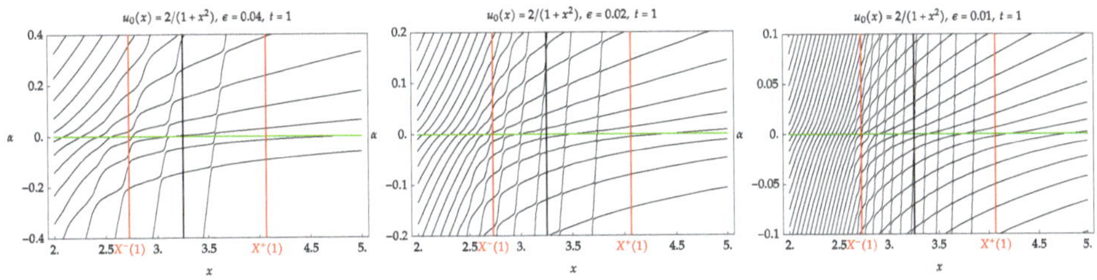

FIGURE 9. Numerical plots of small eigenvalues $\alpha_k(x,1)$ versus x, for three different values of ϵ. The value of $t = 1 > t_b$ is selected so that for the range of x in the plots, there is an interval $X^-(t) < x < X^+(t)$ delineated with vertical red lines on which the solution of Burgers' equation is multi-valued. It is clear that for $x \in (X^-(t), X^+(t))$, there are both slow-moving and fast-moving eigenvalues whose trajectories actually form a system of non-intersecting paths. (The vertical line at $x = 3.25$ is for reference only; see Figure 10 below.) COLOR ONLINE.

Since $-L < \lambda_j < 0$ for $j = 1, \ldots, N(\epsilon)$, it is clear that

$$0 < \frac{\partial \alpha_k}{\partial x}(x,t) < 2L, \quad -4L^2 < \frac{\partial \alpha_k}{\partial t}(x,t) < 0. \tag{3.2}$$

We next note that only small values of α actually contribute significantly to $u(x,t)$. Indeed, we know from (3.2) that $0 \leq \partial_x \alpha_k \leq 2L$, hence

$$\sum_{|\alpha_k| \geq \epsilon^r} \frac{2\partial_x \alpha_k(x,t)}{(\epsilon^{-1}\alpha_k(x,t))^2 + 1} \leq 4L \sum_{|\alpha_k| \geq \epsilon^r} \frac{1}{\epsilon^{2r-2}} \leq 4LN(\epsilon)\epsilon^{2-2r} \lesssim \epsilon^{1-2r}. \tag{3.3}$$

This goes to zero as $\epsilon \to 0$ when $r < \frac{1}{2}$.

3.2. Numerical experiments. We first observe the distribution of small eigenvalues of $\mathbf{A}(x,t)$ and their corresponding (normalized) eigenvectors via numerics. See Figure 9. These computations suggest that on subintervals of $x \in (X^-(t), X^+(t))$ where the method-of-characteristics solution of the inviscid Burgers equation with initial data $u_0(x)$ is multi-valued at some fixed time $t > t_b$, most small eigenvalues $\alpha_k(x,t)$ have either a specific large x-velocity $\alpha_{k,x}(x,t)$ or a small x-velocity. The small eigenvalues $\alpha_k(x,t)$ also appear to never coincide for any x; hence there are also numerous near-collisions between fast-moving and slow-moving eigenvalues in which what actually occurs is that a slow eigenvalue rapidly accelerates while a fast eigenvalue decelerates. On the other hand, if there is only one branch of the Burgers solution above (x,t), then all of the small eigenvalues appear to be of the slow-moving type.

The overall distribution of the eigenvalues $\alpha_k(x,t)$ in the limit $\epsilon \to 0$ is known according to [**MX11**]. Indeed, the following limit holds in the weak-$*$ sense:

$$\lim_{\epsilon \to 0} \frac{M}{N(\epsilon)} \sum_{k=1}^{N(\epsilon)} \delta_{\alpha_k(x,t)}(\alpha) = G(\alpha; x, t)\, \mathrm{d}\alpha, \tag{3.4}$$

where

$$G(\alpha;x,t) := -\frac{1}{4\pi}\int_{-L}^{0} \chi_{[-2\lambda(x+2\lambda t-x_+(\lambda)),-2\lambda(x+2\lambda t-x_-(\lambda))]}(\alpha)\frac{\mathrm{d}\lambda}{\lambda}, \tag{3.5}$$

χ_I denotes the characteristic function of the interval I, and $x_-(\lambda) < x_+(\lambda)$ are the two turning points satisfying $u_0(x_\pm(\lambda)) = -\lambda$. So, given $\alpha \in \mathbb{R}$, the limiting density of eigenvalues near α is computed as $-1/(4\pi)$ times the integral of $1/\lambda$ over those subintervals of $\lambda \in (-L,0)$ where the inequalities $-2\lambda(x+2\lambda t - x_+(\lambda)) < \alpha < -2\lambda(x+2\lambda t-x_-(\lambda))$ both hold. In the case that the inviscid Burgers solution with initial data $u_0(x)$ has three branches, $u_0^{\mathrm{B}}(x,t) < u_1^{\mathrm{B}}(x,t) < u_2^{\mathrm{B}}(x,t)$, and that α is small, there are two such subintervals: $[-u_2^{\mathrm{B}}(x,t)+o(1), -u_1^{\mathrm{B}}(x,t)+o(1)]$ and $[-u_0^{\mathrm{B}}(x,t)+o(1), \eta(\alpha)]$, where $\eta(\alpha) \le 0$ is a small quantity that satisfies the implicit equation

$$u_0\left(\frac{\alpha}{2\eta(\alpha)} + x + 2\eta(\alpha)t\right) = -\eta(\alpha). \tag{3.6}$$

For positive rational $u_0(x)$ there is an integer $p>0$ and a constant $C>0$ such that $u_0(x) = Cx^{-2p}(1+\mathcal{O}(x^{-1}))$ as $x \to \pm\infty$. Solutions $\eta(\alpha)$ of (3.6) that are small as $\alpha \to 0$ necessarily satisfy $\eta(\alpha) \ll \alpha$, in which case the above large-x approximation of u_0 yields that

$$\eta(\alpha) = -K|\alpha|^{2p/(2p-1)}(1+o(1)), \quad \alpha \to 0, \quad K := \left(\frac{1}{2^{2p}C}\right)^{1/(2p-1)}. \tag{3.7}$$

Therefore, for small α, the density $G(\alpha;x,t)$ is approximated by

$$\begin{aligned} G(\alpha;x,t) &= -\frac{1}{4\pi}\left[\ln(u_1^{\mathrm{B}}(x,t)+o(1)) - \ln(u_2^{\mathrm{B}}(x,t)+o(1))\right. \\ &\qquad\qquad \left. + \ln(-\eta(\alpha)) - \ln(u_0^{\mathrm{B}}(x,t)+o(1))\right] \\ &= \frac{1}{4\pi}\frac{2p}{2p-1}\ln(|\alpha|^{-1}) + \mathcal{O}(1), \quad \alpha \to 0. \end{aligned} \tag{3.8}$$

Thus, the overall asymptotic density of eigenvalues $\alpha_k(x,t)$ diverges logarithmically as $\alpha \to 0$. There are therefore many small eigenvalues $\alpha_k(x,t)$, and when $x \in (X^-(t), X^+(t))$, the plots in Figure 9 suggest that the majority of these are slow-moving eigenvalues[2]. One may think of the origin $\alpha = 0$ as locating a kind of "traffic jam" of eigenvalues with small positive x-velocities through which a small number of fast-moving eigenvalues repeatedly pass with increasing x.

The x-velocity of an eigenvalue $\alpha = \alpha_k(x,t)$ can be expressed explicitly in terms of its corresponding normalized eigenvector $\mathbf{u}_k(x,t)$ by (3.1). One can then calculate the x-velocities of all the (small) eigenvalues directly from numerically computed eigenvectors. Selecting the value of x at $t=1$ corresponding to the vertical line at $x=3.25$ in the plots in Figure 9, the velocities of the eigenvalues in the range $|\alpha| < \epsilon^{1/3}$ (larger eigenvalues make a negligible contribution to the sum in (1.23) by (3.3)) are plotted against the eigenvalues α for a series of decreasing values

[2]The majority of the intersections with any given vertical reference line such as $x = 3.25$ in Figure 9 are evidently with curves having the smaller of the two slopes. On the other hand, the majority of the intersections with the horizontal line $\alpha = 0$ as x varies in the multi-valued interval are with curves having the larger of the two slopes, i.e. most of the eigenvalues crossing the origin $\alpha = 0$ with varying x are of the fast-moving variety. Note that according to (1.24), we have $\mathbf{A}(x,t) = \mathbf{D}(x\mathbb{I} - \mathbf{B}(t))\mathbf{D}$, and hence $\mathbf{A}(x,t)$ has $\alpha = 0$ as an eigenvalue precisely when x is an eigenvalue of $\mathbf{B}(t)$.

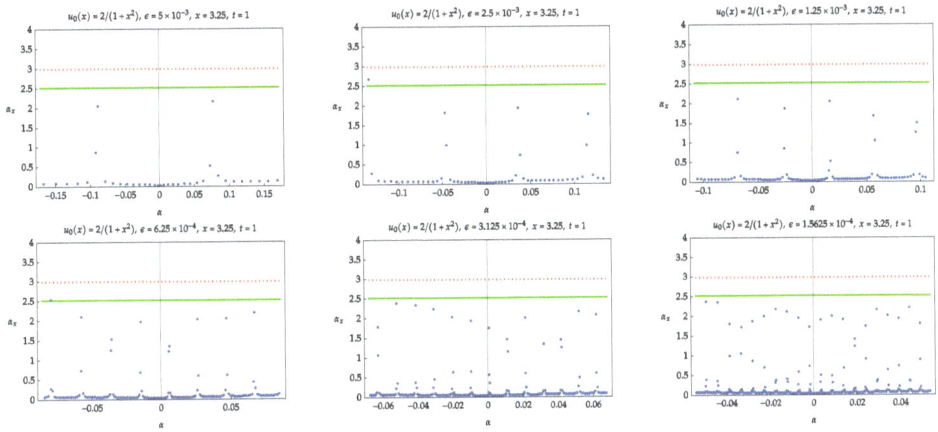

FIGURE 10. The eigenvalue velocity $\alpha_{k,x}(x,t)$ determined from the eigenvector according to (3.1) plotted against the eigenvalues $\alpha_k(x,t)$ in the range $|\alpha| < \epsilon^{1/3}$, for $(x,t) = (3.25, 1)$ and various indicated values of ϵ. For the indicated initial condition, $L = 2$, so all derivatives α_x lie in the range $0 < \alpha_x < 2L = 4$. The dotted red line is $\alpha_x = 2(u_2^{\mathrm{B}}(x,t) - u_1^{\mathrm{B}}(x,t))/(\ln(u_2^{\mathrm{B}}(x,t)) - \ln(u_1^{\mathrm{B}}(x,t)))$, and the green line is $\alpha_x = 2(u_2^{\mathrm{B}}(x,t) - u_1^{\mathrm{B}}(x,t))/(\ln(u_2^{\mathrm{B}}(x,t) - u_0^{\mathrm{B}}(x,t)) - \ln(u_1^{\mathrm{B}}(x,t) - u_0^{\mathrm{B}}(x,t)))$. COLOR ONLINE.

of ϵ in Figure 10. These plots suggest that as $\epsilon \to 0$, the "fast" small eigenvalues are in the minority, they have a regular spacing proportional to ϵ, and (perhaps) their velocities approach a limiting value in the range $(0, 2L)$. Two possible limiting values are shown in the plots as horizontal lines; clearly the solid green line is a better fit than the dotted red line.

To explain the predictions behind the horizontal lines in the plots of Figure 10, we may look at and compare plots of the square modulus of components of an eigenvector $\mathbf{u}_k(x,t)$ for slow and fast eigenvalues. See Figure 11. The plots clearly show that whether a small eigenvalue is "slow" or "fast", the eigenvector is strongly localized in two subintervals of the rescaled index $y_j = \epsilon(j - \frac{1}{2})$, which generally lies in the range $0 < y_j < M$ as $j \in \mathbb{Z}$ varies from $j = 1$ to $j = N(\epsilon)$. One of these intervals abuts the right edge $y_j = M$ but the other is bounded away from this edge. The "slow" eigenvector appears to be supported in the abutting interval while the "fast" eigenvector evidently has some support in both intervals. Moreover, there is evidence in the plots that near the right edge, $|u_{k,j}(x,t)|^2$ is proportional to $(M - y_j)^{-1}$, which is not integrable as a function of y_j. Therefore, if there is any support of $\mathbf{u}_k(x,t)$ in the interval abutting this singularity, the constant of proportionality must be very small for the eigenvector to be normalized. Now, recalling that as j varies from $j = 1$ to $j = N(\epsilon)$ the numbers λ_j increase monotonically from $\lambda_1 \approx -L$ to $\lambda_{N(\epsilon)} \approx 0$, and in fact one can show that for $u_0(x) = 2(1 + x^2)^{-1}$, $\lambda_j = \mathcal{O}(M - y_j)$ near the right edge. Consequently, for such eigenvectors the formula (3.1) predicts a small value of $\partial_x \alpha_k(x,t)$.

This means that in order for an eigenvalue $\alpha_k(x,t)$ to have a velocity $\partial_x \alpha_k(x,t)$ that is not small in the limit $\epsilon \to 0$, it is necessary for the corresponding eigenvector

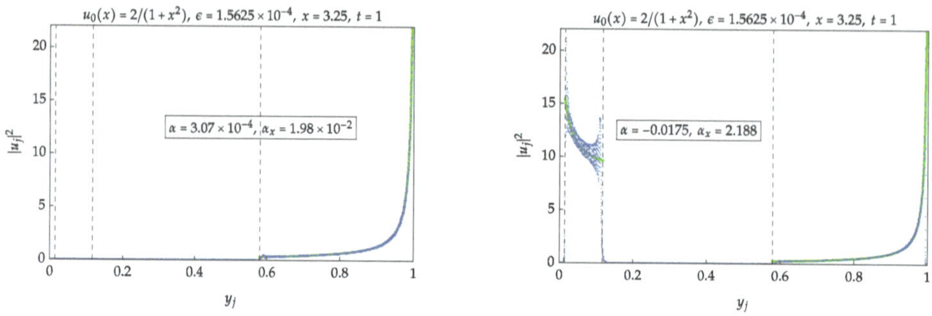

FIGURE 11. Left: an (here, unnormalized) eigenvector of $\mathbf{A}(x,t)$ for a "slow" eigenvalue. Right: an unnormalized eigenvector of $\mathbf{A}(x,t)$ for a "fast" eigenvalue. Values of $|u_j|^2$ are shown with blue points. Also shown with green curves are best fits (choice of $a_0 > 0$) on the intervals bounded by the y-values corresponding to the three values of the multi-valued solution of the inviscid Burgers equation (indicated with dashed vertical lines) to the approximate squared amplitude $a_0^2\Lambda'(y)/(-\Lambda(y))$ (see (3.20) below). For the "slow" eigenvalue, the best fit is for $a_0^2 \approx 6.461 \times 10^{-6}$. For the "fast" eigenvalue, the best fit for the left support interval is $a_0^2 \approx 2.983 \times 10^{-4}$ and the best fit for the right support interval is $a_0^2 \approx 4.873 \times 10^{-6}$; the fraction of the squared ℓ^2 norm in the right support interval is approximately 0.7543. COLOR ONLINE.

to be predominantly localized in the other subinterval of $(0, M)$ that does not abut the right edge. This could happen in two different ways:

(1) In the limit $\epsilon \to 0$, the eigenvector could have *no support* on any interval abutting the right edge $y_j = M$. If this is true, then the plot in the right-hand panel of Figure 11 is misleading in the sense that if ϵ is made smaller the evident support near the right edge should disappear rapidly.
(2) In the limit $\epsilon \to 0$, the eigenvector could have a nonzero limiting fraction of its norm in the subinterval abutting the right edge; since $|u_{k,j}(x,t)|^2$ is proportional to $(M - y_j)^{-1}$ and $M - y_{N(\epsilon)} = \mathcal{O}(\epsilon^{1/2})$ as $\epsilon \to 0$ for $u_0(x) = 2(1+x^2)^{-1}$, this suggests that the proportionality constant on the abutting interval should be small of size $\mathcal{O}(\ln(\epsilon^{-1})^{-1})$.

To see the full implications of these two alternatives requires an asymptotic theory of eigenvectors of $\mathbf{A}(x,t)$ that we will develop next. However, we may point out at this juncture that the dotted red line in the panels of Figure 10 corresponds to the first case, while the green line corresponds to a specific choice of small proportionality constant in the second case, selected to match the formal predictions of Whitham modulation theory [**DK91**].

3.3. Microlocal analysis of eigenvectors of $\mathbf{A}(x,t)$. As suggested by the above numerical observations, we distinguish two families of "fast" and "slow" eigenvectors. An explanation for these two types of eigenvectors may lie in semiclassical analysis, and more precisely, on Toeplitz (or Berezin-Toeplitz) quantization. Indeed, one might notice that the matrix $\mathbf{A}(x,t)$ resembles a generalized Toeplitz matrix, whose entries vary slowly along the diagonals.

To exhibit this structure more clearly, let a monotone increasing function $\Lambda : (0, M) \to (-L, 0)$ (see (1.10) for L, M) be defined as

$$\int_{-L}^{\Lambda(y)} F(\lambda)\,\mathrm{d}\lambda = y. \tag{3.9}$$

It follows that $\Lambda'(y) = 1/F(\Lambda(y))$ and $\lambda_j = \Lambda(\epsilon(j - \frac{1}{2}))$, see (1.15). We can write the off-diagonal elements of $\mathbf{A}(x,t)$ as

$$A_{jk}(x,t) = \frac{2\Lambda(y_j)}{\Lambda'(y_j)} \frac{\mathrm{i}}{k - j} + \epsilon f(y_j, \Delta y_{jk}), \quad k \neq j \tag{3.10}$$

in which

$$y_j := \epsilon\left(j - \frac{1}{2}\right), \quad \Delta y_{jk} := \epsilon(k - j), \tag{3.11}$$

$$f(y, \Delta y) := -2\mathrm{i}\Lambda(y) \frac{\Lambda(y + \Delta y) - \Lambda(y) - \Lambda'(y)\Delta y \sqrt{\frac{\Lambda(y+\Delta y)}{\Lambda(y)}}}{\Lambda'(y)\Delta y[\Lambda(y + \Delta y) - \Lambda(y)]}. \tag{3.12}$$

Both the numerator and denominator of $f(y, \Delta y)$ are smooth ϵ-independent functions of Δy that vanish to second order at $\Delta y = 0$. In fact, we can let $\Delta y \to 0$ and obtain the limiting value

$$f(y, 0) = \mathrm{i}\frac{\Lambda'(y)^2 - \Lambda(y)\Lambda''(y)}{\Lambda'(y)^2} = \mathrm{i}\frac{\mathrm{d}}{\mathrm{d}y}\frac{\Lambda(y)}{\Lambda'(y)}. \tag{3.13}$$

Note that the diagonal elements of $\mathbf{A}(x,t)$ can also be expressed as the sampling of a smooth function:

$$A_{jj}(x,t) = -2\Lambda(y_j)(x + 2\Lambda(y_j)t + \gamma(\Lambda(y_j))). \tag{3.14}$$

Numerics shown in Figure 11 and in Figure 13 below suggest that on ranges of indices j where eigenvector elements $u_{k,j}(x,t)$ of $\mathbf{A}(x,t)$ are not small, they have slowly varying amplitude and rapidly oscillating phase. Heuristically, we therefore propose a wavepacket approximation of the eigenvectors $\mathbf{u}_k$ via a WKB-type expansion. The proof of the following proposition relies on a hypothetical but reasonable estimate; see (3.40) below.

PROPOSITION 3.1 (Wavepacket approximation at non-stationary points). Let J be a finite union of pairwise-disjoint closed subintervals of $(0, M)$, and denote by J_δ the corresponding union of closed intervals each of which is one of the intervals of J extended by δ at both ends, such that the intervals of J_δ are also pairwise-disjoint and contained in $(0, M)$. Suppose that an amplitude function $a : J_\delta \to \mathbb{R}_{>0}$ is of class $C^\infty(J_\delta)$ and strictly bounded away from zero, and that a phase function $S : J_\delta \to \mathbb{R}$ is of class $C^\infty(J_\delta)$ with derivative S' strictly bounded away from $2\pi\mathbb{Z}$. Let χ be a $C^\infty(0, M)$ cutoff function for which $\chi(y) = 1$ for $y \in J$ and $\chi(y) = 0$ for $y \in (0, M) \setminus J_\delta$, assume that $a(\cdot)$ and $S(\cdot)$ are functions independent of ϵ, and define a wavepacket ansatz $\mathbf{u}$ with components u_j given by

$$u_j = \chi(y_j)a(y_j)\mathrm{e}^{\mathrm{i}S(y_j)/\epsilon}, \quad y_j = \epsilon(j - 1/2). \tag{3.15}$$

Then $\mathbf{u}$ is an approximate eigenvector of $\mathbf{A}(x,t)$ with eigenvalue $\alpha \in \mathbb{R}$ in the sense that for each $y_j \in J$,

$$[(\mathbf{A}(x,t) - \alpha\mathbb{I})\mathbf{u}]_j = (\zeta(y_j;\epsilon) + \mathcal{O}(\epsilon^2))u_j \tag{3.16}$$

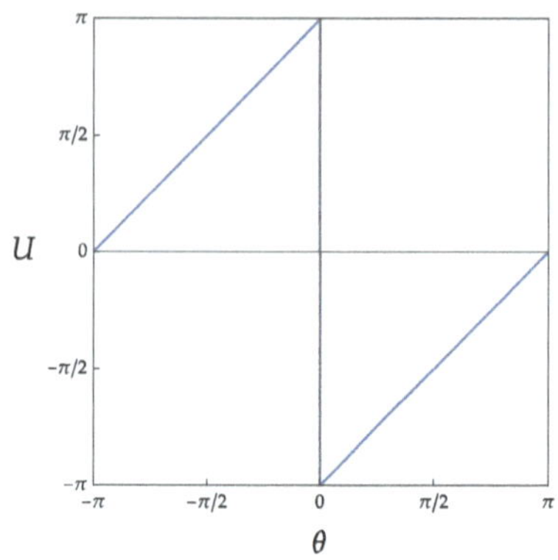

FIGURE 12. The graph of $U(\theta)$ on $(-\pi, \pi)$.

holds with $\zeta(y_j;\epsilon)$ being a uniformly bounded function of $y_j \in J$ that has zero mean with gridscale-wavelength (i.e., $\mathcal{O}(\epsilon)$ in y_j) oscillations wherever $S''(y_j) \neq 0$, provided that

- the phase derivative $S'(y)$ satisfies the eikonal equation

$$-2\Lambda(y)\left(x + 2\Lambda(y)t + \gamma(\Lambda(y)) - \frac{U(S'(y))}{\Lambda'(y)}\right) = \alpha, \quad y \in J, \tag{3.17}$$

  where U is the piecewise-linear function defined on $(-\pi, \pi)$ whose graph is shown in Figure 12;
- the amplitude $a(y)$ is subject to the equation:

$$\frac{\mathrm{d}}{\mathrm{d}y}\frac{\Lambda(y)}{\Lambda'(y)} + \frac{2\Lambda(y)}{\Lambda'(y)}\frac{a'(y)}{a(y)} = 0, \quad y \in J. \tag{3.18}$$

Note that unlike (3.17), the amplitude equation (3.18) is independent of the eigenvalue α.

REMARK 3.2. The constraint that $U(\theta) \in [-\pi, \pi]$ gives a range of admissible values for y for which the eikonal equation (3.17) can be solved for $S'(y)$, and the intervals of J should consist of admissible values only. When $\alpha = 0$, the endpoints of intervals of admissible y can be identified with the branches of the possibly multi-valued solution of Burgers' equation, see Corollary 3.5 below. For the coordinates (x, t) selected for the plots in Figure 11, there are three branches of the solution for the indicated initial condition and the corresponding interval endpoints are indicated on the plots with dotted vertical lines. For such admissible y, we can use the identity $U(U(\theta)) = \theta$ to solve explicitly for $S'(y)$ when $\alpha = 0$:

$$S'(y) = U\left(\Lambda'(y)[x + 2\Lambda(y)t + \gamma(\Lambda(y))]\right). \tag{3.19}$$

From (3.19) we can see that if u_0 is an analytic initial condition so that $\Lambda(y)$ is an analytic function of $y \in (0, M)$, then there are at most finitely-many points $y \in (0, M)$ for which $S''(y) = 0$ and near which the function $\zeta(\cdot;\epsilon)$ fails to be rapidly oscillatory. For the wavepacket ansatz (3.15), the meaning of $S'(y)$ is that it should be the approximate value of the site-to-site phase shift, as can be seen by Taylor-expanding the phase $S(y_j)$ about y_{j-1}, using $y_j = y_{j-1} + \epsilon$. In Figure 13 we illustrate the remarkable accuracy of the prediction of the formula (3.19) for this phase shift.

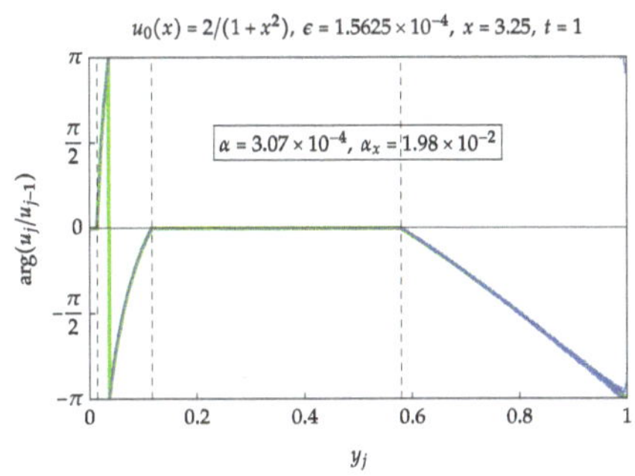

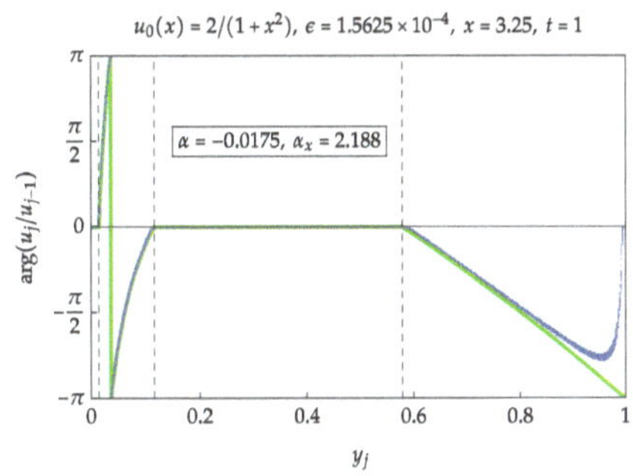

FIGURE 13. Same as Figure 11, except plotting the relative phase of nearest neighbor eigenvector elements in blue, and comparing with $S'(y_k) = U((x + 2\Lambda(y)t + \gamma(\Lambda(y)))/F(\Lambda(y)))$ in green as determined from the eikonal equation for $\alpha = 0$. Note that for the eigenvector for a "slow" eigenvalue (left panel), the phase shift predicted by the eikonal equation is even accurate in the interval of admissible y_j that does not abut $y = M$, where the amplitude is very small. On the other hand, for the "fast" eigenvalue (right panel), some accuracy of the phase prediction is evidently lost near $y = M$, possibly because the eigenvalue α is not sufficiently small. COLOR ONLINE.

REMARK 3.3. Similarly, by explicit integration of the differential equation (3.18), the amplitude $a(y) > 0$ can be obtained for y in the range of admissible values in the form

$$a(y) = a_0\sqrt{\frac{\Lambda'(y)}{-\Lambda(y)}}. \tag{3.20}$$

If there are multiple pairwise disjoint intervals of admissible values of $y \in [0, M]$, then the value of the integration constant $a_0 > 0$ may be different for each. The form (3.20) can be fit by a least-squares computation to determine the value of a_0 in each interval of admissible y_j from given eigenvector data; such fits are shown with green curves in Figure 11; they are also remarkably accurate on the support subinterval of $[0, M]$ that abuts $y = M$. The approximation appears to be slightly less accurate on the support subinterval that is separated from $y = M$ as seen in the right-hand panel of Figure 11, because there are noticeable gridscale-wavelength fluctuations about the mean predicted by (3.20). We may expect that these terms might be captured by a refinement of the ansatz (3.15) to include a highly-oscillatory correction to the amplitude which could perhaps be chosen to removing the oscillatory correction $\zeta(y_j; \epsilon)$ from the residual in (3.16).

Proposition 3.1 essentially shows that $\mathbf{A}(x, t)$ acts microlocally on suitable vectors. Let us now turn to its proof. The following lemma will be useful.

LEMMA 3.4 (Nonstationary phase sums). *Suppose that* $h : \mathbb{R} \to \mathbb{C}$ *is of class* $C^\infty(\mathbb{R})$ *and has compact support* $\operatorname{supp}(h) = [a, b]$, *and suppose that* $S : \operatorname{supp}(h) \to \mathbb{R}$ *is of class* $C^\infty(\mathbb{R})$ *with* $S'(y)$ *bounded away from* $2\pi\mathbb{Z}$ *on* $\operatorname{supp}(h)$. *Then recalling the notation* $y = y_k = \epsilon(k - \frac{1}{2})$,

$$\sum_{k\in\mathbb{Z}} h(y_k)\mathrm{e}^{\mathrm{i}S(y_k)/\epsilon} = \mathcal{O}(\epsilon^\infty), \quad \epsilon \to 0. \tag{3.21}$$

PROOF. By the Poisson summation formula,

$$\begin{aligned}\sum_{k\in\mathbb{Z}} h(y_k)\mathrm{e}^{\mathrm{i}S(y_k)/\epsilon} &= \sum_{m\in\mathbb{Z}}\int_{\mathbb{R}} \mathrm{e}^{-2\pi ikm} h(y_k)\mathrm{e}^{\mathrm{i}S(y_k)/\epsilon}\,\mathrm{d}k \\ &= \frac{1}{\epsilon}\sum_{m\in\mathbb{Z}}(-1)^m\int_{\mathbb{R}} h(y)\mathrm{e}^{\mathrm{i}(S(y)-2\pi my)/\epsilon}\,\mathrm{d}y.\end{aligned} \tag{3.22}$$

Integrating by parts $n > 1$ times, we get

$$\int_{\mathbb{R}} h(y)\mathrm{e}^{\mathrm{i}(S(y)-2\pi my)/\epsilon}\,\mathrm{d}y = \epsilon^n\int_{\mathbb{R}} h_n(y;m)\mathrm{e}^{\mathrm{i}(S(y)-2\pi my)/\epsilon}\,\mathrm{d}y, \tag{3.23}$$

wherein the function $h_n(y;m)$ is defined recursively by $h_0(y;m) := h(y)$ and

$$h_n(y;m) := \mathrm{i}\frac{\mathrm{d}}{\mathrm{d}y}\left[\frac{h_{n-1}(y;m)}{S'(y) - 2\pi m}\right], \quad n \geq 1. \tag{3.24}$$

It is easy to see that for fixed $n > 1$, $h_n(y;m)$ satisfies an estimate of the form

$$\sup_{y\in\mathbb{R}} |h_n(y;m)| \leq \frac{K_n}{\langle m\rangle^n}, \quad \langle m\rangle := \sqrt{1+m^2}. \tag{3.25}$$

Therefore since $\mathrm{supp}(h) = [a,b]$ implies also $\mathrm{supp}(h_n) \subset [a,b]$,

$$\begin{aligned}\left|\sum_{k\in\mathbb{Z}} h(y_k)\mathrm{e}^{\mathrm{i}S(y_k)/\epsilon}\right| &\leq \epsilon^{n-1}\sum_{m\in\mathbb{Z}}\int_a^b \frac{K_n}{\langle m\rangle^n}\,\mathrm{d}y \\ &= K_n(b-a)\epsilon^{n-1}\sum_{m\in\mathbb{Z}}\langle m\rangle^{-n} = \mathcal{O}(\epsilon^{n-1}), \quad \epsilon \to 0\end{aligned} \tag{3.26}$$

holds for each $n > 1$. This completes the proof. □

PROOF OF PROPOSITION 3.1. Again we write $y = y_j$ for a lattice point in J and set $\Delta y = \Delta y_{jk}$. We start by separating the action of $\mathbf{A}(x,t) - \alpha\mathbb{I}$ on the wavepacket vector $\mathbf{u}$ defined in (3.15) according to the diagonal and off-diagonal elements of $\mathbf{A}(x,t)$:

(3.27)

$$\begin{aligned}[(\mathbf{A}(x,t) - \alpha\mathbb{I})\mathbf{u}]_j &= \sum_{k=1}^{N(\epsilon)}(A_{jk}(x,t) - \delta_{jk}\alpha)\chi(y+\Delta y)a(y+\Delta y)\mathrm{e}^{\mathrm{i}S(y+\Delta y)/\epsilon} \\ &= (A_{jj}(x,t) - \alpha)a(y)\mathrm{e}^{\mathrm{i}S(y)/\epsilon} \\ &\qquad + \sum_{\substack{k=1\\k\neq j}}^{N(\epsilon)} A_{jk}(x,t)\chi(y+\Delta y)a(y+\Delta y)\mathrm{e}^{\mathrm{i}S(y+\Delta y)/\epsilon}.\end{aligned}$$

Using (3.10) along with $\chi(y) = 1$, this shows that

$$[(\mathbf{A}(x,t)\mathbf{u} - \alpha\mathbb{I})\mathbf{u}]_j = (A_{jj}(x,t) - \alpha - \epsilon f(y,0))a(y)\mathrm{e}^{\mathrm{i}S(y)/\epsilon} + \frac{2\mathrm{i}\Lambda(y)}{\Lambda'(y)}\sum_{\substack{k=1\\k\neq j}}^{N(\epsilon)} \frac{\chi(y+\Delta y)a(y+\Delta y)\mathrm{e}^{\mathrm{i}S(y+\Delta y)/\epsilon}}{k-j} + \sum_{k=1}^{N(\epsilon)} \epsilon f(y,\Delta y)\chi(y+\Delta y)a(y+\Delta y)\mathrm{e}^{\mathrm{i}S(y+\Delta y)/\epsilon}. \tag{3.28}$$

We use (3.13) and (3.14) to express $f(y,0)$ and the diagonal elements of $\mathbf{A}(x,t)$ respectively. Then, extending the sum on the third line to $k \in \mathbb{Z}$ using compact support of χ and applying Lemma 3.4 with $h(y_k) = f(y_j, y_k - y_j)\chi(y_k)a(y_k)$ individually on each of the intervals of J_δ, we get

$$[(\mathbf{A}(x,t) - \alpha\mathbb{I})\mathbf{u}]_j = \frac{2\mathrm{i}\Lambda(y)}{\Lambda'(y)}\sum_{\substack{k=1\\k\neq j}}^{N(\epsilon)} \frac{\chi(y+\Delta y)a(y+\Delta y)\mathrm{e}^{\mathrm{i}S(y+\Delta y)/\epsilon}}{k-j} + \left(-2\Lambda(y)\left(x + 2\Lambda(y)t + \gamma(\Lambda(y))\right) - \alpha - \mathrm{i}\epsilon\frac{\mathrm{d}}{\mathrm{d}y}\frac{\Lambda(y)}{\Lambda'(y)}\right)a(y)\mathrm{e}^{\mathrm{i}S(y)/\epsilon} + \mathcal{O}(\epsilon^\infty). \tag{3.29}$$

Therefore, to very high accuracy, $\mathbf{A}(x,t)$ acts on the wavepacket $\mathbf{u}$ as multiplication by a sum of a diagonal matrix and the product of a diagonal matrix and an exact Toeplitz matrix with elements $T_{jk} = (k-j)^{-1}$ for $k \neq j$ and $T_{jj} = 0$.

To deal with the Toeplitz part, we again use compact support of χ to write, for $K > 0$ sufficiently large,

$$\sum_{\substack{k=1\\k\neq j}}^{N(\epsilon)} \frac{\chi(y+\Delta y)a(y+\Delta y)\mathrm{e}^{\mathrm{i}S(y+\Delta y)/\epsilon}}{k-j} = \sum_{\substack{k=j-K\\k\neq j}}^{j+K} \frac{\chi(y+\Delta y)a(y+\Delta y)\mathrm{e}^{\mathrm{i}S(y+\Delta y)/\epsilon}}{k-j}. \tag{3.30}$$

To allow $y = y_j$ to range over the full set J and have exact equality in (3.30) for K independent of j, we will assume that $K = N(\epsilon)$. We introduce the notation

$$S_2(\Delta y; y) := S(y+\Delta y) - S(y) - S'(y)\Delta y \tag{3.31}$$

which vanishes to second order as $\Delta y \to 0$. We also use the Taylor expansion

$$\chi(y+\Delta y)a(y+\Delta y)\mathrm{e}^{\mathrm{i}S_2(\Delta y;y)/\epsilon} = a(y) + a'(y)\Delta y + \frac{1}{2}R_2(\Delta y; y, \epsilon)\Delta y^2, \tag{3.32}$$

where the real and imaginary parts of $R_2(\Delta y; y, \epsilon)$ are those of the second derivative with respect to Δy of the left-hand side evaluated at two generally different points between 0 and Δy. Hence we write the sum in (3.30) in the form

$$
\begin{aligned}
(3.33)\quad &\sum_{\substack{k=j-K\\k\neq j}}^{j+K}\frac{\chi(y+\Delta y)a(y+\Delta y)\mathrm{e}^{\mathrm{i}S(y+\Delta y)/\epsilon}}{k-j}\\
&=\mathrm{e}^{\mathrm{i}S(y)/\epsilon}\sum_{\substack{k=j-K\\k\neq j}}^{j+K}\chi(y+\Delta y)a(y+\Delta y)\mathrm{e}^{\mathrm{i}S_2(\Delta y)/\epsilon}\frac{\mathrm{e}^{\mathrm{i}(k-j)S'(y)}}{k-j}\\
&=a(y)\mathrm{e}^{\mathrm{i}S(y)/\epsilon}\sum_{\substack{n=-K\\n\neq 0}}^{K}\left[\frac{\mathrm{e}^{\mathrm{i}nS'(y)}}{n}+\epsilon\frac{a'(y)}{a(y)}\mathrm{e}^{\mathrm{i}nS'(y)}+\frac{\epsilon^2}{2a(y)}R_2(\epsilon n;y,\epsilon)n\mathrm{e}^{\mathrm{i}nS'(y)}\right],
\end{aligned}
$$

where we used $\Delta y=\epsilon(k-j)$ and reindexed by $n=k-j$ in the last line. Since $y=y_j$ is independent of the index n, we just need to examine three summands, $\mathrm{e}^{\mathrm{i}n\theta}/n$, $\mathrm{e}^{\mathrm{i}n\theta}$, and $R_2(\epsilon n;\epsilon)n\mathrm{e}^{\mathrm{i}n\theta}$ for $\theta=S'(y)$, which by assumption is bounded away from $2\pi\mathbb{Z}$. According to [**MX11**, Eqn. 4.36]:

$$
(3.34)\qquad \lim_{K\to\infty}\sum_{\substack{n=-K\\n\neq 0}}^{K}\frac{\mathrm{e}^{\mathrm{i}n\theta}}{n}=-\mathrm{i}U(\theta),\quad \theta\neq 0 \pmod{2\pi},
$$

where the function $U(\theta)$ is periodically extended with period 2π. More precisely, by representing the summand as an integral and exchanging the order of finite summation and integration, we can write

$$
\begin{aligned}
(3.35)\quad \sum_{\substack{n=-K\\n\neq 0}}^{K}\frac{\mathrm{e}^{\mathrm{i}n\theta}}{n}&=\mathrm{i}\int_{\pm\pi}^{\theta}\left[\cos(K\tau)-1+\sin(K\tau)\cot(\tfrac{1}{2}\tau)\right]\,\mathrm{d}\tau\\
&=\mathrm{i}(\pm\pi-\theta)+\frac{\mathrm{i}\sin(K\theta)}{K}+\mathrm{i}\int_{\pm\pi}^{\theta}\sin(K\tau)\cot(\tfrac{1}{2}\tau)\,\mathrm{d}\tau.
\end{aligned}
$$

If we assume that the sign on $\pm\pi$ corresponds to the sign of $\theta\in[-\pi,\pi]\setminus\{0\}$, then $\mathrm{i}(\pm\pi-\theta)=-\mathrm{i}U(\theta)$ and the remaining integral admits repeated integration by parts. In this way we obtain (using also $K=N(\epsilon)=\epsilon^{-1}M+\mathcal{O}(1)$)

$$
\begin{aligned}
(3.36)\quad \sum_{\substack{n=-K\\n\neq 0}}^{K}\frac{\mathrm{e}^{\mathrm{i}n\theta}}{n}&=-\mathrm{i}U(\theta)-\frac{\mathrm{i}}{K}\cdot\frac{\cos((K+\frac{1}{2})\theta)}{\sin(\frac{1}{2}\theta)}+\mathcal{O}(K^{-2}),\quad K\to\infty,\\
&=-U(\theta)-\frac{\mathrm{i}\epsilon}{M}\cdot\frac{\cos((N(\epsilon)+\frac{1}{2})\theta)}{\sin(\frac{1}{2}\theta)}+\mathcal{O}(\epsilon^2),\quad \epsilon\to 0,
\end{aligned}
$$

which holds uniformly for $\theta\in[-\pi,\pi]$ bounded away from zero. Note that the singularity of the correction term at $\theta=0$, which is also the jump point for $U(\theta)$, is related to Gibbs' phenomenon[3]. A more straightforward calculation gives that

[3]This is the main difficulty in extending the theory of Toeplitz quantization to the setting of non-smooth symbols such as $U(\theta)$. See Section 3.4 below for more information.

the sum of $e^{in\theta}$ is exactly a constant shift of the Dirichlet kernel:

$$(3.37)\quad \sum_{\substack{n=-K\\n\neq 0}}^{K} e^{in\theta} = -1 + \sum_{n=-K}^{K} e^{in\theta} = \frac{\sin((K+\frac{1}{2})\theta)}{\sin(\frac{1}{2}\theta)} - 1 = \frac{\sin((N(\epsilon)+\frac{1}{2})\theta)}{\sin(\frac{1}{2}\theta)} - 1.$$

Upon evaluation for $\theta = S'(y) = S'(y_j)$ bounded away from $2\pi\mathbb{Z}$, the terms $\cos((N(\epsilon)+\frac{1}{2})\theta)/\sin(\frac{1}{2}\theta)$ and $\sin((N(\epsilon)+\frac{1}{2})\theta)/\sin(\frac{1}{2}\theta)$ will be highly oscillatory zero-mean functions of $y = y_j$ when ϵ is large, at each point y with $S''(y) \neq 0$. We introduce the notation $\widetilde{\mathcal{O}}(\epsilon^p)$ to denote such a function, whose absolute value is also $\mathcal{O}(\epsilon^p)$ in the usual sense. With this notation, (3.36) and (3.37) respectively imply that

$$(3.38)\qquad \sum_{\substack{n=-K\\n\neq 0}}^{K} \frac{e^{inS'(y)}}{n} = -U(S'(y)) + \widetilde{\mathcal{O}}(\epsilon) + \mathcal{O}(\epsilon^2), \quad \epsilon \to 0,$$

$$(3.39)\qquad \epsilon \sum_{\substack{n=-K\\n\neq 0}}^{K} e^{inS'(y)} = -\epsilon + \widetilde{\mathcal{O}}(\epsilon), \quad \epsilon \to 0.$$

We omit the details of the estimation of the contribution of sum involving the Taylor error term $R_2(\epsilon n;\epsilon)$. However we claim that

$$(3.40)\qquad \sum_{n=-K}^{K} R_2(\epsilon n;\epsilon) n e^{inS'(y)} = \widetilde{\mathcal{O}}\left(\frac{1}{\epsilon^2}\right), \quad \epsilon \to 0.$$

Indeed, keeping one more term in the Taylor expansion (3.32), the first term in $R_2(\Delta y;\epsilon)$ would just be the second derivative of $a(y+\Delta y)e^{iS_2(\Delta y;y)/\epsilon}$ at $\Delta y = 0$, which one can check is of the form $\epsilon^{-1}v(y) + w(y)$ for some smooth and bounded functions v and w. The contribution of this term to the sum in (3.40) is then $-i\epsilon^{-1}v(y) - iw(y)$ times the derivative (from the factor of in in the summand) of the Dirichlet kernel $\sin((N(\epsilon)+\frac{1}{2})\theta)/\sin(\frac{1}{2}\theta)$ evaluated at $\theta = S'(y)$, so one can easily check that the contribution is a term of the form $\widetilde{\mathcal{O}}(\epsilon^{-2})$. Similar arguments apply to the terms obtained by continuing the expansion (3.32) to any finite order. In fact, the kth derivative of the left-hand side of (3.32) evaluated at $\Delta y = 0$ is $\epsilon^{-\lfloor k/2 \rfloor}$ times a polynomial in ϵ with coefficients that are smooth bounded functions of y. This shows that one should not replace $R_2(\Delta y;\epsilon)$ with too many explicit terms, since although their contributions will be oscillatory functions of y, they will also start to grow in size.

Combining (3.29), (3.30), (3.33), and the estimates (3.38)–(3.40) yields

$$(3.41)\quad [(\mathbf{A}(x,t) - \alpha\mathbb{I})\mathbf{u}]_j = \Bigg[-2\Lambda(y)\left(x + 2\Lambda(y)t + \gamma(\Lambda(y)) - \frac{U(S'(y))}{\Lambda'(y)}\right) - \alpha \\ -i\epsilon\left(\frac{d}{dy}\frac{\Lambda(y)}{\Lambda'(y)} + \frac{2\Lambda(y)}{\Lambda'(y)}\frac{a'(y)}{a(y)}\right) + \widetilde{\mathcal{O}}(1) + \mathcal{O}(\epsilon^2)\Bigg] u_j.$$

Neglecting the highly oscillatory term and keeping only the mean yields (3.17) at the leading order in ϵ and (3.18) at second order. □

COROLLARY 3.5 (Admissible values of y and Burgers' equation). Let $\alpha = 0$. The admissible values of $y \in [0, M]$ for which there exists $S'(y) \in (-\pi, \pi)$ such that (3.17) holds are determined by:

- $\Lambda(y) \in [-u_0^{\mathrm{B}}(x,t), 0]$ if (x,t) is a point in the single-valued region for the solution of the inviscid Burgers equation, and
- $\Lambda(y) \in [-u_0^{\mathrm{B}}(x,t), 0] \cup [-u_2^{\mathrm{B}}(x,t), -u_1^{\mathrm{B}}(x,t)]$ if (x,t) is a point in the triple-valued region for the solution of the inviscid Burgers equation.

PROOF. From the eikonal equation (3.17) with $\alpha = 0$, the range condition $U(S'(y)) \in (-\pi, \pi)$ means that $\lambda = \Lambda(y)$ and x have to satisfy the condition

$$\frac{x + 2\lambda t + \gamma(\lambda)}{F(\lambda)} \in (-\pi, \pi), \tag{3.42}$$

where we used the identity $\Lambda'(y) = 1/F(\Lambda(y))$.

It turns out that the curves $x + 2\lambda t + \gamma(\lambda) = \pm\pi F(\lambda)$ in the (λ, x)-plane essentially give the (rotated and reflected) graph of the multi-valued solution $u^{\mathrm{B}}(\cdot, t)$ to the Burgers equation obtained by the method of characteristics. Indeed, at time $t = 0$, assume that

$$\frac{x + \gamma(\lambda)}{F(\lambda)} = \pi. \tag{3.43}$$

From the definitions (1.12) and (1.13) of F and γ, respectively, this means that

$$x = \pm\pi F(\lambda) - \gamma(\lambda) = x_{\pm}(\lambda), \tag{3.44}$$

where $x_-(\lambda) < x_+(\lambda)$ are the turning points satisfying $u_0(x_{\pm}(\lambda)) = -\lambda$, see below (1.10). Therefore $(x, -\lambda)$ belongs to the graph of u_0.

Similarly, at time t, the method of characteristics implies that $(x_{\pm}(\lambda) - 2\lambda t, -\lambda)$ belongs to the graph of $u^{\mathrm{B}}(\cdot, t)$. As a consequence, one sees that $(x, -\lambda)$ is in the graph of $u^{\mathrm{B}}(\cdot, t)$ if and only if $x = x_{\pm}(\lambda) - 2\lambda t$, which is equivalent to

$$\frac{x + 2\lambda t + \gamma(\lambda)}{F(\lambda)} = \pm\pi.$$

Finally, we observe that given x such that $u^{\mathrm{B}}(x,t)$ has only one branch $u_0^{\mathrm{B}}(x,t)$, the range of admissible $(x, -\lambda)$ such that (3.42) holds is the region between the graph of $u_0^{\mathrm{B}}(\cdot, t)$ and the x-axis. This means that the admissible values of y correspond to $\lambda \in [-u_0^{\mathrm{B}}(x,t), 0]$. However, when x is such that there are three branches $u_0^{\mathrm{B}}(x,t) < u_1^{\mathrm{B}}(x,t) < u_2^{\mathrm{B}}(x,t)$, the range of admissible $(x, -\lambda)$ such that (3.42) holds is the union of the region between the graph of $u_0^{\mathrm{B}}(\cdot, t)$ and the x-axis and also the region between the graphs of $u_1^{\mathrm{B}}(\cdot, t)$ and $u_2^{\mathrm{B}}(\cdot, t)$. Hence the admissible values of y correspond to $\lambda \in [-u_2^{\mathrm{B}}(x,t), -u_1^{\mathrm{B}}(x,t)] \cup [-u_0^{\mathrm{B}}(x,t), 0]$. □

3.4. Toeplitz quantization and small eigenvalues of $\mathbf{A}(x,t)$. Proposition 3.1 suggests that the matrix $\mathbf{A}(x,t)$ is represented by a symbol p_x depending on canonical variables $(y, \theta) \in [0, M] \times [-\pi, \pi]$ and given by

$$p_x(y, \theta) := -2\Lambda(y)\left(x + 2\Lambda(y)t + \gamma(\Lambda(y)) - \frac{U(\theta)}{\Lambda'(y)}\right) \tag{3.45}$$

such that the eikonal equation (3.17) takes the form $p_x(y, \theta) = \alpha$. In the following, we mainly consider small eigenvalues satisfying $|\alpha| \leq \epsilon^r$, in view of (3.3). In the case of the zero eigenvalue $\alpha = 0$, the equation $p_x(y, \theta) = \alpha$ can be written as

$$p(y, \theta) = x, \tag{3.46}$$

where a modified symbol is defined by

$$p(y,\theta) := -2\Lambda(y)t - \gamma(\Lambda(y)) + \frac{U(\theta)}{\Lambda'(y)}. \tag{3.47}$$

Unfortunately, we are not aware of results on Toeplitz quantization on rectangles, making the study of p_x and p on $[0,M] \times [-\pi,\pi]$ difficult. In order to gain insight on the eigenvector approximation, one may look instead into the results on Toeplitz quantization on the sphere by using latitude/longitude coordinates (y,θ). The strategy of using Toeplitz quantization on the sphere to study dispersionless PDEs was rigorously implemented in the context of the dispersionless Toda system in [**BGPU03**]. However, this approach is not rigorous in our situation because of the discontinuity of U at the angle $\theta = 0$ and the lack of smoothness of the symbol at the poles $y = 0, M$ of the sphere.

According to Corollary 3.5, the ranges of admissible values of y for (3.46) to hold are determined by the branches of the Burgers solution at the point (x,t). The equation (3.46) for a given (x,t) describes a relation between the latitude coordinate y and the longitude coordinate θ that yields curves generally beginning and ending on the meridian $\theta = 0$ or $\theta = 2\pi$ except for one curve that always emerges from the north pole $y = M$ with longitude $\theta = \pi$. We call different connected components of the curve given by (3.46) *orbits* on the spherical phase space. The connection between vertical slices through the graph of the multi-valued solution of Burgers' equation at different values of x and the orbits on the sphere is illustrated in Figures 14 and 15.

In the context of a smooth symbol on the sphere, the paper [**Cha03**] states that the WKB expansion for the eigenvector $\mathbf{u}$ for a very small eigenvalue α should be valid with amplitude being a discretization of (3.20) on the admissible range of y, and such a small eigenvalue can only occur for x satisfying a certain Bohr-Sommerfeld condition (see also [**Cha06**]). More precisely, we decompose the admissible range of y into one interval if (x,t) is in the single-valued region, and into two disjoint intervals if (x,t) is in the triple-valued region for the inviscid Burgers solution. Then for each interval, there should be an eigenvector with support exactly in that interval, and amplitude given by (3.20). Moreover, the values of x for which $\alpha = 0$ is an eigenvalue follow a Bohr-Sommerfeld condition: at first order, the parallel transport around the orbit with latitude coordinate y confined to a given admissible interval is quantized, see [**Cha03**, Theorem 0.1].

Numerically, however, we observe that the support of the "fast" eigenvector is not localized only in the interval $\Lambda(y) \in [-u_2^{\mathrm{B}}(x,t), -u_1^{\mathrm{B}}(x,t)]$ (the main part), as there is some smaller contribution from the other interval $\Lambda(y) \in [-u_0^{\mathrm{B}}(x,t), 0]$ (the correction), see Figure 11. This may be due to the fact that the symbol is not smooth in our case. Moreover, we observe that this small contribution has a noticeable effect in the calculation. We will choose different values of a_0 in order for both the main part and the correction to have a non-negligible contribution to the Euclidean norm of the eigenvector. More precisely, in Conjecture 6 below, we will assume that the correction has an amplitude of order $a_0 \sim 1/\ln(\epsilon^{-1})$ compared to the main part for which $a_0 \sim 1$. Unlike the scale $\delta(\epsilon)$ in Section 2 which was somewhat flexible (see Conjecture 3 and the preceding paragraph), here the choice of precise scale inversely proportional to $\ln(\epsilon^{-1})$ is important for the proof of Theorem 3.7 below.

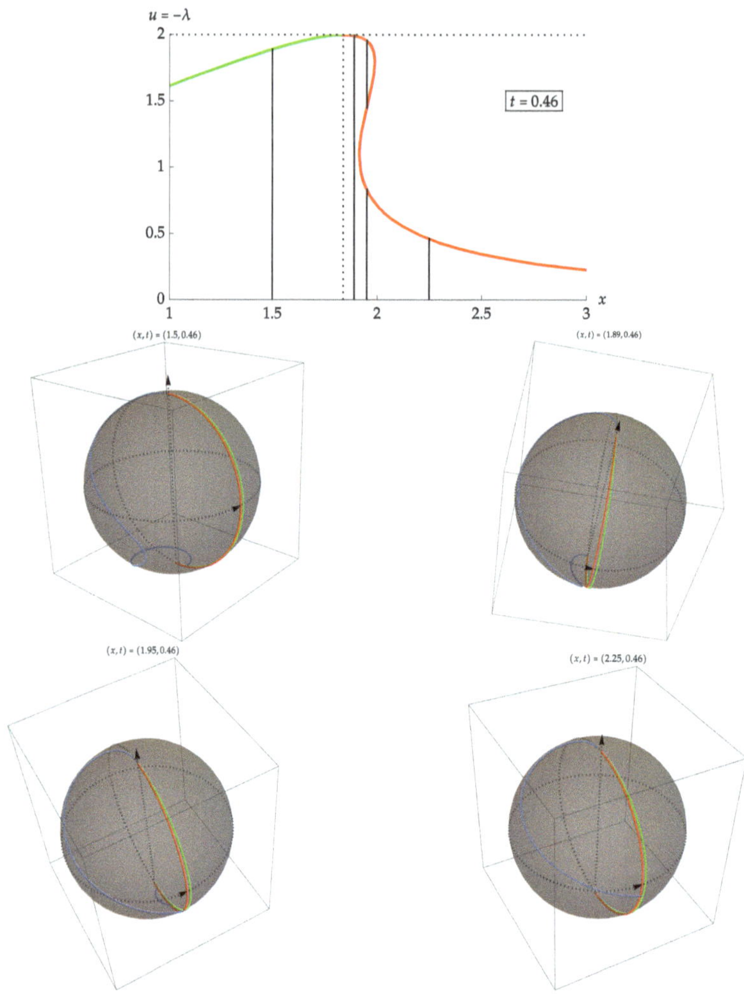

FIGURE 14. Level sets $p(y, \theta) = x$ in the (x, u)-plane and on the sphere with vertical polar axis $y \in (0, M)$ and meridian $0 < \theta < 2\pi$ (green for $\theta = 0$, red for $\theta = 2\pi$) for $t = 0.46$ and various x with $u_0(x) = 2/(1 + x^2)$. Note that when $u = 0$, $y = M$ (north pole) and $\theta = \pi$. As u increases along a vertical line in the top plot, y decreases along one or two components of an orbit on the corresponding sphere plot below. COLOR ONLINE.

3.4.1. *Bohr-Sommerfeld conditions.* We guess the Bohr-Sommerfeld conditions from the case of a smooth symbol on the sphere [**Cha03**] by using the parallel transport in the range of admissible y and $\theta(\Lambda(y)) = S'(y)$ given by (3.19), which we compute using the Levi-Civita connection on the sphere, relative to the metric inherited from the Euclidean one on $\mathbb{R}^3$. At first order, this translates as follows in the case of small eigenvalues $\alpha \approx 0$, using the symbol p given in (3.47). We make

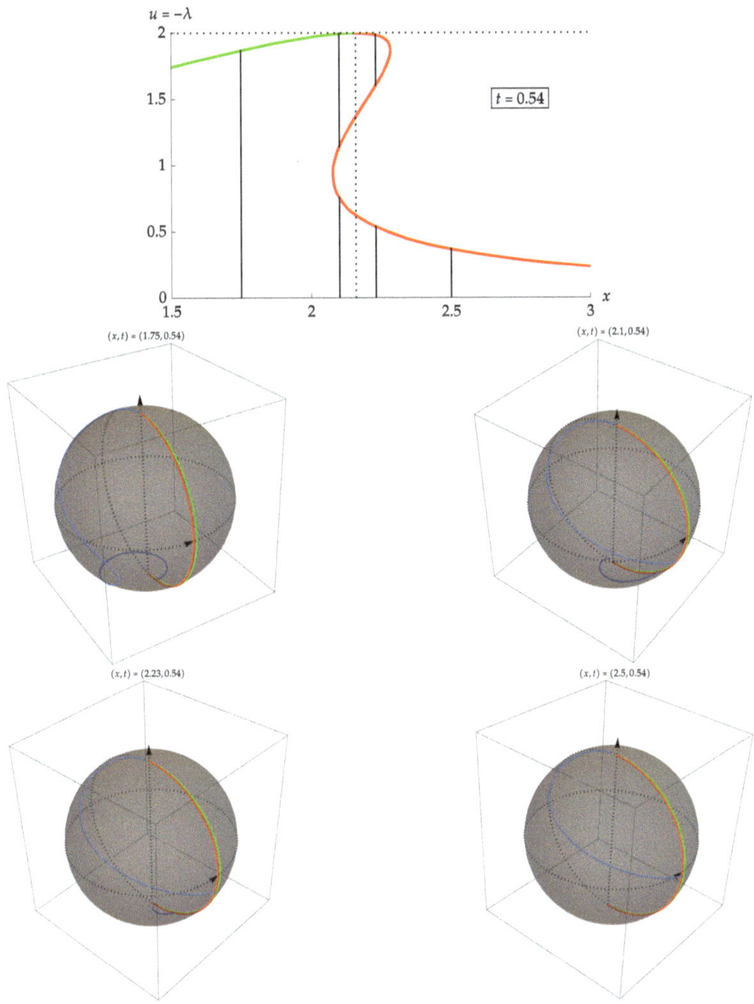

FIGURE 15. Level sets $p(y, \theta) = x$ for $t = 0.54$ and various x with $u_0(x) = 2/(1 + x^2)$.

the change of variable $y = Y(\lambda)$ inverse to $\lambda = \Lambda(y)$ in (3.9), so that

$$Y(\lambda) = \int_{-L}^{\lambda} F(\eta)\,\mathrm{d}\eta. \tag{3.48}$$

Fixing $t > t_b$, we work in the vicinity of a point x_0 in the triple-valued region for the solution of Burgers' equation: $X^-(t) < x_0 < X^+(t)$. Let us define the parallel transport about an orbit corresponding to a fast zero eigenvalue by

$$g_0^+(x, t) = \int_{-u_2^{\mathrm{B}}(x,t)}^{-u_1^{\mathrm{B}}(x,t)} \theta'(\lambda) Y(\lambda)\,\mathrm{d}\lambda, \tag{3.49}$$

and that corresponding to a slow zero eigenvalue by

$$g_0^-(x,t) = \int_{-u_0^{\mathrm{B}}(x,t)}^{0} \theta'(\lambda)Y(\lambda)\,\mathrm{d}\lambda, \tag{3.50}$$

where $\theta(\lambda)$ is determined from the eikonal equation for $\alpha = 0$ in the form $p(Y(\lambda), \theta(\lambda)) = x$. The main idea of Toeplitz quantization is that small eigenvalues of a matrix $\mathbf{A}(x,t)$ associated with[4] a symbol and a corresponding orbit can only occur if, to leading order, the parallel transport around that orbit is an integer multiple of $2\pi\epsilon$, which is known as a *Bohr-Sommerfeld quantization rule*. For smooth symbols, the leading terms $g_0^\pm(x,t)$ are modified by a sum of higher-order corrections $\epsilon^n g_n^\pm(x,t)$, $n = 1,2,3,\ldots$ for which the formulæ written in [**Cha03**] are not expected to be meaningful in the non-smooth setting.

Since $g_0^\pm(x,t)$ should be quantized but also $x \approx x_0$, we define values of $x = x_k^\pm(t) \approx x_0$ corresponding to the two types of orbits by Bohr-Sommerfeld conditions of the form

$$g_0^\pm(x_k^\pm(t),t) + \epsilon g_1^\pm(x_k^\pm(t),t) = 2\pi(k + k_0^\pm)\epsilon, \quad |k| \le C\epsilon^{r-1}, \quad k \in \mathbb{Z}, \tag{3.51}$$

with $k_0^\pm := \lfloor (g_0^\pm(x_0,t) + \epsilon g_1^\pm(x_0,t))/(2\pi\epsilon) \rfloor$.

Assuming moreover that $g_1^\pm$ depends smoothly on x, we replace $g_1^\pm(x_k^\pm,t)$ by $g_1^\pm(x_0,t)$ up to higher-order corrections that we neglect here: with $\rho^\pm(x_0,t) := k_0^\pm - g_1^\pm(x_0,t)/(2\pi)$, the Bohr-Sommerfeld conditions become

$$g_0^\pm(x_k^\pm(t),t) = 2\pi(k + \rho^\pm(x_0,t))\epsilon, \quad |k| \le C\epsilon^{r-1}, \quad k \in \mathbb{Z}. \tag{3.52}$$

The values $x = x_k^+(t)$ (resp., $x = x_k^-(t)$) are expected to approximate the values of $x \approx x_0$ for which $\alpha = 0$ is a fast (resp., slow) eigenvalue, see Conjecture 6 below.

3.4.2. *Simplifying and differentiating the Bohr-Sommerfeld conditions.* We first simplify the expression

$$g_0^\pm(x,t) = \int_{\lambda_{\min}}^{\lambda_{\max}} \theta'(\lambda)Y(\lambda)\mathrm{d}\lambda. \tag{3.53}$$

The integration limits are $\lambda_{\min} = -u_2^{\mathrm{B}}(x,t)$ and $\lambda_{\max} = -u_1^{\mathrm{B}}(x,t)$ in the case of fast eigenvalues (i.e. $g_0^\pm = g_0^+$), but $\lambda_{\min} = -u_0^{\mathrm{B}}(x,t)$ and $\lambda_{\max} = 0$ in the case of slow eigenvalues (i.e. $g_0^\pm = g_0^-$).

Using that $U(\theta) = \theta - \pi$ on $(0,2\pi)$, we parametrize the relevant orbit by $y = Y(\lambda)$, $\theta = \theta(\lambda) \in (0,2\pi)$ and get that because $\Lambda'(y) = 1/F(\Lambda(y))$ the equation $p(Y(\lambda), \theta(\lambda)) = x$ in (3.47) becomes

$$x = F(\lambda)(\theta(\lambda) - \pi) - 2\lambda t \implies F(\lambda)\theta(\lambda) = x + \pi F(\lambda) + 2\lambda t. \tag{3.54}$$

Substituting from (3.48) for $Y(\lambda)$ in (3.53), integration by parts yields

$$g_0^\pm(x,t) = \left[\theta(\lambda)\int_{-L}^{\lambda} F(\eta)\mathrm{d}\eta\right]_{\lambda_{\min}}^{\lambda_{\max}} - \int_{\lambda_{\min}}^{\lambda_{\max}} \theta(\lambda)F(\lambda)\mathrm{d}\lambda. \tag{3.55}$$

Since $g_0^\pm$ is quantized, the positions $x_k^\pm(t)$ at which there is a fast $(+)$ or slow $(-)$ crossing of $\alpha = 0$ are locally regularly spaced if the derivative of $g_0^\pm$ is bounded above and below. This derivative will determine the actual spacing. We note that

[4]The theory of Toeplitz quantization assumes that a sequence of matrices $\mathbf{A}(x,t)$ is constructed systematically from a given symbol, whereas in the present context we start instead with a sequence of matrices $\mathbf{A}(x,t)$ and deduce the symbol from the WKB expansion in Proposition 3.1.

$\theta(\lambda)$ evaluated on the limits of integration is locally independent of (x,t), being equal to 0 or 2π, or in the case $\lambda_{\min}=0$, to π. Hence we compute

(3.56)
$$\begin{aligned}\partial_x g_0^{\pm}(x,t) &= \theta(\lambda_{\max})F(\lambda_{\max})\partial_x\lambda_{\max} - \theta(\lambda_{\min})F(\lambda_{\min})\partial_x\lambda_{\min}\\ &- \partial_x\lambda_{\max}\theta(\lambda_{\max})F(\lambda_{\max}) + \partial_x\lambda_{\min}\theta(\lambda_{\min})F(\lambda_{\min}) - \int_{\lambda_{\min}}^{\lambda_{\max}}\partial_x(\theta(\lambda)F(\lambda))\mathrm{d}\lambda.\end{aligned}$$

Since θF is given by (3.54), we have that $\partial_x(\theta(\lambda)F(\lambda))=1$, hence

(3.57)
$$-\partial_x g_0^{\pm}(x,t) = \lambda_{\max}(x,t)-\lambda_{\min}(x,t) = \begin{cases} u_2^{\mathrm{B}}(x,t)-u_1^{\mathrm{B}}(x,t), & \text{for fast eigenvalues,}\\ u_0^{\mathrm{B}}(x,t), & \text{for slow eigenvalues.}\end{cases}$$

Due to the assumption $|x_l^{\pm}-x_0|\le C\epsilon^r$, one can replace x by x_0 in the integrand of the Bohr-Sommerfeld condition (3.52) rewritten in the form

$$\int_{x_l^{\pm}}^{x_{l+1}^{\pm}} \partial_x g_0^{\pm}(x,t)\,\mathrm{d}x = 2\pi\epsilon, \tag{3.58}$$

up to an error term of order ϵ^r in the integrand. In both cases, we deduce

$$(x_{l+1}^{\pm}-x_l^{\pm})(1+o(1)) = \frac{-2\pi\epsilon}{f^{\pm}(x_0,t)}, \tag{3.59}$$

where

$$f^+(x,t) := u_2^{\mathrm{B}}(x,t)-u_1^{\mathrm{B}}(x,t) > 0 \tag{3.60}$$

and

$$f^-(x,t) := u_0^{\mathrm{B}}(x,t) > 0. \tag{3.61}$$

We conclude that

$$x_{l+1}^{\pm}-x_l^{\pm} = \frac{-2\pi}{\epsilon}f^{\pm}(x_0,t)+o(\epsilon). \tag{3.62}$$

Note that $\partial_x g_0^{\pm}(x,t)<0$, so the sequences $x_l^{\pm}$ are decreasing. More precisely, with the same argument but integrating $g_0^{\pm}$ from x_0 to $x_l^{\pm}$, we can write for some $r^{\pm}(x_0,t)$:

$$x_l^{\pm} = -\frac{2\pi l\epsilon}{f^{\pm}(x_0,t)} + r^{\pm}(x_0,t)\epsilon + o(\epsilon). \tag{3.63}$$

REMARK 3.6. In (3.55), one can make the change of variable $Y(\lambda)=y$ and use (3.48). Since $\theta(\Lambda(y))=S'(y)$, we get

$$\frac{1}{\epsilon}\int_{Y(\lambda_{\min})}^{Y(\lambda_{\max})} S'(y)\mathrm{d}y = \frac{1}{\epsilon}\left[\theta(\lambda_{\max})Y(\lambda_{\max})-\theta(\lambda_{\min})Y(\lambda_{\min})-g_0^{\pm}(x,t)\right]. \tag{3.64}$$

Note that $\theta(\lambda_{\max})$ and $\theta(\lambda_{\min})$ are expected to belong to the set $\{0,\pi,2\pi\}$, and that $Y(\lambda_{\min})/\epsilon$ and $Y(\lambda_{\max})/\epsilon$ should be integers (eigenvector component indices). The Bohr-Sommerfeld conditions (3.52) therefore suggest that between two consecutive eigenvalues, the accumulated phase shift of the corresponding eigenvector as measured by the left-hand side of the above identity changes by 2π at first order.

3.4.3. *Classification of the small eigenvalues according to the Bohr-Sommerfeld conditions.* Conjecture 6 below expresses the following idea. For $t > t_b$ and $x_0 \in (X^-(t), X^+(t))$, we look at the small eigenvalues $|\alpha_k(x,t)| \leq \epsilon^r$ for $x \approx x_0$, where we recall that the parameter $0 < r < \frac{1}{2}$ is chosen so that the large eigenvalues satisfy (3.3). If α_k is "slow", then on Figure 9, one can follow the "slow" line with small slope passing the point (x, α_k). Using the predictions of the Bohr-Sommerfeld conditions, this line crosses the axis $\alpha = 0$ at one of the points x_l^-: in other words $(x_l^-, 0)$ belongs to the same "slow" line. Similarly, if α_k is "fast", then on Figure 9, one can follow the "fast" line with large slope passing the point (x, α_k): this line also passes one point of the form $(x_l^+, 0)$. Then using the predictions from Proposition 3.1 of the wavepacket approximation of the eigenvector when $\alpha = 0$, the eigenvector $\mathbf{u}$ at $x = x_l^\pm$ with eigenvalue $\alpha = 0$ should match the observations from Figure 11 and Corollary 3.5.

We illustrate the predicted position of small eigenvalues in Figure 16 (which may be viewed as the theoretical version of the numerical plots in Figure 9). We model the "slow" and "fast" eigenvalues with two families of parallel lines in the (x, α)-plane with different slopes. True eigenvalues $\alpha_k(x,t)$ are unambiguously slow or fast when they follow a line from one or the other family; however an eigenvalue that is located near a crossing point of a slow and fast line has an ambiguous velocity $\alpha_{k,x}(x,t)$. Note that any eigenvalues for which $\alpha_k(x,t) = \mathcal{O}(\epsilon)$ and for which the value of $\alpha_{k,x}(x,t)$ is ambiguous will make an unpredictable contribution to $u(x,t)$ via the formula (1.23) of order 1. Therefore, to be able prove a convergence result without resolving the ambiguity, i.e., using linear approximations only, we should restrict the analysis to values of x for which all such ambiguous eigenvalues in the small range $|\alpha_k(x,t)| \leq \epsilon^r$ are sufficiently large compared to ϵ.

With this motivation, we introduce a sufficiently small scale $\kappa = \kappa(\epsilon)$ and declare a small eigenvalue $\alpha(x,t)$ with $|\alpha(x,t)| \leq \epsilon^r$ to be *ambiguous* if it lies within a disk of radius κ of a crossing point of the two families of straight lines in the (x, α)-plane near $(x_0, 0)$ given by the leading terms in (3.71) below. Then we pick an exponent s with $0 < r < s < 1$ and define an excluded set $X_o = X_o(x_0, t; \epsilon)$ consisting of values of $x \in \mathbb{R}$ near x_0 such that $x \in X_o$ if there is a crossing point of the straight lines at (x', α') with $|x - x'| \leq \kappa$ and $|\alpha'| \leq \epsilon^s$.

For fixed $x \approx x_0$ not in X_o, we denote by $(\alpha_k^o(x,t))_{|k| \leq C_o \epsilon^{r-1}}$ the ambiguous eigenvalues: for each such $\alpha_k^o(x,t)$, we have $|\alpha_k^o(x,t)| \gtrsim \epsilon^q$. Given that there are $\mathcal{O}(\epsilon^{r-1})$ slow lines intersecting the vertical axis passing the point $(x, 0)$, we roughly estimate the number of such ambiguous crossings to be of order at most ϵ^{r-1}, which explains the index range $|k| \leq C_o \epsilon^{r-1}$. In what follows, we fix parameters $\frac{1}{3} < r < \frac{1}{2}$ and $\frac{1}{2} < q < 1$ to be chosen later.

CONJECTURE 6 (Bohr-Sommerfeld conditions for small eigenvalues). Let $t > t_b$ and fix $x_0 \in (X^-(t), X^+(t))$, and let $\epsilon > 0$ be sufficiently small. Let $0 < r < q < 1$.

For x such that $|x - x_0| \leq \epsilon^r$ and $x \notin X_o$, small eigenvalues α with

$$|\alpha(x,t)| \leq \epsilon^r \tag{3.65}$$

can be split into three pairwise disjoint families:

- a set of "slow" eigenvalues $(\alpha_k^-(x,t))_{|k| \leq C_- \epsilon^{r-1}}$;
- a set of "fast" eigenvalues $(\alpha_k^+(x,t))_{|k| \leq C_+ \epsilon^{r-1}}$;
- a set of "ambiguous" eigenvalues $(\alpha_k^o(x,t))_{|k| \leq C_o \epsilon^{r-1}}$, with $|\alpha_k^o(x,t)| \gtrsim \epsilon^s$.

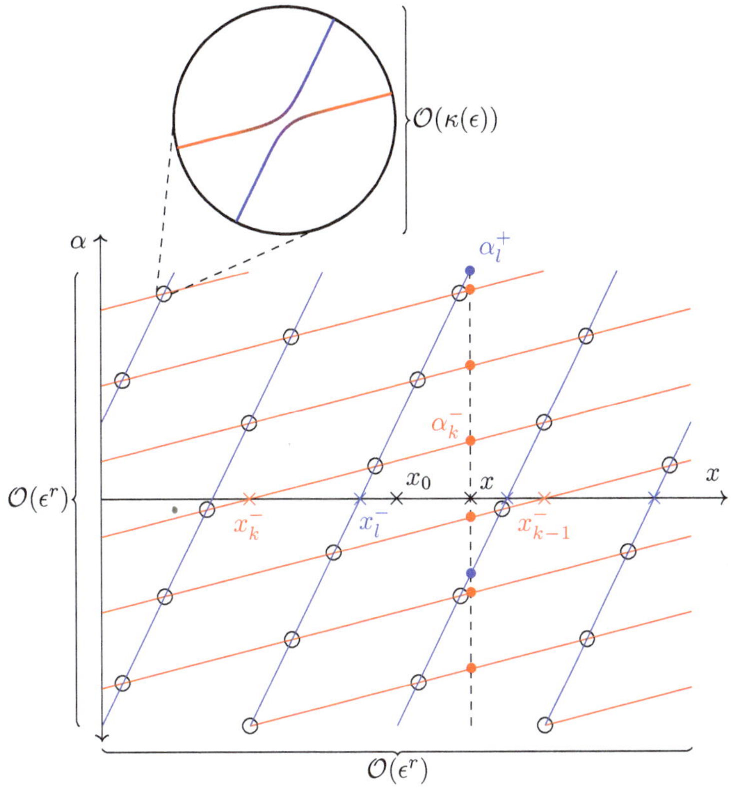

FIGURE 16. Illustration for Conjecture 6. For $x - x_0 = \mathcal{O}(\epsilon^r)$, the small eigenvalues of order ϵ^r are partitioned into two families $\alpha_k^- = \alpha_k^-(x,t)$ ("slow" eigenvalues, in red) and $\alpha_l^+ = \alpha_l^+(x,t)$ ("fast" eigenvalues, in blue), that are modeled by parallel lines in the (x,α)-plane of two distinct slopes. The small black circles in the figure are centered at intersections of slow and fast lines. Eigenvalues within them cannot be reliably characterized as being slow or fast, and for them, the linear approximation is no longer expected to be accurate. The set X_o is the union of projections on the the x-axis of the horizontal diameters of those circles centered on points with $\alpha = \mathcal{O}(\epsilon^s)$. COLOR ONLINE.

Moreover, for $x \notin X_o$, $\alpha_k^\pm(x,t)$ is close to $\alpha_k^\pm(x_k^\pm(t),t)$, in the sense that there is a uniform Taylor expansion in $x - x_k^\pm(t)$:

$$\alpha_k^\pm(x,t) = \alpha_k^\pm(x_k^\pm(t),t) + \partial_x\alpha_k^\pm(x_k^\pm(t),t)(x - x_k^\pm(t)) + o(\epsilon^q), \tag{3.66}$$

$$\partial_x\alpha_k^\pm(x,t) = \partial_x\alpha_k^\pm(x_k^\pm(t),t) + o(\epsilon^q). \tag{3.67}$$

The points $x_k^\pm$ satisfy the Bohr-Sommerfeld conditions in the following sense.

(1) There holds $|x_k^\pm(t) - x_0| = \mathcal{O}(\epsilon^r)$, and $\alpha_k^\pm(x_k^\pm(t),t) = \mathcal{O}(\epsilon^{1+r})$ is a small eigenvalue.

(2) (Eigenvectors for slow eigenvalues.) The amplitude of the normalized eigenvector for the eigenvalue $\alpha_k^-(x_k^-,t)$

$$\mathbf{u} = \mathbf{u}_k^-(x_k^-,t) \tag{3.68}$$

is a discretization of (3.20) restricted to the set $\Lambda(y) \in [-u_0^{\mathrm{B}}(x_k^-,t),0]$, up to a small remainder term.

(3) (Eigenvectors for fast eigenvalues.) We can decompose the eigenvector for the eigenvalue $\alpha_k^+(x_k^+,t)$ as the linear combination of two normalized vectors

$$\mathbf{u} = \mathbf{u}_k^+(x_k^+,t) + \frac{c}{\ln(\epsilon^{-1})}\mathbf{u}_l^-(x_k^+,t) + o\left(\frac{1}{\ln(\epsilon^{-1})}\right). \tag{3.69}$$

The constant c does not depend on k, it is bounded independently of ϵ, x_0, t and:

- (Fast main part.) The component $\mathbf{u}_k^+(x_k^+,t)$ has amplitude that is a discretization of (3.20) with independent variable y restricted to the set $\Lambda(y) \in [-u_2^{\mathrm{B}}(x_k^+,t), -u_1^{\mathrm{B}}(x_k^+,t)]$;
- (Slow correction.) The component $\mathbf{u}_l^-(x_k^+,t)$ is a discretization of (3.20) restricted the set $\Lambda(y) \in [-u_0^{\mathrm{B}}(x_k^+,t),0]$.

Note that in the case of "fast" eigenvectors (as shown in the right-hand panel of Figure 11), the ratio between the parameter a_0^+ tuned for the fast main part and the parameter a_0^- tuned for the slow correction should satisfy

$$\frac{a_0^-}{a_0^+} = \frac{c+o(1)}{\ln(\epsilon^{-1})}. \tag{3.70}$$

Proposition 3.7 (Approximation of α_k and its spatial derivative). Assume there exists a $p > \frac{1}{2}$ such that $u_0(x) \sim Cx^{-2p}$ as $|x| \to +\infty$, and that Conjecture 6 holds. Fix $t > t_b$ and $x_0 \in (X^-(t), X^+(t))$ so that (x_0,t) is within the triple-valued region for the solution of Burgers' equation with initial data u_0. Let k be the index of a small eigenvalue $|\alpha_k(x,t)| \leq \epsilon^r$. For $|x-x_0| \leq \epsilon^q$ and $x \notin X_o$, the following estimates hold, where the remainder terms are uniform over all indices $|k| \leq C_\pm \epsilon^{r-1}$.

(1) One can write

$$\alpha_k^\pm(x,t) = \frac{2}{h_k^\pm}(x_0,t;\epsilon)(xf^\pm(x_0,t) - 2\pi k\epsilon) + R_k^\pm(x,t), \tag{3.71}$$

$$\partial_x \alpha_k^\pm(x,t) = \frac{2f^\pm(x_0,t)}{h_k^\pm(x_0,t;\epsilon})) + Q_k^\pm(x,t), \tag{3.72}$$

$$R_k^\pm(x,t) = \frac{2r^\pm(x_0,t)f^\pm(x_0,t)}{h_k^\pm(x_0,t;\epsilon)}\epsilon + o(\epsilon^q), \quad Q_k^\pm(x,t) = o(1). \tag{3.73}$$

(2) (Fast eigenvalues.) For some bounded $|c_k^+(x_0,t;\epsilon)| \leq C$,

$$h_k^+(x_0,t;\epsilon) = \ln\left(\frac{u_2^{\mathrm{B}}(x_0,t)}{u_1^{\mathrm{B}}(x_0,t)}\right) + c_k^+(x_0,t;\epsilon). \tag{3.74}$$

(3) (Slow eigenvalues.) For some bounded $|c_k^-(x_0,t;\epsilon)| \leq C$,

$$h_k^-(x_0,t;\epsilon) = c_k^-(x_0,t;\epsilon)\ln(\epsilon^{-1}). \tag{3.75}$$

The rest of this subsection is devoted to the proof of Proposition 3.7.

3.4.4. *Spatial derivative of the small eigenvalues.* Choosing an eigenvector $\mathbf{u} = (u_j)_j$ (which is not necessarily normalized) with eigenvalue $\alpha_l^{\pm}(x_l^{\pm}, t)$, the equality (3.1) leads to

$$\partial_x \alpha_l^{\pm}(x_l^{\pm}, t) = \frac{\sum_{j=1}^{N(\epsilon)}(-2\lambda_j)|u_j|^2}{\sum_{j=1}^{N(\epsilon)}|u_j|^2} = \frac{\frac{1}{N(\epsilon)}\sum_{j=1}^{N(\epsilon)}(-2\lambda_j)|u_j|^2}{\frac{1}{N(\epsilon)}\sum_{j=1}^{N(\epsilon)}|u_j|^2}. \tag{3.76}$$

We know that the leading order of the amplitude $|u_{l,j}^{\pm}(x_l^{\pm}, t)|$ is a discretization of a constant multiple (which we take to be 1 for the purposes of this computation as it will cancel between the numerator and denominator of $\partial_x \alpha_l^{\pm}$) of (3.20)

$$a(y) = \sqrt{\frac{\Lambda'(y)}{-2\Lambda(y)}} \tag{3.77}$$

in the range $y \in [y_{\min}(x_l^{\pm}, t), y_{\max}(x_l^{\pm}, t)]$, where we set $y_{\min}(x_l^{\pm}, t) = Y(\lambda_{\min}(x_l^{\pm}, t))$ and $y_{\max}(x_l^{\pm}, t) = Y(\lambda_{\max}(x_l^{\pm}, t))$. Given that $x_l^{\pm}$ is close to x_0, one can replace $x_l^{\pm}$ by x_0 in the bounds of integration, up to a remainder term of order ϵ^r. This implies

$$\lim_{\epsilon \to 0} \frac{-2}{N(\epsilon)} \sum_{j=1}^{N(\epsilon)} \lambda_j |u_{l,j}^{\pm}(x_l^{\pm}, t)|^2 = -2 \int_{y_{\min}(x_0,t)}^{y_{\max}(x_0,t)} \Lambda(y)|a(y)|^2 \mathrm{d}y \in [0, +\infty], \tag{3.78}$$

$$\lim_{\epsilon \to 0} \frac{1}{N(\epsilon)} \sum_{j=1}^{N(\epsilon)} |u_{l,j}^{\pm}(x_l^{\pm}, t)|^2 = \int_{y_{\min}(x_0,t)}^{y_{\max}(x_0,t)} |a(y)|^2 \mathrm{d}y \in [0, +\infty]. \tag{3.79}$$

We first show that the numerator of (3.76) always has a finite limit. Indeed, we compute using (3.77) and the change of variables $\lambda = \Lambda(y)$

$$\begin{aligned} \int_{y_{\min}(x_0,t)}^{y_{\max}(x_0,t)} (-2\Lambda(y))|a(y)|^2 \mathrm{d}y &= \int_{y_{\min}(x_0,t)}^{y_{\max}(x_0,t)} \Lambda'(y)\, \mathrm{d}y \\ &= \lambda_{\max}(x_0, t) - \lambda_{\min}(x_0, t). \end{aligned} \tag{3.80}$$

In the case of fast eigenvalues, one has to sum up the contributions of the main part and of the correction. Since both integrals are finite, the slow correction is a negligible remainder term in the case $x_l^{\pm} = x_l^{+}$. Hence

$$\frac{1}{N(\epsilon)} \sum_{j=1}^{N(\epsilon)} (-2\lambda_j)|u_{l,j}^{+}(x_l^{+}, t)|^2 = u_2^{\mathrm{B}}(x_0, t) - u_1^{\mathrm{B}}(x_0, t) + o(1), \tag{3.81}$$

$$\frac{1}{N(\epsilon)} \sum_{j=1}^{N(\epsilon)} (-2\lambda_j)|u_{l,j}^{-}(x_l^{-}, t)|^2 = u_0^{\mathrm{B}}(x_0, t) + o(1). \tag{3.82}$$

In both cases, we retrieve $f^{\pm}(x_0, t)$.

Let us now study the denominator of (3.76). Again using (3.77) and $\lambda = \Lambda(y)$,

$$\begin{aligned} \int_{y_{\min}(x_0,t)}^{y_{\max}(x_0,t)} |a(y)|^2 \mathrm{d}y &= \int_{y_{\min}(x_0,t)}^{y_{\max}(x_0,t)} \frac{\Lambda'(y)\, \mathrm{d}y}{-2\Lambda(y)} \\ &= \frac{1}{2}\left(\ln(-\lambda_{\min}(x_0, t)) - \ln(-\lambda_{\max}(x_0, t))\right). \end{aligned}$$

The integral is finite and equal to $\frac{1}{2}g^+(x_0,t)$ in the fast case that $\lambda_{\min}(x_0,t) = -u_2^{\mathrm{B}}(x_0,t)$ and $\lambda_{\max}(x_0,t) = -u_1^{\mathrm{B}}(x_0,t)$, but infinite in the slow case because $\lambda_{\max}(x_0,t)=0$.

Let us now use the assumption that there is $p>\frac{1}{2}$ such that $u_0(x)\sim Cx^{-2p}$ as $|x|\to+\infty$. We focus on the values of λ_j which are close to 0, i.e. for which j is close to $N=N(\epsilon)$. Using formula (1.15), and assuming the condition (1.20), we have $\lambda_j=\Lambda(y_j)$, so

$$\int_{\lambda_{N+1-j}}^{0} F(\lambda)\mathrm{d}\lambda = \epsilon\left(j-\tfrac{1}{2}\right). \tag{3.83}$$

Since $F(\lambda)\sim(-C\lambda)^{-1/(2p)}$ as $\lambda\to 0$ by assumption, we get that when $\lambda\to0$,

$$\int_{\lambda}^{0} F(\eta)\mathrm{d}\eta \sim \frac{C^{1/(2p)}(-\lambda)^{1-1/(2p)}}{1-1/(2p)}, \tag{3.84}$$

so that for j close to 1,

$$\Lambda(y_{N+1-j}) \sim -\left(\frac{2p-1}{C^{1/(2p)}2p}\epsilon(j-\tfrac{1}{2})\right)^{2p/(2p-1)}. \tag{3.85}$$

Hence $\Lambda'(y_{N+1-j})/\Lambda(y_{N+1-j}) = [\Lambda(y_{N+1-j})F(\Lambda(y_{N+1-j}))]^{-1}$ has the asymptotic expansion

$$-\frac{\Lambda'(y_{N+1-j})}{\Lambda(y_{N+1-j})} \sim \frac{1}{C^{-1/(2p)}(-\Lambda(y_{N+1-j}))^{1-1/(2p)}} \sim \frac{C^{1/p}2p}{2p-1}\frac{1}{\epsilon(j-\frac{1}{2})}. \tag{3.86}$$

This approximation is valid for $\epsilon j\lesssim 1/\ln(\epsilon^{-1})$ since in this case we can check that uniformly in this range, $|\Lambda(y_{N+1-j})|=o(1)$. Hence the denominator sum has the lower bound

$$\begin{aligned}\frac{1}{N(\epsilon)}\sum_{j=1}^{N(\epsilon)}|u_{l,j}^-|^2 &\gtrsim \frac{1}{N(\epsilon)}\sum_{j=1}^{\mathcal{O}(1/\epsilon\ln(\epsilon^{-1}))}\frac{\Lambda'(y_{N+1-j})}{-2\Lambda(y_{N+1-j})}\\ &\gtrsim \frac{1}{\epsilon N(\epsilon)}\sum_{j=1}^{\mathcal{O}(1/\epsilon\ln(\epsilon^{-1}))}\frac{1}{j-\frac{1}{2}} \gtrsim \ln(\epsilon^{-1}).\end{aligned} \tag{3.87}$$

To get an upper bound, we split between the cases $\epsilon j\lesssim 1/\ln(\epsilon^{-1})$ and $\epsilon j\gtrsim 1/\ln(\epsilon^{-1})$. The small indices j such that $\epsilon j\lesssim 1/\ln(\epsilon^{-1})$ are treated as in (3.87). Concerning the indices j such that $\epsilon j\gtrsim 1/\ln(\epsilon^{-1})$, we use the fact that $t>t_b$ and that $\mathbf{u}_l^-$ is supported on $[-u_0^{\mathrm{B}}(x_l^+,t),0]$, that is, away from the left edge $\lambda=-L$. Then, the maximum of $|\Lambda'(y_{N+1-j})/\Lambda(y_{N+1-j})|$ for $\epsilon j\gtrsim 1/\ln(\epsilon^{-1})$ occurs for the smallest j, and hence $\Lambda'(y_{N+1-j})/\Lambda(y_{N+1-j})=\mathcal{O}(\ln(\epsilon^{-1}))$ holds uniformly on this index range because $\Lambda'(y)/\Lambda(y)\sim(M-y)^{-1}$ for $y\approx M$. This yields the upper bound

$$\frac{1}{N(\epsilon)}\sum_{j=1}^{N(\epsilon)}|u_{l,j}^-|^2 \lesssim \frac{1}{N(\epsilon)}\sum_{j=1}^{N(\epsilon)}\frac{\Lambda'(y_{N+1-j})}{-2\Lambda(y_{N+1-j})} \lesssim \ln(\epsilon^{-1}). \tag{3.88}$$

Consequently, the sum over the "slow" part of the eigenvectors satisfies

$$\frac{1}{N(\epsilon)}\sum_{j=1}^{N(\epsilon)}|u_{l,j}^-(x_l^{\pm},t)|^2 = c_l^-(x_0,t;\epsilon)\ln(\epsilon^{-1}) \tag{3.89}$$

for some bounded constant $c_l^-(x_0, t; \epsilon)$ that may be oscillatory as $\epsilon \to 0$. Regarding the fast eigenvectors, we sum up the contributions $u_{l,j}^+(x_l^+)$ and $u_{l,j}^-(x_l^+)$, which are supported in disjoint intervals,

$$\frac{1}{N(\epsilon)} \sum_{j=1}^{N(\epsilon)} |u_{l,j}(x_l^+, t)|^2 = \frac{1}{2} \ln \left(\frac{u_2^{\mathrm{B}}(x_0, t)}{u_1^{\mathrm{B}}(x_0, t)} \right) + \frac{c}{\ln(\epsilon^{-1})} c_l^-(x_0, t; \epsilon) \ln(\epsilon^{-1}) + o(1).$$

We choose $c_l^+(x_0, t; \epsilon) = c \cdot c_l^-(x_0, t; \epsilon)$ to get the Proposition.

We conclude that spatial derivatives $\partial_x \alpha_l^\pm(x_l^\pm)$ of the fast and slow eigenvalues satisfy the following identities at $x_l^\pm$:

$$\partial_x \alpha_l^+(x_l^+, t) = \frac{2(u_2^{\mathrm{B}}(x_0, t) - u_1^{\mathrm{B}}(x_0, t)) + o(1)}{\ln \left(\frac{u_2^{\mathrm{B}}(x_0, t)}{u_1^{\mathrm{B}}(x_0, t)} \right) + c_l^+(x_0, t; \epsilon)} + o(1), \tag{3.90}$$

$$\partial_x \alpha_l^-(x_l^-, t) = \frac{2u_0^{\mathrm{B}}(x_0, t) + o(1)}{c_l^-(x_0, t; \epsilon)} \ln(\epsilon^{-1})(1 + o(1)). \tag{3.91}$$

We deduce that (3.72) holds. Finally, we write

$$\alpha_l^\pm(x, t) = \alpha_l^\pm(x_l^\pm, t) + (x - x_l^\pm)\partial_x \alpha_l^\pm(x_l^\pm, t) + o(\epsilon^q). \tag{3.92}$$

Then we use (3.63) and the estimate $|\alpha_l^\pm(x_l^\pm, t)| \leq C\epsilon^{1+r}$ to deduce that (3.71) holds.

3.5. Small-ϵ asymptotics of the sum. Assuming now that a stronger form of Conjecture 6 holds, we establish an asymptotic expansion of the soliton ensemble solution to (1.1) for (x, t) near (x_0, t) in the oscillatory region (triple-valued region for the solution of Burgers' equation with data u_0).

We retrieve a result similar to Theorem 2.11 above in the oscillatory region.

THEOREM 3.8. *Assume that Conjecture 6 holds. Fix $t > t_b$ and $x_0 \in (X^-(t), X^+(t))$ so that (x_0, t) is within the triple-valued region for the solution of Burgers' equation with initial data u_0. Let $|x - x_0| \leq \epsilon^q$ and $x \notin X_o$. We assume moreover that uniformly in k, one can write the remainder terms from* (3.73) *as*

$$R_k^\pm(x, t) = \epsilon\varphi^\pm(x_0, t) + o(\epsilon^{2-r}), \quad Q_k^\pm(x, t) = o(\epsilon^{1-r}), \tag{3.93}$$

and that in (3.74) *and* (3.75), $c_k^\pm(x_0, t; \epsilon) = c^\pm(x_0, t)$ *do not depend on k nor on ϵ. Then*

$$u(x, t) = u_0^{\mathrm{B}}(x_0, t) + \frac{f^+(x_0, t) \sinh(\frac{1}{2}h^+(x_0, t))}{\cosh(\frac{1}{2}h^+(x_0, t)) - \cos(f^+(x_0, t)\epsilon^{-1}x + \varphi^+(x_0, t))} + o(1). \tag{3.94}$$

If moreover, in (3.74), *there holds*

$$h^+(x_0, t) = \ln \left(\frac{u_2^{\mathrm{B}}(x_0, t) - u_0^{\mathrm{B}}(x_0, t)}{u_1^{\mathrm{B}}(x_0, t) - u_0^{\mathrm{B}}(x_0, t)} \right), \tag{3.95}$$

then

$$u(x, t) = u_0^{\mathrm{B}}(x_0, t) + \frac{(u_2^{\mathrm{B}}(x_0, t) - u_1^{\mathrm{B}}(x_0, t))(1 - r(x_0, t)^2)}{1 + r(x_0, t)^2 - 2r(x_0, t) \cos(\Theta(\epsilon^{-1}x; x_0, t))} + o(1), \tag{3.96}$$

where

$$\Theta(z; x_0, t) = (u_2^{\mathrm{B}}(x_0, t) - u_1^{\mathrm{B}}(x_0, t))z + \varphi^+(x_0, t), \tag{3.97}$$

$$r(x_0,t)=\sqrt{\frac{u_1^{\mathrm{B}}(x_0,t)-u_0^{\mathrm{B}}(x_0,t)}{u_2^{\mathrm{B}}(x_0,t)-u_0^{\mathrm{B}}(x_0,t)}}\in(0,1). \tag{3.98}$$

REMARK 3.9. The functions φ^+ and φ^- were not present in Proposition 3.7. This is likely because we only considered the principal part of the Bohr-Sommerfeld condition in Conjecture 6, instead of all of its semiclassical expansion. If the symbol p were smooth, then the complete asymptotic expansion of the Bohr-Sommerfeld condition would be given by [**Cha03**, Therorem 3.1], in which the first two terms are explicitly written, and the principal part is precisely the parallel-transport integral $g_0^{\pm}(x,t)$ (see (3.49)–(3.50)).

REMARK 3.10. Aside from details of the phase correction φ^+, the formula (3.94) coincides with (2.73) from Theorem 2.11 upon proper identification of the periodic wave parameters. Note that if one averages (3.96) over the fast variable Θ (requiring an integration since the cosine appears in the denominator), then the result is exactly as expected, namely the weak limit $\overline{u}(x_0,t):=u_2^{\mathrm{B}}(x_0,t)-u_1^{\mathrm{B}}(x_0,t)+u_0^{\mathrm{B}}(x_0,t)$ defined in (1.5).

PROOF. We start from the formula (1.23)

$$u(x,t)=\sum_{k=1}^{N(\epsilon)}\frac{2\partial_x\alpha_k(x,t)}{\left(\epsilon^{-1}\alpha_k(x,t)\right)^2+1}. \tag{3.99}$$

According to (3.3), it is enough to focus on the small eigenvalues α such that $|\alpha_k|\le\epsilon^r$. Each such eigenvalue is either fast, slow, or ambiguous, and the total number of each type in the indicated range of α is proportional to ϵ^{r-1}.

First, let us estimate the contribution of the ambiguous eigenvalues. Using (3.2) and the lower bound $|\alpha_k^o(x,t)|\gtrsim k\epsilon^s$ because $x\notin X_o$, we have

$$\sum_{|k|\le C_o\epsilon^{r-1}}\frac{2\partial_x\alpha_k^o(x,t)}{\left(\epsilon^{-1}\alpha_k^o(x,t)\right)^2+1}\lesssim\sum_{|k|\le C_o\epsilon^{r-1}}\frac{1}{(\epsilon^{s-1})^2+1}\lesssim\epsilon^{1+r-2s}. \tag{3.100}$$

Given that $\frac{1}{3}<r<\frac{1}{2}$ by assumption, one can choose $\frac{1}{2}<s<1$ close enough to $\frac{1}{2}$ so that $1+r-2s>0$, and this sum goes to zero as $\epsilon\to0$. Since the total contribution of the ambiguous eigenvalues in the range $|\alpha_k|\le\epsilon^r$ is negligible, it is harmless to "double count" them as both fast and slow eigenvalues and as this makes the arguments easier going forward we shall do so.

Then, let us tackle the sum over fast eigenvalues. The ideas are remarkably similar to proof of Proposition 2.9, so we omit some of the details. By assumption, we have that

$$\begin{aligned}&\sum_{|k|\le C_+\epsilon^{r-1}}\frac{2\partial_x\alpha_k^+(x,t)}{\left(\epsilon^{-1}\alpha_k^+(x,t)\right)^2+1}\\ &\quad=\sum_{|k|\le C_+\epsilon^{r-1}}\frac{f^+(x_0,t)h^+(x_0,t)}{(-2\pi k+f^+(x_0,t)\epsilon^{-1}x+\varphi^+(x_0,t))^2+\frac{1}{4}h^+(x_0,t)^2}+o(1).\end{aligned} \tag{3.101}$$

In the limit $\epsilon \to 0$, we have $\epsilon^{r-1} \to +\infty$, so the sum on the right-hand side can be computed:

$$\sum_{|k|\leq C_+\epsilon^{r-1}} \frac{2\partial_x \alpha_k^+(x,t)}{\left(\epsilon^{-1}\alpha_k^+(x,t)\right)^2+1} = \frac{f^+(x_0,t)\sinh(\frac{1}{2}h^+(x_0,t))}{\cosh(\frac{1}{2}h^+(x_0,t)) - \cos(f^+(x_0,t)\epsilon^{-1}x+\varphi^+(x_0,t))} + o(1). \tag{3.102}$$

If the expression of $h^+(x_0,t)$ is given by (3.95), one can further simplify

$$\sinh\left(\frac{1}{2}h^+(x,t)\right) = \frac{u_2^{\mathrm{B}}(x,t)-u_1^{\mathrm{B}}(x,t)}{2\sqrt{(u_2^{\mathrm{B}}(x,t)-u_0^{\mathrm{B}}(x,t))(u_1^{\mathrm{B}}(x,t)-u_0^{\mathrm{B}}(x,t))}}, \tag{3.103}$$

$$\cosh\left(\frac{1}{2}h^+(x,t)\right) = \frac{u_2^{\mathrm{B}}(x,t)+u_1^{\mathrm{B}}(x,t)-2u_0^{\mathrm{B}}(x,t)}{2\sqrt{(u_2^{\mathrm{B}}(x,t)-u_0^{\mathrm{B}}(x,t))(u_1^{\mathrm{B}}(x,t)-u_0^{\mathrm{B}}(x,t))}}. \tag{3.104}$$

Hence,

$$\sum_{|k|\leq C_+\epsilon^{r-1}} \frac{2\partial_x \alpha_k^+(x,t)}{\left(\epsilon^{-1}\alpha_k^+(x,t)\right)^2+1} = \frac{(u_2^{\mathrm{B}}(x_0,t)-u_1^{\mathrm{B}}(x_0,t))(1-r(x_0,t)^2)}{1+r(x_0,t)^2-2r(x_0,t)\cos(\Theta(\epsilon^{-1}x;x_0,t))} + o(1). \tag{3.105}$$

Now we consider the sum over slow eigenvalues. Again, a parallel can be made to the previous section with Proposition 2.7, where we obtained the leading order behavior of a sum by turning it into an integral. Given (3.61) of Proposition 3.7, we have that the x-velocities of all slow and small eigenvalues for (x,t) near (x_0,t) are nearly the same, i.e.

$$\partial_x \alpha_k^-(x,t) \approx \frac{2u_0^{\mathrm{B}}(x_0,t)}{h^-(x_0,t)\ln(\epsilon^{-1})}. \tag{3.106}$$

However, at this point we need to approximate the quantity $h^-(x_0,t)$. We will actually proceed more directly to approximate $\partial_x\alpha_k^-(x,t)$, which will automatically produce the logarithmic scaling in ϵ. For this purpose, let $x_-(t)$ denote the largest value of x less than or equal to x_0 for which there is a slow eigenvalue at $\alpha = 0$, let $x_+(t)$ denote the smallest value of x strictly greater than x_0 for which the same is true, and set $\Delta x := x_+(t) - x_-(t)$. By setting the left-hand side of (3.71) to zero we obtain

$$\Delta x \approx \frac{2\pi\epsilon}{f^-(x_0,t)} = \frac{2\pi\epsilon}{u_0^{\mathrm{B}}(x_0,t)}. \tag{3.107}$$

Now as x increases from $x_-(t)$ to x_0, the slow eigenvalue originally at $\alpha = 0$ will increase to the value of the smallest positive slow eigenvalue $\alpha_+(x_0,t)$; likewise as x decreases from $x_+(t)$ to x_0, the slow eigenvalue originally at $\alpha = 0$ will decrease to the value of the negative slow eigenvalue $\alpha_-(x_0,t)$ of smallest absolute value. We set $\Delta\alpha := \alpha_+(x_0,t) - \alpha_-(x_0,t)$, and see that this is the difference between the two closest eigenvalues of opposite signs to $\alpha = 0$ at (x_0,t). We can approximate $\Delta\alpha$ directly using the asymptotic density of all of the eigenvalues given in (3.4). Indeed, the integral of $G(\alpha;x_0,t)\,d\alpha$ between $\alpha_-(x_0,t)$ and $\alpha_+(x_0,t)$ should be approximately

$$\frac{M}{N(\epsilon)} = \epsilon(1 + o(1)). \tag{3.108}$$

Using (3.8), since the limits of integration are small, we get

$$\begin{aligned}\frac{1}{4\pi}\frac{2p}{2p-1}\,[\alpha_+(x_0,t) - \alpha_+(x_0,t)\ln(|\alpha_+(x_0,t)|) \\ - \alpha_-(x_0,t) + \alpha_-(x_0,t)\ln(|\alpha_-(x_0,t)|)] \approx \epsilon.\end{aligned} \tag{3.109}$$

Solving for $\Delta\alpha$ gives

$$\Delta\alpha \approx \frac{2\pi\epsilon(2p-1)}{p\ln(\epsilon^{-1})}. \tag{3.110}$$

Then we approximate the velocity $\partial_x\alpha(x,t)$ for a slow and small eigenvalue by

$$\partial_x\alpha(x,t) \approx \frac{\Delta\alpha}{\Delta x} \approx \frac{(2p-1)u_0^{\mathrm{B}}(x_0,t)}{p\ln(\epsilon^{-1})} \implies g^-(x_0,t) \approx \frac{2p}{2p-1}. \tag{3.111}$$

Using this and the convergence in (3.4), the contribution of the small and slow eigenvalues (which are the majority of the small eigenvalues), is

$$\begin{aligned}\sum_{|k|\le C_-\epsilon^{r-1}} \frac{2\alpha_{k,x}^-(x,t)}{(\epsilon^{-1}\alpha_k^-(x,t))^2+1} &\approx u_0^{\mathrm{B}}(x_0,t)\frac{2(2p-1)}{p\ln(\epsilon^{-1})}\sum_{|k|\le C_-\epsilon^{r-1}}\frac{1}{(\epsilon^{-1}\alpha_k^-(x_0,t))^2+1} \\ &= u_0^{\mathrm{B}}(x_0,t)\frac{2(2p-1)}{p\ln(\epsilon^{-1})}\int_{-\epsilon^r}^{\epsilon^r}\sum_{k=1}^{N(\epsilon)}\frac{\delta_{\alpha_k^-}(x_0,t)(\alpha)}{(\epsilon^{-1}\alpha)^2+1}\,\mathrm{d}\alpha \\ &\approx u_0^{\mathrm{B}}(x_0,t)\frac{2(2p-1)}{p\ln(\epsilon^{-1})}\cdot\frac{1}{\epsilon}\int_{-\epsilon^r}^{\epsilon^r}\frac{G(\alpha;x_0,t)\,\mathrm{d}\alpha}{(\epsilon^{-1}\alpha)^2+1} \\ &\approx \frac{u_0^{\mathrm{B}}(x_0,t)}{\pi\epsilon\ln(\epsilon^{-1})}\int_{-\epsilon^r}^{\epsilon^r}\frac{\ln(|\alpha|^{-1})\,\mathrm{d}\alpha}{(\epsilon^{-1}\alpha)^2+1}.\end{aligned}$$

Scaling by $\alpha = \epsilon z$ and letting $\epsilon \to 0$, the integral above is $\pi\epsilon\ln(\epsilon^{-1})(1 + o(1))$, so we conclude that the leading contribution of the slow and small eigenvalues is

$$\lim_{\epsilon\to 0}\sum_{|k|\le C_-\epsilon^{r-1}}\frac{2\alpha_{k,x}^-(x,t)}{(\epsilon^{-1}\alpha_k^-(x,t))^2+1} = u_0^{\mathrm{B}}(x_0,t). \tag{3.112}$$

□

REMARK 3.11. The proof above shows an interesting connection between the approaches to the strong small-ϵ asymptotic behavior of the BO soliton ensemble $u(x,t)$ based on the eigenvalues of the two matrices $\mathbf{A}(x,t)$ (the approach in this section) and $\mathbf{C}(t)$ (see Section 2). Indeed, we can see that the contributions of the small slow/fast eigenvalues of $\mathbf{A}(x,t)$ respectively correspond precisely to the contributions of eigenvalues of $\mathbf{C}(t)$ with real parts close to $\mu = x$ on the upper/lower branch.

REMARK 3.12. Although for the proof we needed to exclude extremely small ambiguous eigenvalues, we expect that a more accurate modeling of the near-crossings of the eigenvalue curves as seen in Figure 9 would allow such eigenvalues to be included without changing the leading asymptotics.

References

[Ben67] T. Brooke Benjamin, *Internal waves of permanent form in fluids of great depth*, Journal of Fluid Mechanics **29** (1967), no. 3, 559–592.

[BGGM24] E. Blackstone, L. Gassot, P. Gérard, and P. D. Miller, *The Benjamin-Ono equation in the zero-dispersion limit for rational initial data: generation of dispersive shock waves*, arXiv:2410.17405

[BGPU03] A. Bloch, F. Golse, T. Paul, and A. Uribe, *Dispersionless Toda and Toeplitz operators*, Duke Math. J. **117** (2003), no. 1, 157–196, DOI 10.1215/S0012-7094-03-11713-5. MR1962785

[Cha03] L. Charles, *Quasimodes and Bohr-Sommerfeld conditions for the Toeplitz operators*, Comm. Partial Differential Equations **28** (2003), no. 9-10, 1527–1566, DOI 10.1081/PDE-120024521. MR2001172

[Cha06] L. Charles, *Symbolic calculus for Toeplitz operators with half-form*, J. Symplectic Geom. **4** (2006), no. 2, 171–198, DOI 10.4310/jsg.2006.v4.n2.a3. MR2275003

[CLP79] H. H. Chen, Y. C. Lee, and N. R. Pereira, *Algebraic internal wave solitons and the integrable Calogero-Moser-Sutherland N-body problem*, Phys. Fluids **22** (1979), no. 1, 187–188, DOI 10.1063/1.862457. MR516327

[DK91] S. Yu. Dobrokhotov and I. M. Krichever, *Multiphase solutions of the Benjamin-Ono equation and their averaging* (Russian), Mat. Zametki **49** (1991), no. 6, 42–58, 158, DOI 10.1007/BF01156581; English transl., Math. Notes **49** (1991), no. 5-6, 583–594. MR1135514

[DLMF] *NIST Digital Library of Mathematical Functions*, https://dlmf.nist.gov/, Release 1.2.0 of 2024-03-15, F. W. J. Olver, A. B. Olde Daalhuis, D. W. Lozier, B. I. Schneider, R. F. Boisvert, C. W. Clark, B. R. Miller, B. V. Saunders, H. S. Cohl, and M. A. McClain, eds.

[FA83] A. S. Fokas and M. J. Ablowitz, *The inverse scattering transform for the Benjamin-Ono equation—a pivot to multidimensional problems*, Stud. Appl. Math. **68** (1983), no. 1, 1–10, DOI 10.1002/sapm19836811. MR686248

[FFM80] H. Flaschka, M. G. Forest, and D. W. McLaughlin, *Multiphase averaging and the inverse spectral solution of the Korteweg-de Vries equation*, Comm. Pure Appl. Math. **33** (1980), no. 6, 739–784, DOI 10.1002/cpa.3160330605. MR596433

[Gé23] P. Gérard, *An explicit formula for the Benjamin-Ono equation*, Tunis. J. Math. **5** (2023), no. 3, 593–603, DOI 10.2140/tunis.2023.5.593. MR4662323

[Gé24] P. Gérard, *The zero dispersion limit for the Benjamin-Ono equation on the line* (English, with English and French summaries), C. R. Math. Acad. Sci. Paris **362** (2024), 619–634. MR4773745

[Gas23a] L. Gassot, *Lax eigenvalues in the zero-dispersion limit for the Benjamin-Ono equation on the torus*, SIAM J. Math. Anal. **55** (2023), no. 5, 5782–5822, DOI 10.1137/23M154635X. MR4650912

[Gas23b] L. Gassot, *Zero-dispersion limit for the Benjamin-Ono equation on the torus with bell shaped initial data*, Comm. Math. Phys. **401** (2023), no. 3, 2793–2843, DOI 10.1007/s00220-023-04701-0. MR4616654

[GGJM21] M. Girotti, T. Grava, R. Jenkins, and K. D. T.-R. McLaughlin, *Rigorous asymptotics of a KdV soliton gas*, Comm. Math. Phys. **384** (2021), no. 2, 733–784, DOI 10.1007/s00220-021-03942-1. MR4259375

[GR07] I. S. Gradshteyn and I. M. Ryzhik, *Table of integrals, series, and products*, 7th ed., 2007. MR2360010

[Kri91] I. M. Krichever, *The averaging procedure for the soliton-like solutions of integrable systems*, Mechanics, analysis and geometry: 200 years after Lagrange, North-Holland Delta Ser., North-Holland, Amsterdam, 1991, pp. 99–125. MR1098513

[KS21] C. Klein and J.-C. Saut, *Nonlinear dispersive equations—inverse scattering and PDE methods*, Applied Mathematical Sciences, vol. 209, Springer, Cham, 2021, DOI 10.1007/978-3-030-91427-1. MR4400881

[LL83] P. D. Lax and C. D. Levermore, *The small dispersion limit of the Korteweg-de Vries equation. I*, Comm. Pure Appl. Math. **36** (1983), no. 3, 253–290, DOI 10.1002/cpa.3160360302. MR697466

[Mat79] Yoshimasa Matsuno, *Exact multi-soliton solution of the Benjamin-Ono equation*, Journal of Physics A: Mathematical and General **12** (1979), no. 4, 619–621.

[Mat82] Y. Matsuno, *Asymptotic properties of the Benjamin-Ono equation*, J. Phys. Soc. Japan **51** (1982), 667–674.

[Mat84] Y. Matsuno, *Bilinear transformation method*, Mathematics in Science and Engineering, vol. 174, Academic Press, Inc., Orlando, FL, 1984. MR759718

[Mat82] Y. Matsuno, *Number density function of Benjamin-Ono solitons*, Phys. Lett. A **87** (1981/82), no. 1-2, 15–17, DOI 10.1016/0375-9601(81)90603-4. MR638848

[Mil01] P. D. Miller, *Some remarks on a WKB method for the nonselfadjoint Zakharov-Shabat eigenvalue problem with analytic potentials and fast phase*, Phys. D **152/153** (2001), 145–162, DOI 10.1016/S0167-2789(01)00166-X. Advances in nonlinear mathematics and science. MR1837905

[Mol20] A. Moll, *Finite gap conditions and small dispersion asymptotics for the classical periodic Benjamin-Ono equation*, Quart. Appl. Math. **78** (2020), no. 4, 671–702, DOI 10.1090/qam/1566. MR4148823

[MW16] P. D. Miller and A. N. Wetzel, *The scattering transform for the Benjamin-Ono equation in the small-dispersion limit*, Phys. D **333** (2016), 185–199, DOI 10.1016/j.physd.2015.07.012. MR3523500

[MX11] P. D. Miller and Z. Xu, *On the zero-dispersion limit of the Benjamin-Ono Cauchy problem for positive initial data*, Comm. Pure Appl. Math. **64** (2011), no. 2, 205–270, DOI 10.1002/cpa.20345. MR2766527

[MX12] P. D. Miller and Z. Xu, *The Benjamin-Ono hierarchy with asymptotically reflectionless initial data in the zero-dispersion limit*, Commun. Math. Sci. **10** (2012), no. 1, 117–130, DOI 10.4310/CMS.2012.v10.n1.a6. MR2901303

[Ono75] H. Ono, *Algebraic solitary waves in stratified fluids*, J. Phys. Soc. Japan **39** (1975), no. 4, 1082–1091, DOI 10.1143/JPSJ.39.1082. MR398275

[Sun21] R. Sun, *Complete integrability of the Benjamin-Ono equation on the multi-soliton manifolds*, Comm. Math. Phys. **383** (2021), no. 2, 1051–1092, DOI 10.1007/s00220-021-03996-1. MR4239837

[Whi99] G. B. Whitham, *Linear and nonlinear waves*, Pure and Applied Mathematics (New York), John Wiley & Sons, Inc., New York, 1999. Reprint of the 1974 original; A Wiley-Interscience Publication, DOI 10.1002/9781118032954. MR1699025

[Wu16] Y. Wu, *Simplicity and finiteness of discrete spectrum of the Benjamin-Ono scattering operator*, SIAM J. Math. Anal. **48** (2016), no. 2, 1348–1367, DOI 10.1137/15M1030649. MR3484397

[Wu17] Y. Wu, *Jost solutions and the direct scattering problem of the Benjamin-Ono equation*, SIAM J. Math. Anal. **49** (2017), no. 6, 5158–5206, DOI 10.1137/17M1124528. MR3738309

Department of Mathematics, University of Michigan, 530 Church Street, Ann Arbor, Michigan 48104
Email address: `eblackst@umich.edu`

Univ Rennes, CNRS, IRMAR - UMR 6625, F-35000 Rennes, France
Email address: `louise.gassot@cnrs.fr`

Department of Mathematics, University of Michigan, 530 Church Street, Ann Arbor, Michigan 48104
Email address: `millerpd@umich.edu`

Selected Published Titles in This Series

823 **Mikhail Bershtein, Anton Dzhamay, and Andrei Okounkov, Editors,** The Versatility of Integrability, 2025

821 **Bogdan D. Suceavă, Adara M. Blaga, Marian Ioan Munteanu, Cezar Oniciuc, Shoo Seto, Gabriel-Eduard Vîlcu, and Lihan Wang, Editors,** Recent Advances in Differential Geometry and Related Areas, 2025

820 **Sudeb Mitra, Rakesh Kumar Parmar, Saminathan Ponnusamy, and Toshiyuki Sugawa, Editors,** Geometric Function Theory and Related Topics, 2025

819 **Howard S. Cohl, Roberto S. Costas-Santos, and Robert S. Maier, Editors,** Applications and q-Extensions of Hypergeometric Functions, 2025

818 **Howard S. Cohl, Roberto S. Costas-Santos, and Robert S. Maier, Editors,** Classical Hypergeometric Functions and Generalizations, 2025

817 **Mikhail Khovanov, Joshua Sussan, and Anton Zeitlin, Editors,** From Representation Theory to Mathematical Physics and Back, 2025

816 **Luca F. Di Cerbo and Laurenţiu G. Maxim, Editors,** Geometry and Topology of Aspherical Manifolds, 2025

815 **Daniel Orr and Joshua Jeishing Wen, Editors,** Macdonald Theory and Beyond, 2025

814 **Susan Montgomery and Siu-Hung Ng, Editors,** Quantum Groups, Hopf Algebras, and Applications, 2025

813 **Thomas Creutzig, Julia Plavnik, and David Ridout, Editors,** Quantum Symmetries, 2025

812 **Ross Geoghegan, Craig R. Guilbault, and Kim Ruane, Editors,** Topology at Infinity of Discrete Groups, 2025

811 **Krishnendu Gongopadhyay and Shashikant A. Katre, Editors,** Geometry, Groups and Mathematical Philosophy, 2025

810 **Robert J. Berman and Yanir A. Rubinstein, Editors,** Convex and Complex: Perspectives on Positivity in Geometry, 2025

809 **Fabrice Baudoin and Luca Rizzi, Editors,** New Trends in Sub-Riemannian Geometry, 2025

808 **Loring W. Tu, Editor,** Group Actions and Equivariant Cohomology, 2024

807 **Galina Filipuk, Editor,** Recent Progress in Special Functions, 2024

806 **Maximilian Engel, Hildeberto Jardón-Kojakhmetov, and Cinzia Soresina, Editors,** Topics in Multiple Time Scale Dynamics, 2024

805 **Anthony Iarrobino, Pedro Macias Marques, Maria Evelina Rossi, and Jean Vallès, Editors,** Deformation of Artinian Algebras and Jordan Type, 2024

804 **Mahir Bilen Can and Jörg Feldvoss, Editors,** A Glimpse into Geometric Representation Theory, 2024

803 **Peter Gothen, Margarida Melo, and Montserrat Teixidor i Bigas, Editors,** Moduli Spaces and Vector Bundles—New Trends, 2024

802 **Ralph M. Kaufmann, Martin Markl, and Alexander A. Voronov, Editors,** Higher Structures in Topology, Geometry, and Physics, 2024

801 **K. A. Brown, T. J. Hodges, M. Vancliff, and J. J. Zhang, Editors,** Recent Advances in Noncommutative Algebra and Geometry, 2024

800 **Avinoam Mann, Louis H. Rowen, David J. Saltman, Aner Shalev, Lance W. Small, and Uzi Vishne, Editors,** Amitsur Centennial Symposium, 2024

799 **Alberto A. Condori, Elodie Pozzi, William T. Ross, and Alan A. Sola, Editors,** Recent Progress in Function Theory and Operator Theory, 2024

798 **Marat V. Markin, Igor V. Nikolaev, and Carsten Trunk, Editors,** Advances in Functional Analysis and Operator Theory, 2024